D0251526

ARCHAEOLOGICAL GEOLOGY

CONTRIBUTORS

Stanley E. Aschenbrenner is associate professor of anthropology, University of Minnesota—Duluth.

Reuben G. Bullard is chairman of the Arts and Sciences Field and professor of geology and archaeology, Cincinnati Christian Seminary, adjunct lecturer in geology, University of Cincinnati, and consultant to the American Schools of Oriental Research.

Donald A. Davidson is senior lecturer, Department of Geography, University of Strathclyde, Glasgow, United Kingdom.

Robert C. Eidt is director, UW-UWEX State Soils Laboratory and professor of geography, University of Wisconsin—Milwaukee.

Robert L. Folk is D. P. Carlton Professor of Geology, University of Texas, Austin.

John A. Gifford is assistant professor, Department of Anthropology, University of Miami, Coral Gables, Florida.

Fekri A. Hassan is professor, Geoarchaeology Section, Department of Anthropology, Washington State University, Pullman.

Norman Herz is director, Center for Archaeological Sciences and professor of geology, University of Georgia, Athens.

Robert Huggins owns and directs Spectrum Geophysics, Fort Worth, Texas.

Diana C. Kamilli, Ph.D., is an independent consultant based in Denver, specializing in igneous and metamorphic petrography and analysis of ancient ceramic materials.

Doçent Dr. İlhan Kayan is a member of the faculty of Physical Geography and Geology at the University of Ankara, Turkey.

James E. King is head of Scientific Sections at the Illinois State Museum, Springfield.

John C. Kraft is the H. Fletcher Brown Professor of Geology and Professor of Marine Geology at the University of Delaware, Newark.

George Rapp, Jr., is dean and professor of geology and archaeology, College of Letters and Science, University of Minnesota—Duluth.

Virginia Steen-McIntyre is a consulting tephrochronologist.

Arthur Steinberg is associate professor of archaeology in the anthropology/archaeology program at the Massachusetts Institute of Technology, Cambridge.

D. H. Tarling is a reader in palaeomagnetism, Department of Geophysics and Planetary Physics, University of Newcastle upon Tyne, England.

Salvatore Valastro, Jr., is associate director, Radiocarbon Laboratory, University of Texas, Austin.

John W. Weymouth is professor in the Department of Physics and Astronomy with an appointment in the Department of Anthropology, University of Nebraska, Lincoln.

Archaeological Geology

Edited by

George Rapp, Jr., and John A. Gifford

Yale University Press
New Haven and London

LP

Published with assistance from the Mary Cady Tew Memorial Fund.

Designed by Margaret E.B. Joyner
and set in Caledonia type by The Saybrook Press.
Printed in the United States of America by
Thomson-Shore, Inc., Dexter, Michigan.

Library of Congress Cataloging in Publication Data
 Main entry under title:

Archaeological geology.

 Bibliography: p.
 Includes index.
 1. Archaeological geology—Addresses, essays, lectures.
I. Rapp. George Robert, 1930–   . II. Gifford,
John A., 1947–
CC77.5.A73   1985   930.1   84-40201
ISBN 0-300-03142-4 (alk. paper)

10   9   8   7   6   5   4   3   2

7/11/90

To

George R. Gibson

an eminent geologist
whose long-term support and encouragement
of the application of geology to
archaeology helped make this volume possible

Contents

LIST OF FIGURES

LIST OF TABLES

Preface

This volume of fourteen papers, plus an extensive bibliography, illustrating the scope of research at the disciplinary boundary between geology and archaeology is curiously tardy in its appearance. Since the Second World War a considerable number of geologists have, to greater or lesser degrees, focused their research interests on the solution of practical and theoretical problems facing archaeologists in the initial recovery of data and subsequent analysis. The publication of such geologic contributions has most often been limited to appendexes in formal excavation reports of a particular project (*Nichoria I* by Rapp and Aschenbrenner being a notable exception), or as short technical notes, often in highly specialized journals. Consequently, a scholar in either of the two disciplines had no general reference work illustrating the wide range of application of geologic techniques to archaeology.

Many of these techniques were summarized in Brothwell and Higg's monumental *Science In Archaeology* but the second edition of that work is now fifteen years old. A more recent book devoted specifically to archaeological sediments, *Geoarchaeology* by Davidson and Shackley, was published in 1976; by its nature it did not assume or attempt complete coverage.

The editors of this volume also did not attempt complete coverage. Some chapters fell by the wayside, some methodologies were not programmed for inclusion, and additional techniques are being introduced yearly. New thrusts include determination of the thermal history of carbonized wheat grains by electron spin resonance, relative dating by the diffusion of ions into minerals, and the thermal history of cherts. Important and extensive areas such as lithic petrography and petrology are not covered at all; fortunately a new volume, *The Petrology Of Archaeological Artefacts* by Kempe and Harvey, covers this material very well. With the recent establishment of both the Archaeological Geology Division of the Geological Society of America and the Society for Archaeological Sciences, it seems fitting that comprehensive studies are beginning to appear defining the nature and scope of these disciplines.

The authors and editors have benefited from the counsel of colleagues too numerous to mention. We must acknowledge the important contributions of editorial assistants Judith Holz, Carolyn Kueny, and La Donna Harrison; the aid of Avis Hedin, who typed a multiple draft of each chapter; and the long-term support of Bill and Pat King, Bill and Patty O'Brien, Charles McCrossan, and Don Marken as well as George Gibson, to whom this volume is dedicated.

ARCHAEOLOGICAL GEOLOGY

1

History, Philosophy, and Perspectives

JOHN A. GIFFORD and GEORGE RAPP, JR.

ABSTRACT

Nineteenth-century investigations of human remains and artifacts associated with the bones of extinct animals represented a natural combination of the immature sciences of archaeology and geology and resulted in a body of data that was useful to the development of both. After the turn of the century, however, archaeology and geology crystallized into more distinct scientific endeavors, adopting, in the process, more rigorously (if implicitly) defined boundaries. The ease and frequency of collaboration at the boundaries of the disciplines consequently decreased somewhat. A major exception was the transformation, in England, of geological into geographic (physiographic/geomorphologic) studies that were pursued in archaeological projects in both England and the Near East before and after the Second World War.

In the examination and formalization of research paradigms that have characterized American anthropologically oriented archaeology since the 1960s, the place of geological studies in environmental and, ultimately, cultural reconstructions has been addressed by researchers at the archaeological-geological interface.

Archaeological geology as a recognized interdisciplinary field today exhibits characteristics of any learned profession, including a body of active practitioners, a professional organization, outlets for publication, and (to a lesser degree) opportunities for graduate instruction.

> The experience of mankind, not the inquiry of the specialist, has made the primary subdivisions of knowledge.
>
> —Carl O. Sauer, 1925

In this introductory chapter we outline our perception of the history of archaeological geology and its present relationships with its parent disciplines. This is
not an integrated historical narrative of the development of archaeological geology, such as Grayson's study of the establishment of human antiquity (1983), but
rather a short chronicle of events from published primary sources (cited in the
references) and secondary histories such as Willey and Sabloff (1974) and Taylor
(1948, chap. 1). We emphasize that the following chronicle is based only on
published works; it is certain that individuals who published little (or nothing) in
fact practiced what we would call archaeological geology (see the probable
example of G. C. Engerrand, as described in Graham, 1962).

At the end of this chronicle an empirical definition of archaeological geology
will emerge, one based on its relations to archaeology and geology. As Taylor
(1948:25) has stated, this really is a "definition by implication" and to a degree
begs the question of explicit definition, a problem we will address in the chapter
summation.

In the early nineteenth century an ever increasing corpus of information was
being assembled concerning the history of the earth before human existence,
just as antiquarian discoveries illuminating preclassical history were occurring in
Europe and the Near and Middle East. Both geology and antiquarian history
were crystallizing into legitimate scholarly fields whose disciplinary matrices
(Kuhn, 1970) were to be defined by two crucial publications: Lyell's *Principles
of Geology* (original edition, 1830–33) and Thomsen's *Guide to Northern Antiquities* (Danish edition, 1836; English translation, 1848). In both instances, the
theoretical bases presented were not innovative. Lyell wrote in a scholarly
environment that had absorbed Playfair's *Illustrations of the Huttonian Theory
of the Earth* (1802) and William Smith's *Strata Identified by Organized Fossils*
(1816); for Thomsen, numerous Renaissance and earlier writers (all ignored) had
already proposed the framework of a three-age system of stone, bronze, and iron
(Heizer, 1962). However, both Lyell's and Thomsen's works appeared at, and
amplified, critical junctures in the development of their respective disciplines.
What Heizer has written of Thomsen also holds true for Lyell: "The Thomsen
theory was advanced at the proper instant favorable to its acceptance, namely,
when prehistoric archaeology was beginning to be practiced on a scientific basis"
(1962:259). Both works provided a theoretical, temporal, and empirical framework for almost all nineteenth-century geological and archaeological field studies.

In chapter 1 of Lyell's *Principles* (11th revised edition, 1872), he compares
and contrasts at length the physical science of geology with history, a "moral"
science concerned with human affairs. Lyell turns at last to the nature of primary
material evidence for the two sciences:

> The analogy, however, of the monuments [physical remains] consulted in geology,
> and those available in history, extends no farther than to one class of historical
> monuments—those which may be said to be *undesignedly* commemorative of former

events. The buried coin fixes the date of the reign of some Roman emperor; the ancient encampment indicates the districts once occupied by invading armies; and the former method of constructing military defences; the Egyptian mummies throw light on the art of embalming, the rites of sepulture, or the average stature of the human race in ancient Egypt. The canoes and stone hatchets, called celts, found in our peat-bogs and estuary deposits, afford an insight into the rude arts and manners of a prehistoric race, to which the use of metals was unknown, while flint implements of a much ruder type point to a still earlier period, when man co-existed in Europe with many quadrupeds long since extinct. This class of memorials yields to no other in authenticity, but it constitutes a small part only of the resources on which the historian relies, whereas in geology it forms the only kind of evidence which is at our command. For this reason we must not expect to obtain a full and connected account of any series of events beyond the reach of history. [Lyell, 1872:3−4]

It was this distinction between the nature of geological and historical evidence that led Lyell and his geological colleagues to become the central figures in the first great (and controversial) inquiry of archaeological geology: the antiquity of the human race.

Three propositions concerning this question were implicitly tested and confirmed during the nineteenth century in the following order:

1. the "rude stone implements" known for centuries (for example, Dugdale, 1656, col. 1778) through chance finds in Europe were in fact artifacts and not "elf-arrows" or natural conglomerations;
2. fossil remains of extinct mammals found in the "post-Tertiary" (Pleistocene) strata of Europe, as well as evidences of major changes in landforms and drainage since their deposition, indicated that these deposits were of great antiquity;
3. therefore, stone implements directly associated with fossil faunal remains in Pleistocene deposits led to the conclusion that the humans who made them were also of great antiquity.

By even the eighteenth century some natural scientists were ready to admit the probable truth of proposition number 1, owing in large part to ethnographic documentation of stone-tool manufacturing among cultures of North and South America (Plot, 1686:397). Only the less erudite continued to hold that most of the stone implements were produced by lightning striking the ground. However, few scholars of that time took seriously any discoveries involving proposition 3, because the validity of proposition 2 was not completely acknowledged until the 1830s. Lyell's *Principles of Geology* effectively marshaled all the geological and paleontological evidence to prove that post-Tertiary deposits, although the most recent of all recognized European strata, still represented a very remote time.

That the 1830s were an intellectual watershed concerning proposition 3 is illustrated by the contrasting fates of several discoveries involving artifactual-fossil mammal associations. In 1829 J. Christol published a book concerning human and fossil mammal bones he had discovered in a cave in the Provence region of southern France. The book was largely ignored. His collaborator P. Tournal (1833), responding to critics who dismissed the association as due to mixing by the Noachian Flood, presented a carefully documented description of one particular cave's stratigraphic associations. Despite this, their evidence was still disregarded or alternatively explained (for example, by J. Desnoyers, 1849) as artifacts and animal bones of Celtic date.

In 1833, just as the last volume of *Principles of Geology* was appearing, the Belgian paleontologist Schmerling published the first volume of a descriptive compendium of cave deposits in the Meuse River valley. Most of the limestone caverns explored by Schmerling had never been entered purposefully by humans or animals, and he concluded that the fossil bones of both groups found therein had been washed in from the surface. Lyell (1863:68–69) portrays in imaginative but probably accurate detail the labors of Schmerling in securing this information, only to have his work also ignored or disparaged.

Finally, discoveries were made in the drift deposits of the Somme River of northern France that ultimately resulted in the grudging acceptance by European scientists of the coexistence, during a previous geological epoch, of humans and mammals now extinct. This began in 1846 with the publication of J. C. Boucher de Perthes's *De L'Industrie Primitive, ou les Arts et leur Origine* (reissued in 1847 under the title *Antiquités Celtiques et Antédiluviennes*), which documents that renowned amateur antiquary's discoveries (beginning in 1841) of flint "hatchets" with fossil mammal bones. Prestwich (1860:279) noted approvingly: "This work abounds in illustrations; and sections of the beds, drawn by Dr. Ravin, a most competent geologist, are given to show the position in which the flint implements were found. . . ." Boucher de Perthes's work was also largely ignored by the scientific community until Brixham cave in England was excavated and described by a Royal Society–sponsored project in 1858. This discovery was investigated by Joseph Prestwich, then an eminent geologist second only to Lyell. As Prestwich (1860:280) admitted, "It was not, however, until I had myself witnessed that the conditions under which these flint implements had been found at Brixham, that I became fully impressed with the validity of the doubts thrown upon the previously prevailing opinions with respect to such remains in caves."

His investigation led Prestwich to visit the Somme Valley in the spring of 1859, accompanied by the archaeologist (rather than antiquary) John Evans. They both returned to England convinced of the contemporaneity of flint implements and fossil mammals.

Anyone interested in the controversy concerning human antiquity in the mid-nineteenth century should seek out and carefully read Prestwich's article entitled "On the Occurrence of Flint-implements, associated with the remains of animals of Extinct Species in Beds of a late Geological Period, in France at Amiens and Abbeville, and in England at Hoxne." The clarity and detail of the title reflect the nature of Prestwich's entire presentation. In modern scientific terminology, he leaves no implicit assumptions in his hypothesis testing of the associations, and it is worth quoting his general conclusions:

1st.   The flint-implements are the result of the design and work of man.

2ndly.   That they are found in beds of gravel, sand, and clay which have never been artificially disturbed,

3rdly.   That they occur associated with the remains of land, freshwater, and marine Testacea, of species now living and most of them yet common in the same neighborhood, and also with remains of various Mammalia,— a few of species now living, but more of extinct forms.

4thly.   That the period at which their entombment took place was subsequent to the Boulder Clay [till] Period, and to that extent postglacial; and also that it was amongst the latest in geological time,—one immediately anterior to the surface assuming its present form, so far as regards some of its minor features. [1860:308−09]

(As is often the case in scientific developments, Prestwich was correct in these four conclusions, but wrong in a corollary. He deduced that the associations at Abbeville and Hoxne proved the survival of extinct mammals into the more recent prehistoric past, not that human tenure extended back to the "Early" postglacial. The concept of multiple glaciations was just emerging in the late 1850s, and human antiquity was to remain a relative term until well into the twentieth century.)

Evans was primarily responsible for verifying point number 1; he gave a similarly positive assessment of the Somme River associations in a paper delivered in 1859 before the Society of Antiquaries (published 1860). An interesting aspect of this early collaboration between a geologist and an archaeologist is the strong circumstantial evidence that, while at Abbeville, Prestwich and Evans arranged for the taking of a daguerreotype of a flint implement still in situ in a gravel pit face: one of the earliest instances of field documentation by photographic means.

It may be fairly said that the papers presented by Prestwich and Evans redirected the consensus of the European scientific community toward acceptance of a human antiquity older than that allowed in the Scriptures. At least the opponents to human antiquity (as evidenced in the "cave earths") thenceforth had to produce alternative scenarios for the associations that were supportable, within a uniformitarian framework, by geological processes. One such opponent was Charles Babbage, who had collaborated with Lyell in reconstructing the complex sequence of geological events that had affected the so-called Temple of Serapis on the Bay of Naples (Lyell, 1872:168−76). Babbage was a polymath with extraordinary capabilities of observation and analytical reasoning. He objected to the conclusions of Prestwich, Evans, and others on the grounds that "it is certainly premature to assign this great antiquity to our race, as long as the occurrence of such mixtures [of stone implements and fossil mammals] can be explained by known causes admitted to be still in action" (Babbage, 1859:59). He proceeds in his article to account fully for several hypothetical cave deposits containing assemblages of artifacts mixed with fossil mammal bones by assuming that they were deposited at different time intervals, and in doing so he remains strictly within the framework of known geological processes. Babbage's lucid arguments against contemporaneity help to explain why the controversy concerning human antiquity was to continue for several more decades.

Flint implements and bones of extinct mammals found together in a geological deposit: the implication (excepting arguments like those of Babbage) was of a great but indeterminable age for humanity. It points only to a time "beyond the reach of history." However, Lyell and other geologists, in their search to attach rates to the uniformitarian processes they were describing, realized early on that historical monuments could provide at least a key to sedimentation rates. The age of an accurately dated structure partially buried by sediment could be used to arrive at a quantitative estimate of the rate of sedimentation. Some illustrations of this approach are mentioned in early editions of *Principles of Geology*, but it was another geologist, Leonard Horner, who in 1851 first effectively utilized the approach of applying history to geology.

> If in a country in which a certain alteration of the land has occurred, we know that such alteration has taken place in part within historical time, and if the entire change under consideration presents throughout a tolerable uniformity of character, shall we not be justified in holding the portion that has taken place within the historical period to afford a measure of the time occupied in the production of the antecedent part of the same change? If a region existed where such a blending, as it were, of geological and historical time occurs, we may then be able to estimate in definite terms, the time that has elapsed since the change in the form and structure of the land under examination first began. [Horner, 1855:106]

Horner next in this paper identifies rivers as the most dynamic and yet uniform of geological agencies and singles out Egypt as the region containing both the oldest artificial structures and an intimately associated geological agent that has operated (he assumes) uniformly throughout its history: the Nile River and its alluvium. "To investigate the formation of the alluvial land found in the Valley of the Nile in Upper and Lower Egypt is therefore an object of the highest interest to the geologist and the historian" (p. 108).

Horner proposed this project to the Royal Society of London and received substantial monetary support for it (an early instance of interdisciplinary funding); he then arranged to have a number of pits excavated to the base of a standing obelisk at Heliopolis (now in the northern suburbs of Cairo) that had been erected, according to its translated heiroglyphs, by Senwosre I, a pharaoh of the Twelfth Dynasty. Horner's account of the fieldwork and his descriptions of the sedimentary units are worth examining in the original. Though his reasoning and methodological approach to the problem of alluviation rates of the Nile were correct, his final estimate (published in a subsequent report to the Royal Society) was questioned by, among others, Lyell himself (1863:38). Nevertheless, the concept of determining rates of geological processes by associated and datable artifactual material was admitted as a legitimate research strategy by British and Continental geologists, particularly the French.

The tenor of this early period of implicit interdisciplinary research is exemplified in a little-known archaeological journal (one of the first) that was founded in Paris in 1864. Its full title in English is: *Materials for the Primitive and Natural History of Man, A Monthly Report on Work and Discoveries Concerning Anthropology, Prehistoric Times, the Quaternary Epoch, Questions of Species and of Spontaneous Generation (with Illustrations)*. The founder and first editor was Gabriel de Mortillet (1821–1898), an archaeologist who shaped the late-nineteenth-century conception of the Upper Paleolithic of western Europe.

In looking through this prescient publication, some idea is gained of the ease with which geology and archaeology coexisted for several decades, viz.:

Volume 3 (January 1867, pp. 5–13) reported on the program of the International Congress of Anthropology and Prehistoric Archaeology, meeting in Paris that August. The organizing committee of the congress published six major research questions to be considered on successive meeting days in plenary session, of which the first two are:

Sunday 18 August: 1. Under what geological conditions, and with what faunal and floral associations, are to be found the earliest traces of human existence? What are the changes that have occurred since then in the distribution of land and sea?

Monday 20 August: II. Were caves generally inhabited? Was this habitation due to one and the same race and does it belong to one and the same time period? If not, how might the habitation be subdivided, and what are the essential characteristics of each subdivision?

Clearly it was axiomatic in 1867 that geological studies proceeded jointly with archaeological ones in deposits of the prehistoric period.

Volume 3 (November 1867, pp. 486–87) reviewed E. Chantre's recent article entitled "Paleoethnological Studies, or Geologico-Archaeologic Research on the Industries and Customs of Prehistoric Man in the North of Dauphiné and the Vicinity of Lyon," which dealt with the stratigraphy of several caves in the region and their content of flint implements and mammalian fossil bones.

In volume 22 (November–December 1888, pp. 592–93), the following four (of eight) fundamental research topics were proposed by the organizing committee for discussion at the tenth meeting of the above-mentioned international congress, which was held again in Paris in 1889:

I. Cutting and filling of valleys and filling of caves, in relation to the antiquity of man.

II. Periodicity of glacial phenomena.

III. The art objects and stone industries found in caves and alluvium: the value of paleontological and archaeological classifications as applied to the Quaternary Epoch.

IV. Chronological relationships among the civilizations of the Stone, Bronze, and Iron Ages.

In the same issue it is explained that due to continued financial losses *Materiaux* would henceforth be merged with two other journals dealing with "natural human history" (*Anthropology Review* and *Ethnography Review*) to form a new bimonthly journal titled *Anthropology*. In its twenty-four-year lifespan *Materiaux* exemplified nothing so much as a qualitative forerunner of the present interdisciplinary journal *Quaternary Research*.

American archaeologists during the second half of the nineteenth century were, according to Willey and Sabloff (1980:48), spurred to discover traces of human antiquity comparable to the Paleolithic remains being excavated in Europe. As was the case there, such searches in America involved as much geological as archaeological knowledge. Charles Whittlesey, who was identified as a geologist and archaeologist of the Smithsonian Institution in the 1867 membership list of the International Congress of Anthropology and Prehistoric Archaeology, presented a review (1868) of the evidence for human antiquity in the United States. He describes three caves (in Ohio, Kentucky, and New York) in which burials and artifacts were discovered; all evidence pointed to their belonging to the known American Indian cultures. This fact, in conjunction with some early geological estimates of sedimentation rates, suggested to Whittlesey that the "Indian Race" had been present in the eastern states between one and two thousand years ago. He briefly mentions (p. 278) the just developing question of whether human remains associated with bones of extinct mammals actually were to be found in North America.

In his review Whittlesey confronts a fundamental problem of nineteenth-century American archaeology and geology, which was the absolute dating of prehistoric cultures. He recognizes three: the Indian race, the race of the Mound Builders, and "fossil men, contemporary with the elephant, mastodon, horse, megalonyx, and other mammalia of the Quaternary." Several possible instances of "Quaternary Man" are noted by Whittlesey, but he concedes that "unfortunately there is . . . in all the cases quoted in the United States, pertaining to the Quaternary Period, a degree of uncertainty in the evidence, which is fatal to a scientific result" (pp. 285–86). He concludes, "In every instance where we descend below the alluvium in search of human remains and relics we are thus far met by conflicting testimony as to the facts."

What remained lacking throughout the nineteenth and into the early twentieth century was an expertise and sophistication in Quaternary geology that could impose a time frame on the numerous archaeological discoveries then being made. Without such a framework, and some "refinement in geological-archaeological excavation techniques" (Willey and Sabloff, 1980:48), much time and energy were expended by investigators such as G. F. Becker, F. W. Putnam, John Henry Haynes, C. C. Abbott, W. H. Holmes, N. H. Winchell, and A. Hrdlička in either defending or attacking numerous archaeological sites in the United States that were the redoubts of "Glacial Age Man." A long chapter could easily be devoted to any one of these researchers in regard to their embodiment of an interdisciplinary approach (pro or con) to the question of human antiquity in the New World.

One major conclusion that arises from the records of these sometimes acrimonious debates is that whatever detrimental effects the human-antiquity controversy had, it in fact benefited the development of Quaternary geology. It did so in at least two ways: by refining geologists' conception of the absolute duration of the Quaternary epoch and by stimulating investigations of the sedimentology and paleoenvironmental significance of glaciofluvial and periglacial sediment deposits. An editorial comment by N. H. Winchell in the *American Geologist* illustrates this positive effect. It ends with the statement: "The general results of the controversy about the antiquity of man has been to correct the estimates of geological time. Let the discussion go on. Let us not discourage it even by the suggestion that it is beneath the dignity of the true geologist to make estimates of geological time in years." Here we may see, perhaps, an early manifestation of the conceptual schism that exists to this day between Quaternary and pre-Quaternary geologic time (Vita-Finzi, 1973). A second effect was the impetus archaeological studies gave to the interpretation of Quaternary sediments. Such was the case of the human skeleton found in a mixed deposit of loess and slope wash, twenty-five feet beneath the modern surface at Lansing, Kansas, in 1901. Numerous geologists and archaeologists visited the site in the following year, and the several detailed reports on the geological relations of the skeletal remains, particularly that of T. C. Chamberlin (1902), illustrate how observations on sediment deposition and fluvial processes were spurred by such archaeological discoveries. One of Chamberlin's closing remarks reflects attainment of a degree of geological-archaeological refinement that had been lacking a few decades earlier: "I beg to invite the attention of archeologists to the slight grounds for hope of finding really strong evidence of man's antiquity in the fluvial deposits of the glacial rivers, because of the liability of these deposits to deep overworking by scour and fill" (p. 776). In fact, a decade earlier W. H. Holmes had dealt with the same phenomenon—liability of fluvial deposits to reworking—in his definitive refutation of C. C. Abbott's "glacial man" stone artifacts from the gravels of the Delaware River near Trenton, New Jersey (Holmes, 1893).

The decades on either side of the turn of the century evidenced a great interest in archaeological geology. W. H. Holmes, while an archaeologist with the Bureau of Ethnology, served as the "archaeologic geology" editor of the newly formed *Journal of Geology*. His appointment represents an explicit contemporary recognition of the subdiscipline. Further, human antiquity in the New World served as the topic of the Geological Society of America's presidential address (Winchell, 1903), and a subsequent president of that society, Raphael Pumpelly, would lecture on his archaeological-geologic fieldwork in Turkestan (1906).

Pumpelly's survey and excavations at Anau, near Ashkhabad on the north slope of the Kopet-Dag Mountains, between Iran and Soviet Central Asia, were a remarkable manifestation of interdisciplinary field archaeology for their time. Pumpelly was a geologist whose earlier survey work in China and Russian Turkestan led him to an appreciation of the interdependent evolution of prehistoric cultures and the natural environment in a region that has more recently been termed the Dry Belt of Asia (von Wissman, 1956). Pumpelly's appreciation of the value of all excavated remains, rather than just artifacts, is exemplified by statements such as the following:

> In view of the importance that attaches to the question of the origin of our domesti-
> cated animals, I collected systematically, foot by foot from the bottom, all the bones of
> animals found in the older two cultures—that is, in the whole height of the north
> kurgan [mound]—and submitted nearly half a ton of these to Dr. Duerst, comparative
> anatomist and archeological osteologist at Zurich. [Pumpelly, 1906:648]

In the final publication of the Anau excavations (1908), supplementary reports by Pumpelly's collaborators deal with the regional geomorphology of Russian Turkestan, the physiography of habitation mounds, and analyses of metal artifacts, animal bones, human remains, stone tools, and paleoethnobotanical material. (This last report touches on the utility of *phytoliths* for confirming the presence of cultivated grains in certain stratigraphic levels of the excavation!)

Approaching the question of dating the Anau culture (which is now known to have flourished in the fourth millenium B.C.), Pumpelly naturally estimates its age based on rates of sediment accumulation. Although not very accurate, this geological technique represented the only basis for absolute chronologies available to archaeologists before de Geer's varve and Douglass's tree-ring dating techniques.

Pumpelly's survey and excavations (which could be the basis of a fascinating modern history, occurring as they did in the Asiatic borderlands of Imperial Russia just before the first stirrings of the Revolution) mark the apex, in some ways, of archaeological-geological research prior to the 1950s. One is tempted to perceive the history of the subdiscipline in terms of individuals rather than larger social forces and circumstances precisely because of people such as Pumpelly—his work is proof that what we today would call archaeological geology could be and in fact was being done at the turn of the century. We must turn to the question of why others did not follow his lead.

In the United States, and to a lesser degree in Europe, there appears to have been a recession of archaeological-geological collaborative projects around the period of the First World War and into the 1920s. Two possible explanations may be advanced for this:

1. The transformation of archaeology from a natural to a more social science. In Daniel's view (1975:239–42) the increasing manifestation at the turn of the century of human geography and ethnographic anthropology in prehistoric archaeological research meant a diminishing of geology's influence. "It was a complete reorientation of [archaeologists' perception of] prehistoric material and marked the change from the study of man as an animal to the study of him as a human being" (p. 243). He continues: "The new European prehistory which we see coming into being in the mid-twenties of the present century was, quite frankly, not geological in outlook" (p. 248). Taylor (1948:19) draws the same

conclusion regarding prehistoric archaeology's drift from the natural sciences toward anthropology at the end of the nineteenth century.

2. The disciplinary boundaries of archaeology were stabilizing in the last decades of the nineteenth century, with the result that students in Harvard's or Pennsylvania's newly established programs of archaeology and anthropology could not expect to graduate with the eclectic background of the previous generation of archaeologists. The same process may have operated, perhaps to a lesser degree, in graduate studies in geology. It is certainly true that anthropological archaeology as taught in the United States, already heavily influenced by contemporary ethnographic research, widened its divergence from the European concept of the discipline (Willey and Sabloff, 1974:86).

In Paleolithic studies collaboration between physical anthropologists and geologists continued during the first decades of this century, for no one could deny its utility in establishing relative chronologies. The American vertebrate paleontologist H. F. Osborn, in his *Men of the Old Stone Age: Their Environment, Life and Art* (1916), demonstrated that he was capable of explicating the humanistic as well as the natural scientific aspects of Paleolithic archaeology; G. G. MacCurdy's similar review in 1924 was comparably interdisciplinary in its treatment of human prehistory through the Neolithic.

The influence on European archaeology of geological principles was always to be greater than in the United States, due to the Europeans' early adoption and refinement of stratigraphic excavation techniques. Willey and Sabloff (1980:84–88) note that the "stratigraphic revolution" in American archaeology was instigated by two young anthropological archaeologists (M. Gamio and N. C. Nelson) who had been fundamentally influenced by European excavation techniques.

At this point we will digress from the relatively turbulent investigations of human antiquity to a brief consideration of one focus of archaeological geology that has enjoyed a large measure of success and a small amount of controversy. This is the field of petrological examination of artifacts for technical and provenance studies. One of the first geologists to undertake such studies was André Damour, whose reports on the macroscopic characteristics and physiochemical properties of stone axes were published in the Proceedings of the Academy of Science, Paris, in 1865; they later were summarized (presumably for archaeologists who did not belong to the academy) in de Mortillet's *Materiaux*. In his 1865 report Damour shows an appreciation of the ever present problem of determining the source of common lithic materials:

> Because of the abundance and distribution of these siliceous minerals [quartz, agate, flint, jasper] at many locations on the continents, it will always be difficult to specify the source of the majority of specimens of axes or other objects made of these minerals: only for a small number of well-characterized varieties, either by color or by a constant distribution of light and dark tints, could one indicate the sources with some degree of certainty.

(It should be noted that Damour had not adopted H. C. Sorby's recently developed techniques of thin-section preparations and microscopic examination.) Damour also explicitly realizes another current problem of provenance studies:

> One can see . . . that before arriving at specific conclusions regarding Celtic axes and their utilization, in order to aid in resolving problems of human migration, it is necessary to analyze and compare a large number of specimens presently scattered in French and foreign collections. We may, all the same, predict at the present time that the mineral species that will permit us to draw various probable conclusions concerning the movements and contacts of ancient peoples will be limited to a small number, particularly those for which the sources are limited to a few localities on the earth's surface. [Damour, 1865]

With the establishment of microscopic petrography during the last few decades of the nineteenth century, two geologists—G. R. Lepsius in Germany and H. S. Washington in America—devoted a part of their petrographic researches to the provenance of marble statuary of the classical world. (N. Herz gives details of their work in chap. 13.) Here we only record that Washington introduces his 1898 article thus:

> It was suggested to me, who, though a petrologist, have taken, and still take, much interest in archaeological matters, that it would be of value to explain to archaeologists, who, it may be assumed, know little or nothing of petrography, the principles on which such [provenance] conclusions rest, the methods of examination of a given specimen, and how great a degree of confidence may be placed in the identification of the source of the material of a statue.

This he proceeds to accomplish in a very lucid presentation; it is certain that any classical archaeologist who read Washington's article at the turn of the century came away knowing a great deal more about the petrography of marbles.

Several British petrographers became interested in the provenance of the many types of Neolithic stone artifacts and structures to be found around their island. A major achievement growing out of this work was the identification by H. H. Thomas (1923) of the Prescelly Mountains in Wales as being the source of the "blue stone" circle of Stonehenge. This work was systematized in the 1930s under the Council for British Archaeology's Stone Ax Project (cf. Shotton, 1969) and, through its publication, has allowed several sophisticated analyses of economic interaction in the British Neolithic (e.g., Hodder, 1974).

Although archaeologists have occasionally raised some questions about the reliability of petrological examinations (see p. 332), they generally welcome the specific identifications and analyses provided by geologists. This is not so in disagreements concerning excavation, interpretation, and dating, over which geologists and archaeologists often have challenged each others' powers of observation and capabilities of identifying primary versus accidental associations. To return to the central theme of this short history, the decades of the

1910s and 1920s saw American geologists, at least, contributing mostly puzzling and controversial information to the problem of human antiquity in the New World.

The debate concerning "ancient man in Florida"—specifically, the putative associations of human remains and those of extinct vertebrates in Pleistocene deposits at Vero and Melbourne on the east central coast—exemplifies the strained nature of interactions between American geologists and archaeologists during this period. E. H. Sellards, Florida state geologist, was the protagonist (1916, 1919), while Aleš Hrdlička of the Smithsonian Institution led the opposition in this dispute, as well as most others (Hrdlička, 1917, 1918). The drawn-out controversy was a contributing factor in Sellard's move to the Texas Bureau of Economic Geology in 1918. On the positive side, debate about the association of the human and faunal remains in the Melbourne-Vero area generated some careful studies of the vertebrate assemblage (Hay, 1926) and the application of relatively sophisticated excavation techniques (Gidley, 1928) that, by their publication, contributed to the development of American archaeology. (The final verdict concerning this particular association at Melbourne-Vero is not yet in; cf. Rouse, 1951, with Stewart, 1946.)

It was not until 1927 that a discovery of artifacts in association with extinct vertebrate faunal remains was generally acknowledged to support the relatively high antiquity of human entry into the New World: this find was made in northeastern New Mexico near the town of Folsom (Figgins, 1927), and it precipitated a renewed interest in the geological aspects of early archaeological sites. For example, in June of 1931 a joint symposium sponsored by the Geological Society of America and the Anthropology Section of the American Association for the Advancement of Science was held in Los Angeles. It was titled "The Antiquity of Man" and participants included J. C. Mirriam, A. S. Romer, C. Stock, and A. Hrdlička (Mirriam, 1932). In 1935 E. B. Howard, who had excavated at the Burnett cave and Clovis sites in New Mexico, published a review of the evidence for human antiquity in the New World that drew equally upon geological and archaeological data. It is amazing that such a detailed study could appear only seven years after the Folsom discovery.

During the early 1930s Ernst Antevs published extensively on the sequence of ice advances and retreats in North America during the Wisconsin glaciation; his several papers were climaxed in 1935 with the assertion that humans "had reached the Southwest at the age of transition between the pluvial and the post-pluvial epochs, roughly 12,000 years ago." Clearly, once artifacts could be related indisputably to extinct faunal assemblages and paleoenvironments, Quaternary geologists in the United States were ready with the data needed to date and reconstruct the entry of humans into the New World. Geological studies in support (or disputation) of these early finds represented some of the best archaeological-geological fieldwork done on the Quaternary deposits of parts of the United States in the 1930s through the 1950s (e.g., Albritton and K. Bryan, 1939; F. Bryan, 1938; Judson, 1953). After 1935, the topic of human antiquity centered around establishing the age of lithic complexes and human remains older than the Upper Paleolithic "big game hunters," and it remains so today (Simpson, 1978; Adovasio et al., 1978; Bada and Finkel, 1982). Geologists, it is safe to assume, will continue to be deeply involved.

Mention should be made as well of the less controversial environmental reconstructions of Hack (1942) and K. Bryan (1941, 1954), which show a definite geographic influence. In the post–World War II period several long-term interdisciplinary studies were instituted in the Near East and Central America; R. J. Braidwood's investigation of the origins of village-farming communities (Braidwood and Howe, 1960) exemplifies work in the former, and the Tehuacán Archaeological Botanical Project (MacNeish and Nelken-Terner, 1967) in the latter area.

As noted above, prehistoric archaeology in Europe as well as in the United States was beginning to feel geographic and ethnographic influences around the period 1890–1910. Possibly due to Friedrich Ratzel's work, German prehistorians and archaeological geologists working with them seem to have manifested this influence more than others. It is particularly evident in Schmidt's three-volume work (1912) on the diluvial prehistory of Germany and continued to be exemplified in Soergel's work in the 1920s and 1930s on Paleolithic skeletal remains and Quaternary alluvial sequences in central Europe (e.g., Soergel, 1939).

In England, too, emphasis was on an environmental history/geographical approach to prehistoric archaeology even before the mid-1930s, when F. E. Zeuner moved from Germany to London and established the Department of Environmental Archaeology at the Institute of Archaeology. Numerous archaeological geology studies were conducted on British sites in the early twentieth century (see Zeuner, 1959, for references). One of these geologists, K. S. Sandford of Oxford University, was recruited by J. H. Breasted of the University of Chicago's Oriental Institute to direct the interdisciplinary Prehistoric Survey Expedition of the Nile Valley in 1926 (Sanford and Arkell, 1928, 1929). Sandford's major publication of his Nile Valley work (1934) is a good example of the British environmental-geographical approach to archaeological geology. So also are the collaborative studies of the archaeologist G. Caton-Thompson and the geologist E. W. Gardner in the Kharga Oasis (1932), the Fayum Desert (1934), and the Hadramaut region of Arabia (1939). Comparably interdisciplinary projects were undertaken by Garrod and Bate at Mt. Carmel in Palestine (1937), by McBurney and Hay in Cyrenaica (1955), by Leakey at Olduvai Gorge (1965), and by Higgs and Vita-Finzi in Epirus, Greece (1966). Generally, British archaeologists were much more receptive to interdisciplinary field projects, an attitude reflected in articles like that of Boswell (1936) and North (1938). The latter (p. 115) suggested three aspects of basic geology with which, he believed, all archaeologists should have some familiarity:

1. the interpretation of geologic maps,
2. the determination of natural versus artificial formation of large stone structures, and
3. an appreciation of site geology and the range of locally available lithologic materials.

These attainments remain as valid and desirable today as they were almost half a century ago.

In the preceding sections we have attempted to trace the manner in which archaeological geology has developed. Since the early 1970s research at the interface between geology and archaeology has been influenced strongly by the redirection of the conceptual framework of anthropological archaeology in the United States. Butzer (1982:3–4) points out that while the new archaeological paradigm incorporates many new approaches from fields as diverse as ethno-archaeology and computer simulation, the environment remains a poorly articulated variable in the analysis of archaeological context. Building on his own research over the last twenty years, Butzer (1982) has attempted to operationalize what he had termed five years earlier a "fuller understanding of the human ecology of prehistoric communities" (Butzer, 1977). Geology figures in this archaeological endeavor by its contribution of methods, techniques, and concepts. To Butzer this distinguishes *geoarchaeology*—that is, archaeology pursued with the help of geological methodology—from *archaeological geology*, "geology pursued with an archaeological bias or application" (Butzer, 1982:5).

We believe this distinction is a valid one, well reasoned by Butzer, and not as recondite or overprecise as might appear to the outside reader. Butzer defines the ultimate goal of collaboration at the geology-archaeology disciplinary boundary as "to elucidate the environmental matrix intersecting with past socioeconomic systems and thus to provide special expertise for understanding the human ecosystems so defined" (Butzer, 1982:40). He admits that the task is not easy, and it would seem that all collaborative efforts before about 1970, and most since then, do not represent geoarchaeology according to his definition. To the extent that the research presented here does not elucidate Butzer's contextual parameters of space, scale, complexity, interaction, and stability as they apply to human ecology, our volume title is correctly chosen. Though many of the contributors to this volume are "part-time practitioners," we believe the work presented makes significant contributions to the elucidation of past cultures from a disciplinary stance that is more geological than archaeological.

In their history of the Association of American Geographers, James and Martin listed four characteristics of a developed learned profession:

1. a number of scholars actively working on related problems, and in close enough contact with each other that ideas are quickly disseminated, and critical discussion stimulated;

2. departments in universities offering advanced instruction in the concepts and methods of the field;
3. opportunities for qualified scholars to find paid employment in work related to the profession;
4. an organization such as a professional society, to serve the interests of the profession and provide a focus for professional activities [1979:8].

These present a useful perspective from which to investigate archaeological geology.

All scientific disciplines exist to accumulate and disseminate tested knowledge. From the nineteenth century onward there has been apparently no lack of outlets for the publication of archaeological geologists' research. Of the major university presses, that of Cambridge in England has the longest record of such publications. Academic Press, a commercial publisher, has brought out a number of interdisciplinary volumes linking archaeology with natural sciences, including geology (e.g., Sheets and Grayson, 1979; Masters and Flemming, 1982). The publisher also of both the explicitly interdisciplinary journal *Quaternary Research* and the *Journal of Archaeological Science* (which occasionally contains articles on archaeological geology), Academic Press deserves acknowledgment of its role in disseminating research results in this field.

However, most research publications in archaeological geology (and geoarchaeology) appear not in book form but through other publication outlets, to some extent in monograph series but primarily in refereed journals. Serial publications are usually published by smaller university or departmental presses on a regular or occasional basis, as, for a general example, the Geoscience and Man Series of the School of Geoscience, Louisiana State University. Publications in such series are often of high caliber and more than regional interest. One interesting development for archaeological geologists is the commercial publisher British Archaeological Reports (BAR), which produces an international monograph series that has included interdisciplinary research.

Refereed subscription journals, other than the pan-professional *Science*, *Nature*, *Scientific American*, and *American Scientist*, may be categorized as geological, archaeological, or archaeometric by subject content and readership. Those journals in the first category that occasionally may publish articles on archaeological geology include the *Geological Society of America Bulletin*, *Geology*, *Geologie en Mijnbouw*, *Quaternary Research*, and *Environmental Geology*.

English-language archaeology journals include *American Antiquity*, the *American Journal of Archaeology*, *Antiquity*, the *Journal of Field Archaeology*, the *International Journal of Nautical Archaeology*, the *Proceedings of the Prehistoric Society*, and *World Archaeology*. In any of these one may find archaeological geology articles, or sections of articles, relating to site excavations, surveys, or other specific archaeological projects. The archaeological-geological information, as an appended specialist report, often will not be indicated by the article's title.

The third category, of archaeometric journals, would include *Archaeometry*, the *Journal of Archaeological Science*, the *MASCA Journal* (of the Museum of Applied Science Center for Archaeology at the University of Pennsylvania), and the *SAS Research Notes*, a supplement of the Society for Archaeological Sciences quarterly newsletter. This last is a new outlet for rapid publication of current research and interim reports on, among other fields, archaeological geology. By definition, archaeological geology reports appearing in these publications will be skewed toward geochemical and geophysical research.

A final, and labyrinthine, publication outlet for archaeological geology research is through United States government documents, often in the form of reports to the Department of the Interior's National Park Service. Perhaps the most efficient way of keeping track of these is through the National Technical Information Service citations of federally funded research, using appropriate keywords.

This compilation of archaeological geology publications, while not inclusive, indicates the very wide range of printed material that we must review regularly to stay informed. It is a fact of life that, by and large, archaeologists do not read geology publications and vice versa. At least two serious attempts have been made recently to launch a journal of archaeological geology (or geoarchaeology), which, it is assumed, would present the ideal publication outlet for we who are torn in at least two directions for readers of our articles. But perhaps more consideration of a new journal is needed. It may be that neither geologists nor archaeologists would have the time or inclination to peruse such a specialized journal. If we, as interdisciplinary scientists, are trying to reach the mainstreams of both professions, a strictly interdisciplinary publication might compromise the usefulness of our research efforts.

Ultimately, a discipline has the power to prescribe the agenda for university training of new recruits to the profession. Archaeological geology has not yet arrived at this stage of development. There are only a very few universities where archaeological geology may be designated as a focus for graduate study. Some universities allow this focus in their geology departments, a few in their anthropology departments, and one in an interdisciplinary program in ancient studies. While most geology departments are willing to accept interdisciplinary associations with chemistry, physics, or biology, few have established an equivalent relationship with archaeology.

An additional reason for the slowness of academic geology to provide a curricular base for archaeological geology may be the lack of easily identified career positions for graduates of interdisciplinary programs. As an increasing number of researchers in such organizations as the United States Geological Survey devote their efforts to problems at the interface between geology and archaeology, this situation may improve.

Whereas publication outlets have always been plentiful, it is only in the last decade or so that sufficient opportunities existed for the presentation and exchange of ideas at professional meetings. During this time archaeological geology reached a certain "critical mass"—the number of investigators working on related problems became large enough to support an infrastructure of organizations, meetings, and related requirements of a functioning discipline. In 1974 Rapp et al., using both terms *geoarchaeology* and *archaeological geology*, suggested the establishment of an organizational framework within the Geological Society of America for geoscientists engaged in archaeological work. This was the impetus for the establishment in 1977 of the Archaeological Geology Division of the Geological Society of America, which now provides technical sessions and a symposium devoted to archaeological geology at the annual meetings of the society. Since the early 1970s there also have been special sessions devoted to archaeological geology at the annual meetings of the American Quaternary Association, the Society for American Archaeology, the Society for Archaeological Sciences, and the Archaeological Institute of America.

Archaeological geology, in its present stage of development, is undergoing an introspective assessment of what it is and what it should be. In the past eight years at least eight papers have focused on some aspect of the definition or organization of archaeological geology or geoarchaeology.

Butzer's most recent and explicit consideration (1982) has been mentioned above. In a 1975 assessment of the ecological approach to archaeology, Butzer decried the lack of integration of natural scientists working with archaeologists into the planning and the interpretation stages of an excavation. He also called attention to the lack of opportunities within universities for cross-disciplinary research by graduate students.

In the first of a series of articles on the archaeological field staff, Rapp (1975) detailed the wide range of tasks an archaeological geologist may undertake prior to, during, and subsequent to excavation. This paper stresses the uses of geological techniques rather than conceptual problems of integrating the geological with the archaeological studies. In a review of two books that would fall under the category of archaeological geology, Butzer (1977) again addresses the "vital" need for geoarchaeological study in an ecological approach to prehistory. A primary role of geoarchaeology would be the identification of the spatial micro-, meso-, and macroenvironments. (This consideration of the scale of environmental analysis became one of the five central themes of contextual archaeology elucidated in his 1982 publication.) Butzer also lamented the shortage of effective geoarchaeologists.

In a lengthy paper on the contributions of geoarchaeology to environmental reconstruction of archaeological contexts, Gladfelter (1977) narrows the focus of geoarchaeology to the use of geomorphology and sedimentary petrography to elaborate these micro-, meso-, and macroenvironments. His paper presents a detailed exposition of the geomorphological contexts of archaeology and considers in particular the importance of the sedimentary record (materials and textures). Hassan (1979) presents a succinct review of the scope and nature of geoarchaeology and offers a list of nine geoarchaeological topics, illustrating the diverse nature of the techniques earth scientists use in the resolution of archaeological problems.

The most thorough review of the present status of geoarchaeology is given by Gladfelter (1981) in a paper that discusses "what is geoarchaeology," "geoarchaeology in practice," the "conceptual base for geoarchaeology," and "training in geoarchaeology." In this review Gladfelter enlarges somewhat his 1977 view of the scope of the discipline. Finally, a recent addition to the literature on the nature of archaeological geology is an article by the present writers (Rapp and Gifford, 1982) that traces some of the history of the discipline and presents examples of recent research to illustrate the scope of the field.

These papers focus on defining the field. In addition, a number of important statements about the nature and role of archaeological geology or geoarchaeology have been included in publications with a different focus. Notable among these are Renfrew (1976), Bullard (1978), Harris (1979), Shackley (1975), and Butzer (1974, 1982). It might be noted that the book with the title *Geoarchaeology: Earth Science and the Past* (Davidson and Shackley, 1976) is not a survey of the discipline but the publication of an interesting symposium devoted to sediments and what can be learned from them in archaeology.

The emergence of archaeological geology as a specialty has not been the result of a major break with disciplinary tradition, in the Kuhnian sense, but rather a melding at an interdisciplinary boundary. There is some question about whether the term *specialty* should be used. As is apparent from the subsequent chapters in this book, there are many technical specialties involved in the broad discipline of archaeological geology. If we include specialists who would call themselves geoarchaeologists, there are also widely divergent cognitive orientations among practitioners interdisciplinary between geology and archaeology. "The application of geological principles and techniques to the solution of archaeological problems" is perhaps the best definition we can offer for archaeological geology. It is still geology, as are petroleum geology, environmental geology, and glacial geology, but represents a subfield distinct from others in the general science. *Archaeological geology* and *geoarchaeology* do not characterize two ends of a spectrum of techniques, but rather two contrasting and legitimate research goals. We present the following chapters as exemplars of the former.

REFERENCES

Adovasio, J. M., J. D. Gunn, J. Donahue, R. Stuckenrath, J. Guilday, and K. Lord. 1978. Meadowcraft rockshelter. In *Early man in America*, ed. A. L. Bryan, 140–80. Edmonton, Alberta: Archaeological Researches International.

Albritton, C. C., and K. Bryan. 1939. Quaternary stratigraphy in the Davis Mountains, Trans-Pecos Texas. *Bulletin of the Geological Society of America* 50:1423–74.

Antevs, E. 1935. The spread of aboriginal man to North America. *Geographical Review* 25:302–09.

Babbage, C. 1859. Observations on the discovery in various localities of the remains of human art mixed with the bones of extinct races of animals. *Proceedings of the Royal Society of London* 10:59–72.

Bada, J. L., and R. Finkel. 1982. Uranium-series ages of the Del Mar man and the Sunnyvale skeletons. *Science* 217:755–56.

Boswell, P. G. H. 1936. Problems of the borderland of archaeology and geology. *Proceedings of the Prehistoric Society*, n.s. 2:149–60.

Boucher de Perthes, J. C. 1849. *Antiquités celtiques et antediluviennes*, vol. 1. Paris: Treuttel et Wurtz.

Braidwood, R. J., and B. Howe. 1960. *Prehistoric investigations in Iraqi Kurdistan*. Studies in Ancient Oriental Civilization, no. 31. Chicago: Univ. of Chicago Press.

Bryan, F. 1938. A review of the geology of the Clovis finds reported by Howard and Cotter. *American Antiquity* 2:113–30.

Bryan, K. 1941. Pre-Columbian agriculture in the Southwest as conditioned by periods of alluviation. *Annals of the Association of American Geographers* 31:219–42.

———. 1954. The geology of Chaco Canyon, New Mexico, in relation to the life and remains of the prehistoric peoples of Pueblo Bonito. Smithsonian Miscellaneous Collections, vol. 122, pp. 1–65.

Bullard, R. 1978. Geology in field archeology. In *A manual of field excavation*, ed. W. G. Dever and H. D. Lance, 197–235. New York: Hebrew Union College.

Butzer, K. W. 1974. Geo-archaeological interpretation of Acheulian calc-pan sites at Doornlaagte and Rooidam (Kimberly, South Africa). *Journal of Archaeological Science* 1:1–25.

———. 1975. The ecological approach to archaeology: Are we really trying? *American Antiquity* 40:106–11.

———. 1977. Geo-archaeology in practice. *Reviews in Anthropology* 4:125–31.

———. 1982. *Archaeology as human ecology*. Cambridge: Cambridge Univ. Press.

Caton-Thompson, G., and E. W. Gardner. 1932. The prehistoric geography of Kharga Oasis. *Geographical Journal* 80:369–409.

———. 1934. *The Desert Fayum*. London: Royal Anthropological Society.

———. 1939. Climate, irrigation, and early man in the Hadramaut, southwest Arabia. *Geographical Journal* 93:18–38.

Chamberlin, T. C. 1902. The geologic relations of the human relics of Lansing, Kansas. *Journal of Geology* 10:745–77.

Christol, J. de 1829. *Notice sur les ossements humains fossiles des cavernes de département du Gard*. Montpellier: J. Martel.

Damour, A. 1865. Sur la composition des haches en pierre trouvés dans les monuments celtique et chez les tribus sauvages. *Comptes rendus Academie des Sciences, Paris* 63:1038–50.

Daniel, G. 1975. *A hundred and fifty years of archaeology*. London: Duckworth.

Davidson, D. A., and M. L. Shackley, eds. 1976. *Geoarchaeology: Earth science and the past*. London: Duckworth.

Desnoyers, J. 1849. *Grottes ou Cavernes: Dictionnaire Universelle d'Histoire Naturelle*, vol. 6, pp. 342–407. Paris: Renard et Martinet.

Dugdale, W. 1656. *The antiquities of Warwickshire*. London: Warren.

Evans, J. 1860. On the Occurrence of Flint Implements in Undisturbed Beds of Gravel, Sand and Clay. *Archaeologia* 38:280–307.

Figgins, J. D. 1927. The antiquity of man in America. *Natural History* 27 (3):229–39.

Garrod, D. A. E., and D. M. A. Bate. 1937. *The Stone Age of Mt. Carmel*. Vol. 1, *Excavations at the Wady el-Mughara*. Oxford: Clarendon Press.

Gidley, J. W. 1928. Ancient man in Florida: Further investigations. *Bulletin of the Geological Society of America* 40:491–502.

Gladfelter, B. G. 1977. Geoarchaeology: The geomorphologist and archaeology. *American Antiquity* 42:519—38.

———. 1981. Developments and directions in geoarchaeology. In *Advances in archaeological method and theory*, vol. 4, ed. M. B. Schiffer, New York: Academic Press.

Graham, J. A. 1962. George C. Engerrand in Mexico, 1907—1917. *Bulletin of the Texas Archaeological Society* 32:19—31.

Grayson, D. K. 1983. *The Establishment of Human Antiquity*. New York: Academic Press.

Hack, J. T. 1942. *The changing physical environment of the Hopi Indians*. Papers of the Peabody Museum of American Archaeology and Ethnology, vol. 25, no. 1.

Harris, E. C. 1979. The laws of archaeological stratigraphy. *World Archaeology* 11: 111—17.

Hassan, F. A. 1979. Geoarchaeology: The geologist and archaeology. *American Antiquity* 44:267—70.

Hay, O. P. 1926. On the geological age of Pleistocene vertebrates found at Vero and Melbourne, Florida. *Journal of the Washington Academy of Science* 16:387—91.

Heizer, R. F. 1962. The background of Thomsen's Three-Age System. *Technology and Culture* 3:259—66.

Higgs, E. S., and C. Vita-Finzi. 1966. The climate, environment, and industries of Stone-Age Greece: Part II. *Proceedings of the Prehistoric Society* 32:1—29.

Hodder, I. R. 1974. A regression analysis of some trade and marketing patterns. *World Archaeology* 6:172—89.

Holmes, W. H. 1893. Are there traces of glacial man in the Trenton Gravels? *Journal of Geology* 1:15—37.

Horner, L. 1855. An account of some recent researches near Cairo, undertaken with the view of throwing light upon the geological history of the alluvial land of Egypt. *Philosophical Transactions of the Royal Society of London*, pt. 10, pp. 105—38.

Howard, E. B. 1935. Evidence of early man in North America. *Museum Journal* 24: 61—159.

Hrdlička, A. 1917. Preliminary report on finds supposedly ancient remains at Vero, Florida. *Journal of Geology* 25:43—51.

———. 1918. *Recent discoveries attributed to early man in America*. Bureau of American Ethnology, Bulletin, No. 66.

James, P. E., and G. J. Martin. 1979. *The AAG: The first seventy-five years, 1904—1979*. Washington, DC: Association of American Geographers.

Judson, S. 1953. *Geology of the San Jon site, eastern New Mexico*. Smithsonian Miscellaneous Collections, vol. 121, no. 1.

Kuhn, T. S. 1970. Reflections on my critics. In *Criticisms and the growth of knowledge*, ed. I. Lakatos, 231—78. Cambridge: Cambridge Univ. Press.

Leakey, L. S. B. 1965. *Olduvai Gorge, 1951—61*. Cambridge: Cambridge Univ. Press.

Lepsius, G. R. 1890. *Griechische Marmorstudien*. Berlin: Königlich Akademie Wissenschaft.

Lyell, C. 1863. *The geological evidence of the antiquity of man, with remarks on theories of the origin of species by variation*. London: John Murray.

———. 1872. *Principles of geology, or the modern changes of the earth and its inhabitants considered illustrative of geology*. 11th ed., rev. London: John Murray.

McBurney, C. B. M., and R. W. Hay. 1955. *Prehistory and Pleistocene geology in Cyrenaican Libya*. Cambridge: Cambridge Univ. Press.

MacNeish, R. S., and A. Nelken-Terner. 1967. Introduction. In *Prehistory of the Tehuacán Valley*, ed. D. S. Byers, 3–13. Austin, TX: Univ. of Texas Press.

Masters, P. M., and N. C. Flemming. 1982. *Quaternary coastlines and marine archaeology*. New York: Academic Press.

Mirriam, J. C. 1932. Notes. *Bulletin of the Geological Society of America* 43:242.

North, F. J. 1938. Geology for archaeologists. *Archaeological Journal* 94:73–115.

Osborn, H. F. 1916. *Men of the Old Stone Age*. New York: Scribners.

Plot, R. 1686. *The natural history of Staffordshire*. Oxford, at the Theater.

Prestwich, J. 1860. On the occurrence of flint implements, associated with the remains of animals of extinct species in beds of a late geological period, in France at Amiens and Abbeville, and in England at Hoxne. *Philosophical Transactions of the Royal Society of London* 150:277–317.

Pumpelly, R. 1906. Interdependent evolution of oases and civilizations. *Bulletin of the Geological Society of America* 17:637–70.

———. 1908. *Explorations in Turkestan (expedition of 1904): Prehistoric civilizations of Anau, origins, growth and influence of environment*. 2 vols. Washington: Carnegie Institution.

Rapp, G., Jr. 1975. The archaeological field staff: the geologist. *Journal of Field Archaeology* 2:229–37.

Rapp, G., Jr., R. Bullard, and C. Albritton. 1974. Geoarchaeology? *The Geologist, Newsletter of the Geological Society of America* 9:1.

Rapp, G., Jr., and J. A. Gifford, 1982. Archaeological geology. *American Scientist* 70:45–53.

Renfrew, C. 1976. Archaeology and the earth sciences. In *Geoarchaeology: Earth science and the past*, ed. D. A. Davidson and M. L. Shackley, 1–5. London: Duckworth.

Rouse, I. 1951. *A survey of Indian River archaeology, Florida*. Yale Univ. Publications in Anthropology, no. 44.

Sandford, K. S. 1934. *Paleolithic man and the Nile Valley in Upper and Middle Egypt*. Univ. of Chicago, Oriental Institute Publications, vol. 18.

Sandford, K. S., and W. J. Arkell. 1928. *First report of the prehistoric survey expedition*. Univ. of Chicago, Oriental Institute Communications, no. 3.

———. 1929. Paleolithic man and the Nile-Faiyum divide. University of Chicago, *Oriental Institute Publications* 10 (3).

Sauer, C. O. 1925. The morphology of landscape. Univ. of California Publications in Geography, vol. 2, no. 2, pp. 19–53.

Schmerling, P. C. 1833–34. *Recherches sur les ossemens fossiles découverts dans les cavernes de la Province de Liège*. Liège: Collardin.

Schmidt, R. R. 1912. *Die diluviale Vorzeit Deutschlands*. I: *Archaeologischer Teil: Die diluvialen Kulturen Deutschlands*, by R. R. Schmidt. II: *Geologischer Teil: Die Geologie und Tierwelt der paläolithischen Kulturstatten Deutschlands*, by E. Koken. III: *Anthropologischer Teil: Die diluvialen Menschreste Deutschlands*, by A. Schliz. Stuttgart: E. Schweizerbart.

Sellards, E. H. 1916. On the discovery of fossil human remains in Florida in association with extinct vertebrates. *American Journal of Science* 42: 1–18.

———. 1919. Literature relating to human remains and artifacts at Vero, Florida. *American Journal of Science* 47:358–60.

Shackley, M. L. 1975. *Archaeological sediments*. London: Butterworth.

Sheets, P. D., and D. Grayson. 1979. *Volcanic activity and human ecology.* New York: Academic Press.

Shotton, F. W. 1969. Petrological examination. In *Science in archaeology*, rev. ed., ed. D. Brothwell and E. Higgs, 571–77. London: Thames and Hudson.

Simpson, R. D. 1978. The Calico Mountains archaeological site. In *Early man in America*, ed. A. L. Bryan, 218–20. Edmonton, Alberta: Archaeological Researches International.

Smith, W. 1816. *Strata identified by organized fossils.* Pts. 1 and 2. London.

Soergel, W. 1939. Das diluviale System. *Fortschritte der Geologie und Palaeontologie* 39:155–292.

Stewart, T. D. 1946. A reexamination of the fossil human skeletal remains from Melbourne, Florida, with further data on the Vero skull. *Smithsonian Miscellaneous Collections* 106(10):1–28.

Taylor, W. W. 1948. A study of archeology. *American Anthropologist, Memoir*, no. 69.

Thomas, H. H. 1923. The source of the stones of Stonehenge. *Antiquaries Journal* 3:239–60.

Tournal, P. 1833. Considerations Generales sur le Phénomène des Cavernes à Ossemens. *Annales de Chemie et de Physique*, vol. 52, pp. 161–81.

Vita-Finzi, C. 1973. *Recent earth history.* London: Macmillan.

Washington, H. S. 1898. The identification of the marbles used in Greek sculpture. *American Journal of Archaeology* 2:1–18.

Whittlesey, C. 1868. On the evidence of the antiquity of man in the United States. *Proceedings of the American Association for the Advancement of Science*, 17th Annual Meeting, Chicago.

Willey, G. R., and J. A. Sabloff. 1980, 2nd ed. *A history of American archaeology.* San Francisco: Freeman.

Winchell, N. H. 1885. The antiquity of man; some incidental results of the discussion. *American Geologist* 2:51–54.

———. 1903. Was man in America in the Glacial Period? *Bulletin of the Geological Society of America* 14:133–152.

Wissman, H. von. 1956. On the role of nature and man in changing the face of the dry belt of Asia. In *Man's role in changing the face of the Earth*, ed. W. L. Thomas, Jr., 278–303. Chicago: Univ. of Chicago Press.

Zeuner, Frederick E. 1959. The pleistocene period: Its climate, chronology and faunal successions. London: Hutchinson.

2

Geomorphology and Archaeology

DONALD A. DAVIDSON

ABSTRACT

This chapter focuses on the contributions that a geomorphologist can make to an archaeological project. The application of geomorphological mapping at various scales is illustrated, and emphasis is given to the problems associated with postulating the resource significance of various geomorphic units to former societies. Most geomorphological research in collaboration with archaeology has concentrated on seeking evidence for environmental change; this is exemplified by considering archaeological sites associated with fluvial, coastal, and desert landforms and with caves and rockshelters. Attention then turns to the effects of geomorphic processes on archaeological sites. Research in the Mediterranean Basin shows the effects of erosion on site-distribution patterns. Geomorphic processes also influence the spread of artifacts within a site, again emphasizing the need to incorporate geomorphic analysis into archaeological interpretations. Also discussed is the relevance of studying active geomorphic processes on archaeological sites to aid in planning conservation schemes. Two recurring themes evident in this review are identified: the contextual approach of geomorphology to archaeology and the problems of explaining landform change. The geomorphologist, now possessing the techniques for tackling many problems common to archaeological sites, still faces many difficulties in integrating geological results with those of the archaeologist.

In a literal sense, geomorphology is the study of the form of the earth's surface, focusing on description, analysis, and explanation. Such explanation is possible only if the geomorphologist understands the processes, both past and present, that have influenced the form of the land surface. Specific processes result in distinctive landforms—for example, the incision of a river channel into alluvial sediments produces river terraces. A distinction must be drawn between *land form* and *landform*; the former refers in a descriptive sense to the topography of an area, whereas the latter is a distinct surface feature resulting from a particular genesis. Any explanation of land forms must incorporate analysis of the geomorphic processes that gave rise to them. The tradition in geomorphology was to infer process from the nature of the land form, but the modern emphasis is on process-response models, a development made possible because of the increasing amount of data available on individual processes. The frequency and magnitude of geomorphological processes affecting landforms must also be considered, rather than viewing landforms just as expressions of spatial variation in processes.

Geomorphology began to emerge as a distinct subject only in the final quarter of the nineteenth century, despite some pioneering research in the previous century. In Europe James Hutton, the first to establish the principle of uniformitarianism, stressed in his writings in the late eighteenth century that landscapes evolve as a result of past and present processes. His work was developed by J. Playfair and in the nineteenth century by C. Lyell. The first European scientists who might be termed geomorphologists rather than geologists are F. von Richthofen and A. Penck, who were also the first to write geomorphic texts. In the United States, pioneering research was done by J. W. Powell, G. K. Gilbert, and C. E. Dutton, but the writings of W. M. Davis had a major influence on the approach to geomorphology during the first half of the twentieth century. Part of the blame for geomorphology's failure to develop models based on detailed field data must be given to the school of thought promoted by Davis. Despite many important contributions to geormorphology during the first half of the twentieth century, the discipline has flourished in terms of practitioners and research publications only since about 1950. As an example, the British Geomorphological Research Group was formed in 1961, and the journals *Revue de géomorphologie dynamique* and *Zeitschrift für Geomorphologie* were initiated in 1950 and 1956, respectively. During this most recent phase, geomorphology, like any evolving science, has been increasingly divided into subdisciplines such as fluvial geomorphology, glacial geomorphology, hillslope geomorphology, Quaternary geomorphology, and the like. Another trend has been the increasing involvement of geomorphologists with other disciplines, which is well demonstrated in the developing relations between geomorphology and archaeology.

Given the very recent flourishing of geomorphology, it is not surprising that collaborative research between geomorphology and archaeology has evolved only in recent years. In a review of this research, Gladfelter (1977) lists 108 references (only 22 of which are dated before 1960), but most of these demonstrate relevant techniques per se rather than illustrate specific geomorphic studies in archaeology. Of course, there is a long tradition of collaboration between geology and archaeology, the main focus of which has been on the stratigraphic context of archaeological materials (see chap. 1). It is only when archaeological sites began to be related to their local landform settings that a geomorphic input to archaeology became evident. Thus geomorphology initially had a descriptive role in archaeology, since the physical setting of archaeological sites has always been deemed important. Its transition to an analytical role can be traced to the realization that geomorphological analysis can shed light on environmental change. Research by Hack (1942) on the changing physical environment of the Hopi Indians in Arizona is an early example of an analytical approach. As another example, Jacobsen and Adams (1958) relate the increases in salinity and siltation to changes in ancient civilizations in Mesopotamia. Credit must be given to Karl Butzer for stating explicitly the types of geomorphic questions relevant to an archaeological investigation of ancient Egypt (Butzer, 1960:1617). Since they are of relevance to many similar studies, they can be summarized as follows:

1. What is the immediate geologic-geographic setting of the late prehistoric sites in the Nile Valley?
2. What are the relations of such settings to the surficial deposits of the valley margins? What situations are likely to have been deliberately selected by man or accidentally preserved from natural obliteration?
3. What regional generalizations can be made about the likelihood of occurrence of sites?
4. What proportion of the late prehistoric sites has been preserved? Or, rather, are the known sites representative of the density of actual settlement?
5. What physical conditions were dominant during the period of Neolithic—Chalcolithic settlement?

Another trend can be discerned in the 1970s. Geomorphology, like the other environmental sciences, has no neat boundaries but fuses into such subjects as geology, sedimentology, and soil science. Increasingly, geomorphologists have found it necessary to develop expertise in these related disciplines, and it has become progressively more difficult to identify a "true" geomorphologist on an archaeological project. This problem is neatly sidestepped by the rise of *geoarchaeology*, a subject recently reviewed by Gladfelter (1981). The term, taken to mean the study of archaeology using the principles and techniques of the earth sciences, was first used by Butzer in 1973; it was used by Renfrew (1976) at virtually the same time for the introduction to a conference volume. The rise in interest in geoarchaeology is expressed in the results of a survey in 1977 by the Society for Archaeological Sciences: of the 117 responses obtained for North America, 36 percent specialized in geomorphology and historical geology.

This introduction to the increasing collaboration between geomorphology and archaeology would be incomplete without stressing the massive contribution to the subject by Karl Butzer. His fieldwork areas include Majorca (Butzer, 1962), Spain (Butzer, 1965, 1967, in press), Egypt and the Nile Basin (Butzer and Hansen, 1968), East Africa (Butzer, 1970, 1980c), South Africa (Butzer, 1974a, 1976a, 1978b, 1979), and Illinois in the United States (Butzer, 1977, 1978a). His text, *Environment and Archaeology* (Butzer, 1971a), is a detailed synthesis of environmental analysis relevant to prehistoric archaeology whereas his more recent book, *Archaeology as Human Ecology* (1982), provides a contextual framework for environmental archaeology.

GEOMORPHOLOGICAL MAPPING

The production of a geomorphological map of the area around a site should be one of the first tasks on an archaeological project. Such a map describes morphology, landforms, drainage patterns, nature of superficial deposits, tectonic features, and active geomorphological processes. The advantages of producing a geomorphological map are fairly clear. The task enables the geomorphologist to examine systematically the terrain in the study area, which results in an impression of the geomorphic context of the site. Such fieldwork has to be paralleled with examination of exposures, and coring may also be needed to investigate superficial deposits. Not only is a spatial pattern of land types obtained, but also evidence indicating how the geomorphic landscape has changed through time. Thus the archaeological benefits of a geomorphological map and associated stratigraphic investigations are that some pattern in site distribution may become evident and aspects of environmental change may be revealed.

The relevance of a simplified geomorphological map at the regional scale is illustrated in figure 2.1, which indicates the broad geomorphic features of the Plain of Drama in northern Greece. The plain is a limestone-floored graben that has sunk relative to the surrounding uplands. The geomorphological map allows the recognition of five general landform units—foothills, alluvial fans, other alluvial areas, lowland limestone area, and peat and marshland (Davidson, 1971). Superimposed on the map is the distribution of Neolithic and Bronze Age tells. The concentration of the tells on or near alluvium suggests that settlements are preferentially distributed on areas of light alluvial soils with a good water supply. The geomorphological map in combination with data on the chronology of alluviation allow the mapping of alluvial deposits according to age (fig. 2.2). Tells could have been eroded or buried on the areas of younger alluvium and on the secondary fans. Sites on the old lake area may also have been lost.

A spatial relationship is to be expected among the five landform units of figure 2.1, vegetation, and soils. Thus, the landform units could be considered in broader environmental terms. In effect an ecological approach to the landform units is being suggested, a methodology employed by Coe and Flannery (1964), who, in Mesoamerica, found it useful to identify *microenvironments*—landform/soil/vegetation assemblages of differing resource value to communities in the past. Figure 2.3 is an idealized cross section for the Tehuacán Valley, in which the resource characteristics of the different microenvironments are indicated. The geomorphic framework of the valley's subdivision is clear.

An example of geomorphological mapping at a more detailed scale is given by van Zuidam (1975). He has produced such maps, supplemented by aerial-photo stereomodels, for study areas in the Zaragoza region of Spain. One result of his

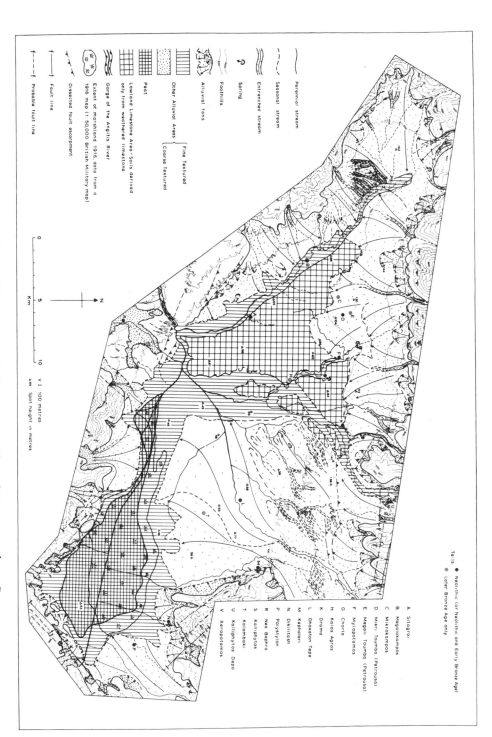

Fig. 2.1. Geomorphology and distribution of tell sites in the Plain of Drama, northern Greece.

Perennial stream

Seasonal stream

Entrenched stream

Spring

Alluvial fans

Foothills

Other Alluvial Areas { Fine Textured / Coarse Textured }

Peat

Lowland Limestone Area – Soils derived only from weathered limestone

Gorge of the Angitis River

Extent of marshland 1916, data from a 1916 map (1: 50,000 British Military map)

Dissected fault escarpment

Fault line

Probable fault line

Tells
● Neolithic (or Neolithic and Early Bronze Age)
⊕ Later Bronze Age only

A   Sitagroi
B   Megalokampos
C   Mikrokampos
D   Mikri Toumba (Petrousa)
E   Megali Toumba (Petrousa)
F   Mylopotamos
G   Choria
H   Kalos Agros
K   Drama
L   Dhoxaton Tepe
M   Kephalari
N   Dikilitash
P   Polystylon
R   Nea Baphra
S   Kalliphytos
T   Kalambaki
U   Kalliphytos Depo
V   Xeropotamos

0   5   10   Km

V I 100 metres

a.ss   Spot height in metres

N

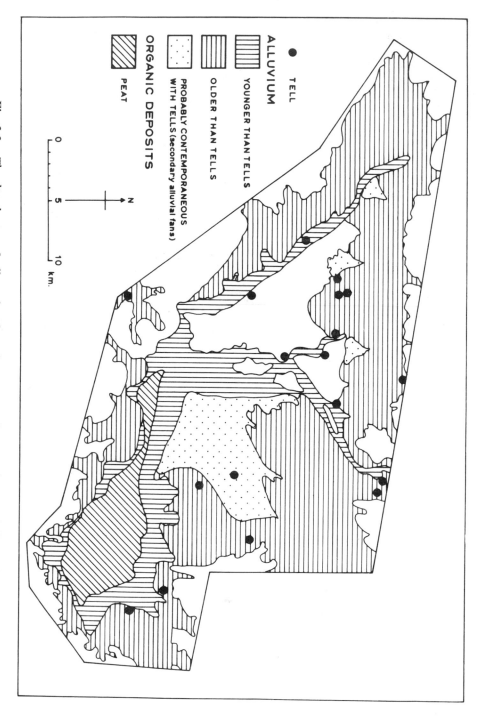

● TELL

ALLUVIUM

⊟ YOUNGER THAN TELLS

⊟ OLDER THAN TELLS

⊡ PROBABLY CONTEMPORANEOUS
WITH TELLS (secondary alluvial fans)

ORGANIC DEPOSITS

▨ PEAT

0   5   10
                km.

N

Fig. 2.2.   The distribution of tells in the Plain of Drama in relation to the age of alluvial deposits.

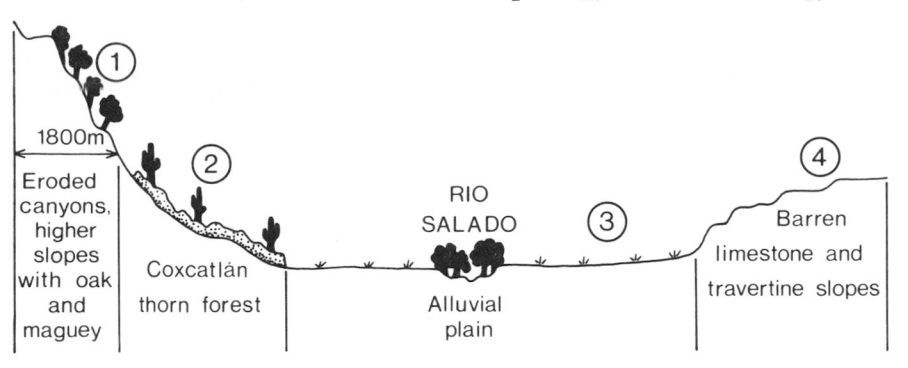

1. Abundant deer and acorns (Autumn); maguey (year-round); wild avocado (rainy season)

2. Abundant deer and peccary (Autumn); cottontails, doves, skunks, (year-round); cactus fruits (Spring)

3. Mesquite pods (rainy season); cottontails, jackrabbits, gophers, quails (year-round)

4. Small numbers of woodrats and doves (year-round); gophers and cottontails in widest ravines (year-round)

Fig. 2.3. An idealized east-west cross section of the central part of the Tehuacán Valley, Puebla, Mexico, showing microenvironments and the seasons in which food resources are exploited. The length of the section is about 20 km (Coe and Flannery, 1964:652. Copyright 1964 by the American Association for the Advancement of Science).

investigation is the identification of an intricate erosional-depositional history: van Zuidam postulates at least two older fills of Pleistocene age, with Holocene valley deposition during the period between 700 B.C. and A.D. 100, reaching a maximum between 500 and 100 B.C. He established the latest possible date for alluvial landforms on the basis of the age of archaeological sites located on their surfaces—a demonstration of how archaeology can help to date geomorphic features. In addition, van Zuidam relates land-use changes from ca. 500 B.C. to accelerated erosion, a theme also developed by Harvey (1978) in his investigation of the evolution of alluvial fans in southeast Spain. Recent work in Spain by Wise, Thornes, and Gilman (1982) has used geomorphological mapping around archaeological sites (fig. 2.8); reference will be made to this in a later section.

In addition to the benefits of geomorphological mapping for an archaeological project, there are a number of practical problems. If the detailed mapping system of Verstappen and van Zuidam (1968), Demek (1972), and Demek and Embleton (1978) is followed, the resultant map becomes too crowded with symbols and shading systems: it is difficult for the nongeomorphologist to make use of it. In part this problem may be solved by producing from the basic map a series of maps showing specific attributes—drainage conditions, alluvial landforms, etc.—that make the data easier to assimilate. A more fundamental problem is that geomorphological mapping forces at least a partially genetic description of the landscape. Areas of river terraces, alluvial fans, or eskers, for example, are thus identified, though often in conjunction with more descriptive attributes, such as boulder-strewn slopes or seepage zones. The archaeological need is to produce one or more maps showing variations in land characteristics that might have been of significance to communities in the past. People do not think of land in terms of geomorphic genetic units, but rather in terms of specific properties, such as deep, well-drained soil, low flood risk, soil of high natural fertility, and so on. Inferences about these characteristics may be possible from a geomorphological map, but for investigations focusing on former ecological relationships between people and their environment, maps of such features as soil depth, stone content, or slope are most relevant.

This approach is illustrated by Kirkby's (1977) environmental analysis of the Deh Luran area of Iran, carried out in collaboration with the archaeological project focused on the excavation of Chagha Sefid. Considering river and alluvial fan patterns, surface stoniness, depth of dissection, degree of salinity, and vegetation zones the prime environmental attributes, Kirkby produced a series of maps describing these properties. He then scored each variable on a rank basis, which permitted the calculation of an overall factor score according to mean rank. The results when mapped present an overall environmental pattern based on attributes thought to be of human significance.

The approach so far has been to view geomorphological mapping as providing a static backdrop on which the evolution of human communities and societies can be placed. Butzer (1980a:418) argues strongly that "environment should not be synonymous with a body of static and descriptive background data. The environment can indeed be considered as a dynamic factor in the analysis of archaeological context." He uses Axum, a first millennium A.D. civilization of the upper Nile, to show how spatial and temporal variations in resources are intimately associated with the evolution of the society. Consideration has thus to be given not only to the changing nature of the physical environment, but also to its varying potential for utilization. The diffusion or local development of new technology may make it possible to exploit areas that previously could not be utilized. Thus it emerges that any assessment of the former significance of land types is possible only when the past physical environment is evaluated with close reference to the technological and economic level of the culture using it.

This problem of identifying former significant elements in the physical landscape may be illustrated by considering the terrain around a Neolithic chambered cairn on the island of Arran off the west coast of Scotland (fig. 2.4a). This tomb, called East Bennan Cairn, is the only indicator of Neolithic activity in this area. It may be viewed as the territorial marker for a small-scale autonomous society practicing a rather varied economy based on land and sea resources: cereal growing; herding cattle, sheep, and pigs; and fowling, fishing, and hunting. At first forest was extensive in the area, but decreased with time.

The form of the basin in which the cairn is located is shown in figure 2.4b, and the soil types are given in figure 2.4c. The basin is floored with a glacial till, so that its lower and bottom slopes are dominated by gleyed soils. To the north the rise in elevation is paralleled by the development of an upland soil sequence—peaty podzols and complexes of peaty gleys, lithosols, and peat. To the west and east of the cairn an irregular ridge with thin soils separates the floor of the basin from the sea. Present-day farming is concentrated in the area mapped as surface water gleys because these are the best soils: they have been improved in terms of drainage and fertility. Thus it would be tempting to propose this as the area of agriculture for the Neolithic group, even accepting that soil conditions would have been different under the previous forest cover. This interpretation would locate the cairn at the edge of the better land.

An alternative view is that the areas of arable cultivation would have been on the steeper slopes—the areas to the immediate east and west of the cairn with rock at or near much of the surface. This interpretation becomes feasible when it is noted that the main agricultural problem on Arran is one of limited drainage—rainfall of the order of 1500 mm annually combined with soils on glacial tills. Before the advent of modern drainage systems, farmers were attracted to steep slopes, where better drainage was possible. This is reflected in the occurrence on steep slopes of remnants of the run-rig system that was practiced before the nineteenth century. The rigs were formed by the ridging up of soil to improve soil depth and drainage. At one locality to the immediate west of the cairn, remnants of rigs occur on a slope of 73 percent. Thus steep and irregular terrain, which today is used only for grazing, was of agricultural value because of a different agricultural system and technological level. Such a view can then be translated back to the Neolithic to suggest that the cairn was in the center of the area of agricultural significance.

These contrasting interpretations demonstrate the need for a contextual approach as argued by Butzer. Consequently the geomorphologist requires a great deal of information about the nature of a former society before being able to evaluate the significance of former terrain types. An alternative contextual approach is for the geomorphologist to carry out a multivariate spatial analysis, illustrated by a research project to examine the distribution of chambered cairns in the Orkney Islands north of Scotland (Davidson, Jones, and Renfrew, 1976; Davidson, 1979). In this study a geomorphological map was first produced, which permitted the compilation of an environmental data bank on a grid basis. Next, a computer-simulation model was used to determine which environmental characteristics led consistently to a simulated site pattern having the closest similarity to the distribution of known sites. The results suggested that angle of slope and the nature of the coastline were the factors of most importance, but the problem still remains of ascertaining how these factors affected site distribution.

Another problem associated with evaluating geomorphic spatial data is that the archaeologist is often unable to specify, at any scale, the territory of a former group. The geomorphologist must know in terms of human significance the area to be assessed. A frequent assumption is that the land around a habitation site

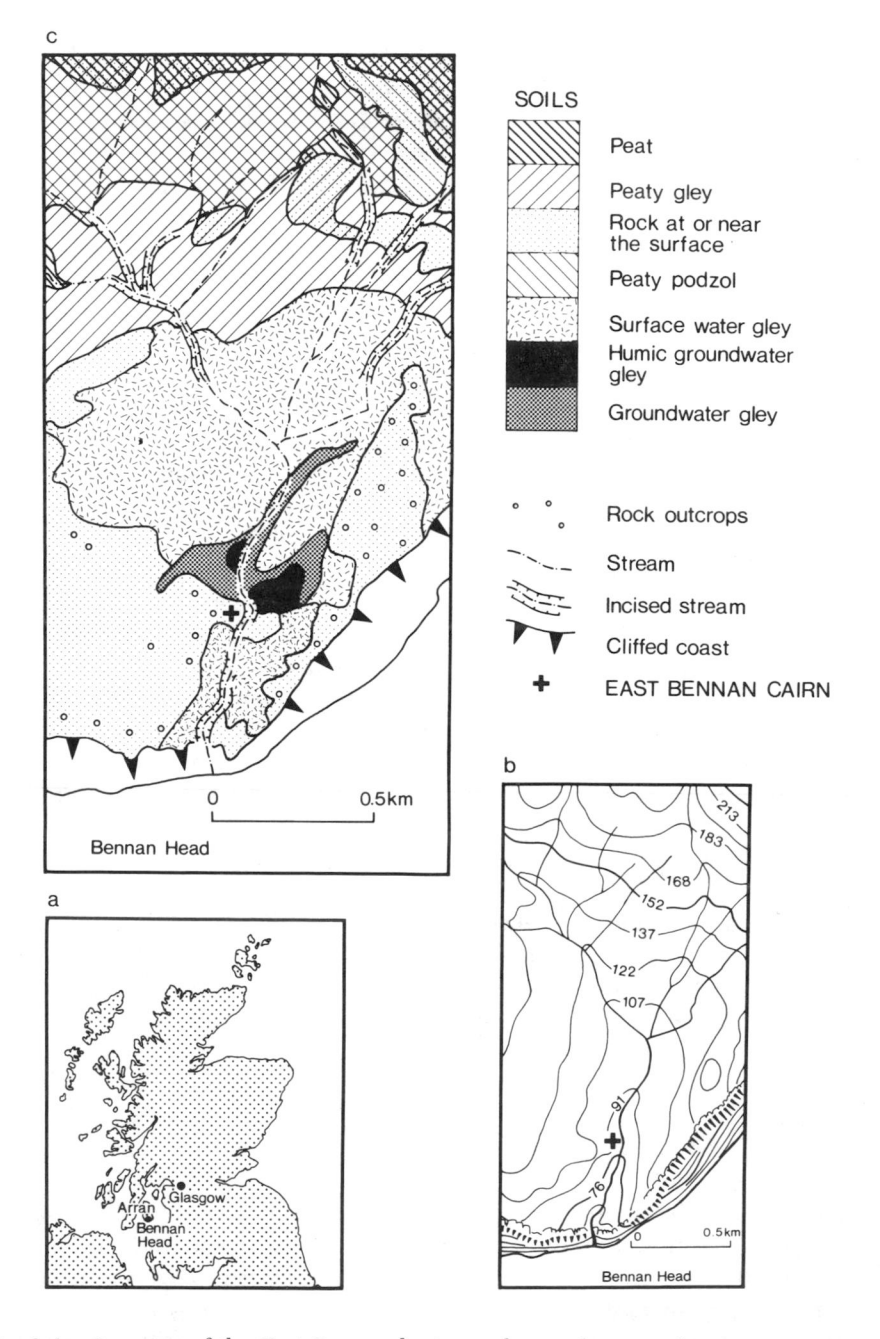

Fig. 2.4. Location of the East Bennan basin on the south coast of Arran, in western Scotland (a); topography of the basin (b); and the setting of the chambered cairn site in relation to the distribution of general soil types (c).

would have been intensively exploited, while land at greater distances would have been less utilized. This approach is fundamental to site-catchment analysis, whereby resources are evaluated in terms of concentric zones at increasing distances from a settlement (Higgs and Vita-Finzi, 1972). Such a geometrical approach is often the only possible one, but the need is for archaeologists to try to delineate areas of varying resource potential.

GEOMORPHOLOGICAL CHANGE

The dominant focus for geomorphologists involved with archaeological projects is the reconstruction of former environments. To ensure that former elements of behavioral significance are identified, this environmental analysis needs to be executed with close reference to the archaeological and nongeological evidence. The closest association between archaeology and geomorphology is evident in geographical areas that have developed only in Pleistocene and Holocene times. Such association is well expressed in the Netherlands, a country formed to a large extent by Pleistocene- and Holocene-age processes (de Jong, 1967). The Saale ice sheet that covered the region resulted in the formation of ice-pushed ridges. During the last glaciation (the Weichsel), the Netherlands was ice-free, but the tundra conditions of the time encouraged active eolian processes. Periglacial outwash sands were thus reworked to produce deposits called coversands. Areas of loess resulted when finer material (dominantly silt) was also wind-deposited. Extensive areas of Holocene clays in the Netherlands owe their origin to marine or fluvial sedimentation. The result is an intricate geological pattern reflecting geomorphic change that has continued up to the present day as a result of Dutch land-reclamation schemes. In consequence, the stratigraphic and spatial occurrence of archaeological sites is intimately associated with Holocene landform development.

The usefulness of this approach is illustrated by the work of Dekker and de Weerd (1973) for the Midden-Westfriesland district to the north of Amsterdam. A sequence of Holocene transgressions over a basement of Pleistocene materials is represented by marine sediments up to 15 m thick. Figure 2.5a shows the surficial distribution of these deposits and figure 2.5b illustrates the stratigraphy. These figures also include known archaeological sites; their spatial and stratigraphic patterning is immediately evident. There are implications in these patterns for further archaeological site surveys. It would be pointless, for example, to search for sites of Roman, Bronze, or Neolithic age on materials that had been deposited between A.D. 800 and 1400. Landform patterns are also of importance. As an example, level areas of coarse sediments, given their drier and elevated nature, were frequent choices for habitation sites, which is reflected in the occurrence of sites on creek ridges in figure 2.5b.

To convey the scope of geomorphic research designed to elucidate environmental history, it is helpful to consider several general geomorphic settings of archaeological sites: fluvial, coastal, and desert landforms, and caves.

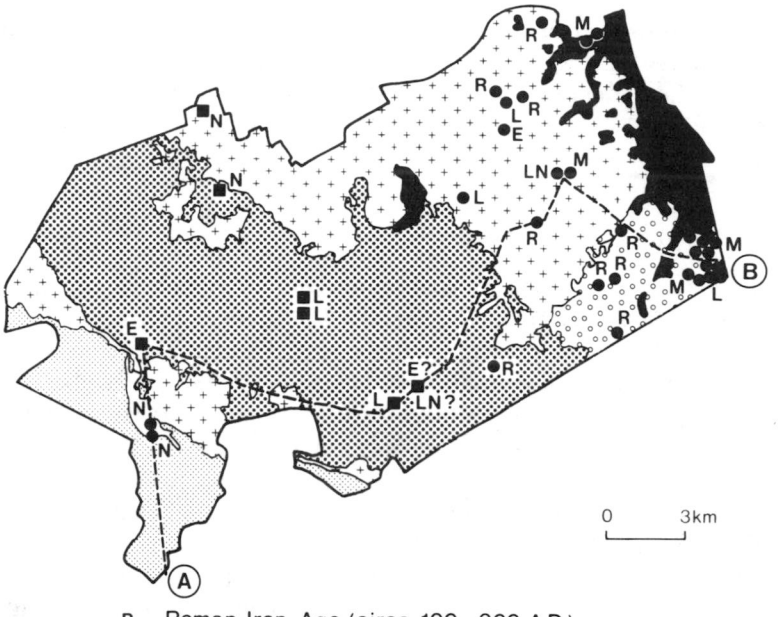

R Roman Iron Age (circa 100 - 300 A.D.)

L Late Bronze/Early Iron Age (800 - 500 B.C.)

M End of Middle Bronze Age (1200 -1000 B.C.)

E Early Bronze Age: Barbed Wired Group (1700 -1500 B.C.)

LN Late Neolithic: Bell Beaker Culture (2000 - 1700 B.C.)

N Late Neolithic: Vlaardingen Culture + Protruding
   Foot Beaker Culture (2500 - 2000 B.C.)

Fig. 2.5.a. Distribution of surface deposits and settlement traces in the Midden-Westfriesland of the Netherlands (Dekker and de Weerd, 1973:171. By permission of Elsevier Scientific Publishing Co.).

FLUVIAL LANDFORMS

The greatest geomorphic input to archaeology has concerned sites whose evolution is linked with fluvial landforms. In Britain, for example, artifacts in river gravel constitute some 95 percent of the evidence for Lower Paleolithic human activity (Wymer, 1976), and thus archaeological interpretation must lean heavily on geomorphic analysis. Such an approach is also demonstrated by Helgren's (1978) study of the lower Vaal Basin in South Africa. Much of Butzer's geo-archaeological research in Africa has focused on the elucidation of Holocene alluvial sequences. For example, the lower Omo Basin at the northern end of Lake Turkana (Lake Rudolf) in Ethiopia has an intricate series of deltaic and alluvial formations, including piedmont alluvial fans and terraces (Butzer, 1971b,

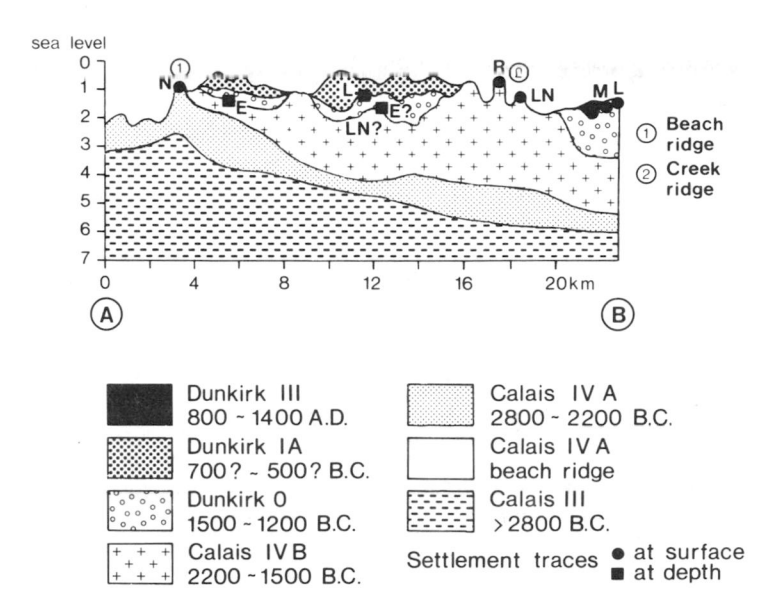

Fig. 2.5.b. Geological cross section along line AB as indicated above; locations of archaeological finds are shown (Dekker and de Weerd, 1973:173).

1976b, 1980c; Butzer, Brown, and Thurber, 1969; Butzer et al., 1972). Outstanding from the Omo-Turkana record is the amplitude and rate of hydrological and geomorphic changes (Butzer, 1980b). These linked and apparently synchronous changes are related by Butzer to wetter and drier phases: alluviation is linked directly to high lake levels, dissection to low levels. In contrast to the Omo-Turkana Basin, the chronological framework for alluvial sequences in the Vaal-Orange Basin of South Africa is less clear (Butzer, 1974b; Butzer et al., 1973, 1978, 1979). In many small basins, an alluvial fill broadly dated to 4,500–1,300 years ago has remained undissected until a century or two ago (Butzer, 1980b).

The relative simplicity of the South African Holocene alluvial record is also expressed along the Nile Valley. Aggradation of a sandy fill began there about 11,000 years ago and continued until 6,000 years ago. The deposition of this alluvium is closely linked with the wadis that feed into the Nile alluvial tract from the adjoining desert areas, as is demonstrated by the geomorphic record from Egyptian Nubia (Butzer and Hansen, 1968). Table 2.1 summarizes this evolution from the beginning of Holocene times, and figure 2.6 is a schematic representation (in simplified form) of the Nile Valley sedimentary sequence. The Pleistocene deposits constitute river-terrace gravels—called the low desert. The most recent deposits are the clayey silts that overlie Early Holocene sandy fill. The diagram neatly demonstrates the geomorphic context of different archaeological sites.

Table 2.1. Latter Phases of the Geomorphic Evolution of Egyptian Nubia during the Holocene

Historical	38.	Wadi dissection and redeposition of fill. Very limited wadi activity.
	37.	Alluviation of 50 cm or more of wadi wash, possibly equivalent to Member III, Shaturma Formation. At this time or earlier, first accumulation of modern Nile silts.
	36.	Dissection of wadi fill (as much as 5 m or more), with Nile floodplain level no higher than it is today. Local accumulation of eolian sands (Seiyala).
	35b.	Alluviation of Shaturma Formation, Member I, by wadis, as a result of significant winter sheetflooding (thickness over 5 m). Contemporary with last part or all of Kibdi unit, phase 35a.
	35a.	Alluviation of Kibdi Member, Gebel Silsila Formation, by Nile (thickness over 6 m, floodplain elevation +6 to +7 m). Slightly more vigorous and possibly higher summer floods. Terminal stage ca. 3000 B.C.
Holocene	34.	Dissection of Nile (to below modern floodplain, vertical differential at least 15 m) and wadi fill (total cutting since beginning of phase 32b over 8 m). Local accumulation of eolian sands (Seiyala).
	33.	Biochemical weathering with red paleosol (Omda Soil Zone). Frequent, gentle rains.
	32b.	Alluviation of Upper Member, Ineiba Formation, by wadis (thickness over 8 m). Accelerated wadi activity. Possibly subdivided into two fills, separated by over 8 m of downcutting. Earlier fill contemporary with phase 32a, later fill terminating ca. 6000 B.C.

SOURCE: Butzer and Hansen, 1968:328.

Vita-Finzi (1969b) has investigated the alluvial sequence in the Mediterranean Basin; he postulates two major periods of fill, one dating to the Pleistocene (the Older Fill) and the other from Roman times to almost the present day (the Younger Fill). A similar two-phase alluvial sequence has been identified in Iran (Vita-Finzi, 1969a). The general difficulties of dating and correlating Holocene alluvial sequences are reviewed by Butzer (1980b), who stresses that the record of each alluvial sequence is complex and unique to each area. These difficulties, as well as the frequent lack of a detailed chronological framework, make regional correlation of alluvial sequences very difficult. This theme is propounded by Davidson (1980a) for the erosional record in Greece over the first two millennia B.C. The synchroneity of erosion there is questioned, as well as single-factor explanations in terms of such variables as climatic change or anthropogenic influence.

COASTAL LANDFORMS

If riverine areas were a prime focus of early settlement, then coastal or lakeside localities are strong challengers for equal importance. The attraction of a riverine, coastal, or lakeside locality lies in the range of resources available within a short distance; in addition the river, sea, or lake provided early settlers a means of contact with other groups. A classic area of early hominid remains, often

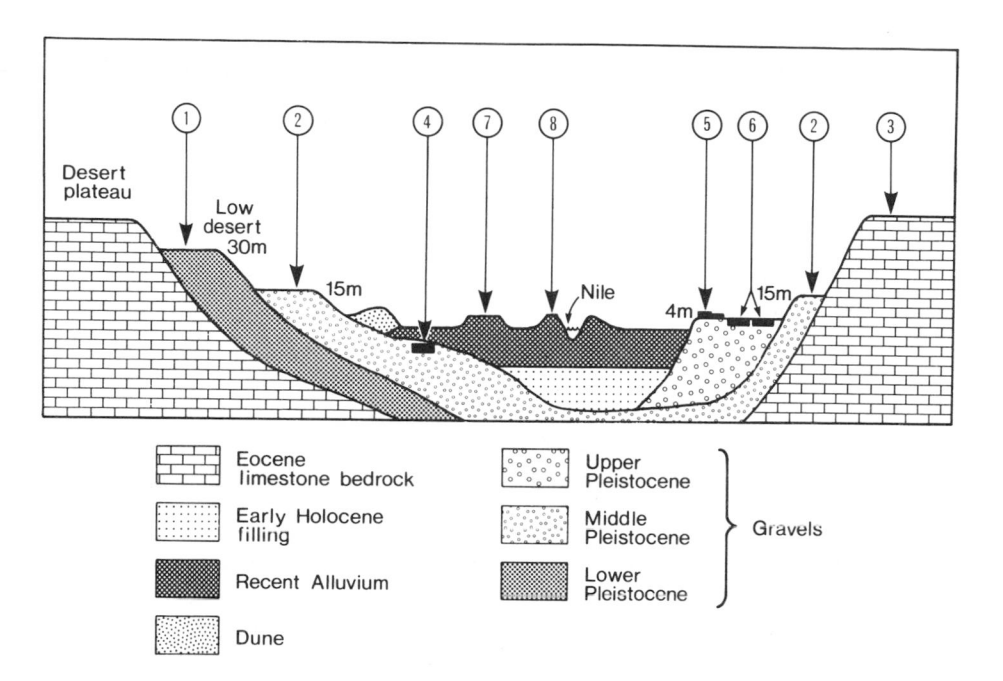

Fig. 2.6. Idealized cross section (not to scale) showing the relationship of archaeological sites to landform elements of the Nile Valley. 1 and 2: Nile gravels containing Lower Palaeolithic implements with scattered Middle Palaeolithic artifacts on the surface; 3: Predynastic flints scattered over desert surface; 4: possible "buried" Predynastic cemetery, under subsequent silt deposits; 5: remains of Predynastic settlement; 6: Predynastic burials; 7: modern village on cultural mound (ancient site at base?); 8: roads on levee embankments bordering low-water channel of the Nile (Butzer, 1960:1618). Copyright 1960 by the American Association for the Advancement of Science.

associated with lakes, is the Eastern Rift of East Africa. In the East Turkana Basin, for example, remains of hominids and occupation sites occur in deltaic and floodplain deposits. Any archaeological investigation of a coastal area ought to include a geomorphic analysis. Two important and interrelated themes are the evolution of landforms at the coast and the changes in base level. Changes in base level, including eustatic sea-level change, have marked implications with respect to the behavior of river channels.

The most obvious areas to demonstrate the importance of changing sea level with respect to archaeology are coastal regions that have emerged only recently as a result of isostatic recovery. In Finland, for example, gradual isostatic recovery resulted in a sequence of receding shorelines that permitted the human exploitation of new areas at different times. A correlation between marine regression and the age of archaeological sites is also evident in northern Canada. In fact, Andrews, McGhee, and McKenzie-Pollock (1971) have investigated the relative emergence of Canadian Arctic shorelines using data from 84 archaeological sites, 71 of which were related to a contemporaneous sea level. The result is a series of five maps showing emergence since 4,000, 3,200, 2,400, 1,600, and 800 years ago. These maps can be used to delimit areas of search for cultural remains of specific ages.

For studies in coastal geomorphology with an emphasis not only on sea-level changes, but also on landform evolution, reference must be made to the work of Kraft (1976, 1977) in Delaware, and Kraft, Rapp, and Aschenbrenner (1975, 1977, 1980) and van Andel et al. (1980) in the Aegean. Their approach is very much focused on detailed stratigraphic investigations of coastal localities. They demonstrate that the changing pattern of coastal settlement is intimately related to variations in sedimentation. This theme is considered in chapter 3 of this book, and for the present purposes only one point is emphasized. The effect of sea-level and coastal change is not limited to the coast but is expressed in a number of inland geomorphic changes. The consequence of base-level changes on rivers has already been mentioned, but little attention has been given to hillslope hydrological considerations. For example, most of the islands comprising the Orkney archipelago are particularly rich in prehistoric monuments, best known of which are the megalithic sites of Neolithic age. The islands are low-lying and have been subjected to slight submergence in the recent geological past, though the precise details have yet to be determined. This means that low-elevation archaeological sites presently located in poorly drained localities may not appear so anomalous, as a slightly lower sea level would have produced better drainage there. Coastal change was mirrored in a transformation of inland resources, at least on lower slopes.

DESERT LANDFORMS

Geomorphic analysis of many desert landscapes has provided evidence of recent environmental change. In part, the impetus for such research has come from archaeological projects, since the prehistories of major river basins such as the Nile, Tigris-Euphrates, and Indus are intimately related to the evolution of the adjacent desert regions. Evidence of environmental change in desert regions can be obtained from lake-shore sequences (Butzer et al., 1972), deflation basins and lunettes (Bowler, 1976), and former river courses and fluvial terraces (Butzer and Hansen, 1968; Allchin, Goudie, and Hegde, 1978), but the most characteristic desert landforms are sand-dune fields. Allchin, Goudie, and Hegde (1978) and Goudie (1977) summarize the evidence for extensive relict sand-dune fields (fossil ergs) in the American Midwest, Brazil, sub-Saharan Africa, the Sudan, Australia, and the margins of the Thar Desert. The extent of environmental shift can be estimated by comparing the distribution of relict dune fields with the distribution of active dunes; shifts of up to 900 km have been recorded. Allchin, Goudie, and Hegde (1978) also collate a range of dates for these relict fields. Their late-Pleistocene age (most were formed between 30,000 and 12,000 years ago) indicates a broad synchroneity with the maximum of the last ice sheets in northern latitudes.

The geomorphology and sedimentology of desert regions is a major topic of study. Mention must be made of the pioneering work on the movement of sand by Bagnold (1941). A specialized text on desert geomorphology is provided by Cooke and Warren (1973). Although there has been much geomorphic research in desert regions, there is a dearth of such investigations done in close collaboration with archaeologists. One major exception is the project on the Great Indian Desert to the east of the lower Indus Plain (Allchin, Goudie, and Hegde, 1978), the aim of which was to investigate the interrelationships between the prehistoric cultures and the changing environment of this region, primarily through study of the distribution of relict sand dunes. Their evidence presented to support the relict form of these features may be summarized as follows:

1. Dunes are being subjected to fluvial incision or gullying.
2. Dunes have a thick acacia scrub with grassland.
3. Dunes have well-developed soil profiles indicative of surface stability.
4. The nature of archaeological materials indicates that little or no deposition is taking place today.

The researchers detected within the fossil sand-dune area the presence of calcium carbonate cemented sand (calcrete), known locally as *kankar*. The development of such a deposit would have been of marked significance to human groups, but unfortunately the chronological details remain to be elucidated.

The most detailed geomorphological-archaeological investigation in an arid area was undertaken by Butzer and Hansen (1968) on the Kom Ombo Plain and lower Nubia, in the Nile Valley, as well as at Kurkur Oasis, in the Libyan Desert, and the coastal plain of Mersa Alam, on the Red Sea. Reference has already been made to the geomorphic evolution of Egyptian Nubia during the Holocene (table 2.1). At the Kurkur Oasis, Butzer and Hansen gave emphasis to the occurrence of tufa on a plateau and in wadis. The plateau tufa is attributed to a long period of complex alluviation of stream and spring deposits prior to dissection and denudation of the area. This dissection is expressed in the formation of wadis within which a younger sequence of tufas was deposited. In summary, the authors elucidate an intricate geomorphic evolution of pedimentation, chemical weathering, tufa formation, dissection, and eolian processes from Oligocene to Holocene times (pp. 363, 389) that culminated in the formation of the last tufa at 8300 B.C. or later, followed by limited fluvial activity and the accumulation of eolian quartz sands.

Regarding the geomorphic evolution, Butzer and Hansen stress the variability of Pleistocene climate in the hyperarid locality of Kurkur Oasis. In this area, archaeological surveys revealed a number of Middle and Late Paleolithic sites located on terrace surfaces adjacent to modern wadi channels. Butzer and Hansen state that "under more humid conditions, vegetation would have been abundant here, and there was possibly some surface water or groundwater. Vegetable foods and game were certainly available" (p. 393). They propose that small hunting-gathering groups began to visit Kurkur Oasis from about 30,000 years ago and lived there over extended periods, exploiting better ecological conditions than exist today. Another example of a desert region that has under-gone considerable geomorphic change over the last 30,000 years is given by Goldberg (1977) for Gebel Maghara, in northern Sinai.

The inception and spread of saline soils on perennially irrigated land is a particular problem in some desert areas. Jacobsen and Adams (1958) recognize three major periods of salinization in Mesopotamia: first from 2400 B.C. to 1700 B.C. in southern Iraq, a second period from 1300 B.C. to 900 B.C. in central Iraq, and a third period after 1200 A.D. in the area east of Baghdad. An example is given in figure 2.7 of the distribution of irrigation channels and settlements for the Post-Sāmarrān period (ca. 883–1150 A.D.) in the lower Nahrawan region (about 70 km southeast of Baghdad). One view is that the spread of irrigation led to salinization, but this is not the whole story. Hardan (1971) demonstrates that salinization was present in Mesopotamia before or at the beginning of irrigation, though intensive irrigation may have accelerated natural processes of salin-ization. Another possible factor is clearance of vegetation in the less arid areas, causing a rise in water tables and concomitant salinization. An increase in throughflow rather than a rise in water-table level could induce the same result.

CAVES AND ROCKSHELTERS

Caves and rockshelters offer particular geomorphic environments and the analy-sis of their sediments can yield much information on paleoenvironmental condi-tions. The study by Tankard and Schweitzer (1976) of the Die Kelders I cave in Cape Province, South Africa, demonstrates how sediment analysis can be used to postulate regional environmental change as early as 120,000 years ago. Shackley (1972) discusses the application of textural parameters to the analysis of cave sediments. Examples of detailed analysis of rockshelter deposits are given by Farrand (1975) and Goldberg (1977).

The present discussion emphasizes that the geomorphic setting of caves or rockshelters also needs to be investigated. In other words, the analysis of a sedimentary sequence ought to be paralleled with an examination of the local geomorphology. The integration of a cave's sedimentary sequence with the geomorphic evolution of the hillslope is demonstrated by Butzer (1976a) for the Swartkrans formation in Transvaal, South Africa. He proposes that the evolution of Swartkrans Hill must be visualized in terms of alternating morphodynamic and morphostatic conditions, patterns of which are recorded in the cave's litho-stratigraphic units. Six periods of pedogenesis and nonsedimentation are interca-lated with phases of deposition linked with accentuated hillslope erosion.

GEOMORPHIC PROCESSES AND ARCHAEOLOGICAL SITES

The point has already been made that some pattern in distribution of archaeolog-ical sites in a region may become apparent when a geomorphic analysis has identified the spatial and chronological evolution of its landforms. Very obviously

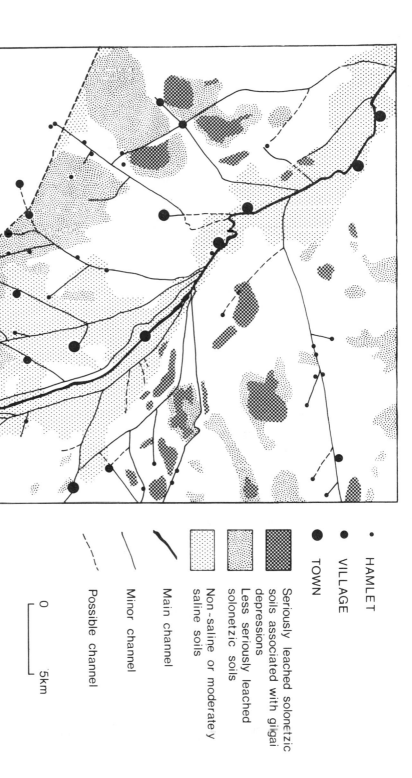

Fig. 2.7. Settlements and irrigation canals in the lower Nahrawan region of Mesopotamia, ca. A.D. 883–1150. Also shown are areas of solonetzic soils (Adams, 1965, fig. 9). Reprinted from *Land Behind Baghdad* by R. M. Adams, by permission of the University of Chicago Press.

● TOWN

● VILLAGE

• HAMLET

Seriously leached solonetzic soils associated with gilgai depressions

Less seriously leached solonetzic soils

Non-saline or moderately saline soils

Main channel

Minor channel

Possible channel

0   5km

a surface site cannot predate the underlying landform. Once a site distribution is established, geomorphic processes can exert another influence, in that the evolving spatial patterns of erosion and deposition can lead to the erosion or burial of sites. This is a major problem in areas that have undergone substantial geomorphic change during the course of their occupation.

The first reaction of a geoarchaeologist to a landscape of apparently high erosion rates might be to assume that many sites will have been lost. In practice this may not be the case. For example, Tasker (1980) examined the results of an archaeological site survey on Melos, in the Aegean, and found that most preserved sites occurred on slopes of less than 18 percent. It could be implied that sites on steeper slopes had been lost by erosion, but it is more likely that these areas were avoided, given their physical difficulty. People tend to select flat sites for occupation—even on a steep slope they will search for small flat areas of land and such sites subsequently will be less subject to complete erosional removal. Most of the sites on Melos were sherd scatters, and Tasker stresses the importance of surface irregularities in trapping them.

Wise, Thornes, and Gilman (1982) have studied the geomorphic setting of some 40 archaeological sites in southeast Spain. They are struck by the fact that on many of these sites the deposits and structures have survived for 4,000 years. In part this must be due to the megalithic nature of the sites, but the authors stress the importance either of flat sites or of the location of sites on the flanks of ridges in severely eroded badland terrain. This latter observation leads the researchers to propose that sheetwash and headward gully erosion have been less active than the badland terrain would suggest. The geomorphic setting of one of their sites (Cuesta del Negro) is shown in figure 2.8. Wise, Thornes, and Gilman comment that the in situ archaeological structures indicate little change in the form of the hill and its superficial deposits and slopes since Argaric times (about 4,000 years ago). These observations are quantified by estimates of erosion rates giving low results. The authors conclude that the evidence points to a landscape demonstrating comparative stability over the last 4,000 years. The important point here is that even in areas that appear to be subject to considerable erosion, there may well be localities of long-term geomorphic stability.

Although the preceding discussion tries to create a more balanced perspective on the effect of erosion on archaeological site occurrence, it is clear that such processes may be of particular importance with respect to the redistribution of artifacts. Rick (1976:133) comments that "very little attention has been paid to the human and natural processes which can destroy or change culturally created patterns of artifact distribution." Rick has examined a large surface collection of artifacts spread downslope from the Ccurimachay rockshelter in Junín, Peru. A sample frame was established in order to collect bone, chipped-stone, and ceramic fragments. When he related the average weights of these artifacts to slope angle, he found a tendency toward negative correlation. In other words, smaller pieces are found on steeper slopes. Bones were found to be in greatest numbers near the top of the site (large slope angle), whereas lithic materials were

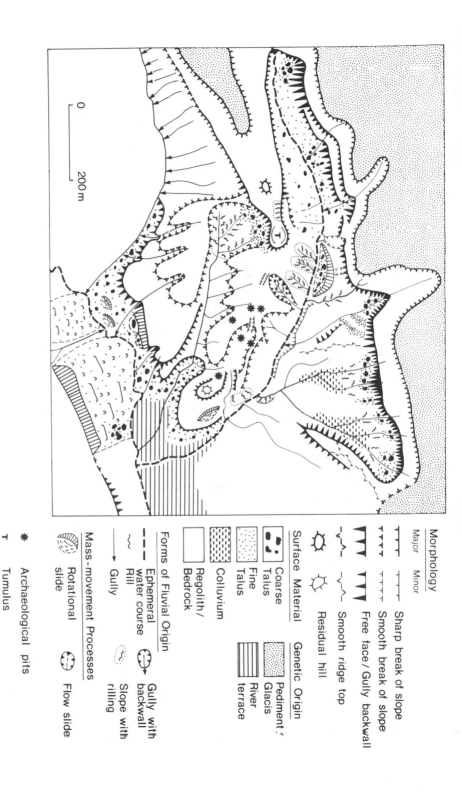

Fig. 2.8. Geomorphological map of the archaeological site of Cuesta del Negro, which lies on a long spur descending from the flat plain east of Diezma down to the Fardes River in southeast Spain (Wise, Thornes, and Gilman, 1982).

Morphology

Major	Minor	
⊥⊥⊥	⊤⊤⊤	Sharp break of slope
⊤⊤⊤⊤	⊤⊤⊤⊤	Smooth break of slope
⋎⋎⋎	⋎⋎⋎	Free face / Gully backwall
⌄⌄⌄	⌄⌄⌄	Smooth ridge top
✳	✧	Residual hill

Surface Material

Coarse Talus

Fine Talus

Colluvium

Regolith / Bedrock

Genetic Origin

Pediment / Glacis

River terrace

Forms of Fluvial Origin

Ephemeral water course

Rill

Gully

Gully with backwall

Slope with rilling

Mass-movement Processes

Rotational slide

Flow slide

✳ Archaeological pits

⊤ Tumulus

0   200 m

found in greater numbers toward the base of the slope. A regression analysis demonstrated that slope value is a good predictor of the average weight of lithic and ceramic fragments, but estimates of bone weight using the same approach are much less satisfactory. Rick is thus able to recognize a spatial patterning of artifacts on the slope below the rockshelter as a result of geomorphic processes. His words of caution are directed at archaeologists who might fail to identify such effects.

The consequence of erosion is deposition: artifacts can be redeposited in different contexts and archaeological sites can become buried. In particular, the surface distribution of sherds will vary according to the interaction of processes of accumulation, breakdown, and transportation. The loss of surface sherds through alluviation/colluviation and continued cultural accumulation over the original site poses problems for the interpretation of site-survey data. These themes are examined by Kirkby and Kirkby (1976), and it is appropriate to summarize here some of their archaeologically significant conclusions, which are based on the geomorphic study of sites in Mexico and Iran. They note that through time, transportational processes are the least important in explaining the redistribution of surface sherds. Surface-sherd concentrations increase over the first 50–100 years after site abandonment, but then decrease exponentially to comparative densities of 0.056 after 500 years and 0.0074 after 1,500 years (for 4–8 cm sherds). For Oaxaca in Mexico, Kirkby and Kirkby estimated an average annual sediment deposition of 0.25 cm, from which they calculated that in 1,000 years only 0.004 percent of the sherds from the buried site will be lying on the present surface. At tells in southwest Iran they found that surface-sherd concentrations were related more to tell height than to the proportion of old buildings reused or the number of site-occupation levels. In summary, the Kirkbys stress "a severe attenuation of site recovery for early periods" (p. 252). The highly selective discovery of sites in geomorphic regions akin to those studied in Mexico and Iran means that site distributions have to be approached with greater care. The use of ratios between surface-sherd counts of different archaeological periods to imply changes in population numbers or settlement patterns must be evaluated with some caution.

Geomorphological techniques may also be applied to the study of individual archaeological sites. For example, tells can be considered as artificial landforms. They accumulate primarily as a result of the collapse of mud-brick houses (Davidson, 1973; Goldberg, 1979), but the effects of erosion can markedly change their form. Investigations at the Neolithic tell of Sitagroi in northern Greece revealed that its summit at abandonment was at least twice its present-day areal extent (Davidson, 1976). Such a finding is of obvious relevance to the archaeological interpretation of the site. Kirkby and Kirkby (1976) have modeled the evolution of the form of tells (fig. 2.9), which, with appropriate calibration, may provide a method for dating mounds solely on the basis of form. Tells are not restricted to the eastern Mediterranean or the Middle East; variants occur in the Netherlands, Arctic Norway (Bertelsen, 1979), and the Orkney Isles north of

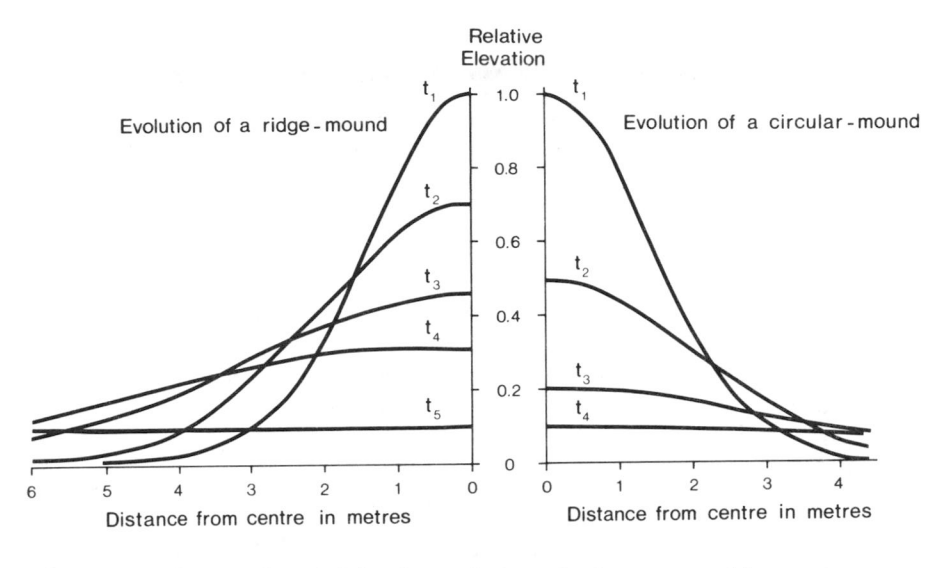

Fig. 2.9. Mathematical model for the evolution of a house-mound by erosion over various times for a ridge-shaped and for a circular mound, showing a mound profile that lowers and spreads out through time as a normal curve (Kirkby and Kirkby, 1976:232).

Scotland (Lamb, 1981; Davidson, Lamb, and Simpson, 1983). The geomorphic evolution of these northern-latitude tells remain to be investigated, though it is to be expected that a combination of anthropogenic and natural processes will be found to be relevant.

ARCHAEOLOGICAL SITE CONSERVATION

The preceding section has demonstrated how geomorphological analysis of sites can aid in their archaeological interpretation, but there are other reasons such analysis is helpful. If sites are subject to serious erosion, there is first the need to identify the threatened sites and second the need to carry out rescue excavations or to provide some site-protection program.

A problem in the western and northern isles of Scotland is the coastal erosion of sites, especially those on sand-dune areas (Crawford, 1974). The total or partial destruction of archaeological sites means the loss of irretrievable information. In certain areas tourism is strongly dependent upon the attraction of archaeological sites, which is why great effort is being made in Greece to minimize the deterioration of classical sites. The Acropolis, for example, is subject to serious pollution-induced weathering and a major attempt is being made to stabilize this decay.

Coastal erosion, the accelerated weathering of the stonework on some Greek sites, and salt weathering, as at Mohenjo-Daro, in the Indus Valley, are obvious problems; less obvious are the gradual effects of geomorphic processes on sites mantled with soils or cultural debris. This is a topic investigated by Tasker (1980) in a study of 40 earthwork sites in Wales and southern England. At each site she mapped its morphology and the evidence for different types of geomorphic processes. Figure 2.10 is an example of one of her maps—in this case for North Poorton, an Iron Age univallate fort in Dorset. Erosional features are indicated by the areas of bare soil, animal burrows, scars, poaching, loose debris, and terracettes. Careful mapping on all the sites allowed Tasker to measure the areas with those features. The overall results are summarized in table 2.2. The figures demonstrate the high incidence of sites with some form of damage. In terms of areal extent, poaching and terracettes are the most extensive. Poaching is caused by hooves of animals sinking into the soil when it has a high moisture content. Although terracettes may form naturally, their development can certainly be aided by livestock or people. Tasker did identify at certain sites the effect of visitors in inducing footpath erosion, but the results give greater emphasis to the damage done by livestock. Her study points to the need for better land-management programs at earthwork sites, with priorities being given to a reduction in livestock numbers and the removal of all livestock from sites during periods when the soil-moisture content is high.

CONCLUSION

This review has focused on the types of approaches geomorphologists can take when they are working in collaboration with archaeologists. A decision was made not to review relevant geomorphological techniques, since these are discussed by Gladfelter (1977) and Hassan (1978). It can be concluded that a geomorphologist ought to be an integral member of any archaeological project, as he or she can tackle a range of questions, from environmental reconstruction and change, site distribution, and site development to conservation evaluation. From this review, two recurring themes emerge: the contextual approach to geoarchaeology and the problems of explaining landform change.

First, emphasis is given to stressing the need for evaluating the significance of physical landscapes with close reference to the nature of the appropriate human society. Perhaps geomorphologists involved with archaeological projects ought also to become associated with present-day anthropological enquiries so that they are in a better position to interpret man-land relationships of the past. Underlying this theme is the lack of a consistent methodology for the analysis and evaluation of environmental conditions relevant to archaeology, a point also emphasized by Gladfelter (1981). One avenue for investigation is the adoption of methodologies and techniques as used in applied soil-survey research for present-day land-use problems (Davidson, 1980b).

The other clear theme in this review is geomorphic change. The identification of geomorphic change over timescales of the order of the Holocene epoch is

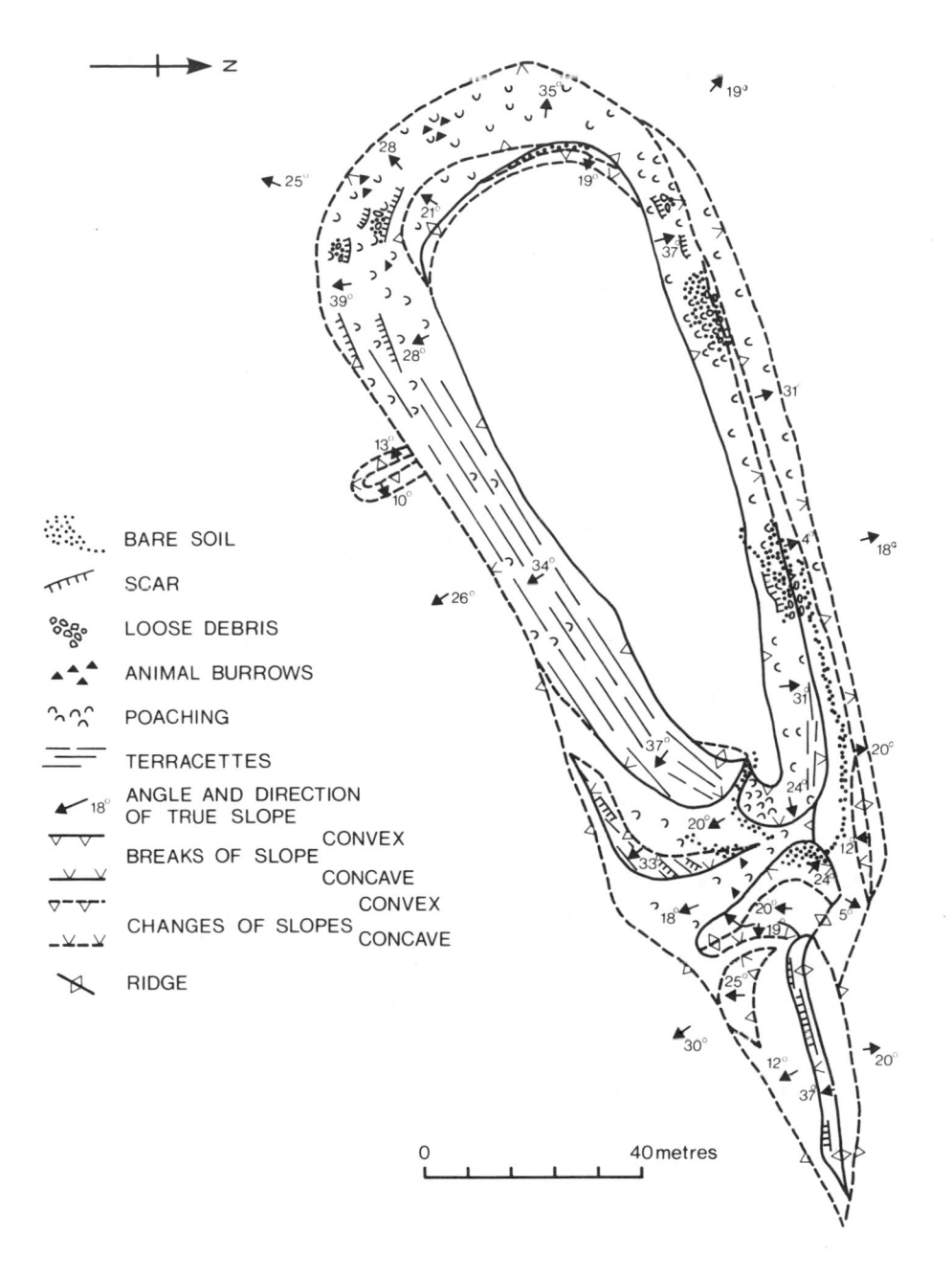

N

BARE SOIL

SCAR

LOOSE DEBRIS

ANIMAL BURROWS

POACHING

TERRACETTES

ANGLE AND DIRECTION
OF TRUE SLOPE

BREAKS OF SLOPE CONVEX
CONCAVE

CHANGES OF SLOPES CONVEX
CONCAVE

RIDGE

0   40 metres

Fig. 2.10.   Geomorphological map of North Poorton hillfort in Dorset, England (Tasker, 1980:174).

Table 2.2. Incidence of the Various Forms of Erosion at the Sample Earthwork Sites in Cardiganshire and Dorset, United Kingdom

Features indicating erosion	Number of sites in sample displaying features (total of 40)	Average percent of area of sites affected by these features
Localized areas of bare soil associated with paths and other concentrated trampling	35	3.4
Poaching on slopes	25	12.9
Terracettes	10	16.9
Sheep hollows	28	0.8
Animal burrows	22	n/a

SOURCE: Tasker, 1980:160.

necessarily very difficult when the landform, stratigraphic, and chronological record is fragmentary. Furthermore, a quantum leap is involved in any attempt to explain a *sequence* of changes. Geomorphologists are only beginning to recognize the nature of constraints, system controls, thresholds, and temporal and spatial responses within individual geomorphic systems; thus they are frequently unable to predict the geomorphic consequences if one or more variables are changed in a system for which process data are available. The implications for explaining geomorphic change in the past are clear. Cooke and Reeves (1976) carried out a study on the development, on a historical timescale, of arroyos in the American Southwest that has important implications for any study trying to explain geomorphic change: they conclude that "apparently similar arroyos can be formed in different areas as a result of different combinations of initial conditions and environmental changes" (p. 189). It is the problem of equifinality—different geomorphic trajectories can result in the same situation. That rivers in a region begin to aggrade at about the same time, for example, should not require a search for one overall explanatory cause.

Arguments on the basis of synchronous change can also be questioned on another point. Thornes (1980) comments that the explanation of soil erosion over the historic timescale in terms of particular causes must be treated with extreme care. He argues that erosional pulses need time to travel across space. This is especially valid when the total timescale may be of the order of one or two hundred years and indeed raises another broad problem—the applicability of geomorphic conclusions drawn from one timescale, say, a process study over ten years to a historical timescale or to a timescale of the Holocene epoch (10^4 years). The importance of varying spatial scales in geomorphology is widely accepted, but fundamental research still remains to be done on the nature of timescales.

Geomorphologists investigating geomorphic change relevant to particular archaeological periods must remember that the archaeological need is for the postulation of environmental conditions. Human activities can then be interwoven with the changing geomorphic patterns so that a fuller picture can be obtained of societies in the past.

Acknowledgment

I am grateful to Karl Butzer for comments on a draft of this chapter.

REFERENCES

Adams, R. M. 1965. *Land behind Baghdad*. Chicago: Univ. of Chicago Press.
Allchin, B., A. Goudie, and K. Hegde. 1978. *The prehistory and palaeography of the Great Indian Desert*. London: Academic Press.
Andrews, J. T., R. McGhee, and L. McKenzie-Pollock. 1971. Comparison of elevations of archaeological sites and calculated sea levels in Arctic Canada. *Arctic* 24:210–28.
Bagnold, R. A. 1941. *The physics of blown sand and desert dunes*. London: Methuen.
Bertelsen, R. 1979. Farm mounds in north Norway: A review of recent research. *Norwegian Archaeological Review* 12:48–56.
Bowler, J. M. 1976. Aridity in Australia: Age, origins and expressions in aeolian landforms and sediments. *Earth Science Review* 12:279–310.
Butzer, K. W. 1960. Archaeology and geology in Ancient Egypt. *Science* 132:1617–24.
———. 1962. Coastal geomorphology of Majorca. *Annals of the Association of American Geographers* 52:191–212.
———. 1965. Acheulian occupation sites at Torralba and Ambrona, Spain: Their geology. *Science* 150:1718–22.
———. 1967. Geomorphology and stratigraphy of the Paleolithic site of Budino. *Eiszeitalter und Gegenwart* 18:82–103.
———. 1970. Contemporary depositional environments of the Omo delta. *Nature* 226: 425–30.
———. 1971a. *Environment and archaeology*. 2d ed. Chicago: Aldine.
———. 1971b. *Recent history of an Ethiopian delta: The Omo River and the level of Lake Rudolf*. Univ. of Chicago Department of Geography Research Paper, vol. 136.
———. 1973. Spring sediments from the Acheulian site of Amanzi (Uitenhage District, South Africa). *Quaternaria* 17:299–319.
———. 1974a. Geo-archaeological interpretation of Acheulian calc-pan sites at Doornlaagte and Rooidam (Kimberley, South Africa). *Journal of Archaeological Science* 1:1–25.
———. 1974b. Geology of the Cornelia beds. *Memoirs van die Nasionale Museum Bloemfontein* 9:7–32.
———. 1976a. Lithostratigraphy of the Swartkrans formation. *South African Journal of Science* 72:136–41.
———. 1976b. The Mursi, Nkalabong and Kibish formations, Lower Omo Basin, Southwest Ethiopia. In *Earliest man and environments in the Lake Rudolf Basin*, ed. Y. Coppens, F. C. Howell, G. L. Isaac, and R. E. F. Leakey, 12–23. Chicago: Univ. of Chicago Press.
———. 1977. *Geomorphology of the lower Illinois Valley as a spatial-temporal context*

for the Koster Archaic Site. Reports of Investigations, no. 34. Springfield: Illinois State Museum.

———. 1978a. Changing Holocene environments at the Koster Site: A geo-archaeological perspective. *American Antiquity* 43:408–13.

———. 1978b. Sediment stratigraphy of the Middle Stone Age sequence at Klasies River mouth, Tsitsikama coast, South Africa. *South African Archaeological Bulletin* 33:141–51.

———. 1979. Geomorphology and geo-archaeology at Elandsbaai, Western Cape, South Africa. *Catena* 6:157–66.

———. 1980a. Context in archaeology: An alternative perspective. *Journal of Field Archaeology* 7:417–22.

———. 1980b. Holocene alluvial sequences: Problems in dating and correlation. In *Timescales in geomorphology*, ed. R. A. Cullingford, D. A. Davidson, and J. Lewin, 131–42. New York: John Wiley and Sons.

———. 1980c. The Holocene lake plain north of Lake Rudolf, East Africa. *Physical Geography* 1:42–58.

———. 1982. *Archaeology as human ecology.* Cambridge: Univ. of Cambridge Press.

———. In press. Archaeological sediments of Cueva Morin. In *Life and death in Cueva Morin*, ed. L. Freeman et al. Chicago: Univ. of Chicago Press.

Butzer, K. W., and C. L. Hansen. 1968. *Desert and river in Nubia.* Madison: Univ. of Wisconsin Press.

Butzer, K. W., F. H. Brown, and D. L. Thurber. 1969. Horizontal sediments of the Lower Omo Basin: The Kibish formation. *Quaternaria* 11:15–30.

Butzer, K. W., G. L. Isaac, J. L. Richardson, and C. K. Washbourn-Kamau. 1972. Radiocarbon dating of East African lake levels. *Science* 175:1069–76.

Butzer, K. W., D. M. Helgren, G. J. Fock, and R. Stuckenrath. 1973. Alluvial terraces of the Lower Vaal River, South Africa: A re-appraisal and re-investigation. *Journal of Geology* 81:341–62.

Butzer, K. W., R. Stuckenrath, A. J. Bruzewicz, and D. M. Helgren. 1978. Late Cenozoic paleoclimates of the Gaap Escarpment, Kalahari margin, South Africa. *Quaternary Research* 10:310–39.

Butzer, K. W., G. J. Fock, L. Scott, and R. Stuckenrath. 1979. Dating of rock art: Contextual analysis of South African rock engravings. *Science* 203:1201–14.

Coe, M. D., and K. V. Flannery. 1964. Microenvironments and Mesoamerican prehistory. *Science* 143:650–54.

Cooke, R. U., and A. Warren. 1973. *Geomorphology in deserts.* London: Batsford.

Cooke, R. U., and R. W. Reeves. 1976. *Arroyos and environmental change in the American South-West.* Oxford: Clarendon Press.

Crawford, I. 1974. Destruction in the highlands and islands of Scotland. In *Rescue archaeology*, ed. P. A. Rahtz, 183–207. Harmondsworth: Penguin Books.

Davidson, D. A. 1971. Geomorphology and prehistoric settlement of the Plain of Drama. *Revue de géomorphologie dynamique* 20:22–26.

———. 1973. Particle size and phosphate analysis: Evidence for the evolution of a tell. *Archaeometry* 15:143–52.

———. 1976. Processes of tell formation and erosion. In *Geoarchaeology: Earth science and the past*, ed. D. A. Davidson and M. L. Shackley, 255–65. London: Duckworth.

————. 1979. The Orcadian environment and cairn location. In *Investigations in Orkney*, ed. C. Renfrew, 7–20. London: Society of Antiquaries.

————. 1980a. Erosion in Greece during the first and second millennia B.C. In *Timescales in geomorphology*, ed. R. A. Cullingford, D. A. Davidson, and J. Lewin, 143–58. New York: John Wiley and Sons.

————. 1980b. *Soils and land use planning*. New York: Longman.

Davidson, D. A., R. L. Jones, and C. Renfrew. 1976. Palaeoenvironmental reconstruction and evaluation—a case study from Orkney. *Transactions of the Institute of British Geographers* n.s. 1:346–61.

Davidson, D. A., R. Lamb, and I. Simpson. 1983. Farm mounds in Orkney: a preliminary report. *Norwegian Archaeological Review* 16:39–44.

de Jong, J. 1967. The Quaternary of the Netherlands. In *The Quaternary*, vol. 2, ed. K. Rankama, 302–426. New York: John Wiley and Sons.

Dekker, L. W., and M. D. de Weerd. 1973. The value of soil survey for archaeology. *Geoderma* 10:169–78.

Demek, J., ed. 1972. *Manual of detailed geomorphological mapping*. Prague: Academia.

Demek, J., and C. Embleton, eds. 1978. *Guide to medium-scale geomorphological mapping*. Stuttgart: E. Schweizerbartsche Verlagsbuchhandlung.

Farrand, W. R. 1975. Sediment analysis of a prehistoric rockshelter: The Abri Pataud. *Quaternary Research* 5:1–26.

Gladfelter, B. G. 1977. Geoarchaeology: The geomorphologist and archaeology. *American Antiquity* 42:519–38.

————. 1981. Developments and directions in geoarchaeology. In *Advances in archaeological method and theory*, vol. 4, ed. M. B. Schiffer, 343–64. New York: Academic Press.

Goldberg, P. 1971. Analyses of sediments of Jerf 'Ajla and Yabrud rockshelters, Syria. *VIIIe INQUA Congrès, Etudes sur le Quaternaire dans le Monde*, pp. 747–54.

————. 1977. Past and present environment in prehistoric investigations in Gebel Maghara, Northern Sinai. In *Quedem No. 7*, ed. O. Bar-Yosef, and J. L. Phillips, 11–31. Jerusalem: Hebrew Univ. Institute of Archaeology.

————. 1979. Geology of Late Bronze Age mudbrick from Tel Lachish. *Journal of Tel Aviv Univ. Institute of Archaeology* 6:60–67.

Goudie, A. 1977. *Environmental change*. Oxford: Clarendon Press.

Hack, J. T. 1942. *The changing physical environment of the Hopi Indians of Arizona*. Papers of the Peabody Museum of American Archaeology and Ethnology, Harvard University, vol. 35, no. 1.

Hardan, A. 1971. Archaeological methods for dating of soil salinity in the Mesopotamian plain. In *Paleopedology: Origin, nature and dating of paleosols*, ed. D. H. Yaalon, 181–87. Jerusalem: Israel Universities Press.

Harvey, A. M. 1978. Dissected alluvial fans in southeast Spain. *Catena* 5:177–211.

Hassan, F. A. 1978. Sediments in archaeology: Methods and implications for palaeoenvironmental and cultural analysis. *Journal of Field Archaeology* 5:197–213.

Helgren, D. M. 1978. Acheulian settlement along the lower Vaal River, South Africa. *Journal of Archaeological Science* 5:39–60.

Higgs, E. S., and C. Vita-Finzi. 1972. Prehistoric economies: A territorial approach. In *Papers in economic prehistory*, ed. E. S. Higgs, 27–36. Cambridge: Cambridge Univ. Press.

Jacobsen, T., and R. M. Adams. 1958. Salt and silt in ancient Mesopotamian culture. *Science* 128:1251−58.

Kirkby, A., and M. J. Kirkby. 1976. Geomorphic processes and the surface survey of archaeological sites in semi-arid areas. In *Geoarchaeology: Earth science and the past*, ed. D. A. Davidson, and M. L. Shackley, 229−53. London: Duckworth.

Kirkby, M. J. 1977. Land and water resources of the Deh Luran and Khuzistan plains. In *Studies in the archeological history of the Deh Luran Plain*, ed. F. Hole, 251−88. Ann Arbor: Univ. of Michigan Press.

Kraft, J. C. 1976. Geological reconstructions of ancient coastal environments in the vicinity of the Island Field archaeological site, Kent County, Delaware. *Transactions of the Delaware Academy of Science for 1974 and 1975*, pp. 83−118.

————. 1977. Late Quaternary paleographic changes in the coastal environments of Delaware, Middle Atlantic Bight, related to archaeological settings. *Annals of the New York Academy of Sciences* 288:35−69.

Kraft, J. C., G. Rapp, and S. E. Aschenbrenner. 1975. Late Holocene paleography of the coastal plain of the Gulf of Messenia, Greece, and its relationships to archaeological settings and coastal change. *Bulletin of the Geological Society of America* 86:1191−1208.

Kraft, J. C., S. E. Aschenbrenner, and G. Rapp. 1977. Paleographic reconstructions of coastal Aegean archaeological sites. *Science* 195:941−47.

Kraft, J. C., G. R. Rapp, and S. E. Aschenbrenner. 1980. Late Holocene paleogeomorphic reconstructions in the area of the Bay of Navarino: Sandy Pylos. *Journal of Archaeological Science* 7:187−210.

Lamb, R. G. 1980. *An archaeological survey of two of the North Isles of Orkney*. Edinburgh: Royal Commission on the Ancient and Historical Monuments of Scotland.

Renfrew, C. 1976. Introduction. In *Geoarchaeology: Earth science and the past*, ed. D. A. Davidson and M. L. Shackley, 1−5. London: Duckworth.

Rick, J. W. 1976. Downslope movement and archaeological intrasite spatial analysis. *American Antiquity* 41:133−44.

Shackley, M. L. 1972. The use of textural parameters in the analysis of cave sediments. *Archaeometry* 14:133−45.

Tankard, A. J., and F. R. Schweitzer. 1976. Textural analysis of cave sediments: Die Kelders, Cape Province, South Africa. In *Geoarchaeology: Earth science and the past*, ed. D. A. Davidson and M. L. Shackley, 289−315. London: Duckworth.

Tasker, C. M. K. 1980. Archaeological site erosion: Studies from Britain and the Aegean. Ph.D. thesis, Univ. of Strathclyde, Glasgow.

Thornes, J. B. 1980. Erosional processes of running water and their spatial and temporal controls: A theoretical viewpoint. In *Soil erosion*, ed. M. J. Kirkby and R. P. C. Morgan, 129−82. New York: John Wiley and Sons.

van Andel, T. H., T. W. Jacobsen, J. B. Jolly, and N. Lianos, 1980. Late Quaternary history of the coastal zone near Franchthi Cave, southern Argolid, Greece. *Journal of Field Archaeology* 7:389−402.

van Zuidam, R. A. 1975. Geomorphology and archaeology: Evidences of interrelation at historical sites in the Zaragoza region, Spain. *Zeitschrift für Geomorphologie*, n.s. 19:319−28.

Verstappen, H. Th., and R. A. van Zuidam. 1968. I.T.C. system of geomorphological survey. *I.T.C. textbook of photo-interpretation*. Delft: International Training Center.

Vita-Finzi, C. 1969a. Late Quaternary alluvial chronology of Iran. *Geologische Rundschau* 58:951–73.

———. 1969b. *The Mediterranean valleys*. Cambridge: Cambridge Univ. Press.

Wise, S. M., J. B. Thornes, and A. Gilman. 1982. How old are the badlands? A case study from South-East Spain. In *Piping and badland erosion*, ed. A. Yair and R. Bryan, 259–77. Norwich, England: Geo Books.

Wymer, J. J. 1976. The interpretation of Palaeolithic cultural and faunal material found in Pleistocene sediments. In *Geoarchaeology: Earth science and the past*, ed. D. A. Davidson and M. L. Shackley, 327–34. London: Duckworth.

3

Geological Studies of Coastal Change Applied to Archaeological Settings

JOHN C. KRAFT, İLHAN KAYAN, and
STANLEY E. ASCHENBRENNER

ABSTRACT

Sedimentological and geomorphological theories, models, and field techniques may be used to yield information on the coastal paleogeographies of archaeological sites. Surface and subsurface reconstructions of late-Quaternary coastal change may be correlated with the historical and archaeological records of an area, thereby clarifying human-environmental interaction and subsistence strategies. This chapter presents several examples of coastal geomorphological and geological studies related to archaeological site settings in the American mid-Atlantic coastal zone and Aegean Sea area. The concepts and methods can be applied to archaeological coastal-zone exploration worldwide.

In recent years the subject of human coastal adaptations has received much attention in the archaeological literature. Ever since Binford's (1968) article on post-Pleistocene adaptations and the work of Kraft (1972), Kraft, Rapp, and Aschenbrenner (1975), and Bintliff (1975), publications on archaeological sites related to coastal paleomorphology and maritime adaptations have taken on a new importance. Although some of the papers generated from this emerging interest are of a theoretical nature (Schalk, 1977; Osborn, 1977), many newer works are based on field surveys of coastal sites in particular environmental zones (Stark and Voorhies, 1978). The recent work of R. S. MacNeish in Belize (pers. com., 1982) has indicated that many statements concerning the distribution of archaeological sites along the Caribbean coast of Central America are at best hypotheses for testing, especially when the statements preclude actual knowledge of the existence of these sites. Even in the field of cultural-resource management, the consideration of paleocoastal site settings has become an important issue in light of surveys of the Gulf of Mexico outer continental shelf initiated by the Minerals Management Service of the Department of the Interior (Gagliano, 1977). Despite an abundance of investigations, many archaeologists remain mystified by the processes of coastal geomorphology. Hassan (1979) has stressed the need for the integration of geomorphological processes and archaeological site distributions, particularly in coastal areas.

The potential for archaeological site preservation and discovery in world coastal zones is at best tenuous (Kraft, Belknap, and Kayan, 1983). Eustatic sea-level change (of 100 m or more) has been dramatic over the last 15,000 years. Transgression and regression of the shorelines have caused many archaeological sites on coastal-plain and continental-shelf areas to be buried and/or destroyed. Sites on the present shoreline may have been far inland when occupied. Conversely, inland sites frequently can be identified as ancient ports or coastal sites. In tectonically active coastal zones, slow upwarp or downwarp of the land surface has caused cities to erode and disappear into the sea. Catastrophic events such as earthquakes and volcanic episodes have destroyed and buried cities. Dependent upon these processes (and possible climatic change), variable amounts of sediment aggrade and/or prograde on the lower floodplains and deltas of the world's coastal zone. Many geologic processes can be used by the archaeologist in evaluating both present archaeological site locales and changes in geomorphic environments since the time of their occupation.

COASTAL SEDIMENTARY SEQUENCE ANALYSIS

Academic geological research on the evolution of coastal systems, supported by massive research into the formation of coastal sedimentary facies models by various major petroleum companies of the world, has generated hundreds of important papers on the construction of coastal sedimentary facies or environmental models from both the hypothetical and empirical viewpoints. As a relatively large number of geological processes are involved in the evolution of any coastal system, it behooves the geologist or archaeologist to consider many factors when interpreting the geology of a coastal system as related to an archaeological site, particularly when attempting to reconstruct paleogeographic settings of archaeological sites. Figure 3.1 illustrates the general forces and processes impinging on a shoreline and potential responses that might affect its configuration. Sedimentary erosion and deposition are the two major elements. As noted in figure 3.1, it is not logical that a single factor will determine the geomorphic configuration of a coast and its evolution; however, depending upon the local geological setting, one geological factor may overwhelm many of the other processes. This is particularly true in the complex interplay of physical forces involved in coastal change over the short term of the Holocene epoch (past 10,000 years) or the longer-term Pleistocene epoch. Sometimes the most important factor affecting coastal change may occur in the interior or in the subsurface as a tectonic event not immediately visible at a coastal archaeological site. Diagrams similar to figure 3.1 can be constructed for all types of coastal settings;

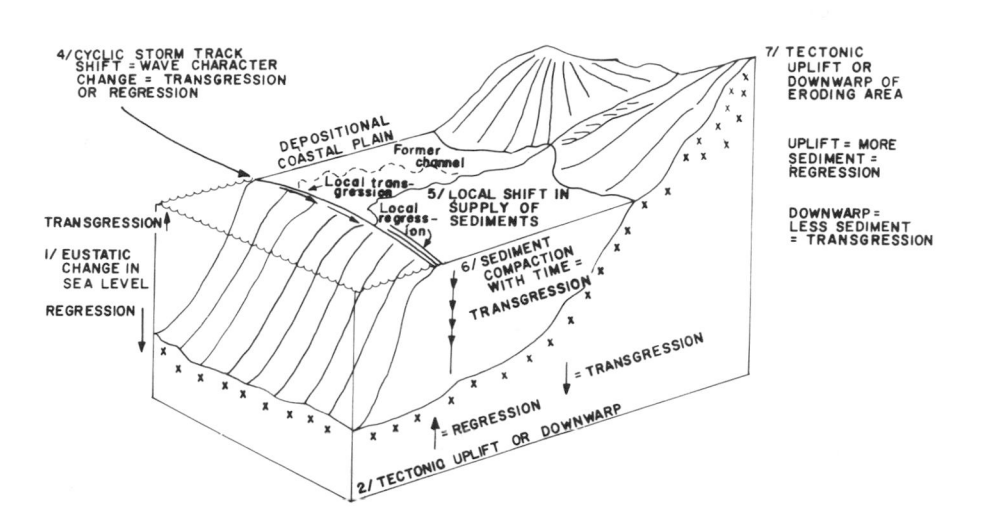

Fig. 3.1. A conceptual diagram of forces and processes impinging upon a sedimentary coastal zone, showing the potential elements of a marine transgression or regression and responses to them (after Kraft, 1972).

the model provides an inventory of the factors that must be considered in any attempt to understand the geology of a coastal setting and its potential for evolving over a period of time.

Geological studies of a coastal environment may be divided into three areas: physical processes and responses, including the types, movement, and provenance of sediments in the coastal zone; the sequence of sediments that are deposited, both vertically and laterally; and the sedimentary environmental *lithosome*, or three-dimensional shape of the sedimentary deposit representing each environment undergoing erosion or deposition in the coastal zone. These three types of analysis might be considered an order of magnitude apart from each other, yet all are important to the study of coastal-zone geology. Most important of all, however, is that studies be directed toward forming three-dimensional models of coastal morphology and of the lithosomes resulting from the physical processes. In addition, it is important to consider the fourth dimension, time. Ultimately, time, plus changes in relative and absolute sea levels locally and worldwide, controls whether a coastline is eroding and transgressing landward, or depositing and regressing seaward, or remaining in a roughly "neutral" configuration.

Since the waning of the latest (Wisconsin or Würm) glaciation, from approximately 18,000 to 14,000 years ago, the world's sea level has risen rapidly toward its present position. There is much argument and discussion among workers on relative and absolute (eustatic) sea levels at any particular time over the past 10,000 to 15,000 years, but a general consensus has been reached regarding this period: during the early portion of this time, sea level rose rapidly and then slowed its rate of rise. After this, one school of thought believes, sea level rose to its present position approximately 6,000 years ago and has since fluctuated above and below this level. Another theory holds that sea level rose relatively rapidly in early Holocene time, with a slower rise from approximately 6,000 to 4,000 years ago and then a still slower rise from about 4,000 years ago to its present position (which is still rising). Figure 3.2 shows a number of curves illustrating the diversity of opinion concerning late-Quaternary eustatic sea-level change. A major controversy involves whether sea level has risen smoothly to its present position or whether there have been wide fluctuations during the past 12,000 to 15,000 years. Obviously, all the curves shown cannot be true eustatic sea-level curves for the world's oceans. It is known that presently there are slight differ-

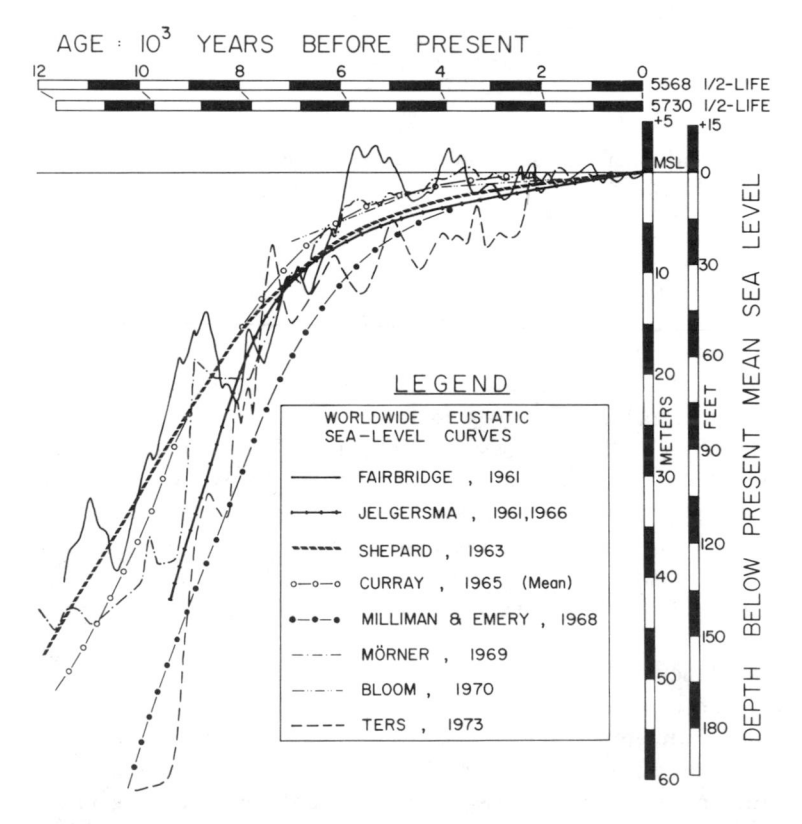

Fig. 3.2. Opinions of various geologists as to the world's sea level over the Holocene epoch. The dominant cause of change is, of course, climatic; however, tectonics and compaction effects are surely also involved. Curves assembled by D. F. Belknap (from Kraft et al., 1976).

ences in eustatic sea level, but not of the magnitudes shown in figure 3.2. Therefore, we are left with the obvious conclusion that most of the curves are not eustatic sea-level curves, but rather *relative* sea-level curves based on data gained from local and regional studies around the world. It is possible that one or two of the curves may closely resemble the as yet unknown "true" eustatic sea-level curve for the Holocene epoch, but, as agreement has not and cannot yet be reached, the geologist and archaeologist cannot simply select a curve that best fits their data and use it to make paleogeographic reconstructions. The only absolute requirement is that *all paleogeographic reconstructions of coastal archaeological settings must be based on relative sea-level curves from local data.* Reconstructions based on presumed "eustatic" sea-level curves must be considered to be conceptual or hypothetical interpretations, certainly not factual in nature.

For specific details on techniques of studying coastal sedimentary facies or sedimentary environments, the reader is referred to the excellent text by Friedman and Sanders (1978). Walker (1979) summarizes the study of sedimentological sequences, techniques, and schematic methods of presentation that might be used by the archaeologist and geologist in attempting to interpret a coastal setting. Details of technique are discussed and illustrated in Kraft, 1976a; Kraft et al. (1976, 1979); Kraft, 1978; and Kraft and John (1978, 1979). Critical to the entire concept of coastal reconstruction is the capability to predict, based on surficial geomorphic and sedimentologic data and subsurface drilling or geophysical data, what coastal environment should be in what position at what time. The conceptual framework for modern stratigraphic-sedimentologic reasoning in this area is summarized by Walther's Law, formulated approximately one century ago, which states:

> The various deposits of the same facies areas and similarly the sum of the rocks of different facies areas are formed beside each other in space though in cross section we see them lying on top of each other. As with biotopes, it is a basic statement of far-reaching significance that only those facies and facies areas can be superimposed primarily which can be observed beside each other at the present time. [Middleton, 1973:979]

Walther's Law provides us with a guide whereby we may study the lateral and vertical facies changes of both biological and sedimentological coastal environments. If we can understand these changes physically and historically, we can, with enough data, produce empirical reconstructions and even project potential coastal geomorphic change into the future.

Probably the most important element of any coastal study related to archaeology is to determine whether a marine transgression or regression is occurring. A regression will leave a former coastal archaeological site somewhere inland, possibly buried by fluvial sedimentary lithosomes. A transgression will result in a former inland site being located in a coastal setting, potentially resulting in a false interpretation of the site's original location. Furthermore, transgressions,

accompanied by erosion and local relative sea-level rise, may inundate archaeological sites and leave them stranded in the submarine zone. Frequently
the marine transgression of a coastal archaeological site will destroy it. However,
if the site is transgressed rapidly, as in a catastrophic event such as an earthquake, it may be preserved in the submarine environment. Many examples of
submerged settlements are known in the Mediterranean area. Figure 3.3 is a
schematic diagram showing examples of transgressions and regressions in the
coastal zone of Delaware and in the coastal zone of Greece. Although the geologic-
tectonic settings of these areas are extremely different, similar processes and
Walther's Law still apply, so that one can note a common sequence of events
that has occurred over the past 9,000 years. The transgressions and regressions
have been reconstructed using data from a large number of drillings and radiocarbon dating of organic peats, shells, etc. Despite the similarities that may be
observed in the two transgressions and the two regressions shown in the figure,
each is an individual case under different climatic and geologic-tectonic conditions. Thus paleogeographic maps showing the geomorphology of these areas
at particular times in the past must be based on a proper understanding of the

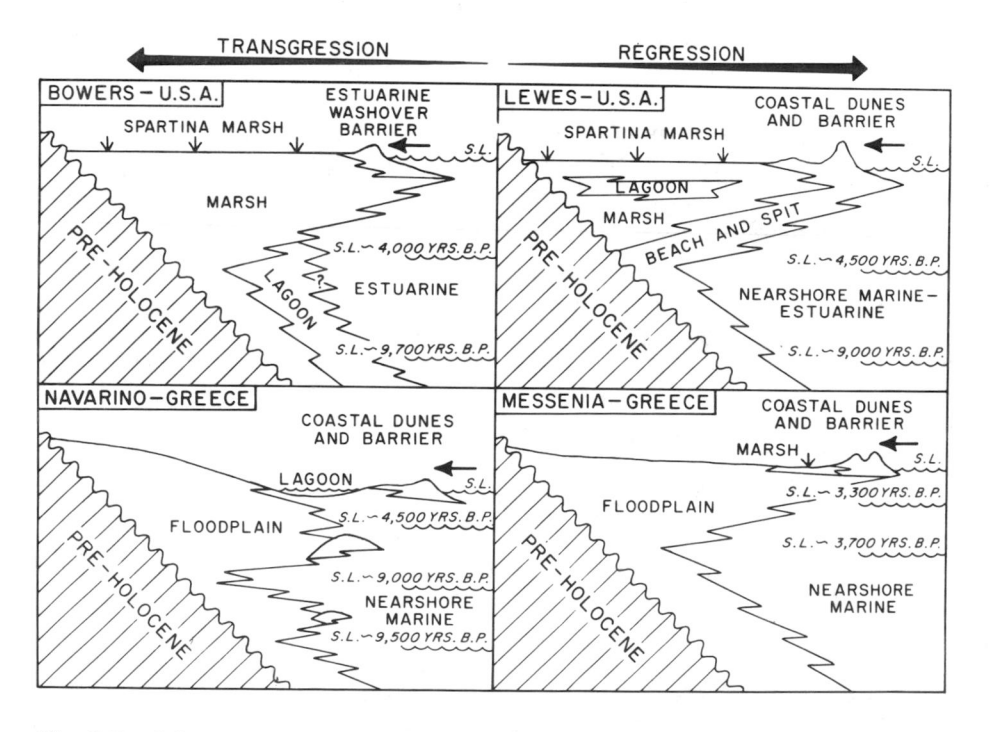

Fig. 3.3. Schematic interpretations of transgression and regression along the Atlantic
coast of Delaware (a sandy coastal plain) and along the Mediterranean coast of Greece (a
tectonically active coast). Note that although the geological settings are very different a
similar analysis, Walther's Law, may be applied. Note also that the sea levels shown are a
local relative sea level and therefore are not the same for the Greek and Delaware coasts.

nature of local relative sea level and of time lines as they might be drawn through the diagrams.

Figure 3.4 presents a geological cross section along the axis of the northern coastal plain of the Bay of Navarino, Peloponnesus, Greece (Kraft, Rapp, and Aschenbrenner, 1980). A time-depositional line has been drawn through the diagram. Note that such time lines in general are based on an understanding of variations in coastal sedimentary facies and on radiocarbon dates to support the actual time of the depositional events. Also note that the time line of figure 3.4 is not straight; rather, it follows the morphology of the depositional environments at the time indicated, about 5,000 years ago. With a large number of radiocarbon

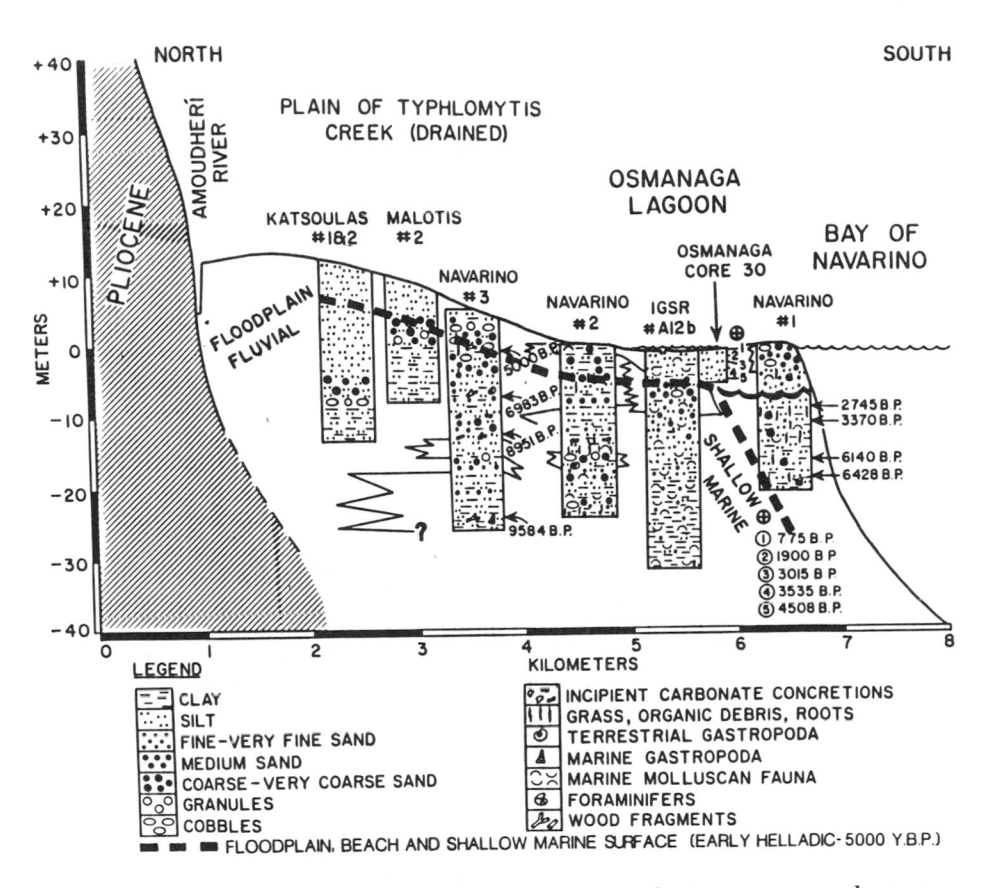

Fig. 3.4. A north-south geologic cross section interpreting the transgressive and regressive sedimentary environmental lithosomes in the alluvial plain and lagoon-barrier area north of the embayment at Navarino, Greece. Radiocarbon dates are shown for various environments. Note that time lines used in paleogeographic interpretations should be roughly parallel to the present depositional surface and thus can span a large topographic elevation range—in this case approximately 45 m (modified from Kraft, Rapp, and Aschenbrenner, 1980).

dates and a large number of drill cores, fairly precise time lines may be drawn for any coastal locality, from the historic past back through early Holocene time. The more and better the data, the more accurate are the geological-paleogeographical interpretations that can be made. Figure 3.5 is an example of a paleogeographic reconstruction made from data and interpretations, shown in figure 3.4, of the embayment at Navarino, Greece.

A local relative sea-level-rise curve (fig. 3.6) has been constructed for the Delaware coastal zone, based on data from over 100 drill tests in a limited shoreline area that is believed to be slowly subsiding tectonically (Kraft, 1976b). With this knowledge and with a knowledge of where sea level was at various times over the last 10,000 years, information from borings can be used to determine the shape of the sedimentary lithosomes of the coastal environments, as related to past positions of sea level. From these data precise paleogeographic

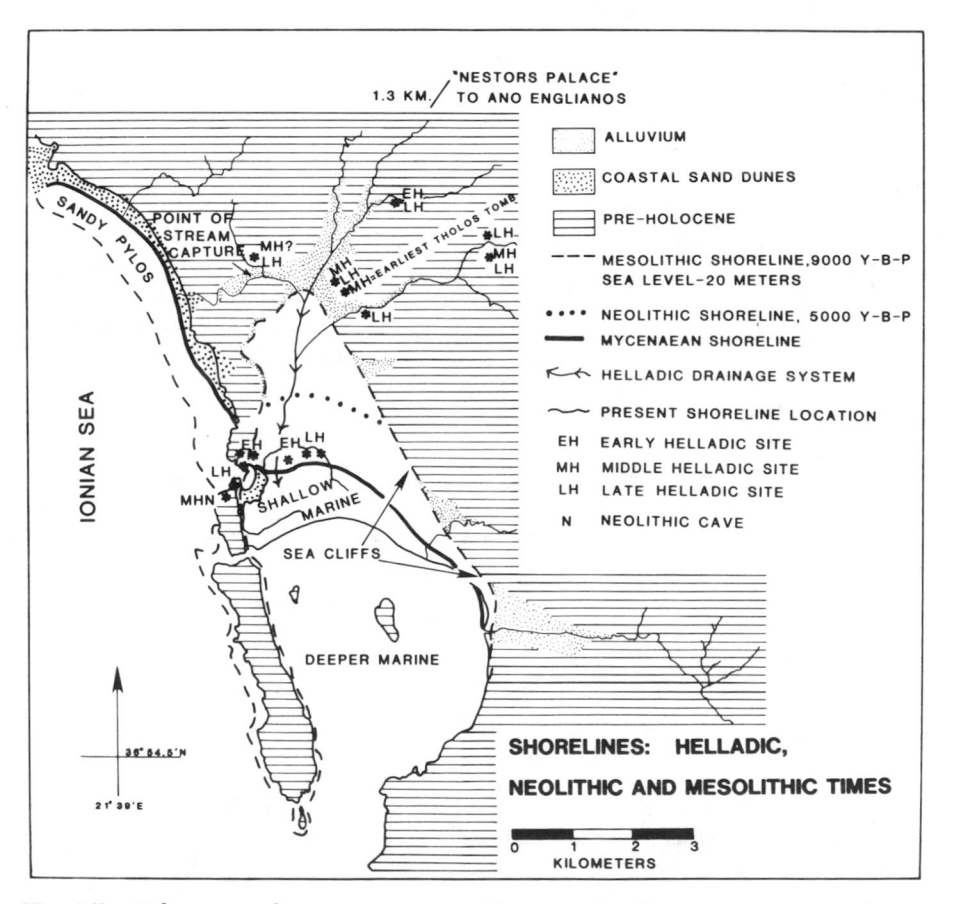

Fig. 3.5. Paleogeographic reconstructions of ancient shorelines in the embayment at Navarino (Sandy Pylos), based on data presented in figure 3.4.

maps can be reconstructed. Figure 3.7 is a typical subsurface cross-sectional interpretation of coastal environments along the west side of Delaware Bay, on the Atlantic coast of the United States. Note the radiocarbon dates showing time and position of earlier marshes that grew in an area below the present sea surface. Dates for salt-marsh peat are ideal in locating sea levels, as they indicate the high sea-level position, occurring in spring, for the time of the living marsh; basal dates are preferable for absolute positions of sea level because they are less subject to compaction of the sedimentary section. The cross section in figure 3.7 shows a deeply incised pre-Holocene tributary of the ancestral Delaware River at a time when sea level was approximately 100 m below the present level. As may be seen from the figure, about 9,700 years ago sea level was slightly more than 25 m below its present position.

Data from a large number of borings similar to those shown in figure 3.7 were used in making paleogeographic reconstructions for this coastal zone in the vicinity of Bowers Beach and Island Field, a very important Woodland period—

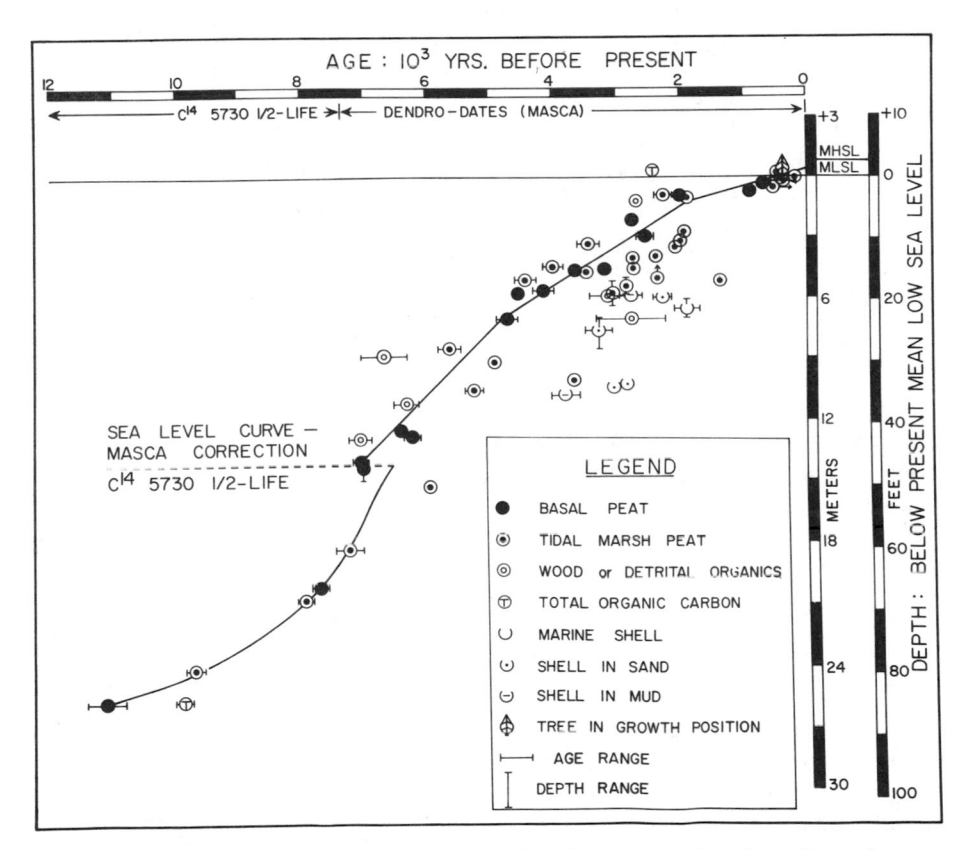

Fig. 3.6. A local relative sea-level-rise curve based on coring data from the Delaware coastal zone (after Kraft, 1976b).

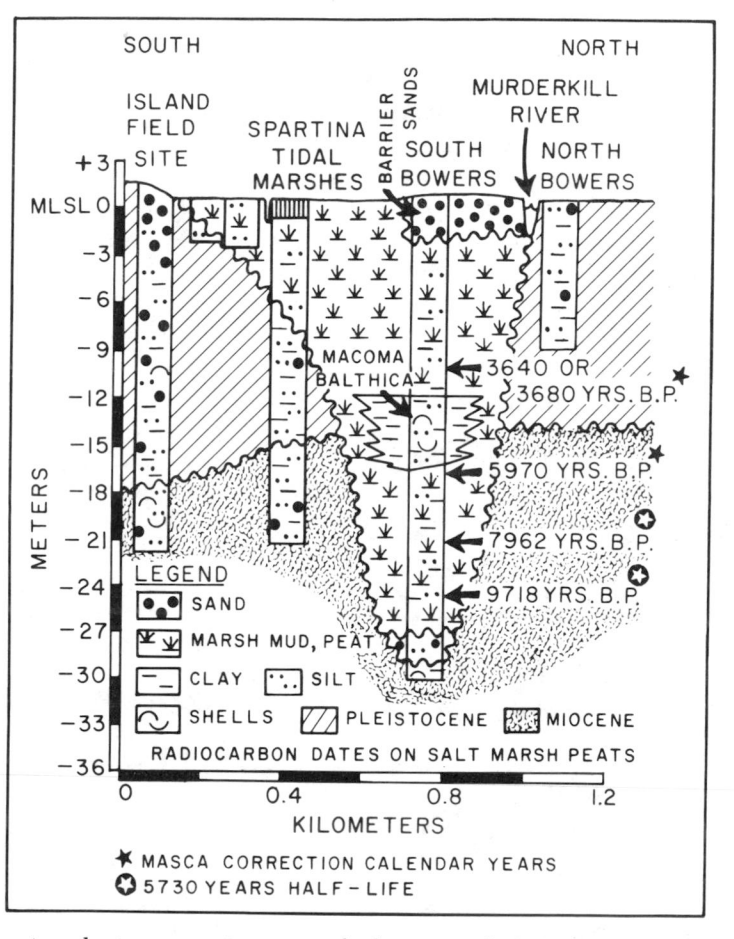

Fig. 3.7. A geologic cross section across the lower Murderkill River valley in Delaware, adjacent to the Island Field archaeological site. Note the numerous radiocarbon dates in salt-marsh peats; they provide a geographically accurate position for an environment although not necessarily a precise relative sea-level position, as they may be subject to some compaction. However, the basal date of 9718 B.P. might be considered to be unaffected by compaction because it impinges upon earlier Pleistocene sediments that are more strongly compacted. Cross sections of this type have been used to reconstruct the paleogeographies shown in figure 3.8 (from Kraft and John, 1978).

Archaic period archaeological site in coastal Delaware. Figure 3.8, based on a number of cross sections similar to that shown in figure 3.7, shows the present coastal configuration and a series of paleogeographic reinterpretations of the area at different periods in the past. At the Island Field archaeological site, a large and important settlement was occupied by Amerinds in the transitional early Woodland to middle-late Woodland periods, from approximately 1,000 to 3,000 years

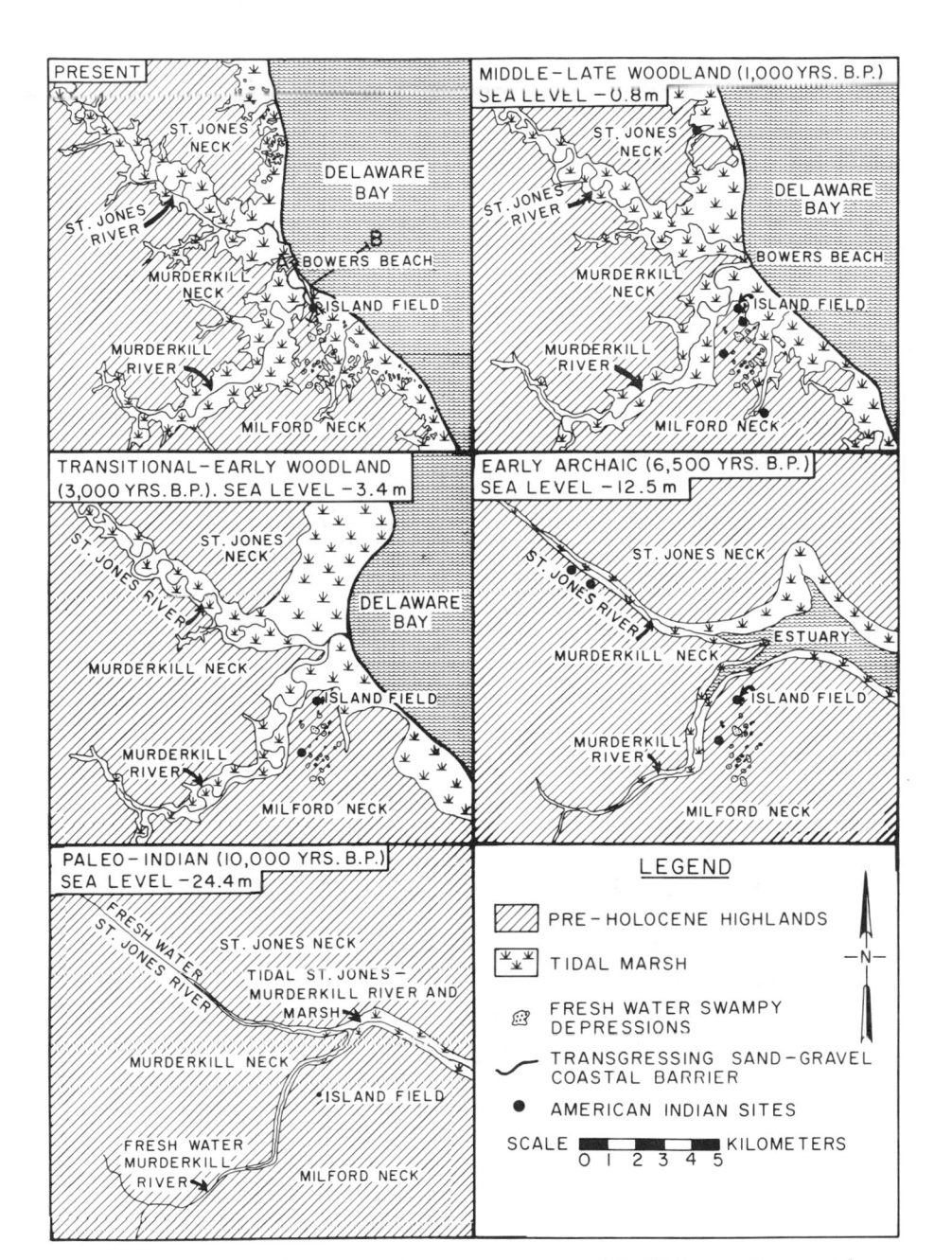

Fig. 3.8. A series of paleogeographic reconstructions of the Delaware Bay coastal zone in the vicinity of the Island Field archaeological site (after Kraft et al., 1976; Kraft, 1976a).

ago. There are also indications of earlier Archaic sites in the area. Finally, a map is presented for Paleo-Indian times (approximately 10,000 years before the present), although no archaeological sites of this age have been identified as yet in the area.

One might observe that the Island Field site at present is very close to Delaware Bay, an important shellfish and seafood resource; however, it was much further from Delaware Bay at the time of site occupancy. Probably, the Island Field site was occupied because it lay near a series of freshwater ponds and along the edge of a tidal river with easy access to the sea. The implications of this example of paleogeographic reconstruction are profound in that they suggest, as seen from the maps, that Delaware Bay, a large marine estuary, at one time lay much further to the southeast on the Atlantic continental shelf. Thus, sites dating from earlier Archaic and Paleo-Indian times might occur in abundance on the submerged Delaware Bay bottom and adjacent shelf. Here, of course, they would be extremely difficult to discover, for they were very likely buried or destroyed by the transgression that has pushed the shore landward nearly 100 km and the sea level upward over 25 m during the past 10,000 years.

PRESERVATION OF ARCHAEOLOGIC SITES

Preservation or destruction of archaeological sites in coastal settings in Greece and Turkey may depend on a number of factors. For instance, the aggradation of alluvial surfaces buries sites as floodplains evolve and prograde seaward (fig. 3.9). An example of this is noted in a study by Kraft, Rapp, and Aschenbrenner (1980) of the embayment of Navarino in the Peloponnesus. Any early Helladic sites that were constructed on the floodplain at the head of the Bay of Navarino are now buried by up to 5 m of alluvium (fig. 3.4). The chances of their discovery are almost nil. On the other hand, classical- to Roman-period and later sites are buried only 1–2 m deep. Thus, wells dug on the plain and footings for modern buildings frequently encounter such archaeological sites.

Coastal erosion is another and pervasive factor in the Mediterranean. Although many of the cliffs of the Mediterranean are eroding at very slow rates, those formed of Neogene sediments—often softer silts, clays, and sands—are eroding at very rapid rates. Therefore, any archaeological sites that were located on Neogene formations near the shore zone may have been completely destroyed by erosion. In addition, many cases of catastrophic events are known in which all or part of a site on Neogene bedrock has disappeared in an earthquake. Helice, on the southern shore of the Gulf of Corinth, is one such case; it is said to have disappeared in 373 B.C. during an earthquake along the southern fault of the graben (Schwartz and Tziavos, 1979). Helice presently is covered by a thick alluvial fan of sands and gravels. Attempts to locate this ancient city have been and will prove to be very difficult. Other cases of earthquakes destroying cities and dropping them below sea level are known. A part of the ancient city of Gythion in the Gulf of Laconia dropped into the sea in A.D. 374–375 (Scoufopoulos and

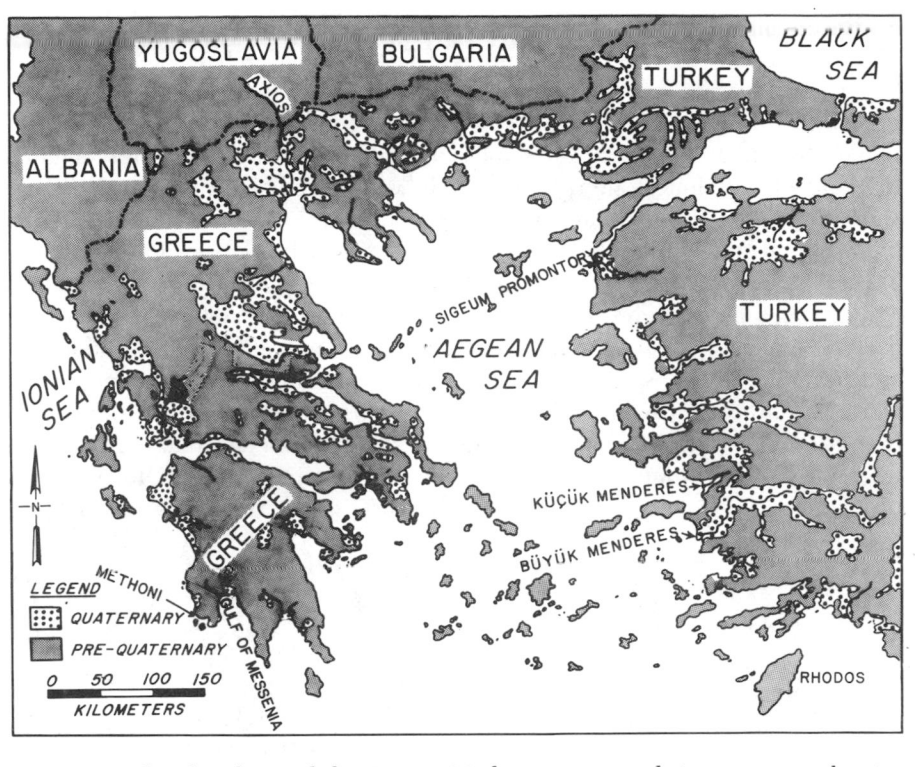

Fig. 3.9. The shorelines of the Ionian, Mediterranean, and Aegean seas, showing Quaternary sedimentary embayments. The majority of the Quaternary sedimentary embayments are presently undergoing alluviation and progradation, resulting in burial of archaeological sites. Many of the clifflike shorelines of the map area are eroding at very slow rates; little change can be noted over the past three to five millennia. However, in those areas where the cliffs are comprised of Neogene silts, sands, and limestones, wave-cut cliff retreat is in some cases very rapid. In these areas large numbers of sites are being destroyed or have disappeared. Furthermore, the deep embayments of the shorelines are in many cases down-faulted grabens and are therefore subject to catastrophic dropping of the land surface, resulting in the disappearance of archaeological sites under the sea.

McKernan, 1972). Flemming (1971) discusses the site at Elaphonisos at the southern end of the southeast tip of the Peloponnesus. Here a settlement dating from early to late Helladic period (ca. 5,000–3,000 years B.P.) lies in 3–5 m of water. It remains unknown whether the submergence of Elaphonisos was cataclysmic or a gradual inundation due to relative sea-level rise. More recent work in Greece indicates that traces of Neolithic coastal settlements may have survived dispersal by the Holocene worldwide sea-level rise (Flemming, 1982; Gifford, 1982).

Reconnaissance and detailed studies of the shorelines of Greece and Aegean Turkey have shown numerous examples of destruction of archaeological sites and depositional events that could have buried sites (fig. 3.9). A few of these examples will be discussed here. Kraft and Aschenbrenner (1977) studied the evolution of the embayment at Methoni, 12 km south of Navarino Bay. Here a hard, resistant Eocene limestone ridge forms the western edge of the Peloponnesus, and rates of shoreline erosion along this coast are very slow. Farther eastward, in the Methoni embayment, Neogene silts occur. As sea level rose with the waning of the last major ice age, these Neogene silts were exposed to wave erosion, resulting in rapid shoreline retreat (fig. 3.10). In the embayment at Methoni, the islet of Nisakouli is known to have been occupied from about 2100 to 1580 B.C. and in medieval times. Ten years ago remains of walls from both occupations were still present (McDonald and Hope Simpson, 1969). By 1973, Aschenbrenner found that the medieval walls had disappeared and that only two tile fragments and a stone with lime-rich mortar adhering attested to medieval occupation. The islet is continuing to decrease in size through wave erosion.

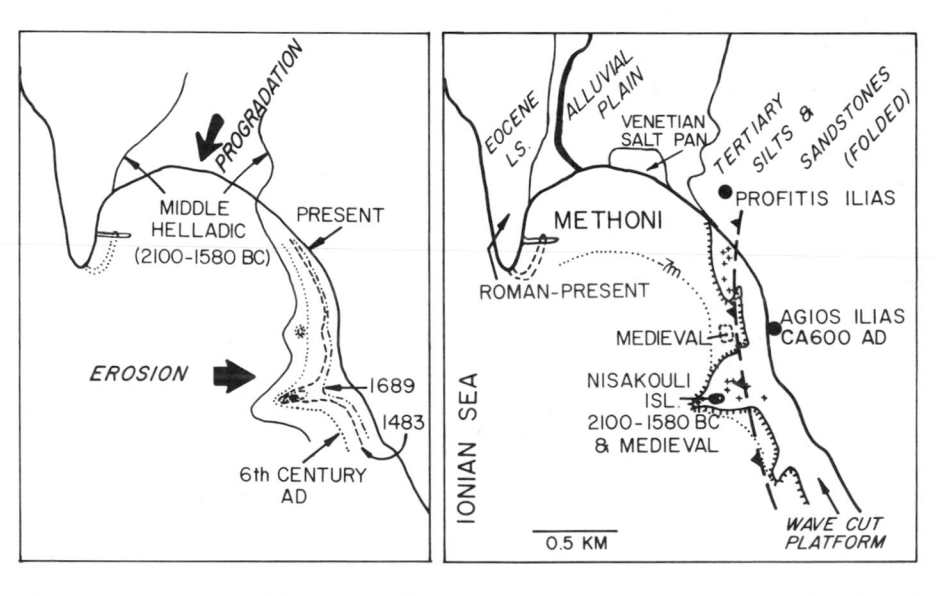

Fig. 3.10. Wave-cut cliff retreat and paleogeographic changes related to archaeological settings at Methoni, in the Peloponnesus. The diagram on the left shows shoreline progradation at the head of the embayment and erosion by wave-cut cliff retreat on the eastern side of the embayment since 3000 B.C. The diagram on the right shows some of the historic and prehistoric sites that are presently being eroded or destroyed. (The writers thank Dr. Tsugo Sunamura, Institute of Geoscience, University of Tsukuba, Tokyo, for his comments and ideas incorporated in the diagram.) This figure was originally published as figure 10 in John C. Kraft and Stanley E. Aschenbrenner, "Paleogeographic Reconstructions in the Methoni Embayment in Greece," *Journal of Field Archaeology* 4 (1977):33. It is reprinted here with the permission of the publisher.

Furthermore, an illustration from 1574 shows the fortifications and harbor of Methoni and two islands in the embayment (fig. 3.11). The northern islet has now disappeared. However, examination of air photographs and bottom topography shows the base of this island to be still evident as a part of the submerged wave-cut shelf (fig. 3.12). Figure 3.10 also shows progressive erosional cliff retreat on the eastern edge of the Methoni embayment. From this interpretation it can be seen that when the site on Nisakouli was occupied in middle Helladic times it was located on a promontory or peninsula. The site of Agios Ilias, a sixth-century A.D. chapel that is falling into the sea, is also shown.

In Helladic times (ca. 3,000–5,000 years ago) the Methoni embayment extended far to the north of the present shoreline. Yet for medieval times a map from the archives of Venice identifies the location of a salt pan on the lower part of the present alluvial plain. Air-photographic studies show part of the periphery of the salt pan. The shoreline in this area is now eroding and the Venetian salt-pan remnants are disappearing. Thus, accretion is occurring by alluviation with burial of archaeological sites in the northern part of the Methoni embayment, while wave-cut cliff retreat and total destruction of archaeological sites is occurring on the northeastern and eastern sides of the embayment.

Elsewhere in the southwestern Peloponnesus, at the head of the Gulf of Messenia, lies the plain of the Pamisos River (Kraft, Rapp, and Aschenbrenner, 1975). Presently it is a drained agricultural plain of alluvial silts with occasional channels of rivers with their sand and gravel beds (fig. 3.13). A few low-lying swamps remain undrained behind the high sand-gravel coastal accretion barriers. Although it appears that the coast currently is undergoing erosion, the longer-term trend in this area is one of seaward progradation of the deltaic coastal plain. Study of geologic maps and air photographs shows smooth arcing lines between the older pre-Quaternary terrain and that of the Holocene depositional plain. These arcing lineaments are highly suggestive of ancient shorelines. On the Pamisos River plain two Helladic sites, at Bouxas in the west and Akovitika in the east, are known. Bouxas lies at the edge of a highland that was a former shoreline.

Akovitika remained a problem, as it includes occupation from approximately 2500 B.C. through Geometric times (about 800 B.C.). Ruins of walls at Akovitika are buried under a coastal swamp. This is a case of destruction of archaeological sites by progradation of a delta-floodplain swamp. Subsurface stratigraphic studies in this area enabled the reconstruction of the coastal geography as it was in middle Helladic times. The shoreline position approximately 5,000 years ago is shown in figure 3.13. Figure 3.14 is a geologic cross section interpreting the relationship of the ruins of Akovitika buried under the swamp to the present beach-accretion plain. The Helladic remains at Akovitika may be interpreted as a coastal site similar to Bouxas to the west.

Similar processes of coastal change occur in the Aegean Sea (Kraft, Aschenbrenner, and Rapp, 1977). Several examples undergoing study are next presented. In the northwest corner of the Biga Peninsula of the Troad, near the

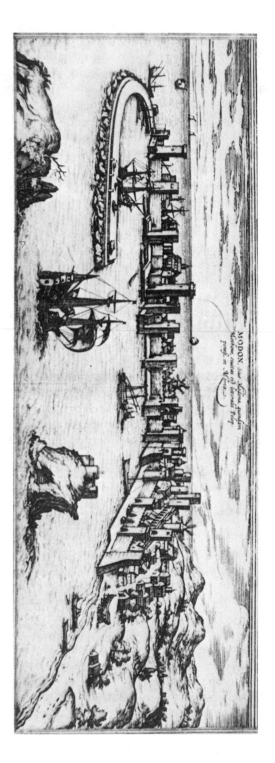

Fig. 3.11. A sixteenth-century schematic panorama by Hohenberg of fortress Methoni as viewed from the cliffs on the east side of the embayment (Braun and Hohenberg, 1574).

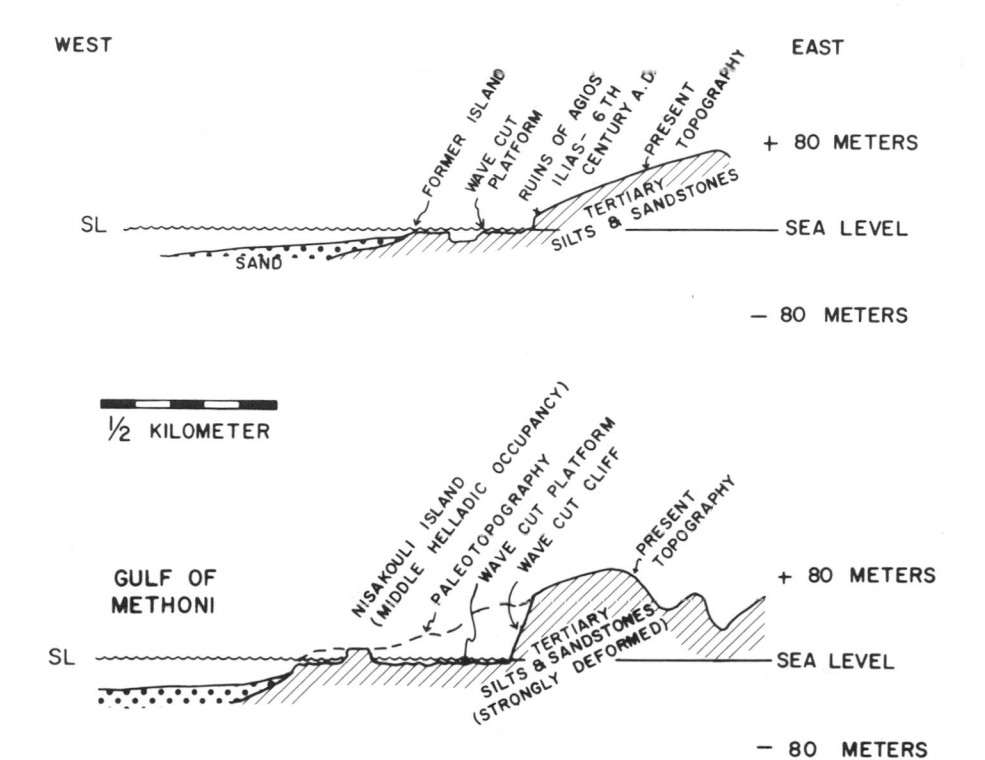

Fig. 3.12. Topographic profiles along the eastern flank of the Gulf of Methoni. Topographic details are shown in sections running east-west through the ruins of the sixth-century A.D. chapel, Agios Ilias and through Nisakouli Island. The wave-cut cliffs are well developed. It is hypothesized that a former island existed to the west of the ruins of Agios Ilias (Kraft and Aschenbrenner, 1977; reprinted with permission from the *Journal of Field Archaeology*).

mouth of the Dardanelles, lies the ancient site of Troy (fig. 3.15). Kraft, Kayan, and Erol (1982) demonstrated that a 15 km marine embayment occurred south of the present shoreline of the Dardanelles approximately 7,000 years ago. Since that time, sediments of the Kara Menderes (Scamander) River and Dümrek Sü (Simois) River have infilled this embayment as relative sea level rose and sediments were eroded from the surrounding highland plateaus. Accordingly, ancient shoreline and floodplain sites have been buried, at depths of up to 10 m below the present alluvial plain, since first occupation here in early Helladic times, about 5,000 years ago.

More important, however, is the Sigeum Promontory lying to the west of the plain of Troy. Here a Neogene plateau, partly incised, is undergoing intense wave-cut cliff retreat and erosion. The Neogene sands, silts, and limestones are relatively horizontal and unconsolidated. The .cliff is steep and undergoing

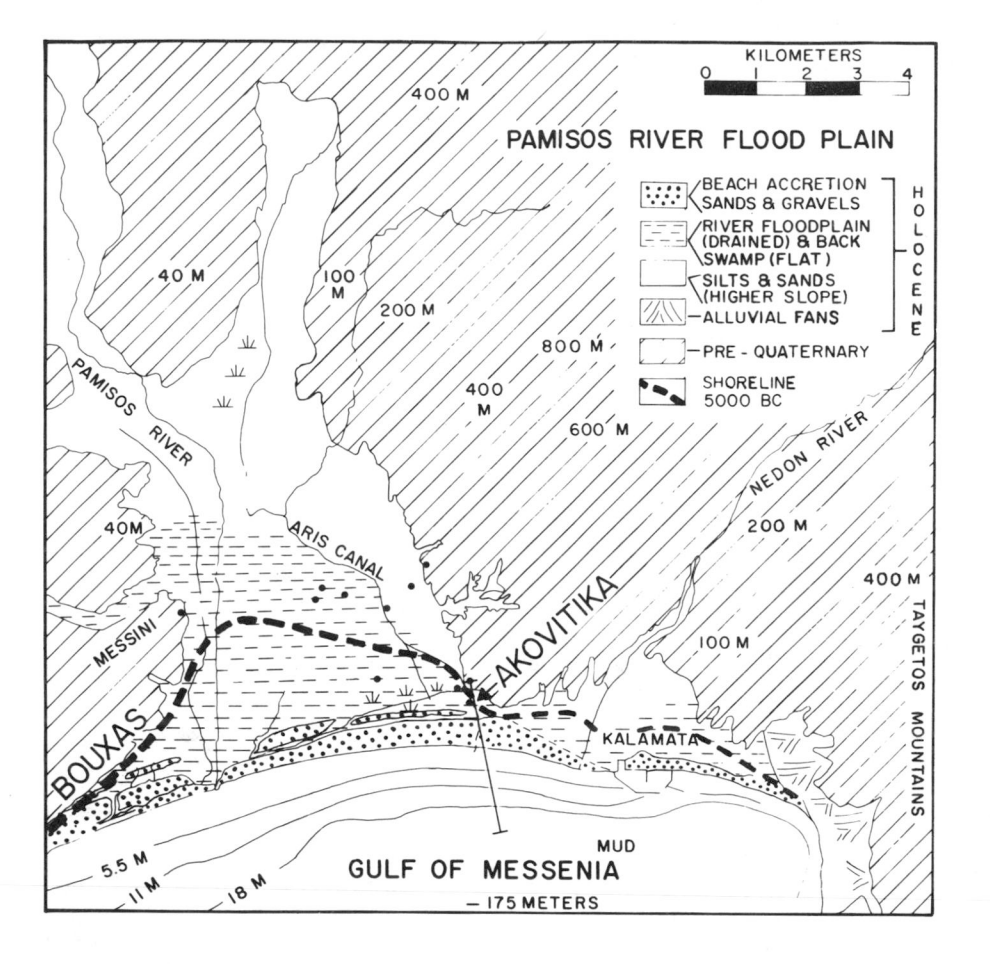

Fig. 3.13. The Pamisos River floodplain and its depositional environments at the head of the Gulf of Messenia. Deposition has been infilling a drowned embayment of the Gulf of Messenia over the past 9,000 to 10,000 years. Based on subsurface evidence from drill-hole studies a shoreline position for approximately 5000 B.C. has been identified. This shoreline reconstruction identifies the Early Helladic sites at Bouxas and Akovitika to have been along a shoreline. The site at Akovitika is presently buried under the marshes of the Pamisos floodplain. It is probable that a large number of occupation sites are buried under the plain of the Pamisos River. Only those that are by chance encountered by foundation studies or building activities will be discovered. However, the possibility for use of geophysical methods to discover other sites remains (Kraft, Rapp, and Aschenbrenner, 1975; reprinted with the permission of the Geological Society of America).

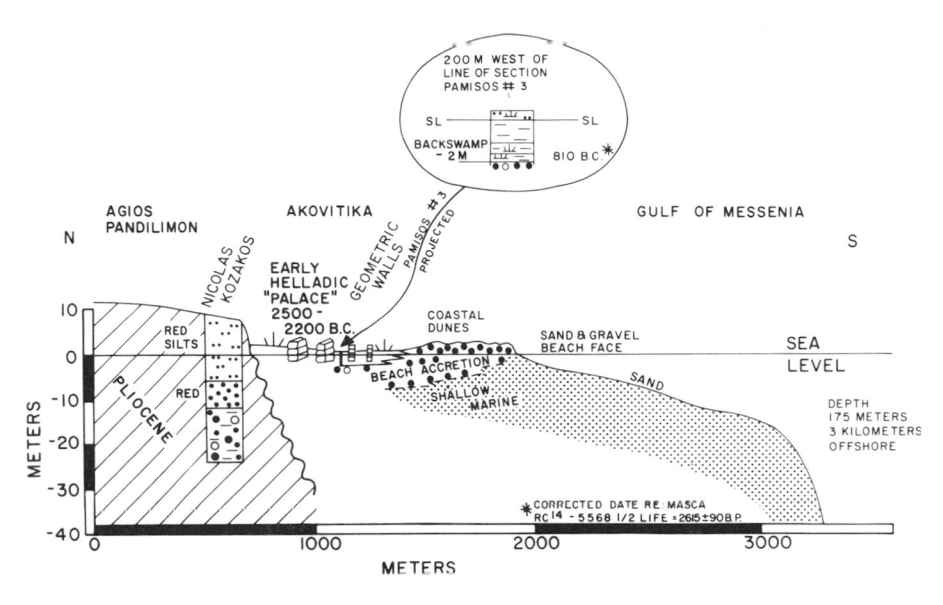

Fig. 3.14. Geologic cross section interpreting the site at Akovitika. An Early Helladic "palace" and Geometric walls are buried under the marsh silts of the accreting Pamisos River plain. Their discovery and excavation led to a number of speculations as to their reason for being in a marsh. Subsurface geologic studies of the floodplain of the Pamisos River and its tributaries show that the site was clearly a coastal or shoreline site at the time of occupancy (Kraft, Rapp, and Aschenbrenner, 1975; reprinted with permission of the Geological Society of America).

continual erosion, as indicated in figure 3.16. Littoral transport of sediment accompanying this erosion is tending to smooth the shoreline and fill in former embayments such as existed at Bewsik Düzü, which was formerly 2 km broad. Trench studies by Mey and others in 1926 showed that the shoreline of this embayment was occupied in early Helladic times, about 5,000–4,000 years ago. Progradation of the shoreline in Beşik Düzü, therefore, may have buried many other archaeological sites, since the littoral transport has provided sediment from the eroding adjacent cliffs.

The classical city of Sigeum was located somewhere along the promontory of the same name near the confluence of the Dardanelles and the Aegean Sea (fig. 3.16). Although Cook (1973) has identified Sigeum as being located under the ruins of Yenişehir, which is now abandoned, its location is still uncertain. Nevertheless, classical ruins exist in the area and it is the most logical site for this ancient city. Probably the major part of the site has long since been eroded and buried under the sand of the shallow adjacent shelf.

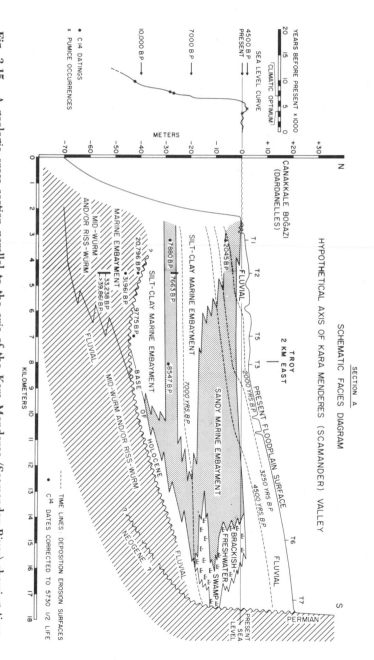

Fig. 3.15. A geologic cross section parallel to the axis of the Kara Menderes (Scamander River) showing time-depositional surfaces of shallow marine-embayment and floodplain environments. Any archaeological sites built on the floodplain surface in Helladic times (3000–1200 B.C.) would be buried approximately 4 to 12 m under alluvium. Their potential discovery using modern site-search techniques is, therefore, unlikely. Refer to Fig. 3.16 for geography of the Trojan plain (Kraft, Kayan, and Erol, 1982; reprinted with permission of Princeton University Press).

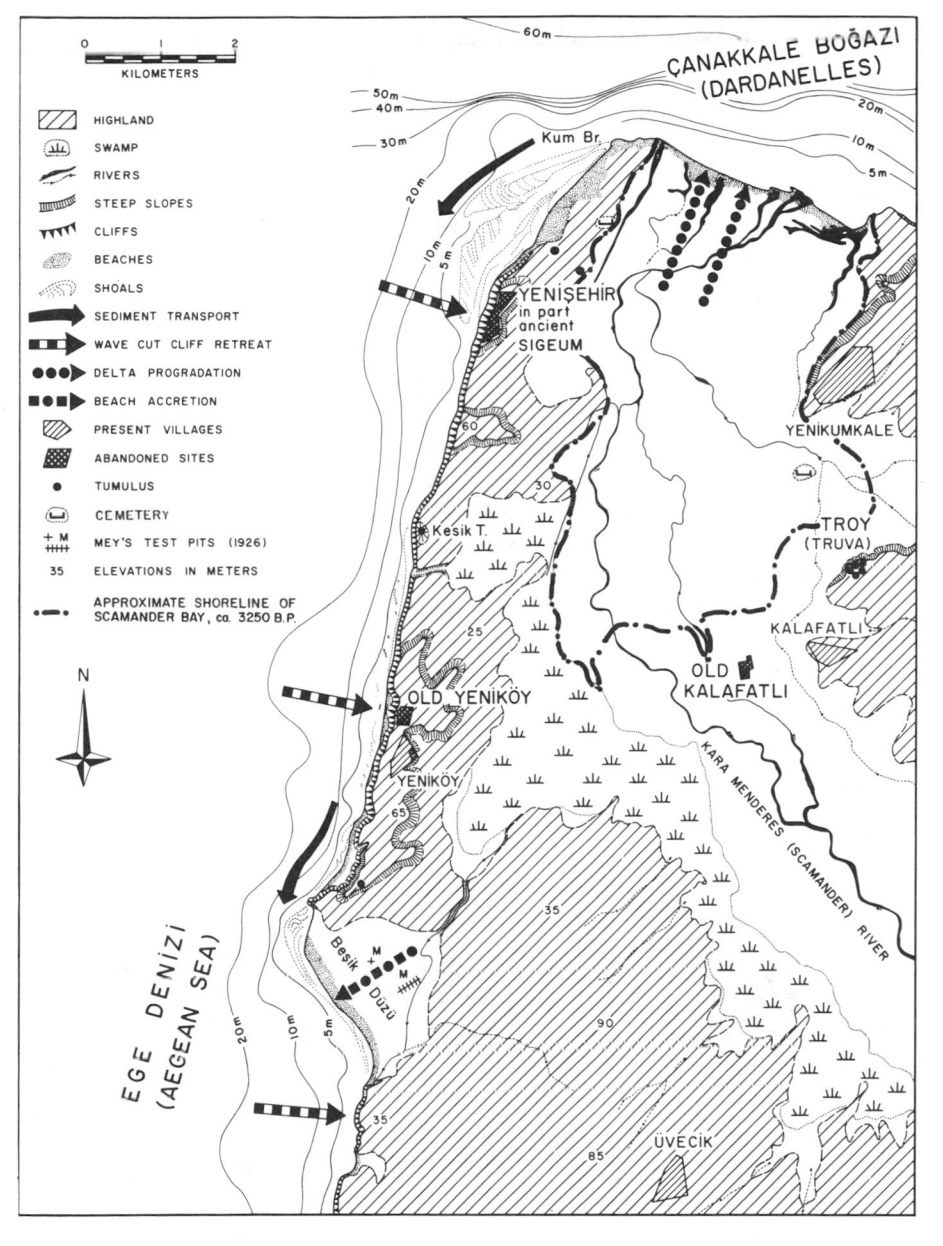

Fig. 3.16. The shoreline of the Sigeum Promontory in the vicinity of ancient Troy. The low-lying dissected plateau surface that forms the cliff shoreline along the Aegean Sea includes soft Neogene sands, silts, and limestones. These sediments are particularly prone to rapid erosion by wave activity and resultant cliff retreat. Further, littoral transport by wave activity tends to straighten or modify the shape of the shoreline, such as the infilling of an embayment at Beşik Düzü. The site of ancient Sigeum is identified as in part buried under the more modern abandoned site of Yenişehir (map drawn by İ. Kayan).

A site on a seventeenth-century sketch of the mouth of the Dardanelles and the adjacent Sigeum Promontory is identified as the ruins of ancient Sigeum (fig. 3.17). Possibly at that time the ruins were still evident to visitors. The erosional process continues along the Sigeum Promontory. About 5 km to the south of Yenişehir occurs the site of "Old Yeniköy," which became uninhabitable because of undercutting of the cliff and subsidence of portions of the town. Accordingly, it had to be moved to its present site. Surely this is a prime example of an event that must have occurred thousands of times around the Mediterranean in cliff-type coasts incised into Neogene sediments.

Farther south along the Aegean shoreline of Turkey lie the deltaic coastal plains of the Küçük Menderes and the Büyük Menderes rivers. These plains are of importance to archaeologists and historians because major ancient cities were situated along their edge. For instance, Ephesus (Efes) lay along the shore of an embayment of the Küçük Menderes. In addition, the ancient site of Artemision occupied a shoreline site in classical times near the hill later occupied by Selçuk castle at the eastern end of the embayment. The record of progradation and aggradation of the Küçük Menderes Delta is well known to historians, as shown in figure 3.18 (Schindler, 1904; Darkot, 1938; Erinç, 1955; Eisma, 1962; Russell, 1967). As the delta advanced, the alluvial plain aggraded and the river meandered across the plain. The shoreline eventually reached the vicinity of the harbor of Ephesus in "late ancient times." Then, because of the importance of the city, the occupants continued to dredge a channel from the Aegean into the

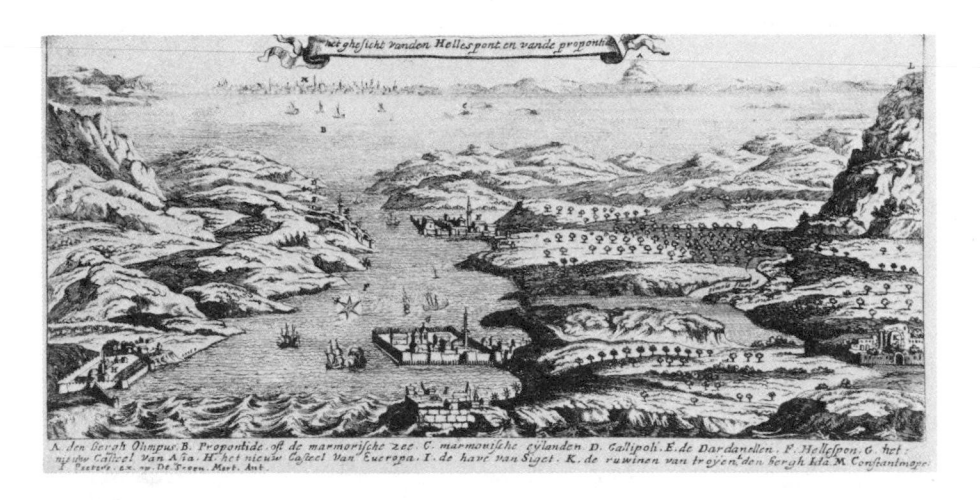

Fig. 3.17. A panorama by Jacques Peeters, ca. 1686, of the mouth of the Dardanelles and the Sigeum Promontory. Wave-cut cliff retreat is evident along the line of the Sigeum Promontory (along right half of lower edge). The ruins of ancient Sigeum are identified (lower center). A coastal defense wall is evident, possibly an attempt to maintain the site of Yenişehir (which was occupied at this time) or possibly representing walls of ancient Sigeum.

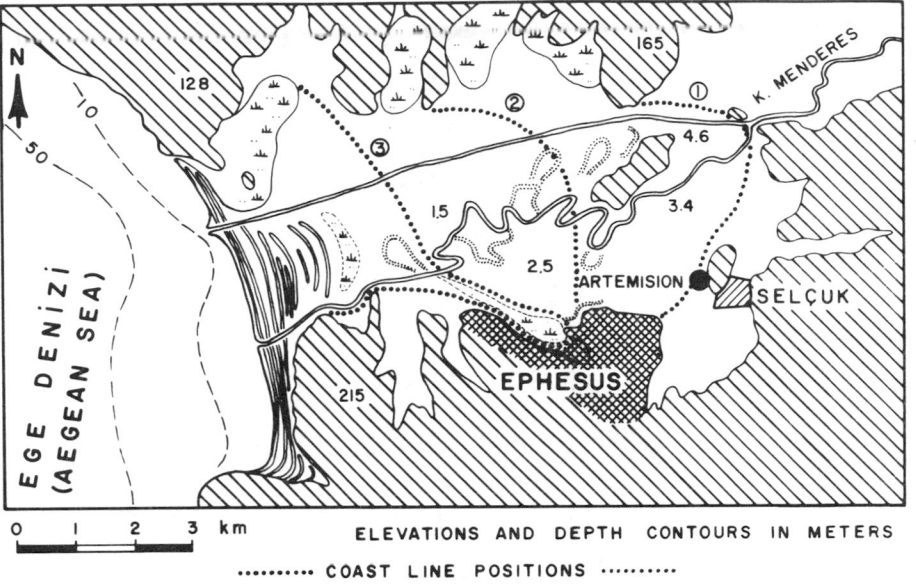

Fig. 3.18. The infilled embayment of the Küçük Menderes River. Lines of shoreline progradation are from Schindler (1904). Note the attempt to keep the channel open to the harbor of the major city of Ephesus. The potential for an extremely large number of sites to be buried here under alluvium is evident in an area of embayment infill and shoreline progradation.

harbor. Finally, with the abandonment of the city the shoreline migrated past the site of Ephesus and the beach-accretion plain continued to move westward to its present position. With two major ancient cities along the shoreline of the embayment, it is most likely that numbers of other sites of the classical-Roman period, such as agricultural villages and/or individual farm sites, occurred on the delta plain itself. All of these are now buried and lost to us by alluviation.

Figure 3.19 shows the relationship of the smaller plain of the Küçük Menderes to the large plain of the Büyük Menderes River. The embayment of the Büyük Menderes River was much larger in classical times. Near the mouth of the embayment on its southern side was Miletus, one of the five largest cities in the Roman Empire and one of great importance. Heracleia, a major seaport in the classical-Hellenistic period, lay on the eastern end of an arm of the embayment. At the site of Priene, located on cliffs overlooking the northern side of the embayment, major ruins are still visible. However, Priene's lower port area is now buried by alluvium. Heracleia is now partly under the waters of Bafa Lake, a

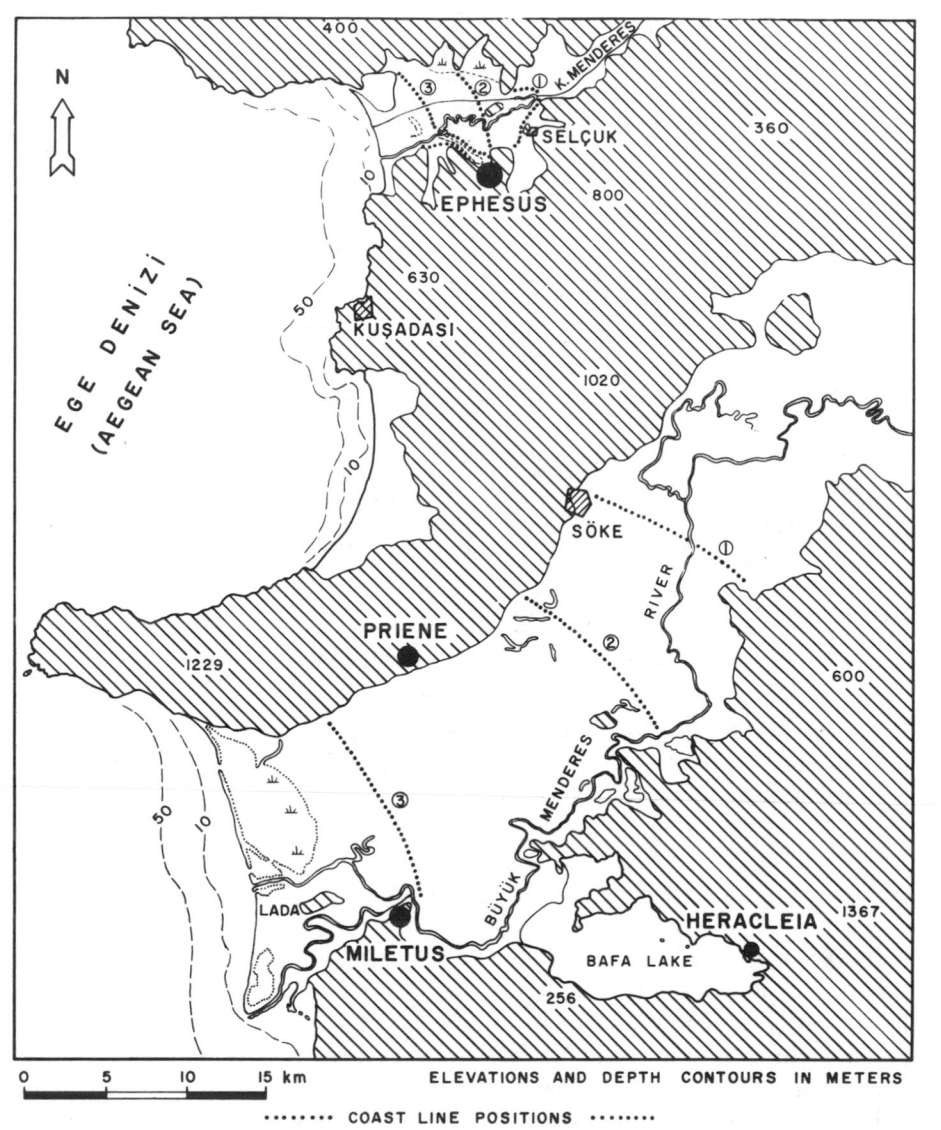

Fig. 3.19. The Büyük Menderes River embayment. Coastal positions are shown for various times as the delta prograded seaward. The potential is very high for site burial beneath this prograding and alluviating surface. The remains of a major naval battle fought between Miletus and the (at that time) Island of Lada about 2,500 years ago are buried under the alluvium.

brackish lake that formed and was filled to a slightly higher level than the sea as the alluvial plain of the Büyük Menderes River dammed the embayment.

Miletus is almost totally destroyed, with much of its building materials obviously removed for construction elsewhere. Here again history tells us of an event in which archaeological remains probably are still buried at great depth under the alluvium of the Büyük Menderes Plain. Between the Island of Lada and Miletus the peoples of Miletus and their allies from Priene fought a major sea battle with the Persians (Kraft, Aschenbrenner, and Rapp, 1977). Somewhere within the 10 square km between Miletus and Lada lie the buried remains of the ships destroyed in 494 B.C. The Büyük Menderes continues to prograde seaward and its channel continues to migrate. The river itself has lent an important term to the geographer and geologist, *meander*.

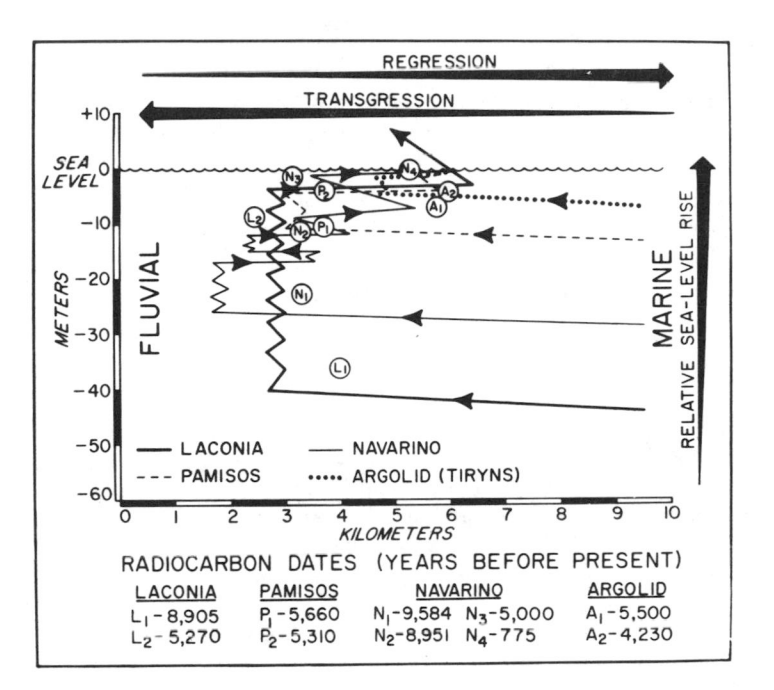

Fig. 3.20. Accompanying the post-Würm (Wisconsin) rise in sea level, a number of valleys in the Ionian-Mediterranean-Aegean region underwent structural deformation and downwarping as grabens. In addition, sediment was eroded off the land surface and transported to the heads of the grabenlike embayments. The net result was a complex of shoreline transgression and regression. This diagram illustrates shoreline transgression and regression at various times as indicated by radiocarbon dates. The potential, therefore, exists for either burial of sites by coastal progradation and floodplain alluviation or destruction by coastal erosion. The rates of subsidence of the embayments varied, showing that each embayment must be studied individually.

SUMMARY

The examples discussed here are but a few of the many known cases of partial destruction or preservation of archaeological sites by alluviation, shoreline accretion, or wave-cut cliff retreat along the shorelines of the Ionian, Mediterranean, and Aegean seas. Figure 3.20 summarizes the coastal transgression and regression accompanied by rise of the sea relative to the land in several marine embayments around the Peloponnesus. Only a beginning has been made in understanding the importance of studies of coastal processes and geology of coastal change and their impact on archaeological sites and the interpretation of the history of coastal settings. Similar concepts and interpretations can and should be applied to the problems of coastal archaeological site preservation or destruction elsewhere in the world's coastal zone.

Acknowledgments

This research summarizes work over the past twelve years partially supported by the University of Minnesota Messenia Expedition (W. A. McDonald, Director); the Office of Naval Research, Geography Programs; the Archaeometry Laboratory, University of Minnesota at Duluth (G. Rapp, Jr., Director); the University of Delaware; and Maden tektik ve Arama Institüsü, Ankara (S. Alpan, former Director). We thank Dr. Jay Custer for his review of portions of the manuscript and Dr. M. L. Schwartz (Kraft, Aschenbrenner, and Kayan, 1980) for his support and suggestions regarding concepts and content.

REFERENCES

Binford, L. R. 1968. Post-Pleistocene adaptations. In *New perspectives in archaeology*, ed. L. R. Binford and S. R. Binford, 313–42. Chicago: Aldine.

Bintliff, J. L. 1975. *Natural environment and human settlement in prehistoric Greece.* British Archaeological Reports, Supplementary Series, vol. 28.

Braun and Hohenberg. 1574. *Theatres des villes*, II. Amsterdam.

Cook, J. M. 1973. *The Troad: An archaeological and topographical study.* Oxford: Clarendon Press.

Darkot, B. 1938. Ege halialerinin mense ve tekamulu. *Cografi Arastirmalar, Istanbul Universitesi Yayinlari*, no. 62, pp. 29–52.

Eisma, D. 1962. Beach ridges near Selcuk, Turkey. *Tijdschrift van het Koninklijk Nederlandsch Aardrijkskundig Genootschap* (Amsterdam) II, 79, 234–46.

Erinc, S. 1955. Gediz ve Kucuk Menderes Deltalarinin Morfolojisi. *Dokuzuncu Cografya Meslek Haftasi, Tebligler ve Konferanslar, Turk Cografya Kurumu Yayinlari*, Sayi 2:33–66.

Flemming, N. C. 1971. *Cities in the sea.* New York: Doubleday and Co.

———. 1982. A submerged Neolithic site in the Sporadhes, Greece. In *Quaternary coastlines and marine archaeology*, ed. P. M. Masters and N. C. Flemming. New York: Academic Press.

Friedman, G. M., and J. E. Sanders. 1978. *Principles of sedimentology.* New York: John Wiley and Sons.

Gagliano, S. 1977. *Cultural resources evaluation of the northern Gulf of Mexico continental shelf.* Vol. 1, *Prehistoric cultural resource potential.* Washington, DC: National Technical Information Service.

Gifford, J. A. 1983. Core sampling of a Holocene marine sedimentary sequence and underlying Neolithic cultural material off Franchthi Cave, Greece. In *Quaternary coastlines and marine archaeology,* ed. P. M. Masters and N. C. Flemming, 269–81. New York: Academic Press.

Goney, S. 1973. Buyuk Menderes Deltasi. *Istanbul Univ. Cografya Enstitusu Dergisi, Sayi* 18–19:339–54.

Hassan, F. A. 1979. Geoarchaeology: The geologist and archaeology. *American Antiquity* 44 (2):267–70.

Kraft, J. C. 1972. *A reconnaissance of the geology of Greece and the sandy coastal areas of eastern Greece and the Peloponnese.* Newark, DE: Univ. of Delaware College of Marine Studies Technical Report no. 9.

———— 1976a. The coastal environment. In *Selected Papers on the Geology of Delaware.* In *Transactions of the Delaware Academy of Science,* vol. 7, ed. J. C. Kraft and W. Carey. Newark: Delaware Academy of Science.

———— 1976b. *Radiocarbon dates in the Delaware coastal zone.* Newark, DE: Delaware Sea Grant College Program, DEL-SG-19-76.

———— 1977. Late Quaternary paleogeographic changes in the coastal environments of Delaware, Middle Atlantic Bight, related to archaeological settings. In *Amerinds and their paleoenvironments in northeastern North America,* ed. W. S. Newman and B. Salwen, 35–69. *Annals of the New York Academy of Sciences,* vol. 288.

———— 1978. Coastal stratigraphic sequences. In *Coastal sedimentary environments,* ed. R. A. Davis, Jr., 361–83. New York: Springer-Verlag.

Kraft, J. C., G. Rapp, Jr., and S. E. Aschenbrenner, 1975. Late Holocene paleogeography of the coastal plain of Messenia, Greece, and its relationships to archaeological settings and to coastal change. *Bulletin of the Geological Society of America* 86:1191–208.

Kraft, J. C., E. A. Allen, B. F. Belknap, C. J. John, and E. M. Maurmeyer. 1976. *Delaware's changing shoreline.* Dover, DE: Technical Report no. 1, Delaware Coastal Zone Management Program.

Kraft, J. C., and S. E. Aschenbrenner. 1977. Paleogeographic reconstructions in the Methoni Embayment in Greece. *Journal of Field Archaeology* 4 (1):19–44.

Kraft, J. C., S. E. Aschenbrenner, and G. Rapp, Jr. 1977. Paleogeographic reconstructions of coastal Aegean archaeological sites. *Science* 195:941–47.

Kraft, J. C., and C. J. John. 1978. Paleogeographic analysis of coastal archaeological settings in Delaware. *Archaeology of Eastern North America* 6:41–60.

———— 1979. Lateral and vertical facies relations of transgressive barriers. *American Association of Petroleum Geologists Bulletin* 63 (12):2145–63.

Kraft, J. C., E. A. Allen, D. F. Belknap, C. J. John, and E. M. Maurmeyer. 1979. Processes and morphologic evolution of an estuarine and coastal barrier system. In *Barrier islands: From the Gulf of St. Lawrence to the Gulf of Mexico,* ed. S. P. Leatherman, 149–183. New York: Academic Press.

Kraft, J. C., S. E. Aschenbrenner, and I. Kayan. 1980. Late Holocene coastal changes and resultant destruction or burial of archaeological sites in Greece and Turkey. In *Proceedings of the Coastal Archaeology Session CCE Field Symposium, Interna-*

tional Geographical Union, ed. M. L. Schwartz, 13–31. Bellingham, WA: Western Washington Univ.

Kraft, J. C., G. Rapp, Jr., and S. E. Aschenbrenner. 1980. Late Holocene paleogeographic reconstructions in the area of the Bay of Navarino: Sandy Pylos. *Journal of Archaeological Sciences* 7:187–210.

Kraft, J. C., İ. Kayan, and O. Erol. 1982. Geology and paleogeographic reconstructions of the vicinity of Troy. In *Troy: The archaeological geology*, ed. G. R. Rapp and J. A. Gifford. Princeton, NJ: Princeton Univ. Press.

Kraft, J. C., D. F. Belknap, and İ. Kayan. 1983. Potentials of discovery of human occupation sites on the continental shelves and near shore coastal zone. In *Quaternary Coastlines and Marine Archaeology*, ed. P. M. Masters and N. C. Flemming, 87–120. New York: Academic Press.

McDonald, W. A., and R. Hope Simpson. 1969. Further explorations in the southwest Peloponnese, 1964–1968. *American Journal of Archaeology* 73:153–54.

Mey, O. 1926. *Das Schlachtfeld vor Troja: Eine Untersuchung*. Berlin: Verlag Walter De Gruyter.

Middleton, G. V. 1973. Johannes Walther's law of the correlation of facies. *Bulletin of the Geological Society of America* 84 (3):979–87.

Osborn, A. J. 1977. Strandloopers, mermaids, and other fairy tales: Ecological determinants of marine resource utilization—The Peruvian case. In *For theory building in archaeology*, ed. L. R. Binford, 157–206. New York: Academic Press.

Russell, R. J. 1967. *River plains and sea coasts*. Berkeley, CA: Univ. of California Press.

Schalk, R. F. 1977. The structure of an anadromous fish resource. In *For theory building in archaeology*, ed. L. R. Binford, 207–50. New York: Academic Press.

Schindler, A. 1904. Umgebung von Ephesos. Map in *Ephesos, Stadt der Artemis und des Johannes*, by F. Miltner. Vienna: Deuticke, 1958.

Schwartz, M. L., and C. Tziavos. 1979. Geology in the search for ancient Helice. *Journal of Field Archaeology* 6 (3):243–52.

Scoufopoulos, N. C., and J. G. McKernan. 1975. Underwater survey of ancient Gythion, 1972. *International Journal of Nautical Archaeology* 4 (1):103–16.

Stark, B. L., and B. Voorhies. 1978. *Prehistoric coastal adaptations: The economy and ecology of maritime Middle America*. New York: Academic Press.

Walker, R. G. 1979. *Facies models*. Toronto: Geological Association of Canada, Geoscience Canada, Reprint Series 1.

4

Paleoenvironments and Contemporary Archaeology: A Geoarchaeological Approach

FEKRI A. HASSAN

ABSTRACT

Contemporary archaeology with its emphasis on cultural processes requires a consideration of paleoenvironments from an anthropological viewpoint. Geoarchaeological investigations provide a means for reconstructing prehistoric and ancient landscapes, depositional environments, and paleoclimatic regimes. These investigations, however, should be regarded as a prelude to further studies. Landforms and depositional environments should be evaluated in terms of their potential contributions and relationship to site-formation processes, subsistence activities, and settlement location. The climatic regime and its temporal variability should also be viewed in terms of its reciprocal link with human cultural activities for an adequate understanding of human adaptation and cultural change.

The association between the earth sciences and archaeology is strong and deeply rooted. It was as a result of the effort of geologists that the notion of human antiquity was established (see chap. 1). The emphasis on geology as a source of methods and principles for age determination, however, seems to have detracted from a fuller utilization of other methods and principles of the earth sciences. The range of potential applications of geology is remarkably wide, as shown by the chapters of this volume and the references listed in Hassan (1978, 1979), but one of its most outstanding contributions to archaeology deals with prehistoric environments.

In contemporary archaeology, the place of prehistoric environments must be understood in light of recent changes in the scope and theoretical orientation of the field. To start with, archaeologists have to contend with a diversity and multiplicity of artifactual data. Before any serious attempts can be made to interpret the archaeological record, these data must be placed in a temporal sequence. This requirement fostered a diachronic approach and a strong identification of archaeology with history. It was this need for temporal and historical order that shaped the early (1900–1920) and middle (1920–1940) phases of archaeological investigation. Stages of cultural development were constructed, classificatory-typological schemes and seriation methods of artifacts were formulated, and methods for precise excavation to discern relative ordering of archaeological strata were devised. These matters have not ceased to be of concern to archaeologists. Elaborate and sophisticated techniques for "absolute" dating, starting with Libby's radiocarbon dating in the 1950s, are still in progress (Berger and Suess, 1979). Typology and seriation are subjects of continuous debate and improvement (Marquardt, 1978), and principles of archaeological stratigraphy are still under review (Harris, 1979).

The early phases of archaeological investigation were also characterized by speculative theories that attempted to explain the emerging differences and similarities of the archaeological cultures and the temporal sequence of cultural stages. Without going into the details of this episode in archaeological thinking (Daniel, 1964), it is necessary in the context of this chapter to note the leading role of the British archaeologist Gordon Childe in shaping contemporary archaeology. Childe articulated views about culture that were crucial for a reorientation of the whole field of archaeology:

> A culture is the durable material expression of an adaptation to an environment, human as well as physiographical, that enabled a society to survive and develop. From this point of view the buildings, tools, weapons, ornaments, and other surviving constituents are interrelated as elements in a functioning whole. [Childe, 1951:16]

Although many would disagree with certain concepts expressed in this statement, the impact of Childe's approach on archaeology is undeniable. Within this paradigmatic framework, archaeologists face the prospect of understanding prehistoric cultures as a result of adaptive processes of a functioning whole. At present, this approach, with numerous elaborations and variations, defines the profile of contemporary archaeology. The major addition in recent years has been the incorporation of the systemic holism of culture (Redman, 1973), which is rooted in the concept of functional holism expressed by Childe. Although concepts such as "adaptation" and "function" and the potential utility of systems theory in archaeology are controversial, archaeologists today work within the framework of adaptation and functional interdependence of cultural components. Ecological archaeology, the study of the interrelationships between human groups and their habitat, is one of the key approaches to archaeology today.

The place and importance of the ecological approach to prehistory has found its spokesman in Karl W. Butzer, who in 1964 (and more forcefully in 1971) provided the first coherent and comprehensive treatise on environment and archaeology. It is within the context of this approach that the contributions from the earth sciences to paleoenvironmental analysis in archaeology may be assessed.

Pioneer works include those by Bryan (1926, 1941, 1954) and Hack (1942) on the impact of climatic change on arroyo cutting and agricultural activities in the southwestern United States and by Huzayyin (1939), Passarge (1940), and Butzer (1959) on the geomorphology and hydrology of the Nile River as they affected agriculture and agricultural origins in Egypt. Other notable contributions include Caldwell's (1958) interpretation of the prehistory of the northeastern United States in terms of environment and economic modes of production. These studies went beyond the reconstruction of ancient landscape and paleoclimate to investigate the *relations* between the geological environment and culture. It may also be noted here that many studies have been aimed at using an environmental approach as a tool for dating, as exemplified in the work by Antevs (1941, 1949, 1955), Bryan and Ray (1940), and Haurey, Antevs, and Lance (1953) in the southwestern United States and by Campbell (1936) in California.

It is my aim in the present chapter to discuss the potential contributions from the earth sciences to some of the major problems in contemporary archaeology within the framework of an ecological paradigm. These problems focus on food-getting (subsistence) activities and on settlements, as well as on culture change.

WHAT IS AN ENVIRONMENT?

An environment is a subdivision of the earth's surface distinguished and delimited on the basis of physical, chemical, and/or biological criteria. Sedimentologists are often concerned with sedimentary or depositional/erosional environments. These are often geomorphological units (e.g., rivers or glaciers) where sediments are formed (cf. Potter, 1967:340; Reineck and Singh, 1975:4–5). There is, however, a certain amount of confusion in the literature, since sedimentary environments often are extended to include climatic zones, such as deserts.

It is perhaps advisable to deal with sedimentary depositional environments as depositional systems (cf. LeBlanc, 1972) characterized and differentiated on the basis of the dominant agent or combination of agents of deposition (ice, water, wind, gravitation). The following depositional systems may be recognized:

Glacial, in which sedimentation is directly or intimately related to glacial ice;

Fluvial, in which sediments are deposited from running stream water;

Colluvial, in which sediments are deposited from surface runoff;

Lacustrine, in which sediments are deposited from bodies of standing water (lakes, peat bogs, marshes, swamps);

Spring, in which material is deposited from spring water, geysers, and water seeps;

Gravitational, in which sediments accumulate primarily as a result of the effect of gravitational force, with or without the presence of moderate amounts of water under freeze-thaw or freeze-free conditions;

Alluvial fan, characterized by predominance or alternation of deposition from running water and debris flows; and

Eolian, in which wind is the agent of deposition.

Obviously these sedimentary environmental systems overlap with geomorphological units, but it is essential to separate the two categories. Geomorphological units, such as deltas, may be characterized by a complex combination of deposition from fluvial, lacustrine, and marine systems. A distinction ought to be made beween a geomorphological unit, such as a river, and a geomorphic landscape, which consists of the association of geomorphological units—an association of rivers with alluvial fans, for example, or ephemeral streams with playas and dunes.

We should also distinguish climatic-geologic systems from either sedimentary or geomorphological environments. Climatic systems involve large tracts (zones) of the surface of the earth that are differentiated primarily on the basis of average annual precipitation and temperature. It may be noted here that climatic conditions influence vegetation and lead to specific associations between climate and vegetation (biomes), such as tropical rainforest, temperate rainforest, temperate forest, woodland, grassland, tundra, and desert (Whittaker, 1970). Classifications of climatic zones include such schemes as those developed by Köppen, Thornwaite, and Miller (Money, 1972). In geomorphology, Peltier (1950) proposed nine climatic-morphogenetic systems. These morphogenetic zones are listed in table 4.1.

The three kinds of environments are closely linked. Climate, with its overwhelming influence on weathering, erosion, transportation, and deposition, directly or via its impact on vegetation influences both sedimentary types and geomorphic landscapes (fig. 4.1). Table 4.2 shows some of the associations of climatic-morphogenetic systems, geomorphic units, and depositional systems. Unfortunately, most textbooks deal with these environments separately, but Butzer (1976b) provides a well-balanced view.

Table 4.1. Climatic-Geologic Systems

Climatic-Morphogenetic Region	Range of Average Annual Temperature (in °C)	Range of Average Annual Rainfall (in mm)
Glacial	$-18--7$	$0-1,150$
Periglacial	$-15--1$	$125-1,400$
Boreal	$19-3$	$250-1,500$
Maritime	$1.6-21$	$1,250-1,900$
Selva	$16-29$	$1,400-2,250$
Moderate	$3-29$	$900-1,500$
Savanna	$-12-29$	$600-1,250$
Semiarid	$2-29$	$250-650$
Arid	$13-29$	$0-350$

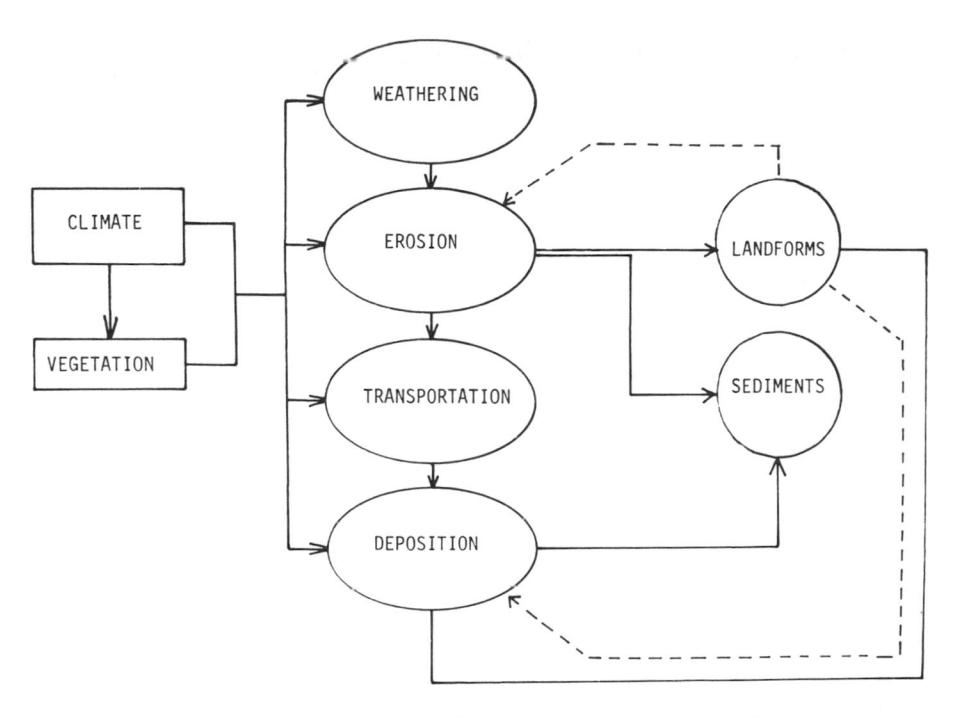

Fig. 4.1. A diagram showing the relationship of the climatic regime with landforms and sediments. Climate influences vegetation, and both control sedimentary processes (weathering, erosion, transportation, and deposition). For example, under tropical humid conditions intensive chemical weathering prevails, and abundant rainfall enhances fluvial transportation and deposition. By contrast, under arid climate, the scarcity of water reduces chemical weathering and permits low to moderate mechanical weathering. Eolian transport and deposition are most prevalent. The landscape is determined by the balance between erosion and deposition. In turn, relief influences depositional processes. A climatic regime may thus be characterized by a set of sediments related to an association of depositional environments with a definite geomorphic expression. For example, in an arid zone, eolian, lacustrine, and gravitational depositional environments provide an association of eolian sand in the form of dunes or sand sheets, lacustrine deposits in playas, and talus or alluvial fans on hillslopes.

In the course of paleoenvironmental analysis for archaeological purposes, reconstruction of both paleoclimate and geomorphic landscape are of primary importance, as will be shown in the following two sections. Reconstruction of both paleoclimate and landscape from sedimentological and stratigraphic studies is considerably improved by a conjunctive analysis of paleosols, fauna, pollen, and plant remains. Independent sources of data also serve as a means for cross-checking results; they may help disentangle natural events from those induced by prehistoric peoples, as well as screen the data from archaeological sites for assessing the magnitude of human interference.

Table 4.2. Associations of Climatic-Morphogenetic, Geomorphic, and Depositional Systems

Climatic-Morphogenetic System	Geomorphic Units	Depositional Systems
Glacial	Moraines	Glacial
	Kames	Subglacial and englacial Fluvial
	Eskers	
Periglacial	Braided streams	Fluvial
	Alluvial fans	Alluvial fan
	Loessial sheets and sand hills	Eolian
	Proglacial lakes	Lacustrine
	Thermokarstic lakes	Lacustrine
	Slopes	Gravitational
Boreal (temperate) Maritime	Rivers	Fluvial
	Slopes	Gravitational
Tropical/Selva	Rivers	Fluvial
	Swamps	Lacustrine
Savanna	Rivers	Fluvial
	Lakes	Lacustrine
	Slopes	Colluvial
Semiarid	Ephemeral streams	Fluvial
	Alluvial fans	Alluvial fan
	Dunes	Eolian
Arid (desert)	Ephemeral streams or none	Fluvial
	Playas	Lacustrine
	Slopes	Colluvial
	Dunes	Eolian

The time and spatial scale at which archaeological research is conducted is in most cases much finer than that dealt with by Quaternary geologists and certainly that of pre-Quaternary geologists. Climatic events over a few hundred years or changes in the geomorphic landscape in a matter of decades can significantly alter the interrelations between people and the environment. It is perhaps the concern for minute details that characterizes the practice of environmental analysis in archaeology. This is often reflected in a program of sampling that is sufficiently extensive to detect geomorphic subunits—e.g., levees, point bars, lake shorelines (see Hay, 1976, and Kraft, Kayan, and Erol, 1982, for case studies)—and to discern spatial shift or evolution of geomorphic units over as small a time span as possible—e.g., the change in the position of a channel bar, the shift in the position of dripline in a cave, or the developmental history of a channel bar (Hassan, 1976; Blinman, 1980). The sampling often takes advantage of available exposures in road cuts, quarries, and scarps. Test pits and trenches are also necessary for a complete understanding of the relationship between sites and their geologic setting. Attention to field characteristics of sediments (bedding, sedimentary structures) and a follow-up quantitative analysis of sedimentary attributes (grain size, mineral content, shape of particles) are vital for an adequate interpretation of the depositional system. The concept of the depositional system (Selley, 1970) is very useful as a heuristic device. It is based on theoretical assumptions of lithological associations in a depositional system (i.e., lithofacies). Models of climatic-geomorphic landscapes are also useful in extending paleoenvironmental analysis from local depositional or geomorphic units to a regional perspective. Allen (1965), for example, presents lithofacies models of fluvial depositional units in various climatic-geomorphic settings. In a semiarid environment alluvial fans may coexist with intermittent streams, and in an arid environment ephemeral streams may be associated with dunes and playas.

Paleoenvironmental analysis in archaeology is by no means an easy task. Without a special commitment and a close association between the earth scientist and the archaeologist, it can become a frustrating and sometimes a fruitless task. The following sections provide insights into some areas of concern for archaeologists—subsistence, settlement, and cultural change—and document ways in which contributions from the earth sciences can meet the challenge of solving key problems in those areas. It will be clear, I hope, that both archaeologists and earth scientists must be aware of the potentials and limitations of their respective fields and that integration of research goals and strategies is essential for a successful interdisciplinary interaction.

ENVIRONMENT AND SUBSISTENCE:
A GEOARCHAEOLOGICAL PERSPECTIVE

The survival of human groups ultimately hinges upon their ability to procure food. There are primarily two modes of food getting: extraction from wild resources (e.g., hunting, plant gathering, fishing, and fowling) and production, the deliberate manipulation of resources by actively controlling their yield through cultivation or stock-raising. For most of the prehistoric past, people subsisted on available wild animal and plant resources. It was not until about 10,000 years ago that agriculture became a viable mode of subsistence (Redman, 1978).

In general, the archaeologist is concerned with the following aspects of subsistence:

1. proportional use of resources;
2. annual scheduling of subsistence activities; and
3. spatial organization (allocation) of subsistence activities.

Proportional use of resources depends in part on the benefits to be gained from using certain resources versus the cost of their extraction, including the cost of travel and transportation of food. In order to determine the proportional use of resources, it is essential to obtain a reliable map of the economic potentials of the exploited region. The role of the geologist at this stage is to provide a reconstruction of the geomorphic landscape and paleoclimate. Reconstructions of this kind

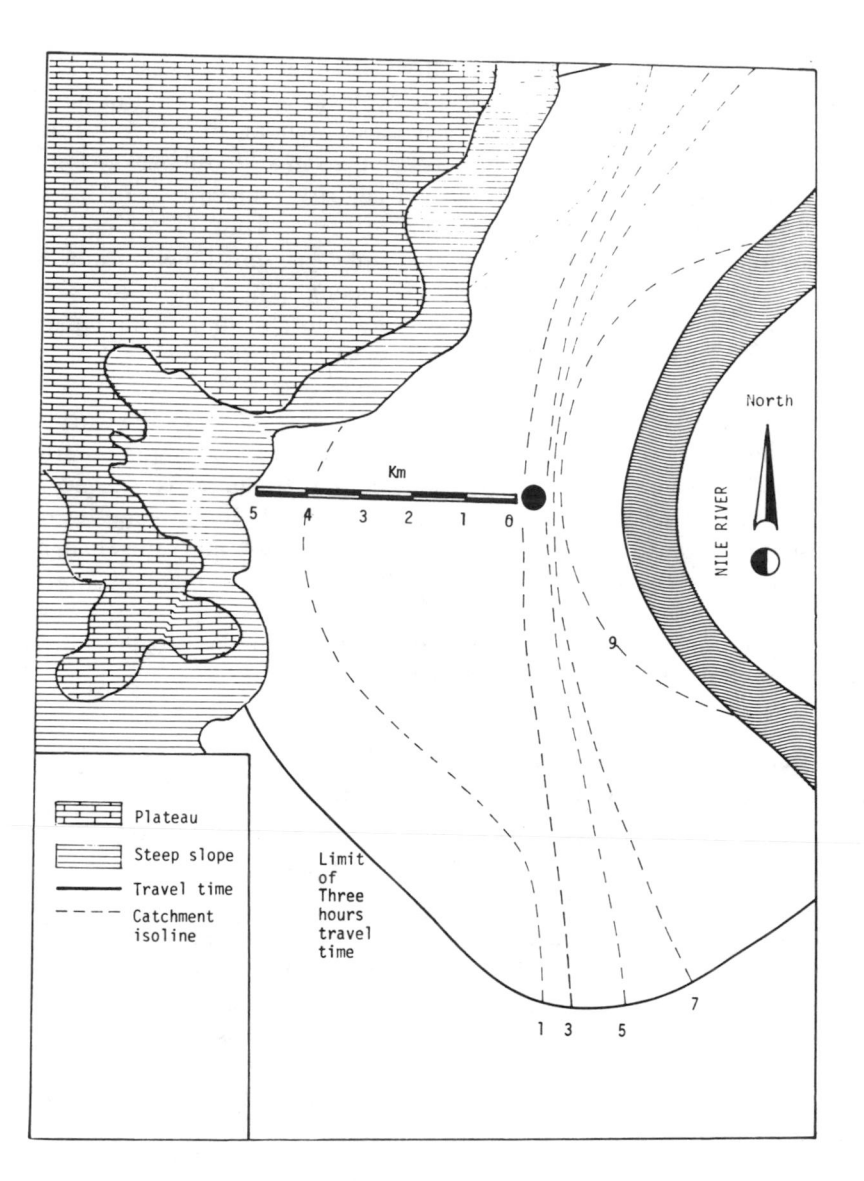

Fig. 4.2. A catchment map of the Nagada sites, Upper Egypt, showing the results of evaluating the geomorphic map, assuming a slightly moist paleoclimate, for travel time and yield of resources. Catchment values were obtained from 1 km × 1 km grid units. For each grid unit the cost of travel from the site (darkened circle) and the yield of resources (given a hunting-gathering mode) were determined in calories, and from these figures the benefit/cost ratio (catchment value) was calculated. The landscape surrounding a site is thus evaluated in terms of its "economic pull." The zone surrounded by the 9:1 isoline is the most economically profitable zone. Areas of low catchment value will seldom, if ever, be utilized. The map is based on geomorphological, sedimentological, and stratigraphic

are often achieved through detailed and intensive geomorphological, strati-graphic, sedimentological, and pedological investigations and use of paleo-geographic maps similar to those prepared by Butzer for the Koster site in Illinois (1977) and Kom Ombo Plain, Egypt (Butzer and Hansen, 1968), and by Hassan (1974) for the Dishna Plain, Egypt. The task consists of identifying the geomor-phic landscape and the depositional systems. For studying hunter-gatherers, these maps, in conjunction with paleoclimatic inferences, provide a basis for determining the biotic zones, as well as the productivity of resources in these zones. In the case of the Dishna Plain, the geomorphic zones that were identified suggested five geomorphic-biotic zones (assuming a paleoclimate wetter than the climate at present): a seasonally inundated floodplain, probably with a wooded or tree-savanna vegetation; periodically flooded wadi floors with thorn-savanna vegetation; an edaphic zone with a high water table supporting grasses; a low desert or piedmont with a desert vegetation; and a plateau with a desert vegeta-tion. One can estimate the biotic productivity of these zones by analogy with similar habitats in Africa (Bourlière, 1963) and from these estimates arrive at the potential caloric yield to humans (Hassan, 1974:152–54).

The energy cost of travel from a habitation site to a resource location can be estimated from the physiographic relief and nature of surficial deposits along the route. For example, if the slope of the cliffs defining the edge of a desert plateau is more than 15 percent, plateau resources may be reached only through major wadis (see fig. 4.2). Variations in slope and surface geology (e.g., sticky mud, boulders, pebble gravel) can be used to calculate travel time or approximate speeds. Knowledge of travel cost and productivity of resources may thus be utilized along with other factors, including cost of food extraction, to create a benefit/cost map (cf. Foley, 1977). Such a map was constructed for the Nagada region, Egypt, starting with a site of known location and showing isolines of equal benefit and cost, called *catchment isolines* (fig. 4.2). The recognition of types of available resources and their spatial loci, along with other factors, can be used to estimate the proportional use of resources (Jochim, 1976).

The spatial allocation of subsistence activities can be profitably approached by considering the geomorphic-biotic map and the *catchment map* (cf. Vita-Finzi and Higgs, 1970; Webley, 1972; Roper, 1979). Additional information can be obtained from a knowledge of the seasonal availability and accessibility of re-sources. During prehistoric times in Egypt, two key seasonal variables are important: precipitation and annual flooding (Butzer, 1976a). Reconstruction of paleoclimatic conditions for the early Holocene (Wendorf and Hassan, 1980), for

---

studies for the identification of geomorphic units. Paleoclimatic inferences allow an estimate of their biomass by analogy with other known environments. The results can be tested by estimating the relative abundance of resources from various zones at archaeolog-ical sites. In addition, the map can be reproduced with a changing site location to test the role of proximity to profitable resources in determining site location.

example, suggests that precipitation in Upper Egypt was higher than it is at present and fell mostly during the summer.

The flooding of the Nile since at least 200,000 years ago (Butzer, 1980) occurred during the summer. During the flood season, most of the floodplain was under water and only its edge could be exploited, but following the water's recession, fish could be obtained readily from remaining shallow pools (Hassan, 1974:148–51). After rains, grasses and desert animals could be exploited in the desert region adjacent to the floodplain. Thus, seasonal variations of the geological environment may provide insights into the seasonal availability and accessibility of resources and allow for a well-grounded interpretation of annual subsistence activities and changes in the spatial organization of activities through the year (fig. 4.3).

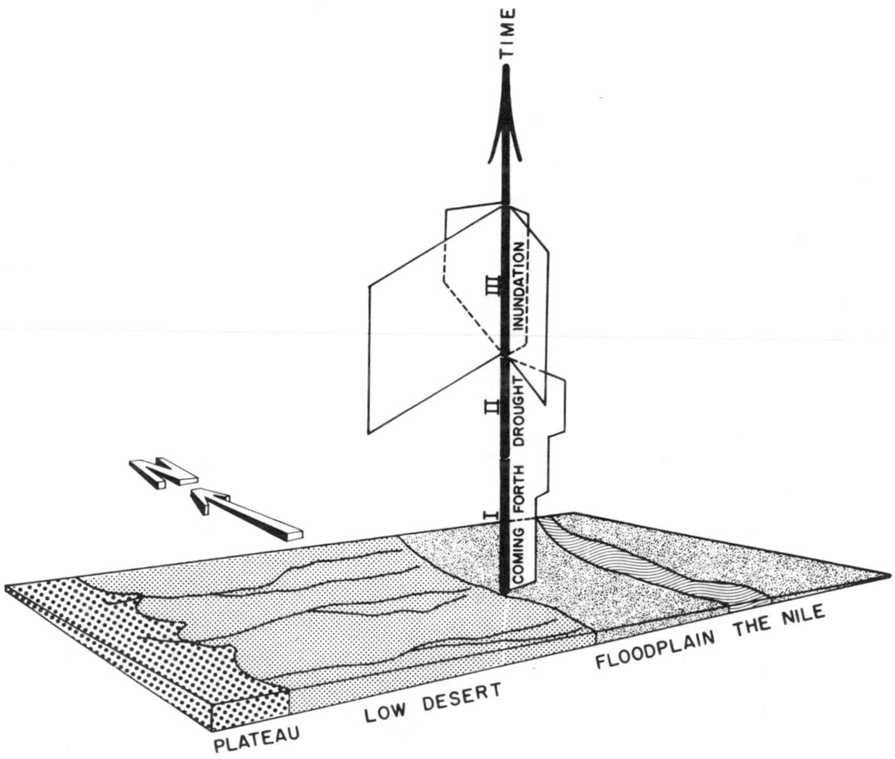

Fig. 4.3. A time-space map showing probable utilization of the Nilotic landscape during the ecological seasons of the year. During the season of inundation (July–October), subsistence activities are concentrated in the low-desert zone, in contrast with a greater utilization of the floodplain during the season of drought (March–June). The nature of subsistence activities also varies depending on available resources—e.g., fishing following the recession of the floodwater.

The agricultural potential of land can be assessed only by considering the nature of the geological environment. Vita-Finzi (1969), for example, brings attention to the study of alluvial sequences in the southwestern United States and its implications for floodwater farming on an undissected valley floor. Given a regional approach (cf. Vita-Finzi and Higgs, 1970), the reconstruction of the landscape can provide a basis for constructing a land-use map (Stamp, 1963) showing the agricultural and pastoral potential of various geomorphic zones under the contemporaneous paleoclimatic conditions. Land can be classified on the basis of elevation, slope, and surface relief (steepness and ruggedness); depth of soil; solid texture; soil fertility; stoniness; and water conditions (drainage, susceptibility to droughts).

SETTLEMENTS IN ARCHAEOLOGY: A GEOARCHAEOLOGICAL PERSPECTIVE

The role of geologists in settlement studies can hardly be underestimated: they can be instrumental in finding archaeological sites and interpreting their environmental setting. The recognition of probable site locations is contingent upon identifying the factors likely to influence human habitation. These include proximity to water and high-yield/low-cost food resources, accessibility, protection from bad weather and environmental hazards, dry ground surface, defensibility, and availability of building materials. After reconstructing the geomorphic landscape and paleoclimate, the geologist can evaluate the various geomorphic zones on a nominal scale and produce a *geoekistic map* showing the settlement potential, from optimal to least desirable. The archaeologist may use this map either for locating sites or for testing the optimal locations of sites by prehistoric peoples. For the Nagada region, Upper Egypt, a geoekistic map was constructed (fig. 4.4) showing the edge of cultivation and the low desert bordering the floodplain as the optimal zones for permanent settlement locations. A systematic archaeological survey confirmed that sites are, in fact, located in that zone.

The spatial pattern of archaeological sites in a region is of concern to archaeologists (Hodder and Orton, 1976) because of its utility in testing the Central Place Theory or equidistant spacing. However, it can be drastically altered by destruction of archaeological sites as a result of erosion or by burial under dunes or alluvial silt (Hamond, 1978:3). It is, therefore, important to reconstruct the landscape not only during the period of occupation, but also in the period between the abandonment of settlements and the discovery of their remains. In Algeria, research on Epipaleolithic Capsian sites, for example, revealed that many sites had been destroyed by torrential stream flow (Hassan, 1975). In Egypt many sites of predynastic age were destroyed or buried by later Nile flood deposits. During the late predynastic and the following Archaic period, the Nile floods were declining (Bell, 1970). Higher floods during the Old and Middle Kingdoms destroyed or buried earlier settlements not sufficiently protected by high levees and artificial dykes.

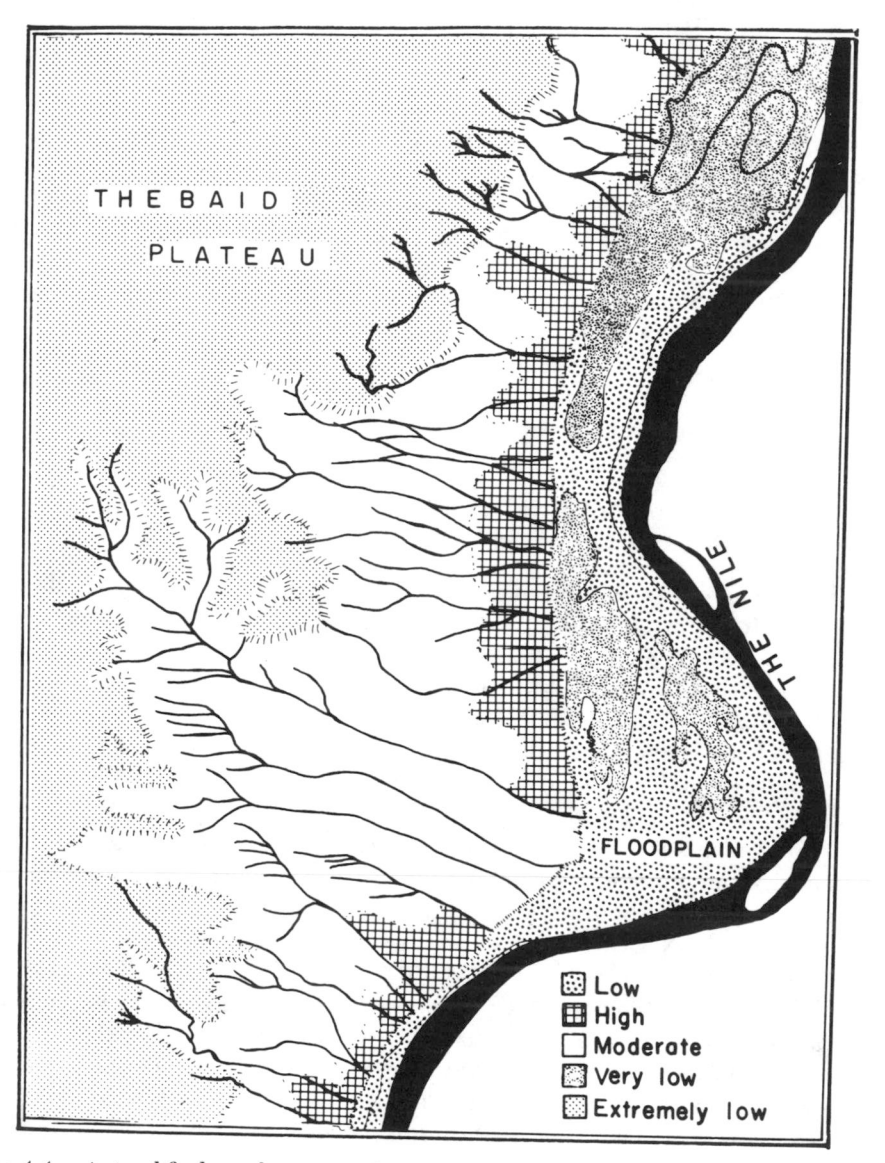

Fig. 4.4. A simplified geoekistic map showing rank order of geomorphic units in terms of their suitability for settlement location. The preparation of the map begins with a definition of a number of geoekistic variables (e.g., proximity to water, suitability of surface materials for building or habitation, probability of geological hazards). The variables are then considered together for each geomorphic zone in the landscape after each variable is weighted on a scale from zero to one and a value for the variable from one to five is determined for each geomorphic zone. The score for each geomorphic zone is then used to generate a geoekistic surface with various degrees of suitability for habitation. In this map for an area along the Nile Valley, Upper Egypt, the low desert adjoining the floodplain was found to provide the best location for predynastic occupation under conditions of early farming and natural basin irrigation. The actual location of predynastic settlements matched the expectations based on this map.

The destruction of sites or their burial by later deposits may bias significantly any interpretation of the regional settlement pattern. Since settlements in a region will consist of different types—cities, hamlets, villages, hunting camps, seasonal farmhouses—and since they may be located in various geomorphic zones, they may be subject to differential preservation. Isolated farm sites on a hillslope, for example, may be more easily eroded than others located in more protected areas.

At the local level, the morphometry and spatial organization of a site are as much a result of geological processes as of cultural activities. Deposition and erosion during the period of site occupation must be taken into consideration because of their impact on duration of occupation and preservation of archaeological material. The geological environment of the site also influences the state of surviving materials. A consideration of the availability of water, drainage, the presence of organic acids, hydrogen-ion concentration (pH), redox potential (Eh), and temperature can be very useful in interpreting the relative frequency of bones, wooden artifacts, and other perishable or alterable substances. For example, the prevalence of acidic conditions may lead to the destruction of bone and mollusk shells but enhances the preservation of delicate basketry, as at the Hoko site on the northwestern coast of the United States (Croes and Blinman, 1980).

CULTURAL CHANGE

The role of the environment in cultural change has been a subject of numerous studies. A synthesis of the interrelationships between environments and culture throughout the Pleistocene, from the emergence of man to the origin of cities, is to be found in Butzer (1971).

It is the task of the archaeologist to provide or construct the explanatory model of cultural change, but if he or she is to assess the role of the environment, the relevant data must be available. The geologist can serve this cause best by making a detailed and thorough analysis of paleoclimatic-environmental changes and their implications for the spatial location, yield, and seasonal or periodic predictability of the key resources.

Climatic change, for example, was considered the key variable in agricultural origins by Gordon Childe (1936). He assumed that a period of desiccation forced people and animals to aggregate near water sources (rivers or oases), where both plants and animals became domesticated. Even if desiccation were shown to be the case, the model is theoretically inadequate, but it motivated a consideration of a changing climate on agricultural origins. Subsequent studies have shown that no major change in the amount of rainfall occurred prior to the Neolithic in the Near East (van Zeist and Wright, 1963). However, the transition from glacial to postglacial global climatic conditions was associated with severe fluctuations (Butzer, 1971:530) that might have dramatically affected the predictability of food resources. These fluctuations apparently motivated Epipaleolithic hunters and food gatherers to include in their range of exploitable resources wild grasses, the progenitors of domesticated cereals. The impact of environmental change on the spread of grasses, as argued by Wright (1977), might have made these resources more desirable by minimizing the cost of their extraction. The trajectory leading to domestication of cereals, however, should be sought in a combination of demographic and cultural transformations (Hassan, 1977). In addition, domestication could have been a result of spatial conflict in subsistence activities. Certain zones where wild cereals were found, such as the highlands of the Near East, were not as suitable as other areas for grazing, collection of legumes, and perhaps hunting. Consequently, relocation to places where both cereals and legumes could be cultivated and where pastures were close by might have provided a viable solution to subsistence problems.

The geologist ought, therefore, to focus on elements of the prehistoric landscape that provided opportunities for exploitation, as well as on elements that may have impinged on the viability of a contemporaneous subsistence system. For the resource base of Epipaleolithic hunter-gatherers, the effects of both climatic fluctuations and the landscape potential for exploitation or cultivation of wild cereals are equally important (cf. Sauer, 1952:5—6). The geologist must also identify the limitations of the paleoenvironment, as, for example, the impact on agricultural development of soils prone to salinization or of the characteristics of tropical soils and their implications for shifting agriculture. The consequences of past human activities for the landscape should be carefully scrutinized. Salinization, as occurred in Mesopotamia (Jacobson and Adams, 1958), the impact of widespread use of fire during the terminal Pleistocene (Butzer, 1971:482), or overgrazing and tree cutting in predynastic Egypt (Hassan, 1980) can have great impact on yield and thereby motivate remedial action, such as agricultural intensification or extensification through expanding agricultural land, fertilizing, or weeding.

SUMMARY

Archaeologists are interested in ancient landscapes because many of their studies focus on the location of settlements and the spatial organization of food-getting activities. They are also interested in past climates because of the potential effects of climatic change on food resources and, hence, patterns of cultural adaptation. Archaeologists often concentrate on sites, but since human activities are distributed over a range or territory surrounding a site, a regional approach to paleoenvironmental reconstruction is necessary.

For conceptual clarity paleoenvironments may be classified according to climatic-geologic systems; geomorphic units or landscapes; and sedimentary (depositional) environments.

Sedimentological and geomorphological investigations are essential for recognizing depositional environments, such as rivers, dune fields, or lakes. The identification of such environments is a prerequisite for reconstructing prehistoric landscapes. Once a landscape is reconstructed it may be evaluated for habitation with the help of geoekistic maps and for subsistence by assessing its suitability for grazing, farming, hunting, etc. Geological processes, such as salinization and frequency of hazardous floods or landslides, may be considered. Relief, slope, and surface geology are also useful in estimating travel time and transportation costs.

The long-term characteristics of an environment (climatic stability, fluctuations, secular changes) are crucial for interpreting certain cultural changes. The role of the environment, however, should not be restricted to its potential causal impact on adaptive responses, but should include an evaluation of its "passive" potential and limits to human exploitation. The damaging impact of human activities on the environment (mostly inadvertent results of unforeseen consequences) is worthy of special consideration, especially in the context of the origins of civilization. Modifications of the environment that are beneficial at one time may entail future dangers and vice versa. The identification of reciprocal interactions between people and environment is crucial in understanding prehistoric cultural change, especially that which has occurred since the terminal Pleistocene.

REFERENCES

Allen, J. R. L. 1976. A review of the origin and characteristics of recent alluvial sediments. *Sedimentology* 5:89–191.

Antevs, E. 1941. Age of the Cochise culture stages. *Medallion Papers* 29:31–56.

———. 1949. Geology of the Clovis Site. In *Ancient man in North America*, ed. H. M. Wormington. Denver, CO: Denver Museum of Natural History.

———. 1955. Geologic-climatic dating in the West. *American Antiquity* 29 (4):317–35.

———. 1959. Geological age of the Lehrner Mammoth site. *American Antiquity* 25:31.

Bell, B. 1970. The oldest records of the Nile floods. *Geographical Journal* 136 (4):569–73.

Berger, R., and H. E. Suess, eds. 1979. *Radiocarbon dating*. Berkeley: Univ. of California Press.

Blinman, E. 1980. Stratigraphy and depositional environment. In *Hoko River: A 2500 year old fishing camp on the Northwest Coast of North America*, ed. D. R. Crops and E. Blinman. Washington State University Laboratory of Anthropology, Reports of Investigations, no. 58.

Bourlière, F. 1963. Observations on the ecology of some large African mammals. In *African ecology and human evolution*, ed. F. C. Howell and F. Bourlière, 43–54. Chicago: Aldine.

Bryan, K. 1926. Recent deposits of Chaco Canyon, New Mexico, in relation to the lives of the prehistoric peoples of Pueblo Bonito. *Washington Academy of Sciences Journal* 16:75–76.

———. 1941. Pre-Columbian agriculture in the Southwest, as conditioned by periods of alluviation. *Annals of the American Association of Geographers* 31:219–42.

———. 1954. *The geology of Chaco Canyon, New Mexico*. Smithsonian Institution Miscellaneous Collected Papers, vol. 122, no. 7.

Bryan, K., and L. L. Ray. 1940. *Geological antiquity of the Lindenmeier Site in Colorado.* Smithsonian Miscellaneous Collections, vol. 99, no. 2.

Butzer, K. W. 1959. Environment and human ecology in Egypt during predynastic and early dynastic times. *Bulletin de la Société Royale de Géographie d'Egypte* 32: 43–87.

———. 1964. *Environment and archaeology.* 1st ed. Chicago: Aldine.

———. 1971. *Environment and archaeology: An ecological approach to prehistory.* 2d ed. Chicago: Aldine.

———. 1976a. *Early hydraulic civilization in Egypt: A study in cultural ecology.* Chicago: Univ. of Chicago Press.

———. 1976b. *Geomorphology from the earth.* New York: Harper and Row.

———. 1977. *Geomorphology of the lower Illinois Valley as a spatial-temporal context for the Koster Archaic Site.* Reports of Investigations no. 34. Springfield, IL: Illinois State Museum.

———. 1980. Pleistocene history of the Nile Valley in Egypt and Lower Nubia. In *The Sahara and the Nile,* ed. M. A. J. Williams and H. Faure. Rotterdam: A. A. Balkema.

Butzer, K. and C. L. Hansen. 1968. *Desert and river in Nubia: Geomorphology and prehistoric environments at the Aswan Reservoir.* Madison: Univ. of Wisconsin Press.

Caldwell, J. 1958. *Trend and tradition in prehistory of the eastern United States.* American Anthropologist, Memoir no. 88.

Campbell, E. W. C. 1936. Archaeological problems in the southern California Deserts. *American Antiquity* 1:295–300.

Childe, V. G. 1936. *Man makes himself.* London: Watts.

———. 1951. *Social evolution.* New York: Henry Schuman.

Croes, D. R., and E. Blinman, eds. 1980. *Hoko River: A 2500 year old fishing camp on the Northwest Coast of North America.* Washington State University Laboratory of Anthropology, Reports of Investigation, no. 58. Pullman, WA.

Daniel, G. 1964. *The idea of prehistory.* Harmondsworth: Penguin Books.

Foley, R. 1977. Space and energy: A method for analyzing habitat value and utilization in relation to archaeological sites. In *Spatial archaeology,* ed. D. L. Clarke. New York: Academic Press.

Hack, J. T. 1942. *The changing physical environment of the Hopi Indians of Arizona.* Peabody Museum Papers, vol. 35. Cambridge, MA: Harvard Univ. Press.

Hamond, R. W. 1978. The contribution of simulation to the study of archaeological processes. In *Simulation studies in archaeology,* ed. I. Hodder. Cambridge, England: Cambridge Univ. Press.

Harris, E. C. 1979. *Principles of archaeological stratigraphy.* London: Academic Press.

Hassan, F. A. 1974. *The archaeology of the Dishna Plain, Egypt: A study of a Late Pleistocene settlement.* The Geological Survey of Egypt, Paper no. 59.

———. 1975. Geology and geomorphology of the Aim Mistehiya locality. In *The prehistoric cultural ecology of Caspian Escargotiere,* ed. D. Lubell et al., 60–70. *Libyca,* vol. 23.

———. 1976. Stratigraphic and geomorphic setting of the Miller Site, Strawberry Island. In *Preliminary archaeological investigations at the Miller Site, Strawberry Island, 1976, a late prehistoric village near Burbank, Franklin County, Washington,* ed. G. C. Cleveland et al., 143–63. Washington Archaeological Research Center, Project Report no. 46. Pullman, WA.

————. 1977. The dynamics of agricultural origins in Palestine: A theoretical model. In *Agricultural origins*, ed. C. Reed. The Hague: Mouton.

————. 1978. Sediments in archaeology. *Journal of Field Archaeology* 5:197–213.

————. 1979. Geoarchaeology: The geologist and archaeology. *American Antiquity* 44: 267–70.

————. 1980. Prehistoric settlements along the main Nile. In *The Sahara and the Nile*, ed. M. A. J. Williams and H. Faure, 421–50. Rotterdam: A. A. Balkema.

Haury, E. W., E. Antevs, and J. F. Lance. 1953. Artifacts with mammoth remains, Naco, Arizona. *American Antiquity* 19:1–24.

Hay, R. L. 1976. *Geology of Olduvai Gorge: A study of sedimentation in a semi-arid basin*. Berkeley: Univ. of California Press.

Hodder, I., and C. Orton. 1976. *Spatial analysis in archaeology*. Cambridge, England: Cambridge Univ. Press.

Huzayyin, S. A. 1939. Some new on the beginnings of Egyptian civilization. *Bulletin de la Société Royale de Géographie d'Egypte* 20:203–73.

————. 1950. Origins of Neolithic and settled life in Egypt. *Bulletin de la Société Royale de Géographie d'Egypt* 23:175–81.

Jacobson, T., and R. M. Adams, 1958. Salt and silt in ancient Mesopotamia. *Science* 128:1251–58.

Jochim, M. A. 1976. *Hunter-gatherer subsistence and settlement—A predictive model*. New York: Academic Press.

Kraft, J. C., I. Kayan, and O. Erol. 1982. Geology and paleogeographic reconstructions of the vicinity of Troy. In *Troy: The archaeological geology*, ed. G. R. Rapp and J. A. Gifford. Princeton, NJ: Princeton Univ. Press.

LeBlanc, R. J. 1972. Geometry of sandstone reservoir bodies. In *Underground waste management and environmental implications*, ed. T. D. Cook, 133–90. American Association of Petroleum Geologists Memoir, vol. 18.

Marquardt, W. H. 1978. Advances in archaeological seriation. In *Advance in archaeological method and theory*, vol. 1, ed. M. Schiffer, 257–314. New York: Academic Press.

Money, D. C. 1972. *Climate, soils and vegetation*. London: Univ. Tutorial Press.

Passarge, S. 1940. Die urlandschaft Agyptens und die Lokalisierung der Wiege der Altagyptischen Kultur. *Nova Acta Leopolding* 9:77–152.

Peltier, L. C. 1950. The geographic cycle in periglacial region as it related to climatic geomorphology. *Annals of the Association of American Geographers* 40:214–36.

Potter, P. E. 1967. Sand bodies and sedimentary environments: A review. *Bulletin of the American Association of Petroleum Geologists* 51:337–65.

Redman, C. L. 1973. Research and theory in current anthropology: An introduction. In *Research and theory in current archeology*, ed. C. L. Redman, 5–20. New York: Wiley-Interscience.

————. 1978. *The rise of civilization*. San Francisco: W. H. Freeman.

Reineck, H. E., and I. B. Singh. 1975. *Depositional sedimentary environments*. Berlin: Springer-Verlag.

Roper, D. C. 1979. The method and theory of site catchment analysis: A review. In *Advances in archaeological method and theory*, vol. 2, ed. M. B. Schiffer, 119–40. New York: Academic Press.

Sauer, C. 1952. *Agricultural origins and dispersals*. New York: American Geographic Society.

Selley, R. C. 1970. *Ancient sedimentary environments*. London: Chapman and Hall.

Stamp, L. 1963. *Applied geography*. Harmondsworth: Penguin Books.

van Zeist, W., and H. E. Wright, Jr. 1963. Preliminary pollen studies at Lake Zeribar, Zagros Mountains, Southwestern Iran. *Science* 140:65–69.

Vita-Finzi, C. 1969. Geologial opportunism. In *The domestication and exploitation of plants and animals*, ed. P. J. Ucko and G. W. Dimbleby, 31–34. Chicago: Aldine.

Vita-Finzi, C., and E. S. Higgs. 1970. Prehistoric economy in the Mount Carmel area of Palestine: Site catchment analysis. *Proceedings of the Prehistoric Society* 36: 1–37.

Webley, D. 1972. Soils and site location in prehistoric Palestine. In *Papers in economic prehistory*, ed. E. S. Higgs, 169–80. Cambridge, England: Cambridge Univ. Press.

Wendorf, F., and F. A. Hassan. 1980. Holocene ecology and prehistory in the Egyptian Sahara. In *The Sahara and the Nile*, ed. M. Williams and H. Faure, 406–19. Rotterdam: Balkema.

Whittaker, R. H. 1970. *Communities and ecosystems*. New York: MacMillan.

Wright, H. E., Jr. 1977. Environmental change and origin of agriculture in the Old and New Worlds. In *Origins of agriculture*, ed. C. Reed. The Hague: Mouton.

5

Sedimentary Environments and Lithologic Materials at Two Archaeological Sites

REUBEN G. BULLARD

ABSTRACT

An archaeological site cannot be said to have been thoroughly excavated if no information concerning its geology was collected in the process. At Tell Gezer (Israel) and at Carthage (Tunisia), studies of the regional geological environment, of local sedimentation and stratigraphy prior to, during, and after occupation, and of the lithologic resources available to the inhabitants aid our understanding of the range of activities performed at these sites. Bedrock geology and geomorphology around both Tell Gezer and Carthage influenced their situation, ground-plans, and construction techniques. If they are correctly interpreted, human activities of an ephemeral nature (such as threshing of grain) may leave some traces in the sedimentary matrix surrounding a buried architectural feature. Sedimentary fabrics, textures, and compositions can be related to activities surrounding the deposition of strata and to occupational hiatuses. Simple field petrographic descriptions of building materials may allow specific identifications of quarry locations.

There is no more complete destruction than archaeological excavation of a site down to preoccupation soils or to bedrock. Whatever the state of the art or science of information gathering utilized at that time, the data not recorded are forever lost.

Contemporary excavations without archaeological-geologic investigations are markedly traditional, stressing as they do architecture, *objets d'art*, and ceramics. One has only to read the volumes of *Hesperia, American Journal of Archaeology*, the *Biblical Archaeologist*, the *Bulletin of the American Schools of Oriental Research*, and other archaeological journals throughout the past fifty years to observe these areas of principal concern. Sedimentological, stratigraphic, and environmental parameters are generally not considered.

Notable exceptions to this situation existed and are to be seen in the publications of some excavators in Britain, of many North American archaeological investigations by anthropology departments, and of French excavators working in cave sites where paleoenvironmental and geological specialists were engaged as staff. These efforts have shown concern for capturing all data pertinent to the fullest possible recovery of human-occupational history.

Recently the American geological community has shown appropriate responsibility for conservation and interpretation of the fullest possible sedimentological record from archaeological excavations. This encouraging trend is seen in the emergence of the Archaeological Geology Division of the Geological Society of America, which now embraces more than 400 members. These scientists are applying their expertise in petrology, paleontology, geomorphology, clay mineralogy, pedology, geohydrology, mineralogy, geochemistry, stratigraphy, and structure to archaeological problems.

Not all departments of geology have shown constructive and supportive interest in working in deposits that are post-Pleistocene. An amazing antipathy still exists among some geologists toward archaeologically related research.

Human history should be at least as important as life in the Triassic period; as such, it deserves no less the attention that paleontological and sedimentary depositional environment models, for instance, command in the effort to recover Jurassic or Cretaceous geological history. The issue, of course, concerns the expenditure of geological energies (and monies) on historical research. Since excavation entails near-total destruction of the sedimentary matrix containing the cultural residue of the occupants of a site, the discovery/recovery process is a nonrepeatable venture that should yield maximum data about the site's occupational environment. It should embrace "ecofacts" as well as artifacts (Shackley, 1981). The objective is the fullest possible restoration of human-activity patterns in connection with the site.

THE REGIONAL AND LOCAL GEOLOGICAL ENVIRONMENT OF A SITE

Certain preliminary information is necessary for a geologist to function in an excavation, the most important being a thorough knowledge of the bedrock and surficial geology of the area. A very effective archaeological framework for this geological conceptualization lies in its relationship to local geomorphic provinces, which form an environmental setting that strongly affects the character of a site's development.

The examples most effective in the writer's work are the geomorphic divisions of Israel and Jordan, Cyprus, and northern Tunisia.

Israel and Jordan
  The Coastal Plain
  The Shephelah (Foothills Belt)

The Ephraim-Judean Mountains
Mount Carmel
The Plain of Esdraelon (Jezreel)
Galilee
The Rift System of the Sea of Galilee, Jordan River, Dead Sea, and Araba
    Valley (includes the Beqa of Lebanon)
The Negev
The Transjordanian Plateau

Cyprus
    The Kyrenia Range
    The Mesoria Plain
    The Troodos Mountains (including Foothills Belt)

Tunisia
    The Folded Cretaceous and Early Tertiary Mountain Belt
    Triassic Evaporite Areas
    Jurassic Incipiently Marmorized Mountains
    Late Tertiary/Pleistocene Basins
    Holocene Fill Areas and Coastal Modifications

These designations are unrefined and pragmatic. They have served, however, to synopsize areas of affinity for the purpose of ascertaining their influence on the selected site studied and are more useful archaeologically than, for instance, divisions by tectonic provinces.

Relationships between surface morphology and the stratigraphy and structure of underlying geological formations are well understood by earth scientists. A geomorphic or physiographic province possesses character that has significant influence upon the physical attributes of the environment of an archaeological site. The example of the site of Tell Gezer will serve to illustrate the range of this influence.

Tell Gezer is situated upon an outlier of Eocene chalk on the northern limits of the Shephelah, the western foothills of the Judean Mountains of Israel. A north-south erosional valley on Paleocene and Senonian (Late Cretaceous) marls separates the Shephelah from the higher Judean ranges to the east. The Coastal Plain of Israel bounds the foothills area to the west. Figure 5.1 illustrates this east-west relationship. Other sites occupying the strategic region of the Shephelah are Beth Shemesh, Azekah, and Lachish.

Judging from the remains, Gezer was highly dependent upon the local geological environment. Its topographic situation gave it command over the lower regions to the west, north, and east. Residents of Tell Gezer dug into the *nari*-encrusted chalks and chalky limestone formations for their cisterns, water tunnel, tombs, and wine presses. They employed the indurated *nari* crust (the C-Ca calichified zone of the weathering-soil profile upon the chalk) for threshing floors, mortar or mill basins, and architectural stone. The marls underlying the Eocene chalks were quarried for use in floors, courtyards, and wall plasters.

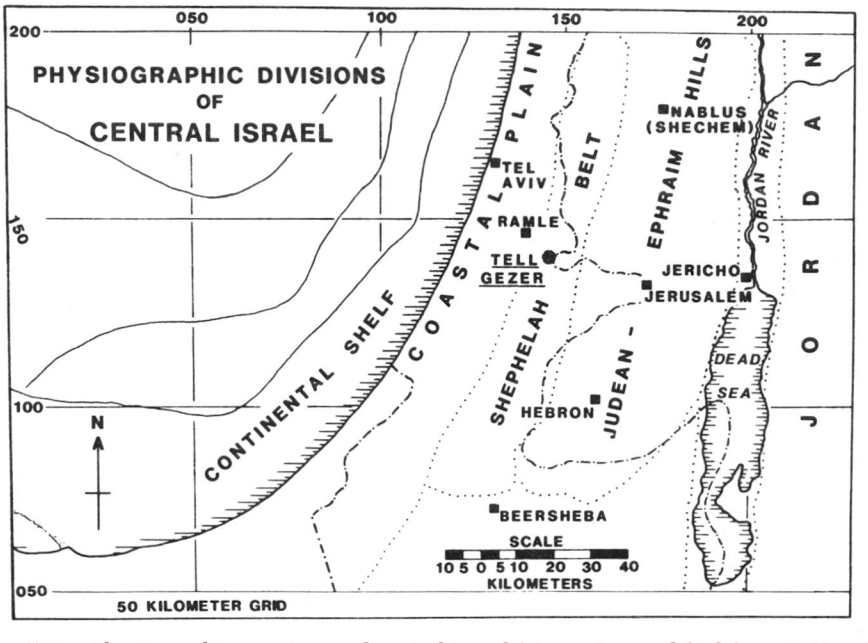

Fig. 5.1. Physiographic provinces of central Israel (Base map modified from Ball and Ball, 1953).

Even the soft, unweathered chalk material was quarried for the construction of a sloping defensive rampart against the Middle Bronze IIC wall structures (fig. 5.2). The *nari* crust apparently was quarried easily with the use of chisels. It appears in house and courtyard walls as ashlar blocks in both Bronze Age and Iron Age outer defensive walls, towers, and gate works. It was quarried in long subrectangular pillar forms for erection in the "High Place," where a rectangular basin, cut from the same source, was placed. Additionally, a subrectangular "horse trough" of cut *nari* was emplaced in the Persian-phase levels of the former Iran Age I Solomonic gate works. The occupants of all the phases of the city found use for the palygorskite-rich soils and soil clays formed on both the chalks and marls of the area. Highly tempered and bonded ceramic and terra-cotta vessels and installations, in part mixed with the kaolinite-illite alluvial clays of the Wadi Aijalon, were fabricated from the local soil clays.

Mud bricks, both sun-dried and fired, came from a variety of sources. During the construction of the Middle Bronze IIC gate works, brick-making parties drew raw materials from the alluvial wadi clays, Paleocene and Upper Cretaceous marls, rendzinate soils, and even from occupational debris from former structures. A highly colorful "mosaic" of the various bricks forms a checkered facade on the eastern side of the gate works.

Nearby geologic sources furnish the bulk of the constructional requirements at most archaeological sites. More remote sources, however, are of special

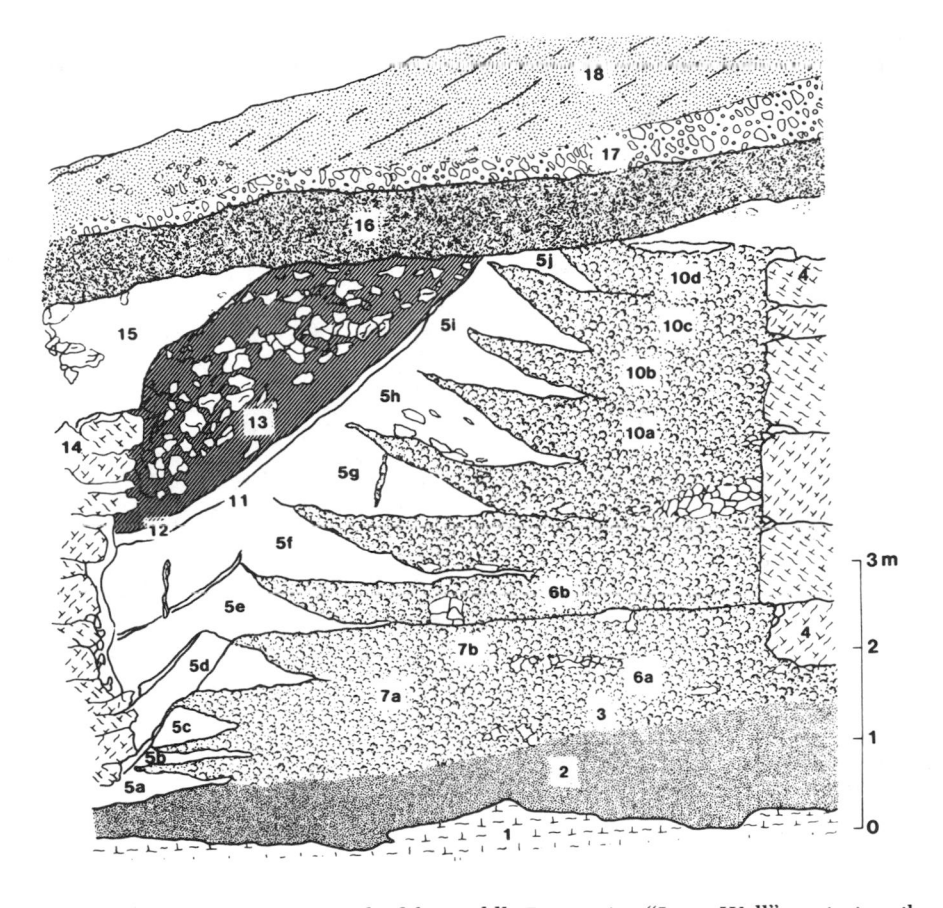

Fig. 5.2. Glacis stratification south of the Middle Bronze Age "Inner Wall" on virgin soil with a post-Roman incipient soil developed on the occupation debris. Key: 18: Macallster excavation spill (1902–1919); 17: field rock; 16: soil (weathered Tell debris); 15: fill on wall 14; 14: early late Bronze wall (hewn *nari*); 13: *nari* rubble—angular boulders, cobbles, and pebbles; 12: sesquioxide deposition zone; 11: indurated, impermeable compacted chalk paste; 10: brown lenses of rendzinate soil and occupational debris; 9: brown lens of rendzinate soil with minor gray-green marl and occupational debris; 8: brown lens of rendzinate soil, angular *nari* cobbles, and occupational debris; 7: brown horizons of rendzinate soil, packed with large potsherds, and *nari* pebbles and debris; 6: gray-green angular marl slabs and conglomerate; 5: indurated *nari* chalk lenses; 4: Middle Bronze IIC massive wall (15.5 m thick), of hewn (face only) *nari*; 3: soil-occupational debris—mingled zone—Paleo "A" zone relict; 2: virgin rendzina soil with minor occupation-debris intrusions (upper 10 cm); 1: *nari* bedrock, jointed and irregular.

interest to archaeologists and historians and require accurate description on the part of the staff geologist. A knowledge of the geographic locations of lithic occurrences, based on geologic maps, contacts with local geologists, and additional fieldwork, is necessary. Theories of trade routes, political alliances, and captured booty may depend upon such provenance studies. At the Gezer site, imported lithic materials stood in marked contrast to those available from local sources. Examples of imported materials and in what form they were found are:

Alkaline olivine basalt	Mills, bowls, weights, whorls, door sockets, thresholds, pounders;
Porphyritic red granite	Fragments without any worked boundaries;
Alabaster	Bowl and jar fragments, mace heads, fragments of figurines;
Malachite	Pieces of ore;
Fossiliferous biomicrite	Shaped fragments of *meleke*, a hard architectural stone occurring in the Middle Cretaceous formations of the Judean Hills;
Specular and sedimentary hematite	Burnishing pieces, pigments, and ore;
Amethyst	Unworked crystal fragments and scarab seals;
Steatite	Seals, whorls, and beads;
Kurkar Formation	A quartz and bioclastic conglomeratic sandstone from the Coastal Plain that found use in mills and loom weights;
Marble	White, coarsely crystalline, probably Aegean.

There is no doubt concerning the exotic nature of the above; they are not indigenous to the Shephelah of Israel. Each was brought to Gezer for a specific purpose that was important to the occupants.

SITE SEDIMENTOLOGY AND STRATIGRAPHY

Consideration of sediments as sources of useful archaeological information came only recently in Mediterranean excavations. Typically, excavators were uninformed about the nature of site sediments, simply regarding them as the "dirt," "soil," or "earth" matrix of the prime objectives of their work: architecture and artifacts. This attitude was a function of the education of Mediterranean and Middle Eastern archaeologists: classics, theology, history, or sometimes social science.

For example, K. M. Kenyon has received justified praise for the descriptive notations recorded in her balk-stratigraphic sections (e.g., 1957). She not only drew carefully the shape of the sediment bodies in the profiles of her trenches into the Jericho site, but also noted some distinguishing characteristic about each. However, expressions such as "sticky black burnt," "dark brown stoney," "green bricky," and "brown ashy" carry little useful descriptive, much less genetic, information. They serve primarily as notes verifying the stratigraphic discrimination and separation that the excavator observed, but apart from the sequential context their information value is low.

More penetrating questions must be asked about the total context of archaeological remains. Examination of the composition, particle size, shape, orientation and color, depositional texture and fabric, compaction, cementation, scour and fill, and any grading or sorting of the bedding should be conducted by a geologist on the site at the time of excavation. Texture, fabric, particle orientation, compaction variations, and sorting will be lost if an excavator merely bags the sediments and ships them to a laboratory.

Most traditionally trained geologists will require a little adjustment to the kind of preparation essential to developing sufficient confidence in making inferences concerning the genesis of a given sedimentary body in an archaeological site. In general, the use of multiple working hypotheses is the most objective framework. The probability of being able to "read" correctly an origin for a sediment increases in proportion to the degree to which a geologist can observe and document the spectrum of modern human-occupational activities and their associated sedimentary records. The present remains they key to the past. Depending upon the geographic location of the archaeological site and the culture of the modern inhabitants of the area, ethnographic analogy may be an important aspect of the investigation.

One example of a quasi-sedimentary depositional activity that is foreign to most geologists is harvesting wheat: the separation of the grain from the husks and stalks on the threshing floor, the winnowing process of casting with a fork the threshed products into currents of air to remove the chaff from the grain, and its storage in bins or cylindrical pits dug into the floor of the place of residence. All the depositional products of these activities were encountered at Tell Gezer.

During the last two years of excavations at Tell Gezer, we encountered in the early Iron I Philistine levels some unusual stratiform deposits that were usually designated "industrial deposits." These were cyclical or recurring layers made up of triplets in sequence. Each triplet was composed first of occupational debris with pebble- to sand-sized particles in a matrix of light gray-brown dust (silt- and clay-sized). The significant fraction of coarse particles had a high frequency of chert flakes or chips. Above this unit was a black, charred band of scorched or smothered chaff and grain residue, the flotation of which yielded wheat, barley, and lentil fragments with some grape seeds. The uppermost unit of each triplet contained light- to medium-gray oxidized ash. The units varied in thickness from a few mm to a few cm and were truncated locally with the next triplet sequences above such erosional contacts, in angular unconformity with the lower sequences. Moreover, the occupational context was problematic in these levels because they were deposited in an open space between buildings in the city. There was no known "industry" to be found *within the defensive walls* of such sites described by historical sources. The agency (or agencies) of deposition was cyclical, with at least 14 sets of triplets (locally lenticular, tonguing and/or truncated in shape) being noted. The overall morphology of the deposits was concave downward slightly and was also pitted from above. What hypothesis could possibly explain these deposits?

The middle unit of the triplets offered the best evidence: it apparently originated in some kind of grain-processing activity. The threshing floors for Gezer lay on abraded *nari* surfaces in the topographic saddle separating the mound of Gezer from the next hill in the Shephelah to the south. Moreover, wind-shadow effects would adversely affect efficient winnowing within the walls of the city. And why was repeated burning connected with these sedimentary events?

Although fraught with many problems, the threshing-floor hypothesis was the most logical. The political situation at this time in Philistine history seems to reflect pressure from the Israelites: the former could have been forced to protect their crops from raids by intramural threshing, despite the inefficient winnowing obtained in this method. Additionally, the scouring activity that locally truncated lower deposits would fit the movements of a threshing sled, the bottom of which could have been studded with chert flakes for better grain separation. (Caches of dressed and undressed chert blades or "teeth" were found in the same levels on the eastern side of this field of excavation.) But how could the grain- and chaff-rich sediments also show consistent burning as a sequel to the harvesting activities?

This question remained unsolved until the summer of 1971, when the writer was studying the site of Heshbon in Jordan. It was just after harvest time, and he observed in the nearby village of Hisban a blackened area in an open space on the windward side of the settlement. The charred sediment covering the area of about 7 by 8 m was rich with chaff and grain. Upon inquiring of the villagers about this burned deposit, he was told a fascinating story: one of the senior citizens of the community had endured a plague of gnats for several evenings and his patience was exhausted. He tied rags to a stick and lighted them to produce a smoke screen that he thought would be effective as a deterrent against the pests. In the process of smoking the area near his house, some remnants of the burning rags fell to the ground and the threshing floor was burned over in a small flash fire. This last piece of the puzzle was now in place. In Iron Age cities, fire for hearth and lamp would be carried from a public hearth or that of a neighbor in the form of lighted faggots or torches, the droppings from which could have ignited the grain on the threshing floor, which would account for the uppermost layer of our triplet sequences. The Gezer staff accepted this hypothesis, and we now assume that similar deposits found at Shechem, Dothan, and sites in southern Judea and Shephelah and in the northern Negev are of the same origin.

The expediency of using such "ethno-sedimentary" models is supported by similar studies carried out at ten other archaeological sites where the writer has been involved. One of the most useful criteria is the physical appearance or texture and fabric of the sediments in a fresh section or in microsection exposures. Hand-lens (10–30X) study of these in the immediate area of excavation, with both macro observations in situ and close-up work, is always fruitful. Both black-and-white and color photography (for documenting freshly exposed sedimentary fabrics) will prevent the loss of information that comes with drying and disaggregation.

Laboratory facilities in the excavation workspace should include a stereo-zoom (usually 30–45X) binocular microscope for examination of objects and sediments. A wet-surface inspection with this microscope is equivalent to the use of acetate peels, polished-section, and even thin-section analysis without, of course, polarizing functions. Flotation screenings may also require microscopic inspection as well. A set of standard sieves of either the 3- or 9-inch diameter in 4-, 16- (or 20-), 50-, and 100-mesh sizes is necessary equipment, and requires water for wet-sieving methods. Other items of excavation lab equipment are listed below.

magnet	porcelain spot-test plate
microscope probes	porcelain evaporating dish
tweezers	chemical reagents for spot tests
single-edge razor blades	petri dishes
plastic water buckets	stainless steel spatulas
plastic sample bags	HCl (37%, in tight containers, used
15 cm ruler in mm divisions	in 1:10 dilution)
3 m steel tape measure	eye-dropper bottle for the HCl
Munsell Soil Color Chart	black felt-tip pens
bright light source for close work	alcohol

Excavation staffs, especially area supervisors, welcome descriptive analyses of the loci (levels) being exposed and removed. Such descriptions are especially important when unusual sedimentary bodies are encountered. If a geologist cannot be present during excavation, a large block of undisturbed sediment should be dug out, set aside out of the direct sunlight, and covered with plastic to conserve the moisture content. Although inferior to examination in situ, study of the sedimentary fabric in this way may produce information vital to determination of the source of the sediment and the environment and agency of deposition.

EXAMPLES OF ARCHAEOLOGICAL SEDIMENTARY ENVIRONMENTS AT
TELL GEZER

Unpaved Streets and Roads

A prominent feature of unpaved streets and roads is the "metalling" effect on the used surface. In dry weather the fine fractions are deflated, leaving behind a lag deposit of the coarse fraction in areas where no effort has been made to add aggregate to the surface. As in aqueous environments, where natural bottom deposition of concavo-convex objects is affected by currents, particle orientation of objects of similar geometry (such as shells and potsherds) is influenced by the forces of movement on the street surface. A convex-up positioning of such particles is evidence of surface movement. Additionally, tabular particles are deposited parallel to the bedding fabric of an active-use surface and furnish confirmation of the existence of such a surface in problematic contexts. Moreover, these deposits are generally highly compacted or indurated.

Dumps and Fills

In contrast to the slow accumulation of sediment on street and passageway surfaces, environments of dumps and fills, of levelling activities, and of destructive events (both artificial and natural) stand at the opposite end of the depositional-rate spectrum. In these environments random, "chaotic," and nonsystematic sedimentation occurs, producing a fabric not unlike that of colluvium or till (with the obvious exception of drumlin and ice-contact surface-particle alignment). Coarse and fine particles composed of lithics, mud-brick and mud-roof fragments, mud-mortar and stucco fragments, and indurated sediment agglomerates may occur in a matrix of sand, silt, and clay-sized fine particles from any of these sources. Courtyard and hearth ash (either dung and chaff or wood, if charcoal fragments are large and occur with high frequency) may be admixed.

A hallmark of rapid deposition is the preservation of large agglomerates of highly friable and easily crushed sediments or unfired mud bricks that still show the molds of the bonding chaff. Such materials would be destroyed by traction of any sort across a surface in existence for any extended period of time. I have documented not only the composition of brick fragments in a Gezer wall collapse, but also structural evidence for the direction in which the defensive feature failed. In areas that were filled systematically by the dumping of sediment from containers, clinoform units marked off by tip lines and having a quasi-graded bedding fabric were laid down. (These are somewhat analogous to prograding delta-front depositional units.) This is a specific fabric type very diagnostic of such fill agencies (fig. 5.2).

Industrial Activity Sites

Many crafts produce debris, such as:

1. chert flakes and cores from knapping in the manufacture of blades, knives, and augers;
2. lithic flakes and chips from stonemasonry;
3. murex-shell fragments, evidence of crushing large quantities of the gastropod for dye or food;
4. ore, slag, and clay-tuyere fragments from smelting, refining, or melting of metals;
5. dung ash from cooking or baking;
6. loom weights and fragments (clay, terra-cotta, and lithic) from weaving locations;
7. unfired, fired, and sometimes fused ceramic fragments, with or without the remaining kiln installation of the potter.

Running Water and Ponding

Running water and ponding are familiar to the sedimentologist. The evidence for water action exists in an ancient urban environment on surfaces where runoff and local ponding occurred. Such deposits may furnish insight for correlation of occupational surfaces and gradients across a site that is being excavated in randomly selected areas. If a drain was not efficiently designed, sediment accumulation eventually may have ended its service.

A running-water system of an archaeological site usually will not have had sufficient time or energy to round the coarse-sediment fraction, but localized sorting and grading often occurred as a function of current velocity. Except where sheet-wash events predominated, some channel scouring may be discernible. Careful attention to these sedimentary features may enable the geologist and the architect to work out the archaeohydrodynamics of the city (e.g., Wilkinson, 1976).

Ponded conditions, analogous to deposition in lakes and reservoirs, give rise to three types of sediment bodies: deltaic forms at the inlets (unless a waterfall type exists); graded bedding arising from high surface-runoff discharge rates; and laminar silts and clays in basins where undisturbed, quiet water conditions exit. Cisterns usually exhibit structures of the second and third type, whereas surface pools may show characteristics of the first two types.

Occupational Hiatuses

In the course of time most major sites of the Near East have experienced periods of abandonment. Gezer and Carthage were examined in sufficient detail to recognize such events in the sedimentary record. Destructional horizons are always of major archaeological interest but require little special consideration because of their usually obvious fabric.

At Tell Gezer we encountered 15 cm of well-sorted silt (ca. 0.3−0.8 mm) across the Solomonic-period street surfaces. Archaeological evidence demanded no more than half a century (more likely about a quarter) of occupational discontinuity. This loessic unit had a random distribution of very minor pebbles and granules that probably represents the gravity waste from adjacent structures and the effects of animal movements over the site. The fabric was uniform without any distinctive depositional fabric.

One of the best models of site abandonment I have examined anywhere during 14 years of work was the nearly abandoned Arab refugee village just south of the Wadi Kelt and modern Jericho. Mud-brick structures there exist under various states of disintegration. The agencies of erosion were mass wasting and the few winter rainstorms. My studies were made during 1973; the abandonment of the village occurred in 1967. This six-year hiatus had been of sufficient duration for wedges of mud (banded with chaff), stucco, and mud-brick-wall gravity wastage to produce a maximum of about 1 m of deposition adjacent to structures. This ghost town provided numerous excellent examples of architectural features and their weathering and erosional behavior.

Courtyard Sediments

A large cow, several sheep and goats, and some chickens completely filled the courtyard of a house in the village of Balata, in Samaria, West Bank. The owner took us to see an architectural block of marble surfacing through the sediments in the courtyard. The experience helped to explain the deposits most commonly observed in domestic occupational quarters of Bronze and Iron Age sites in the Levant: dung, chaff, and dung ash. During the first year of my research at Tell Gezer, I went on a survey traverse across the topographic saddle to the slopes of the hill immediately south of the mound. The air carried a strong odor as I advanced up the hill and followed the increasing pungency to its source. An abandoned quarry had served as a corral for flocks of sheep and goats for years. The accretion of dung on the floor of this area was about 20 cm in places, and a portion of this deposit was burning with the speed and character of a lighted cigar. I carefully examined both the unburned- and the burned-dung sedimentation and sampled them. Sheep, goat, donkey, cow, horse, and camel dung universally has a texture of plant stalks and blades, usually finely comminuted, in a matrix of fine organic paste. In ancient sediments only the fibers remain, and in many instances only traces of the mineral content of the fibers remain as surfaces of molds. The ash generally preserves these mineral-fiber residues, perceptible, at least, under magnification.

The identification of such deposits in an unburned state indicates courtyard and stable environments. Compaction is proportional to the depth of burial.

Other aspects of this domestic sedimentation involve burning contexts. Domestic ovens usually were fueled by sheep dung. This commodity had wide and general application, especially from the time of the Middle Bronze Age, and is a product of the keeping of flocks. Modern Bedouin use this fuel, along with that obtained from camels, in the preparation of coffee, the boiling and roasting of meats, and the heating of washwater.

Shepard (1965) observes that American Indians used dung for another purpose:

> Dung, a fuel widely used by potters who had domestic animals, has both advantages and disadvantages. It burns easily and rapidly because it is open in texture and composed of fine material. It also retains its form after the more combustible material has been burned out, and thus maintains a warm blanket around the pottery, but it gives a short firing because it burns out quickly. [p. 77] By the time she [the Zia potter] had laid the fuel, a temperture of 415°C was reached. The slabs burned more slowly than chips; consequently the rate of heat increased to the first peak was less than half that of the other 3 Pueblo firings, and since there was more fuel, a high temperature was maintained much longer. [p. 85]

Mud-brick Building and Roofing Debris

Mud bricks may be observed in various states of existence: (1) unfired mud brick with vugs taking up as much as 35 percent of total volume, representing the molds of straw and chaff used as a binder in the brick manufacture; (2) reduction-fired mud brick; and (3) mud brick fired above 900°C (no endothermic-reaction peaks remain).

One of the best examples of this last type of mud bricks occurs in the Middle Bronze Age II gate-house structure in the southwestern wall area at Tell Gezer; it is a massive construction of fired bricks. Trace-element fingerprinting led us to the inference that sediment sources utilized in this defensive feature were (a) the rendzinate soils formed on the Eocene chalk above the *nari*; (b) similar soils formed on the Taqiya marl (Paleocene), stratigraphically the next unit below the chalk and hence outcropping on the more gentle slopes that apron the base of the mound of Tell Gezer; and c) the Wadi Aijalon colluvial, alluvial, and residual soils (Holocene), which exhibit much higher kaolin content in tests on unfired materials and afford a brick that has more structural integrity. The compositional variation of the gate's eastern facade is vividly evident in the multicolored brickwork. Based upon this visual differential and the trace-element discrimination, we assumed that at least six brick-making parties were engaged in the construction of this building phase of the city's defenses.

THE SITE OF CARTHAGE

There are few other sites in the Mediterranean that offer a greater challenge for environmental studies than ancient Carthage. Founded at the interface between the land and the sea, Carthage developed from the interplay of earth materials and forces. The site generally affords excellent examples for observing the setting and its natural and anthropogenic change.

The modern physiographical shoreline configuration of northern Tunisia is a consequence of geological development. The topograhic highs in this region are a result of tectonic forces that have produced local structural folding and uplift in the North African crust. Differential lithic resistance to weathering and erosion further punctuate this morphology. These striking massifs occasionally feature ore concentrations that were exploited by both Punic and Roman inhabitants. The massifs delineate valleys, plains, and depressions (which some workers regard as past embayments of the sea) where active alluvial and eolian sedimentation provide new land surfaces for human development.

Geological conditions in this part of the North African continent were favorable for exploitation by the early Punic settlers. The easternmost arm of Tunisia, the Cap Bon Peninsula (fig. 5.3), forms the Gulf of Tunis, on the southwest shore of which Carthage would eventually develop as a major maritime state.

Carthage, like all city-sites, was not only influenced by its physiographic setting but in many ways was also strongly molded and influenced by the geological conditions and resources from which the city and its harbors arose. Indeed, perhaps there is no other archaeological city-site more exemplary in the study of the effects of environmental geology on the life-style of an urban settlement. The land around Carthage is isolated from areas that provided various essential lithological materials for the construction of Punic and Roman buildings and for the essential functions of an urban way of living. For example, there is not a suitable lithic source for building stone anywhere in the immediate region of Carthage. Both Punic and Roman inhabitants had to travel many kilometers from the city in quest of this essential resource. Figure 5.4 highlights

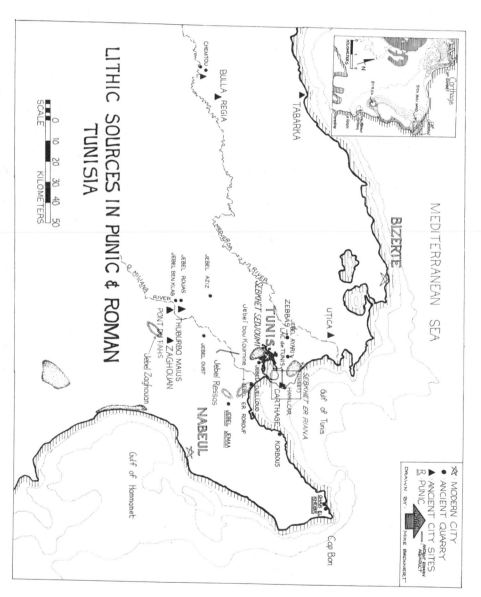

Fig. 5.3. Lithic sources in Punic and Roman Tunisia.

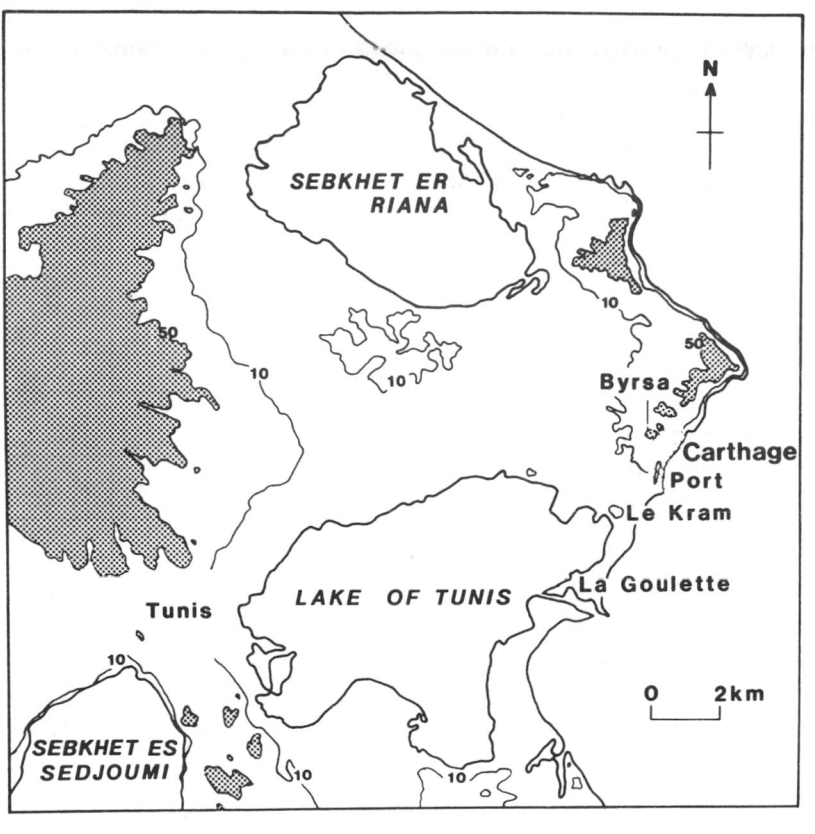

Fig. 5.4. Location of the ancient city of Carthage in relation to the modern coastline (map data from the British and American excavations at Carthage).

the topographic and lithologic isolation of the Carthage highlands and headlands (see also fig. 5.3, inset).

THE GEOLOGICAL ENVIRONMENT OF CARTHAGE AND ITS ECONOMIC EXPLOITATION BY PUNIC AND ROMAN PEOPLES

Basement rocks of the continental crust of North Africa are not exposed in the area around Carthage, although they have been observed in well cores from the southern region, under exploration by oil companies (Baird, 1967). Continuing up through the stratigraphic section, we shall note only those units that are archaeologically significant for the purposes of this discussion.

Early Mesozoic (Triassic) strata, highly deformed, outcrop in several areas of northern Tunisia. They are usually composed of clays and silty clays, locally vuggy dolomite, and (most important) anhydrite and gypsum. Our lithic-source map (fig. 5.3) locates a surface occurrence with very ancient workings centrally placed in the mountainous terrain of Jebel Amar, northeast of the city of Zebbas. Here large selenite crystals form a mosaic along the sides of the Triassic-age deformed strata. This site is the closest probable source of gypsum for Roman (and perhaps Punic) Carthage. Many hundreds of cubic yards of gypsum rock have been removed, not only pitting the slopes, but also creating an artificial topographic saddle. The purity and abundance of gypsum in this area make it a potential source for plaster, stucco (on *cippi* and buildings), and building blocks of Roman "African" architectural style (Bullard and Vann, in preparation).

Some of the most spectacular mountain masses in northern Tunisia are composed of Jurassic-age strata. Those nearest Carthage are Jebels bou Kournine and Ressas, while to the south, Jebels Zaghouan, Oust, Aziz, Rouas, and Ben Klab are all prominent topographically (fig. 5.3). Although this study does not feature a detailed analysis of Roman lithic sources, further research in these areas will be noted as it bears on the lithic repertoire of the merchant-harbor portion of Carthage.

The lithics utilized by the Roman inhabitants of Tunisia, especially the material found in the well-built private houses and in the public buildings, were quarried principally from the Jurassic sources mentioned above. These were not by any means the only sources for Roman architects and contractors, but because of their accessibility and desirable properties these particular resources were exploited heavily. The attractive mountains to the west of Thuburbo Maius (fig. 5.5) reveal the sculpture of the telltale Roman step-quarry activity. Material from these sources appeared in the harbor-side sediment sequences from the early

Fig. 5.5. View of Jurassic Jebel Ben Klab (left) and Jebel Rouas from Thuburbo Maius.

part of the second century A.D. onward. The delightful travertines of Rouas are not only found at Thuburbo Maius but also at Utica and throughout the Roman strata of several excavations in the Carthage area (fig. 5.3). An excellent study in the contrast between Roman and modern quarry techniques can be observed presently in the quarries at Jebel Aziz. Nowhere else in quarry sites throughout Tunisia were examples of Roman step-ledge quarry technique more clearly shown than as they appear in figure 5.6, slightly higher and to the right and left of the modern, wire-cut exposures in the view. This Jurassic facies is intermittently punctuated by white veins of calcite and occasional mylonized zones of hematite-rich cataclastics.

Perhaps the most famous Roman quarries in Tunisia are those of Smitthou (named Chemtou in fig. 5.3), which are cut into three differently colored facies of the Jurassic mountain mass north of the Medjerda floodplain west of Bulla Rhegia. The widely used golden yellow variety of this submarble was prominent not only in this area of North Africa, but also in Italy in the construction of the Pantheon in Rome. The dominant feature on the skyline in the area of Carthage is imposing Jebel bou Kournine, across the Bay of Tunis to the south, the double-pronged peak of which is composed of the same characteristically massive Jurassic submarmorized carbonate. Petrographic examination of these lithics show them to be recrystallized, dolomitic limestone, with local pelletal micro-structures. The recrystallization is incipient and in some instances there are

Fig. 5.6. Jebel Aziz; Roman (step cut) and modern (slab-cut with wire saw) quarry techniques are shown.

reports of diagnostic fossil remains that have not been destroyed—for example, algae and foraminifera (Bonnefous, 1967:110).

In contrast to the nature of the Jurassic-age sediments, Cretaceous through Eocene sediments in the eastern area of the Mediterranean basin include abundant microfauna, especially foraminifera. In northeastern Tunisia, thin to moderately thick bedded limestone sequences comprise the topographic highs. Locally, these resistant carbonates may be replete with foraminifera-forming biosparites. More clay-rich, sometimes colorfully iron-stained biomicritic limestones may occur interbedded in the same outcrop sequence. Carthage is flanked on the west and south by these microfaunal-rich carbonate mountain resources. Figure 5.7 illustrates the moderate topographic relief of these mountains, in contrast with the heights of Jebel Ressas in the distant background. Examination of the lithic scree capping the ridge in the foreground reveals that it is not the consequence of physical weathering. The elongate and occasionally tabular nature of these rock fragments is a consequence of the bedding characteristics of this Late Cretaceous (Senonian) biomicrite. The scree is composed of Punic lithic rejects, among which one may observe numerous unfinished obelisks. This limestone contains microfauna typical of that illustrated in figure 5.8. Similar carbonate rocks are exposed along another Punic quarry face to the northeast of the area in figure 5.7. Numerous artificial "talus" cones of rejects or quarry waste rock may be observed, the consequence of intensive quarrying operations on this part of the Jebel Jemaa Range. This carbonate material was exploited by the

Fig. 5.7. Jebel Jemaa, capped with the quarried Punic funeral-stelae rejects, observed by Jean Le Maresquier. Jebel Ressas is the distant peak to the left.

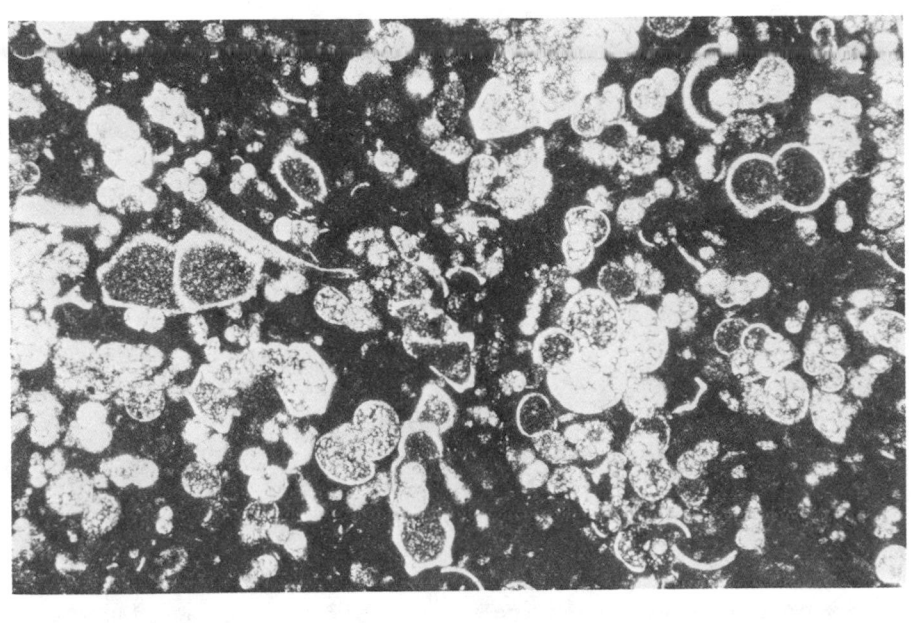

Fig. 5.8. Photomicrograph of a foraminiferal biomicrite, typical of the Nepheris lithic repertoire. The original sediment was a moderately clayey globigerina ooze of the Cretaceous period.

Punic Carthaginians for use as funerary or burial stelae *(cippi)*, in, for example, the Tophet area. The Gulf of Tunis afforded easy transport by boat from this quarry to Carthage.

Lithic resources of Upper Cretaceous to Eocene age from Jebel er Rorouf, a foothill along the northern slopes of Jebel bou Kournine, and Jebel Djelloud, a gentle topographic swell just southeast of Tunis (fig. 5.3), afforded paving stone for streets, plazas, and walkways (such as those illustrated in fig. 5.9) from the merchant-harbor site at Carthage.

Stratigraphically higher than most of the biomicritic carbonates available to the inhabitants of Carthage was a prominent sandstone sequence at Korbous (fig. 5.10). Increased tectonic activity in the Oligocene epoch resulted in the deposition of thick sandstone sequences in parts of Tunisia. In the Korbous area, sediment of this age is known as the Fortuna formation; its upper part is composed of massive sandstone layers, frequently cross-bedded and locally coarse-grained, and cemented by a silica that imparts to it a variable character. This lithic resource has been observed in Roman strata of Carthage at both the American merchant-harbor excavation site and that of the German excavations just south of the Antonine baths.

The most exciting quarry source studied in the Carthage region was observed in a Pleistocene (Tyrrhenian) deposit near the northern tip of the Cap Bon Peninsula at Ghar el Khebir (fig. 5.3). In this area, sand and granules composed

Fig. 5.9. Fourth-century Roman flagstone paving and curbing composed of micritic limestone derived largely from the Hammam-Lif Jebel er Rorouf quarries on the southeast shore of the Gulf of Tunis. This lithic paving shows the abrasion of curbing by Roman vehicles. The street surface shows a highly indurated compaction of occupational lithics and detritus. Some compaction is due to the presence of iron-oxide cementation from slag droppings.

Fig. 5.10. Korbous. Famous for its hot springs, this village is set in the midst of resistant orthoquartzitic sandstone units of the Oligocene age (Fortuna formation).

Fig. 5.11. Ghar el Khebir quarry on the north shore of Cap Bon Peninsula, about 5 km from the northeastern terminus and northwest of the village of El Haouaria. Initially exploited by the Punic inhabitants of Carthage, the coarse bioclastic material (mostly wave-abraded shell material) was employed in urban structures and in the larger, sometimes massive, Tophet stelae. Later, the Roman builders of Carthage utilized the same quarry site.

of bioclasts and minor quartz were blown inland from an active beach to cover the land surface with several tens of meters of coarse colianite. Figure 5.11 shows the locale of this deposit and quarry site, which is now undergoing active wave erosion. Two different types of quarrying activity were observed at the Ghar el Khebir site. In the Punic technique, shafts (apparently) were sunk, from which cut blocks were lifted, whereas the Romans utilized the step-ledge quarrying technique. The latter activity broke into shafts made by the older methods of extraction in several places in the quarry site (fig. 5.12).

The Ghar el Khebir bioclastic sandstone was utilized extensively by the Punic occupants of the city of Carthage, not only in the harbor quay wall (fig. 5.13), but also in the *cippi* found in the Tophet area (fig. 5.14). The Romans extensively reused this material from destroyed levels of the Punic city. The massive character of the Ghar el Khebir sandstone not only afforded easy workability in the quarry, but also in the subsequent preparation of the stone for use in various structures within the city and harborside installations. It was an excellent, light-weight architectural stone not unlike modern-day cinder blocks.

Fig. 5.12. Ghar el Khebir quarry cuttings showing the technique of lithic removal. In some areas of the site, vertical shafts were cut to give access to the more suitable rock below (center). Caliche layers (left) were unsuitable for economic use at Carthage; hence, the shafts may have provided access to the earliest Punic material shipped from the quarry.

Fig. 5.13. Harbor excavation area GH.2: fully exposed quay wall as viewed from the east. Boring mollusks and barnacles decorate the surface of the cut and trimmed lithics composing this harbor wall. The large, well-dressed blocks are composed of Ghar el Khebir sandstone.

Fig. 5.14. Stele in the Tophet area excavated by Kelsey, University of Michigan, 1925. The combination of Ghar el Khebir sandstone in contrast to the fine Nepheris biomicrite illustrates the diverse character of these lithic materials.

ENVIRONMENTAL GEOLOGY OF CARTHAGE

Studies of the Pleistocene and Holocene sedimentation patterns in and around Carthage reveal the interesting paleogeography of the area. Work in the area of the depositional basin of the Medjerda River has revealed the influence of the sediments of this active stream on the shoreline configurations, especially on the western side of the Gulf of Tunis (Warmington, 1967; Harbridge et al., 1976). Originally a Pleistocene island, the headlands of Caps Gamart and Carthage and the area southward to Byrsa (fig. 5.3, inset) were linked to the low-lying mountains to the west (such as Jebel Ayari) by a tombolo deposited by near-shore processes reworking sediment from the Medjerda River. It was this shoreline configuration that the early Punic settlers found when they arrived in the Carthage area. The continuing activity of the Medjerda River (especially after Roman deforestation came to climax in the second century after Christ) resulted in further shoreline accretion around Carthage. Not only was the basin of the Sebkhet er Riana cut off by long shoreline sedimentation, but currents around the Gamart and Carthage headlands deposited sediments in an ever-growing spitlike projection from the southern shoreline of what is today the area of Le Kram (fig. 5.4). The new land now occupied by Khereoline and La Goulette, and the peninsula extending from Maxula Rades to the west of Tunis were deposited in post-Roman times.

The area occupied by the rectangular commercial and circular naval harbors and their connecting channels (fig. 5.15) is of concern to the history of the Punic development of Carthage and later Roman modifications. Air photos (figs. 16a and b) show the nearly contemporary and mid-twentieth-century configuration, respectively, of these artificial lagoons. Comparison of these two photographs reveals the increasing shoreline activity. Channels cut to freshen these artificial lagoonal areas have been widened by breaker and wave erosion and have introduced tidal inlet-type sedimentation to the extent that these bodies of water are being infilled. Already the filling of the southernmost lagoonal pond (fig. 5.16b) has given rise to speculation that there is no evidence for a precise understanding of the direction of the entrance channel by which ancient ships entered both harbors.

These shoreline processes are quite dramatic: a comparison of the quantities of sand available along the shoreline between the time of the air photo in figure 5.16b and that of figure 5.16a reveals striking erosion of the beach material. Such geological activities help to explain the nature of the rapidly changing environment of the harbor area of Carthage.

PREHABITATION AND OCCUPATIONAL ENVIRONMENTAL CONDITIONS OF THE HARBOR AND TOPHET AREAS

The first indication of truly undisturbed rock appeared in the merchant-harbor excavation site beneath the quay wall (fig. 5.13), where large Ghar el Khebir ashlars were set into a semiplastic Pleistocene argillaceous coquina. The abun-

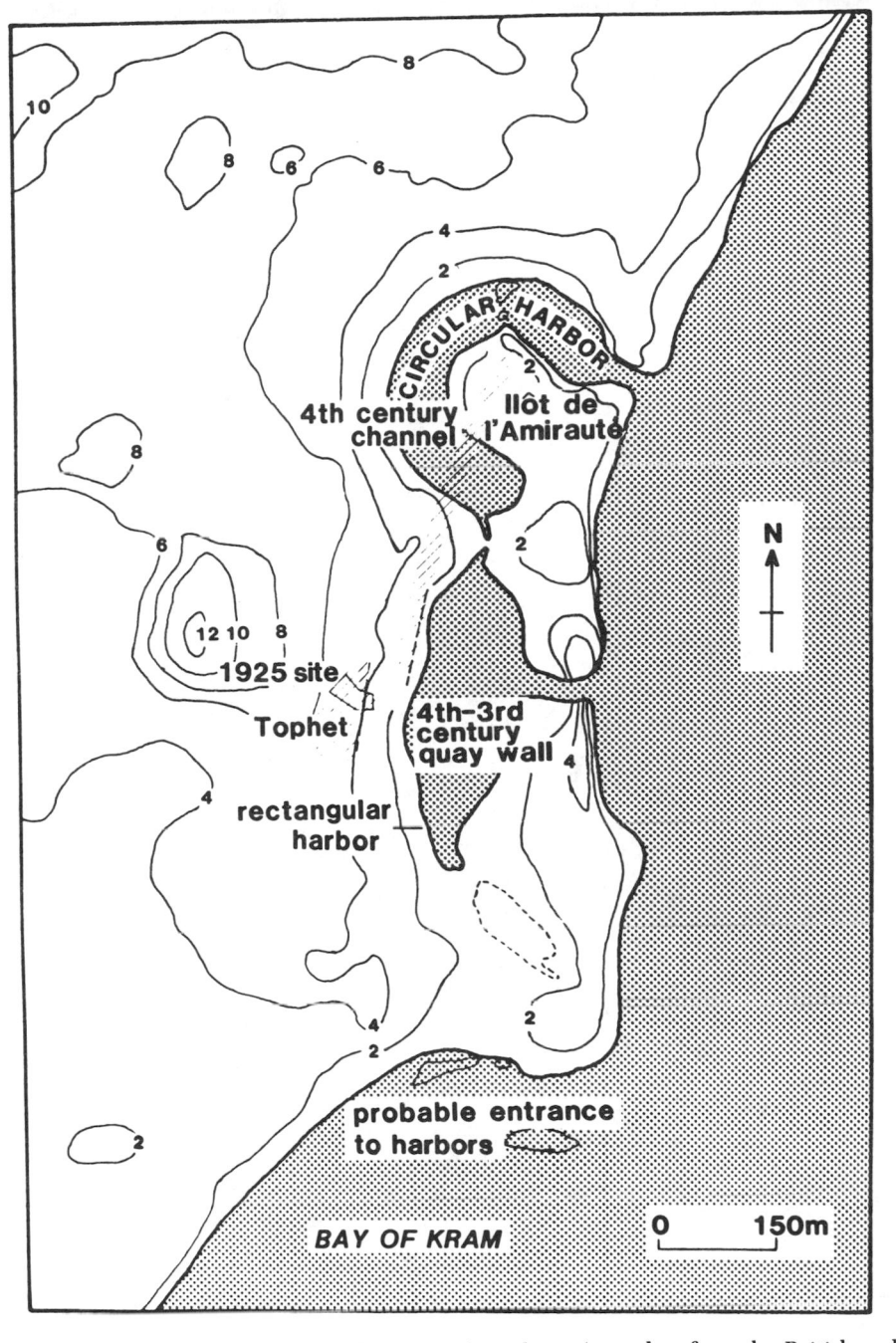

Fig. 5.15. Topography of the harbor area of Carthage (map data from the British and American excavations at Carthage).

(a)

Fig. 5.16(a). Pre-1975 photograph of the Carthage harbor area, courtesy of the government of Tunisia. Note the extensive erosion in the marine-water access channels leading into the naval and merchant harbors. Note also the higher density of structures along the streets.

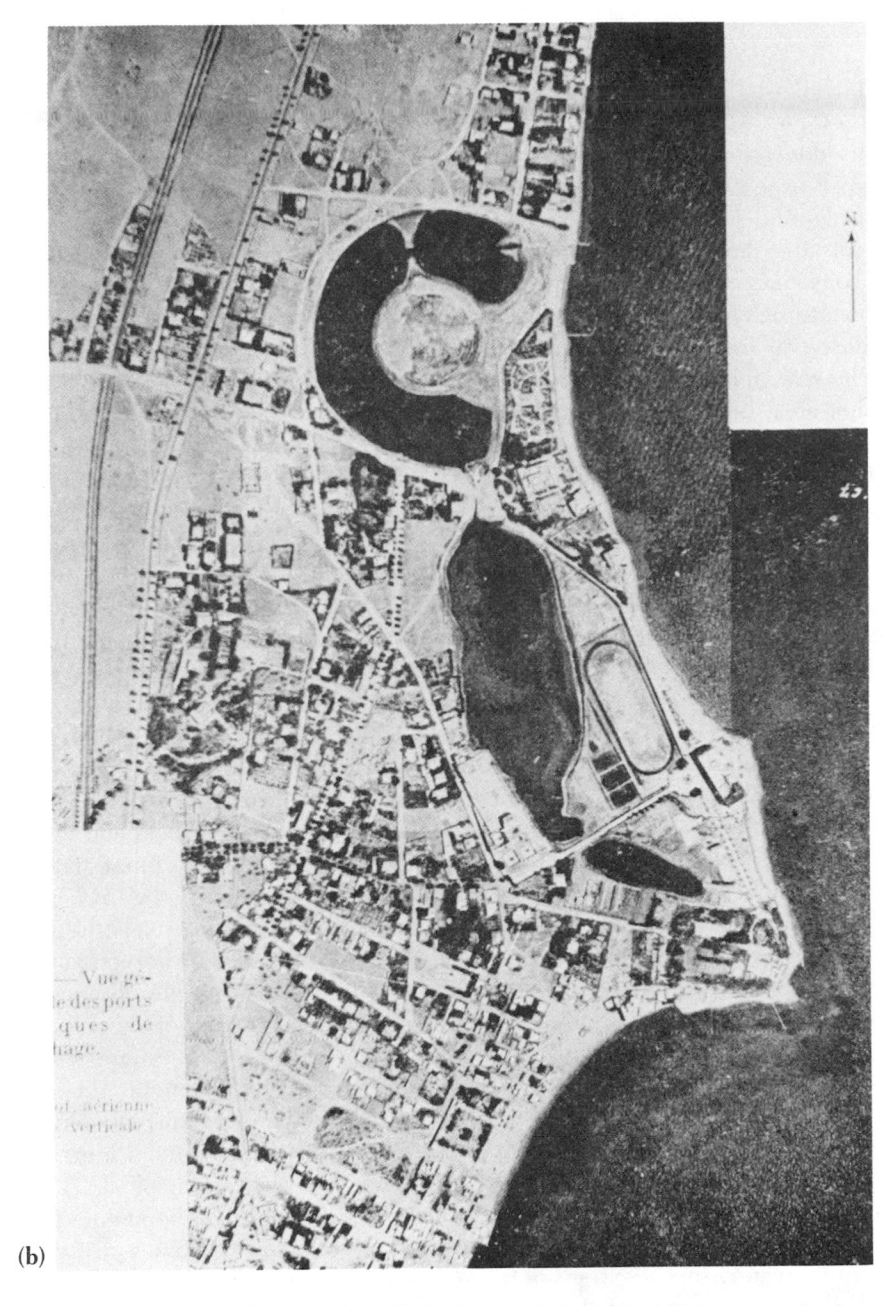

(b)

Fig. 5.16(b). Aerial photograph published by Jean Baradez in "Mission archéologique française en Tunisie," *Karthago, Revue d'archéologie africaine* 9 (1958). This photograph offers a dramatic view of environmental alteration by man. There is no channel cut between the sea and the merchant harbor. The opening to the sea from the naval harbor is a covered culvert. A remnant of the eastward-trending "access channel" exists as an unfilled lagoon.

dant whole-shell remains of marine pelecypods, gastropods, and annelids represent a typical shallow, near-shore faunal assemblage of Plio-Pleistocene age.

Evidence for this land mass appeared in two parts of the excavations. The bedrock exposed in the Tophet investigations and in the harbor-area research offers helpful information. Unlike the argillaceous marine coquina observed beneath the quay wall, this lithic material is a deposit known as *caliche*. It occurs as a consequence of groundwater solution and precipitation activities of calcium carbonate-rich material. It is not a submarine depositional feature. Thus we pass in these two areas from a marine to a subaerial depositional environment. The caliche was quarried not only to provide burial space in the shallow soils of the Tophet area, but also to provide wall-rock filling for structures that were plastered or stuccoed. The texture of the caliche does not show any uniform bedding or layering, but rather an almost travertine-like and highly variable solution-deposition fabric.

Awareness of the nature of caliche and its sedimentary characteristics enabled us to determine which side of a fragment was "up" and provided the basis for several observations concerning the dredging activities that were part of the construction and maintenance of the Punic harbor. A cobble-size caliche fragment with marine fauna encrusted on its bottom surface was observed in the clay deposits of the lowermost lagoonal harbor. Other caliche fragments were observed in these relatively pure, reduced bluish-gray lagoonal clays, which are composed primarily of montmorillonite, kaolinite, and illite. These clays, with their observed fauna, are typical of a stressed lagoonal environment having limited access to the sea. The clays were nearly a meter in thickness and were covered by undulating areas of contorted sands and oxidized clays immediately underlying the layer of the Punic city destroyed by the Romans (fig. 5.17).

Significant for our understanding of the environment of the area is the soil sediment overlying the calichified bedrock; it is especially well preserved in the Tophet area, where more than 20 cm of gray-black, highly reduced soil remains as an undisturbed deposit. Locally, slightly angular fragments of caliche are suspended in the first 5 cm of this soil above the bedrock, the residue of the soil-forming process. These soils were formed in a marshy environment, not only from the sedimentary material present in the bedrock undergoing calichification, but also from wind-blown sources that still affect the local environment.

In the harbor bounded by the quay wall on the west, ashlar blocks have settled into the Pleistocene argillaceous marl to a depth of nearly 30 cm. The offset blocks reveal that the method of placement of the lower courses probably was under water. Investigations of the sediments adjacent to the quay wall and further out in the area of the present merchant-harbor lagoon revealed an important chapter of Carthaginian history (fig. 5.18). Immediately above the Pleistocene argillaceous coquina, another 10 cm of sediment underlies quiet-water lagoonal deposits. The 10–20 cm level is composed of highly contorted, light- to medium-gray, mottled, medium-sand-size quartz within which are agglomerates of medium- to dark-gray lagoonal clays containing Punic ceramic

Fig. 5.17. Merchant-harbor site, excavation area, E.1.070, showing the massive, reduced, blue-gray Punic harbor clays. The upper left and right areas of the photograph show debris from the Roman destruction of Carthage.

fragments. These prelagoonal sediments with their undulating upper surface are not typical of any natural activity. They most likely arose from dredging activities. If so, most if not all of the Punic harbor lagoonal sediments to the east of the quay wall were dredged away, and the sedimentary history there begins with late-Roman cultural deposition.

Investigation of the merchant-harbor sediments included six cores taken along a north-south transect in the center of the contemporary body of water. These cores reveal a depositional record that shows very clearly the Punic sediments resultant from the Roman destruction overlain by stressed quiet-water lagoonal sediments with marine fauna. The lagoonal sediments subsequently were buried by a series of cyclic sedimentary laminae that indicate disuse of the harbor. These are overlain in turn by highly saline and evaporitic sediments, which are in turn overlain by gypsum- and carbonate-rich clays, reflecting a highly eutrophic, closed (or nearly closed) depositional basin. Finally, these are capped by indications of a renewed marine connection, with living gastropods on top of the uppermost layer.

One interesting feature of the marine fauna observed in the harbor waters adjacent to the quay wall is the encrustation on the stone courses of the quay wall. A barnacle assemblage differing from the modern one is seen to encrust oysters and the exposed ashlar quay surface. This barnacle assemblage has been displaced recently by a more aggressive variety that entered the Mediterranean

Fig. 5.18. Merchant-harbor bottom sediments. The lower section is below dredge level. This argillaceous coquina contains clay-lined water pipes and conduits. The block footing of the quay wall is set into this lower section. The middle section is made up of highly convoluted sand, medium gray in color, with inclusions of gray clays, of up to small-cobble size, which contain ceramic fragments. The contact at the 20 cm level with the overlying (and very low energy) harbor clays is undulating and appears in slight relief. The massive upper section is composed of nearly pure, gray clays. Colonies of pelecypods and barnacles are intrastratified within this upper unit, thinning away from the quay wall.

after the opening of the Suez Canal (J. Zaouali, pers. com., 1976). Wedges composed of barnacle detritus up to 30 cm thick were observed at various levels of the sedimentary sequence against the quay wall. These were interstratified with the clays described above and were not significant in the sediments at the time that coarse detritus began to enter the harbor.

Acknowledgments

This chapter is dedicated to those who have made possible years of exciting and fruitful archaeological geology: Lynn Y. Bullard, Richard H. Durrell, Frank L. Kouchy, Nelson Glueck, George Ernest Wright, William G. Dever, Cedric Boulter, Kenneth Caster, Leonard H. Larsen, and Lawrence Stager. Opportunity, funds, and research facilities and publication assistance were forthcoming as a consequence of the interest, assistance, and vision of these friends and associates and my wife, Lynn.

REFERENCES

Baird, D. W. 1967. The Permo-Carboniferous of southern Tunisia. In *Guidebook to the geology and history of Tunisia*, 85–107. Petroleum Exploration Society of Libya, Ninth Annual Field Conference.

Ball, M. W., and D. Ball. 1953. Oil prospects in Israel. *American Association of Petroleum Geologists Bulletin* 37(1):127–28.

Bonnefous, J. 1967. Jurassic stratigraphy of Tunisia: A tentative synthesis. In *Guidebook to the geology and history of Tunisia*, 109–30. Petroleum Exploration Society of Libya, Ninth Annual Field Conference.

Bullard, R. G., and L. Vann. In preparation. The occurrence and use of gypsum in Roman Tunisian architecture.

Harbridge, W., O. H. Pilkey, P. Whaling, and P. Swetland. 1976. Sedimentation in the Lake of Tunis: A lagoon strongly influenced by man. *Environmental Geology* 1:215–25.

Kelsey, F. W. 1926. *Excavations at Carthage, 1925: A preliminary report.* New York: Macmillan.

Kenyon, K. M. 1957. *Digging up Jericho.* London: Ernest Benn.

Shackley, M. L. 1981. *Environmental archaeology.* London: George Allen and Unwin.

Shepard, A. O. 1965. *Ceramics for the archaeologist.* Washington, DC: Carnegie Institution Publication no. 609.

Warmington, B. H. 1967. Phoenician Carthage. In *Guidebook to the geology and history of Tunisia*, 5–16. Petroleum Exploration Society of Libya, Ninth Annual Field Conference.

Wilkinson, T. J. 1976. Soil and sediment structures as an aid to archaeological interpretation: Sediments at Dibsi Faraj, Syria. In *Geoarchaeology*, ed. D. A. Davidson, and M. L. Shackley, 275–88. London: Duckworth.

6

Palynological Applications to Archaeology: An Overview

JAMES E. KING

ABSTRACT

Palynological studies as part of archaeological investigations can yield valuable information on the relationship of prehistoric people and their environments. Identification of cultigens, dietary information, site seasonality, relative age placement, and anthropologically related questions can frequently be addressed by the analysis of pollen grains preserved in archaeological sites. By correlating archaeological sequences with stratigraphically continuous pollen chronologies from within the same region, paleoenvironmental reconstructions can be developed for the site. Changes in the regional vegetation and climate may be of significance in interpreting the archaeological evidence.

Palynology is broadly defined as the study of pollen and spores from living and fossil vascular plants, including their dispersal and application in stratigraphy and paleoecology (Gray, McAfee, and Wolf, 1974). The usefulness of pollen and spores as tools in both stratigraphic and paleoecologic studies is one of the main reasons that this subdiscipline, encompassing elements of geology, ecology, and botany, has enjoyed such remarkable growth since its development early in this century. Paleoecological reconstructions based on pollen have been especially valuable in archaeological studies.

Palynology originated in nineteenth-century Scandinavia with the recognition that pollen grains are taxonomically identifiable and that they are preserved in lake sediments and peat deposits (Erdtman, 1943). However, it was not applied until 1916, when Lennart von Post, a young Swedish geologist working on stratigraphic correlations of peat deposits, realized that changes in pollen composition of a sediment stratum represent temporal changes of plants on the landscape. An important corollary to von Post's discovery was that these changes in the plant community reflect changes in climate of that locality. Since the recognition of this relationship, palynology (the term was not introduced until 1944) has developed in many divergent directions; in particular, the study of Quaternary-period pollen grains has had a great impact on archaeology. Although spores are also included in pollen analysis, it is the pollen grain, the conveyor of the male gametophyte in the higher plants, that is of primary importance in vegetative reconstructions and a major tool in Quaternary paleoecology.

Early pollen studies, including those of archaeological sites, employed pollen analysis solely as a dating technique to determine relative chronologies within a single site as well as between sites. This application worked especially well in northwestern Europe, where the regional vegetational history since the end of the Pleistocene epoch is well known and has been described as a series of distinct successional communities, the Blytt-Sernander sequence, each with a characteristic pollen flora. By identifying the pollen from a particular stratigraphic level of a site, its age relative to other levels or sites could be determined through correlation with this sequence. With the development of the Scandinavian varve chronologies by Gerard de Geer and others during this time (see Flint, 1971, for a discussion), approximate ages in years could be assigned to the Blytt-Sernander vegetational stages. Thus pollen assemblage zones, including those from archaeological sites, could be dated with reasonable confidence. The development of radiocarbon dating during the 1950s displaced pollen analysis as the primary dating technique; since then Quaternary palynologists have turned their attention more toward paleoecological investigations. In the past 20 years, with the introduction of techniques for measuring absolute pollen influx, of newer radiometric dating methods, such as the use of ^{210}Pb, and of numerical analysis of pollen data, palynology has evolved from a qualitive to a quantitive method for vegetational reconstruction.

The field of modern palynology now can include not only the study of pollen and spores but also a host of microorganisms that have little relationship to either Quaternary vegetation or archaeology. Many of these forms are marine and have direct application primarily in stratigraphic correlation and petroleum geology. Throughout this chapter references to palynology, pollen analysis, and paleoecology therefore refer specifically to Quaternary-age sediments and plants. Most of the archaeological applications are drawn from North American examples, not as a slight to other areas but as a result of the familiarity and biases of the author.

APPLICATIONS OF POLLEN ANALYSIS TO ARCHAEOLOGY

Palynological studies as a part of archaeological investigations include both intrasite analyses and extrasite reconstructions of regional vegetation and climatic history. Studies within an individual site may include stratigraphic correlations and relative age placement of sedimentary units, seasonality of site occupation, local paleoecological reconstruction, and studies related to subsistence, such as the identification of cultigens and other economically important plants. In addition, the analysis of pollen preserved in prehistoric human fecal matter and the delineation of specialized cultural activities represent new applications of palynology to archaeological site studies.

In North America, pollen analyses of cultural deposits have been most successful in the semiarid environments of Mexico and the southwestern United States. The dry climate of this region facilitates pollen preservation by retarding the destructive effect of the predominately alkaline sediments on the outer layer (exine) of pollen grains. Alkaline conditions degrade pollen even more in the presence of moisture, as these two factors combine to produce a highly oxidizing environment (Tschudy, 1969).

Major archaeological pollen studies in the southwestern United States have dealt with sites ranging in age from Paleo-Indian to late prehistoric and have included syntheses from many sites (Bohrer, 1970; Hall, 1977; Martin and Byers, 1965; Mehringer and Haynes, 1965; Schoenwetter, 1962, 1966, 1970, 1974b). Jelinek (1966) demonstrated correlations between major pollen groups and artifact types at several North American sites, suggesting that the relative quantities of certain artifacts contained within a site might be useful in explaining and correcting the abnormal pollen compositions frequently found in archaeological samples, which result from cultural activities and disturbance. Other uses of pollen data from archaeological sites have included internal dating within a complex site and delimiting areas of specialized activities (Hill and Hevly, 1968). Models of predicted carrying capacity and population equilibrium for prehistoric peoples have also been based in part on pollen data (Zubrow, 1971).

CULTURALLY RELATED PLANTS

The presence of cultigens or wild plants gathered for food or raw material at a site is frequently evidenced only by their pollen grains. In cases where rarely preserved soft plant tissues—flowers, fleshy fruits, leaves, and stems—had been used prehistorically, pollen may provide the only present evidence for their use. The earliest evidence for flowers placed with a human burial is based on highly localized pollen concentrations of brightly colored spring flowers associated with a Neanderthal burial in Iraq (Leroi-Gourhan, 1975). The presence of large amounts of *Cleome* (bee-weed) pollen in archaeological samples from Mesa Verde, Colorado, suggests prehistoric use of this plant (Martin and Byers, 1965), and prehistoric maize fields have been identified by pollen preserved in the sediments of such former fields (Martin and Schoenwetter, 1960). Recently, previously unknown prehistoric agricultural fields, first observed on aerial thermograms as unusual striate surface features, have been confirmed by pollen studies (Berlin et al., 1977).

The effects of prehistoric cultural activities on the contemporary pollen deposition at a site have long been recognized as a problem in archaeological pollen studies, a problem that must be solved by separating the cultural and natural factors that produced these pollen records. The activities of prehistoric peoples in occupying a site often radically altered the local environment by displacing the natural vegetation and encouraging the growth of disturbance (ruderal) plants, such as ragweed. In many places the local tree cover was cleared for building materials, firewood, and agricultural fields, reducing the arborescent pollen component of the associated sediments to an unknown extent. Distinguishing these cultural modifications of the pollen rain from changes brought about by climate and by natural plant succession is one of the main problems in interpreting pollen data from terrestrial samples. In order to attempt to correct for cultural biases, some authors have excluded the known disturbance species, like ragweed, from the pollen data to produce an adjusted pollen sum (Schoenwetter and Eddy, 1964). Others have considered the entire suite of species but attempted to balance the suspected culturally related taxa against more ecologically significant types in making interpretations (Martin and Byers, 1965). Hevly (1981) examined pollen production, dispersal, and preservation in the natural environment as well as pollen source areas as a means of assessing the effects of cultural influences on the pollen record.

Although there is as yet no satisfactory solution to these problems, thus making pollen interpretation still something of an art, many of the cultural effects on pollen deposition at a site can be detected by determining the pollen influx for the species in question. Whereas pollen percentages in a diagram are measures of increase or decrease relative only to those grains identified in the samples, a species' pollen influx is an absolute measure of the number of pollen grains deposited per unit volume of sediment per unit time (Davis, 1966; Davis and Deevey, 1964). The influx value for a particular species will thus fluctuate independently of the other species, and large increases or decreases in pollen of suspected cultural origin will have a higher visibility than in a percentage diagram.

POLLEN IN DESICCATED FECAL REMAINS

Since the first report that pollen grains were preserved in prehistoric human feces (Martin and Sharrock, 1964), this specialized source of archaeologically related pollen has yielded important cultural information. Desiccated fecal remains have been found not only in the semiarid West, but also in dry caves in eastern North America and the Old World. The pollen preserved in human specimens represents plants consumed as food as well as accidental types— pollen grains deposited naturally on food plants that were later ingested— having little or no cultural affiliation. The principal emphases of pollen studies on desiccated fecal remains have been dietary reconstruction and food preferences, seasonality of food consumption, and paleoenvironmental reconstruction (Schoenwetter, 1974a). In addition, pollen from human feces have demonstrated the consumption of flowers and green, leafy parts of wild plants not indicated by the seeds contained in the desiccated feces. The prehistoric inhabitants of Mammoth Cave, Kentucky, apparently ate flowers of dandelion-like plants (Compositae, Liguliflorae) and sweet flag *(Acorus calamus)* (Bryant, 1974a).

In contrast to the abundant dietary information derivable from pollen in desiccated fecal remains, the amount of paleoenvironmental information is less certain. At Hogup Cave in Utah, Kelso (1970) correlated variations in pollen types found in feces with the changing cultural strata over an 8,000-year period. This correlation, however, was related more to changes in cultural activities and food preferences than to changes in the local natural vegetation. Studies of desiccated human fecal remains from Mammoth and Salts caves (Bryant, 1974a; Schoenwetter, 1974a) suggest that the remains contained insufficient environmental pollen for local vegetational reconstruction. The exception may be when a large number of samples are available from a single site. For example, Bryant (1974b) analyzed 43 specimens from a site in southwest Texas and was able to separate the pollen of intentionally ingested plants from that of the normal background, thus permitting limited paleoenvironmental interpretations.

The primary studies attempting to correlate pollen of desiccated fecal remains with the natural local vegetation have been conducted with animal dung, particularly that of large herbivores. This work was spurred by the discovery of the large deposits of fossil ground sloth *(Nothrotheriops shastense)* dung in Rampart Cave in the Grand Canyon (Long, Hansen, and Martin, 1974; Martin, Sabels, and Shutler, 1961) and in Nevada's Gypsum Cave (Laudermilk and Munz, 1934). However, pollen studies of large-animal dung have shown rather convincingly that their pollen composition reflects primarily what was last consumed by the animal and not the regional vegetation (F. King, 1977; Martin, 1961). This same lack of a regionally significant pollen assemblage was also found in pack rat *(Neotoma* sp.) dung preserved in their fossil middens (Van Devender and J. King, 1971). The cemented matrix of the pack rat middens does, however, reflect the regional pollen rain (J. King and Van Devender, 1977). From these studies it appears that the greatest potential for pollen analysis of desiccated fecal remains is in the dietary and cultural information they contain and that their paleoecological significance is limited.

ENVIRONMENTAL RECONSTRUCTION

Quaternary palynology's quintessential problem is climatic history (Faegri and Iversen, 1975), and probably the most important application of pollen analysis to archaeological studies is as a method of reconstructing vegetation and climatic change. The atmospheric-circulation patterns that determine the climate of any given locale affect vast areas. Any climatic change altering the vegetation in one location will be recorded in pollen diagrams from throughout the affected region. For example, the changes in Holocene climate that resulted in the development of the Prairie Peninsula are recorded in pollen diagrams from Iowa (Van Zant, 1976), Missouri (J. King and Allen, 1977), Illinois (J. King, 1981), and elsewhere in midcontinental North America (Wright, 1968). By correlating vegetational successions on a regional basis, their interrelationships with changing cultural adaptation become apparent. Wright (1976), through such an analysis of climatic events of the late Pleistocene and early Holocene from Near Eastern and Mediterranean pollen diagrams, has suggested a probable time for the domestication of the wild cereal grains.

The replacement of the late Pleistocene spruce forests in eastern North America by the oak-dominated forests of the Holocene illustrates another feature of vegetational and climatic changes directly relevant to the archaeological perspective: that they are frequently time-transgressive (see maps by Bernabo and Webb, 1977). Environmental change affecting one site may occur at a slightly different time from its occurrence at another site, or it may be expressed unevenly in the palynological record. This is especially true when looking at sites from different ecological situations, such as uplands versus river valleys.

Palynology's greatest contribution to archaeology, as well as to paleoecology, is expressed in the stratigraphically continuous pollen records that have been recovered from natural sedimentary basins, such as lakes, ponds, and bogs. Such basins have remained free of natural and cultural disturbance and the pollen diagrams produced from these sedimentary records provide the vegetational history against which other lines of evidence—plant macrofossil, faunal, geochemical, and artifactual—can be compared and interpreted. The basic strength of these data lies in the large number of samples having high pollen counts and numerous taxa, all correlated by radiocarbon and stratigraphic controls. Records from these samples are replicable, in contrast to the low number, small size, and few taxa that are characteristic of many archaeological pollen-sample suites. It is probably a fair assessment that most Quaternary palynologists put little faith in these kinds of samples. However, if archaeological pollen data can be correlated with regional, stratigraphically continuous pollen chronologies, then they may take on paleoenvironmental significance.

A sufficient number of pollen diagrams is now available from most of North America, Europe, and the Near East so that the major outline of paleoenvironmental change in these regions is known and its probable effects upon cultural history can be assessed (Wright, 1977). Although local influences and conditions affecting any one archaeological site must be accounted for, the regional climate, through its effect upon vegetation, controls the limits within which the biotic as well as cultural communities exist. Recent studies on prehistoric subsistence demonstrate the abilities of people to alter their life-ways to accommodate environmental shifts (Smith, 1978; Wood and McMillan, 1976). The large midwestern river valleys of the United States, with their multitude of resources and ecological niches, provide an example of a fertile area for this type of archaeological investigation. The challenge for the paleoecologist is to refine the paleoenvironmental models in order to identify the microhabitats exploited by prehistoric peoples. A good illustration is the separation of the midwestern upland floras, usually composed of prairie and species-poor forests, from the mesic bottomland forests often dominated by the same tree species but with much greater floral diversity. These bottomland river valley habitats may have continued producing the resources necessary to support human settlement during periods of unstable climate, when the uplands no longer could. Such models are refined primarily through the analysis of stratigraphically continuous pollen samples from natural sedimentary basins. When regional paleoenvironmental history is developed in conjunction with the archaeological records, their interrelationship can greatly enhance the final environmental reconstruction (Bryant, 1969; Hall, 1977; Martin, 1963; Mehringer, 1967; Wendorf, 1961; Wendorf and Hester, 1975; and others). Principal-component analysis, a multivariate statistical technique, has been used successfully with a large suite of pollen samples from Antelope House ruin in Arizona (Fall, Kelso, and Markgraf, 1981). Climatic and paleoenvironmental influences were separated from culturally induced changes by identifying groups of covariant taxa. This technique should find wide application at sites with large palynological data bases.

Another contribution of pollen analysis to archaeology has been in documenting floral disturbance by prehistoric human settlement practices. The extent of such disturbance as reflected in pollen diagrams was first recognized in Denmark by Iversen (1941), who noted increases in early successional trees, such as birch, alder, and hazel, following the appearance of cereal-type grasses and herbaceous weedy species. This pattern was apparent in numerous pollen diagrams, and, when combined with the archaeological evidence, indicated that Danish Neolithic farmers practiced a sequence of forest clearance, grazing, cultivation, and abandonment. Iversen applied the name *landnam*, after the Danish term for land settlement, to this phenomenon. Palynology thus confirmed the historically documented practice of primitive slash-and-burn agriculture in a prehistoric setting. The landnam sequence has been documented elsewhere in Europe and radiocarbon-dated to about 4,000 years B.P. (Pilcher et al., 1971).

Although numerous archaeological sites in North America are extensive in size, and some, like the Mississipian-period city-state at Cahokia (near St. Louis, Missouri), may have contained as many as 50,000 inhabitants about A.D. 1000, the natural record of what must have been extensive environmental modifications by prehistoric people is generally incomplete. As an example of what can be done given the proper situation, however, recent pollen studies from small lakes in southern Ontario have established a North American equivalent of landnam practiced by the late-prehistoric Iroquois (McAndrews, 1976). These pollen diagrams record a sequence of forest clearance and Indian agriculture followed by abandonment and forest succession in the area surrounding the lakes. As was the case in the European experience, the species involved in the Ontario landnam are the forest trees and small numbers of herbaceous weeds. The main characteristic of the pollen record is the indication of renewed forest succession following an apparent decline in forest trees. The presence of *Zea mays* pollen indicates that maize agriculture was practiced at one of these sites. Interestingly, prehistoric Iroquois agriculture is not accompanied by an increase in ragweed pollen such as is associated with the introduction of large-scale land clearance and agriculture in the later Euro-American period.

For the historic period, many pollen studies document Euro-American alteration of the landscape not only of local areas (Davis, Brubaker, and Beiswenger, 1971; Janssen, 1967; Brugam, 1978), but also of large drainage basins like those emptying into Lake Superior (Maher, 1977) and Lake Michigan (J. King, Lineback, and Gross, 1976). The effects of modern human disturbance are also recorded in lake sediments by the changes in microorganisms, such as diatoms and cladocera (Bradbury and Waddington, 1973).

PROBLEMS IN ARCHAEOLOGICAL POLLEN ANALYSES

As previously noted, pollen investigations of American archaeological sites have been most successful either in the dry Southwest or in unusual situations like the Itasca, Minnesota, bison kill site (Shay, 1971), unique in that it was located in a bog. The reasons for this are twofold: the mechanics of pollen deposition and the environments of pollen preservation. Clearly, in order for pollen to be present in a sample it must have been deposited in the sediment initially and then have been preserved until the present. Many, if not most, pollen samples from archaeological sites do not meet these two criteria; this is especially true in humid eastern North America (J. King, Klippel, and Duffield, 1975).

Although pollen grains are relatively resistant to natural decay, they can be destroyed by oxidation processes and rapidly digested by soil microbes. These are the main factors responsible for pollen destruction. Unfortunately, during deposition the surfaces of archaeological sites were generally oxidizing environments, so little or no pollen was preserved. In addition, such deposits are often subjected to the destructive factors of oxidation or microbial action at some time between site abandonment and the present. As a result, whatever pollen accumulated in these cultural contexts often has not survived. Another major factor is that many of the cultural features often sampled for pollen—floors and refuse pits, for example—were exposed to active pollen fallout from the atmosphere for too brief a period during their construction or use and as a result accumulated little or no pollen.

This discouraging assessment, however, should not deter pollen research at archaeological sites. Rather it should make us all think carefully about what we are trying to achieve and how best to accomplish it. Sampling contexts that have frequently been productive are those special situations in which pollen has been preserved with cultural remains due to unusual depositional circumstances. Sediment in proximity to copper artifacts, tree bark, deeply buried trash pits, the contents of sealed containers, leather bundles, prehistoric dirt piles, and other unique microenvironments have all shown the potential to yield palynological information. Undoubtedly, many more of these special situations exist and need only to be ferreted out.

In pollen research, as in any type of archaeological study, the method of investigation should be determined by the questions under consideration. Questions concerning the regional vegetation and its history will be better answered by stratigraphic pollen studies, whereas questions pertaining to intrasite problems, such as presence of economic plants and depositional history, may be addressed by site-oriented studies if suitable pollen-bearing sediments are present. One of the greatest uses of pollen analyses is in allowing a better understanding of the relationship between humans and their changing environments. The following example is but one of many that emphasizes this relationship.

PREHISTORIC SETTLEMENT AND PRAIRIE DEVELOPMENT IN CENTRAL ILLINOIS

The archaeological evidence from central Illinois suggests that prehistoric peoples occupied and utilized the prairie areas much less than the forested valleys along the major rivers. These major river valleys and valley edges have been the location of major human occupations throughout much of the Holocene. The archaeological record for the upland areas, on the other hand, indicates predominantly short-term or special-purpose occupations, especially following the early archaic period, which ended approximately 8,000 years ago.

The Illinois State Museum has been conducting archaeological surveys in central Illinois for the past decade. These surveys, centered on the Sangamon River drainage basin, have included both forested and prairie-covered river bottoms and upland areas. One small tributary of the Sangamon River, Willow Branch, was chosen for a detailed survey, and the results yielded an interesting pattern of prehistoric settlement (Klippel and Maddox, 1977).

Willow Branch is a small third-order stream (Strahler, 1957) with a drainage basin of approximately 150 km^2. It originates on the rolling Shelbyville moraine near the southern limit of the Wisconsinan glaciation in central Illinois (Miller, 1973; Willman and Frye, 1970). The stream flows roughly southwest across the flat outwash plain, then into the low relief of the Illinoian end moraine before entering the Sangamon River (fig. 6.1). The local relief along Willow Branch rarely exceeds 12 m and the valley slopes are less than 9 m above the floodplain. There is no terrace system along the stream and the rolling uplands slope gently to the valley floor. The upper reaches of Willow Branch flow intermittently, as do most of its small tributaries.

The Willow Branch drainage basin is located in the Grand Prairie division of Illinois (Schwegman, 1973), that vast expanse of prairie that occupied much of Illinois at the time of the Euro-American settlement (fig. 6.2). During the early nineteenth century, the basin was occupied entirely by mesic tall-grass prairie (Braun, 1950). The only upland forest along Willow Branch recorded by the United States General Land Office survey in 1819–1823 was near its confluence with the Sangamon River, which itself supported rich bottomland forest (F. King and Johnson, 1977). Willow Branch was a small stream flowing entirely through

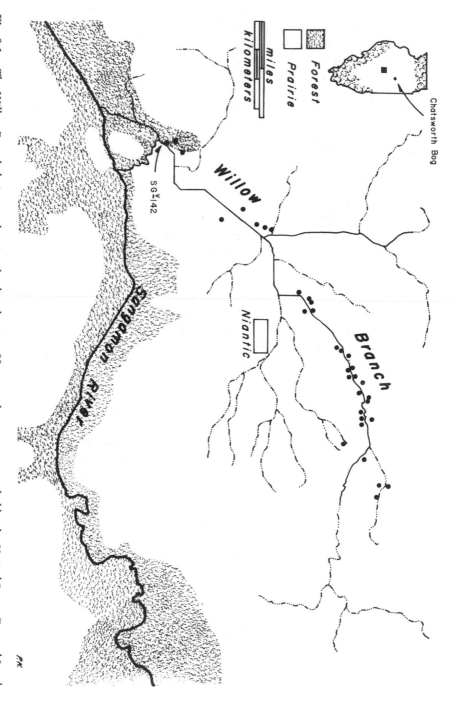

Fig. 6.1. The Willow Branch drainage, showing the distribution of forest and prairie as recorded by the United States General Land Office Survey (1819–1823). The dots represent the location of known archaeological sites (modified from Klippel and Maddox, 1977, by permission of the Kent State University Press).

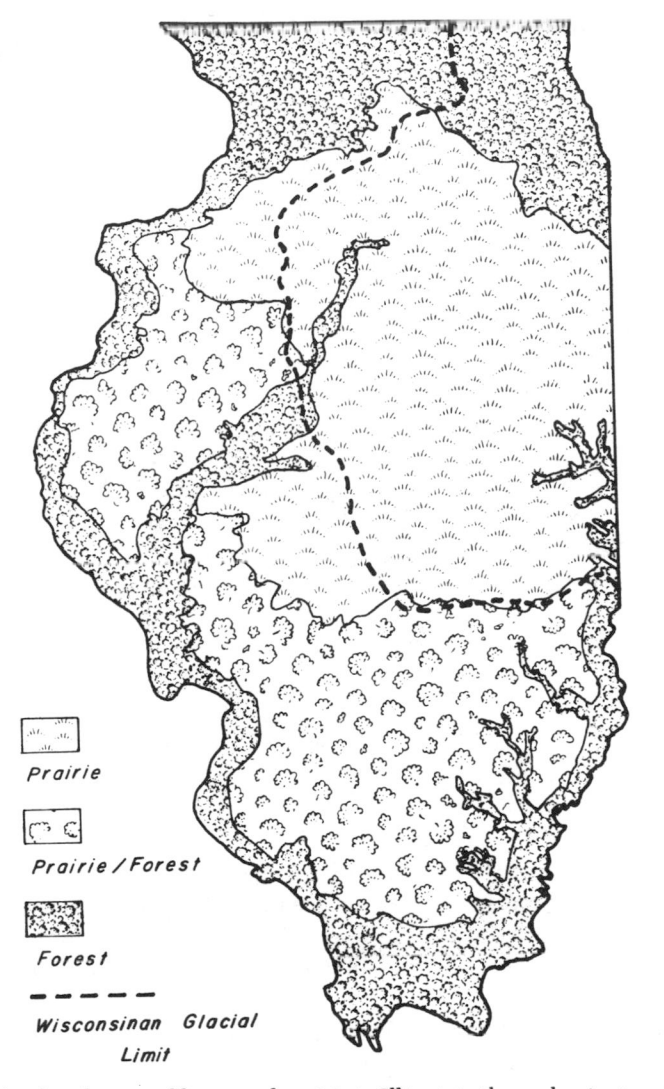

Fig. 6.2. The distribution of forest and prairie in Illinois in the early nineteenth century (modified from Schwegman, 1973).

the flat to gently rolling prairie and encountering forest only at its extreme lower end. Today, the prairie is gone and the land is farmed to the edge of the stream; no areas of riparian vegetation remain and as a result nothing obscured the archaeological surveys that were conducted along Willow Branch during the winter and early spring, when the ground was bare.

The accounts of the early travelers and settlers in central Illinois portray a harsh environment: water-saturated ground throughout the spring and early summer, followed in late summer by drought and awesome prairie fires (Winsor, 1975; Vestal, 1939). In winter the absence of trees meant no wood for cabins or fires and no protection from the cold winds and drifting snow. In summer, stifling heat and malaria were common.

The Willow Branch survey revealed 33 surface sites, almost all of which were located on the summits and shoulders of the gently sloping valley (Klippel and Maddox, 1977). With only one exception, the recovered artifact assemblages indicate that the recorded sites are early Archaic, or more than about 8,000 years old. The exceptional site, SG⌣142, was located near the confluence of Willow Branch and the Sangamon River in the area of nineteenth-century forest (fig. 6.1) and contained artifacts representing younger cultural periods. Projectile points from SG⌣142 included Paleo-Indian through late-Woodland/Mississippian specimens. Although specific dates are sometimes difficult to assign to cultural periods, the early Archaic in central Illinois probably extended from about 10,000 to 8,000 years ago. Thus, on the basis of the survey results, it appears that prehistoric groups no longer extensively inhabited the Willow Branch drainage after about 8,000 years ago. The same pattern of widespread early-Archaic occupations followed by restricted distribution has been documented in other systematic surveys by the museum elsewhere in central Illinois.

CHATSWORTH BOG

Chatsworth Bog is located in what was Grand Prairie on the gently rolling, Wisconsinan-age land surface about 120 km northeast of Willow Branch in Livingston County, Illinois (fig. 6.1). The bog occupies a depression within a late Wisconsinan outwash channel that flowed from the Chatsworth moraine a few kilometers to the northeast; it is approximately 30 acres in size. About half of the area of the bog has been dredged for marl. Although the bog supported a riparian community and was surrounded by a thin edge of forest in the nineteenth century, the primary vegetation of the area was wet prairie, very similar to that of Willow Branch.

Sediments from Chatsworth Bog were originally analyzed for pollen by Voss (1937), who reported abundant spruce (Picea) pollen at a depth of 11 m in the marly sediments. In 1977 a continuous core of bog sediments was collected by the author for pollen analysis, using a modified Livingston-piston sampler 5 cm in diameter. A total of 12.7 m of sediment was recovered; the base of the core was radiocarbon-dated at over 14,000 years before the present. The stratigraphic sequence—clay, gyttja, marls, and surface peat—indicates that the sediments have accumulated under different sedimentary regimes.

The summary pollen diagram from Chatsworth Bog (fig. 6.3) reveals the changes in vegetation that have occurred in central Illinois since the late Pleistocene. It is divided into four pollen assemblage zones corresponding to stages in vegetational succession. Picea pollen dominates the lowest portion of the core (zone I), reflecting the late-Pleistocene spruce forests that were ubiquitous throughout eastern and central North America. Spruce-pollen percentages as high as 76 percent indicate a forest dominated by spruce trees almost to the exclusion of other species. This spruce maximum at a 12.6 m depth is dated at 14,380 ± 150 years ago (ISGS–527). By a depth of 12.2 m, spruce declines abruptly to approximately 10 percent of the total pollen, indicating that the climate had warmed so much that spruce could no longer compete successfully with other species. Zone I ends about 13,800 years ago.

Zone II follows the spruce zone; it is dominated by ash *(Fraxinus)* but also includes the beginnings of the oak *(Quercus)* increase. The peak in ash pollen in zone II apparently reflects an early successional stage during a period of climatic amelioration in which ash replaced spruce around the bog. Pollen-grain morphology indicates that the ash belonged to the subgenus Bumelioides, which includes black ash *(Fraxinus nigra)*. The present habitat of black ash is rich woods, moist bottom lands, and valleys (Braun, 1950); it is also found in peat bogs (Fowells, 1965). Zone II lasts until about 12,000 years ago, when ash declines and oak, elm *(Ulmus)*, and alder *(Alnus)* become prominent. This change marks the transition to the third zone.

Pollen assemblage zone III is dominated by mesic deciduous species and encompasses the period during which the oak-dominated deciduous forest, characteristic of the remainder of the midwestern Holocene, developed. The last of the spruce pollen disappears from the Chatsworth pollen record at this time. Although spruce was generally absent from central Illinois after 13,800 years ago, it remained in northern Illinois until about 10,700 years ago (J. King, 1981). Zone III can be further subdivided into an early alder and birch *(Betula)* phase and a later elm and ironwood *(Ostrya/Carpinus)* phase. During this later phase, hickory *(Carya)* also increases and remains prominent for the rest of the sequence. These phases reflect the continued climatic warming and drying that started with the end of Pleistocene glaciation. The continued increase in temperature and decrease in effective precipitation culminated in the subtle but highly significant changes in pollen frequency recorded at 9.6 m depth, the beginning of zone IV.

With the transition to zone IV, there is an abrupt decline in elm and ironwood and an increase in ragweed *(Ambrosia)*. Within the next 40 cm, the sunflower group *(Tubuliflorae)*, the grasses, and the chenopods *(Chenopodiaceae)* increase. This shift to grass and herbs is dated at 8,300 ± 100 years before the present (ISGS–519) and marks the replacement of the mesic deciduous forest in central Illinois by prairie. Pollen zone IV is the palynological expression of the climatic change that initiated the development of the Prairie Peninsula. Comparable shifts from forest to prairie species are seen in pollen diagrams from Iowa (Van Zant, 1976), Wisconsin (Webb and Bryson, 1972), Minnesota (Wright, Winter, and Patten, 1963), and southeastern Missouri (J. King and Allen, 1977). Although oak continues to dominate the pollen in zone IV, it no longer dominates the vegetation. White ash *(Fraxinus americana)*, a more temperate species, replaces *F. nigra* as the dominant ash species because of the effectively drier environment.

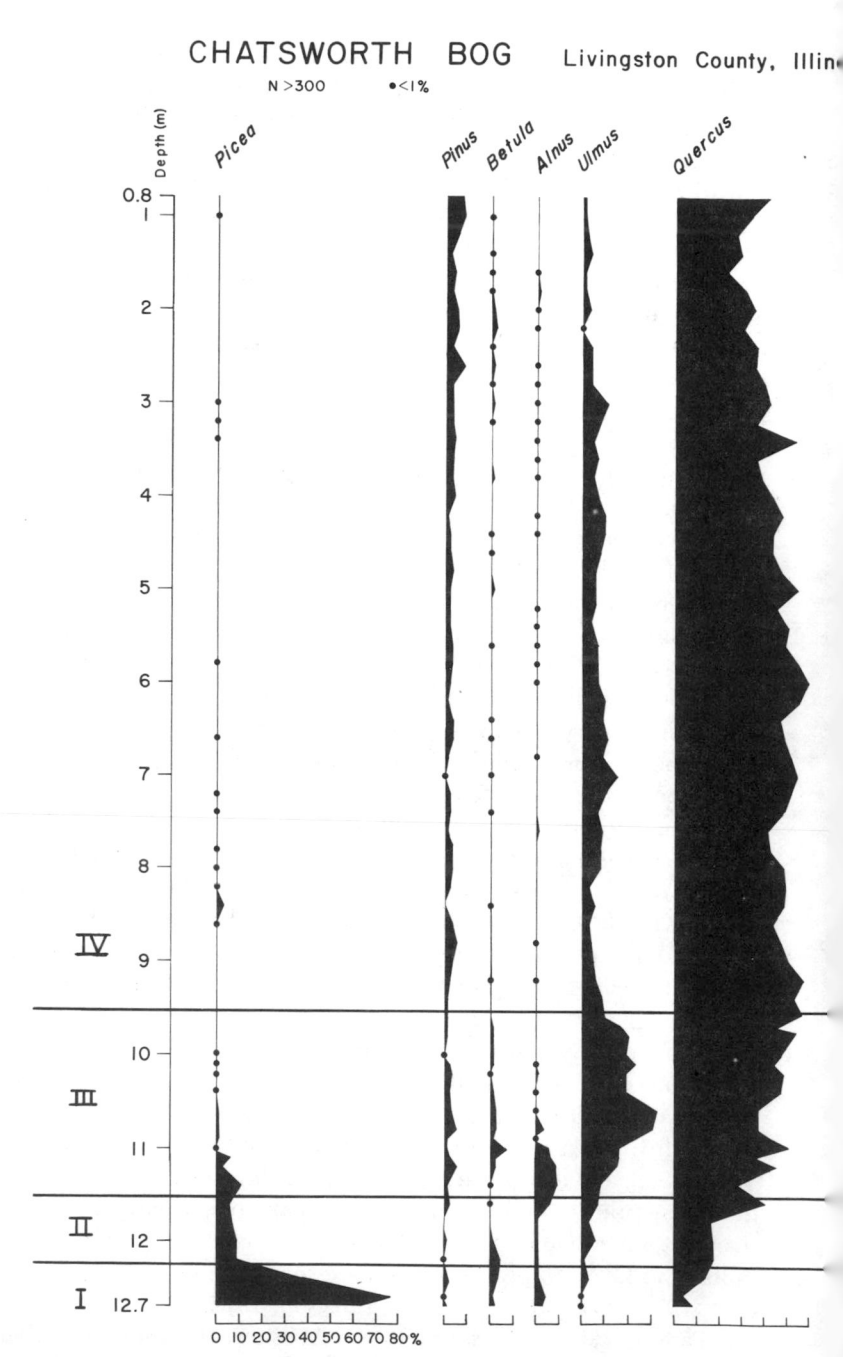

Fig. 6.3. Summary pollen diagram from Chatsworth Bog, Illinois. Individual taxa are shown as a percentage of the total pollen exluding aquatic species (adapted from J. King, 1981).

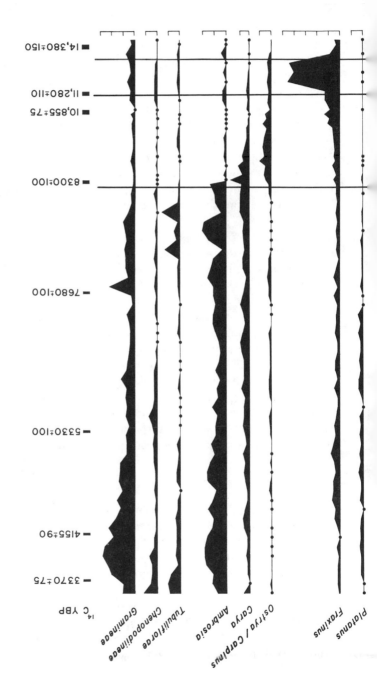

Herbaceous species produce pollen in much smaller quantities than arboreal forms, thus their small increases are of greater significance than the higher tree-pollen percentages. Once ragweed and grass attain their high percentages in zone IV, they remain at these levels to the top of the sequence, indicating that once prairie developed in central Illinois it remained the dominant vegetation type to the present.

Zone IV extends to the top of the pollen diagram, and although some subzones can be defined within it, they will represent only variations of the dominant prairie vegetation. This pollen diagram ends at an 80 cm depth, about 3,000 years ago. The upper section, as well as the complete diagram containing all identified plant taxa, has appeared elsewhere (J. King, 1981).

PALEOECOLOGY OF WILLOW BRANCH

As the late-Pleistocene spruce forest collapsed around Chatsworth Bog about 13,000 to 14,000 years ago, it was followed in turn by an intermediate successional stage dominated by black ash, then by a mixture of mesic deciduous trees about 11,500 years ago. This mesic arboreal flora exhibits an early wet phase followed by a later dry phase as early maxima of alder, birch, and pine are replaced by elm, ironwood/hornbeam, and hickory. By 8,300 years ago, this warming and drying trend had progressed sufficiently to initiate the replacement of the upland forests with herbaceous prairie species and grass. As the prairie developed, the forest retreated to the larger, more mesic river valleys, leaving vast expanses of central Illinois as treeless prairie. Small streams with no appreciable gradient lost their gallery forests. The prairie elements at Chatsworth Bog continued to increase after 8,300 years ago as climatic conditions became progressively warmer and drier. The pollen from Illinois and throughout the Midwest indicates that this warm, dry climate trend peaked about 7,000 years ago (Wright, 1968).

The archaeological data from the Willow Branch survey indicate that starting about 8,000 years ago, coinciding with the decline of the upland forests, the stream was no longer the site of aboriginal occupation. Klippel and Maddox (1977) suggest that climatic change played a crucial role in this cultural shift. The pollen record from Chatsworth Bog confirms their interpretation and provides data on the paleoenvironmental changes that occurred. As the Prairie Peninsula was developing just prior to 8,000 years ago, the vegetation of the glacially formed flat uplands and gently rolling landscape was shifting from forest to prairie. Willow Branch, lacking the protective relief or the water volume of the larger river valleys, became a small stream in the prairie. As its catchment basin lost tree cover, stream flow was probably reduced and many of the animals that depended on the stream and its surrounding forest may have largely disappeared from the area. With the protective cover of the forest gone, the full force of the climate would have been felt more severely. Thus, Willow Branch probably was no longer as hospitable a place for people or animals as it had been before prairie development.

Throughout prehistory, people have utilized and modified the environments in which they lived. Early human dependence on hunting of animals gave way to increased utilization of plants, first by the gathering of wild species and then by horticulture. Pollen analysis can provide information on this relationship from a local-site as well as a regional perspective. Site-specific pollen studies can be of value if suitable pollen-bearing sediments are present, but palynology's main contribution will continue to be in the detailed description of regional vegetational history and related climatic change that it provides. This natural history forms a background for investigating both site-specific questions and larger cultural/settlement problems.

Acknowledgments

Many of the ideas presented here are the result of numerous discussions with Walter E. Klippel, R. Bruce McMillan, Bonnie W. Styles, and Frances B. King; I gratefully acknowledge their help. The palynological research at Chatsworth Bog was supported by National Science Foundation grant #DEB76−20126.

REFERENCES

Berlin, G. L., J. R. Ambler, R. H. Hevly, and G. G. Schaber. 1977. Identification of a Sinagua agricultural field by aerial thermography, soil chemistry, pollen/plant analysis, and archaeology. *American Antiquity* 42:588−600.

Bernabo, J. C., and T. Webb, III. 1977. Changing patterns in the Holocene pollen record of northeastern North America: A mapped summary. *Quaternary Research* 8:64−96.

Bohrer, V. L. 1970. Ethnobotanical aspects of Snaketown, a Hohokam village in southern Arizona. *American Antiquity* 35:413−30.

Bradbury, J. P., and J. C. B. Waddington. 1973. The impact of European settlement on Shagawa Lake, northeastern Minnesota, U.S.A. In *Quaternary plant ecology*, ed. H. J. B. Birks and R. G. West, 289−307. New York: John Wiley and Sons.

Braun, E. L. 1950. *Deciduous forests of eastern North America.* Philadelphia: Blakiston.

Brugam, R. B. 1978. Pollen indicators of land-use change in southern Connecticut. *Quaternary Research* 9:349−62.

Bryant, V. M., Jr. 1969. Late full-glacial and postglacial pollen analysis of Texas sediments. Ph.D. thesis, Univ. of Texas, Austin.

————. 1974a. Pollen analysis of prehistoric human feces from Mammoth Cave. In *Archeology of the Mammoth Cave area*, ed. P. J. Watson, 203−09. New York: Academic Press.

————. 1974b. Prehistoric diet in southwest Texas: The coprolite evidence. *American Antiquity* 39:407−20.

Davis, M. B. 1966. Determination of absolute pollen frequency. *Ecology* 14:310−13.

Davis, M. B., and E. S. Deevey, Jr. 1964. Pollen accumulation rates: Estimates from late-glacial sediments of Rogers Lake. *Science* 145:1293−95.

Davis M. B., L. B. Brubaker, and J. M. Beiswenger. 1971. Pollen grains in lake sediments: Pollen percentages in surface sediments from southern Michigan. *Quaternary Research* 1:450−67.

152 / *James E. King*

Erdtman, G. 1943. *An introduction to pollen analysis.* Waltham, MA: Chronica Botanica.
———. 1969. *Handbook of palynology.* New York: Hafner.
Faegri, K., and J. Iversen. 1975. *Textbook of pollen analysis.* Copenhagen: Munksgaard.
Fall, P. L., G. Kelso, and V. Markgraf. 1981. Paleoenvironmental reconstruction at Canyon del Muerto, Arizona, based on principal-component analysis. *Journal of Archaeological Science* 8:297–307.
Flint, R. F. 1971. *Glacial and Quaternary geology.* New York: John Wiley and Sons.
Fowells, H. A. 1965. *Silvics of forest trees of the United States.* Agricultural Handbook, no. 271. Washington, DC: U.S.D.A. Forest Service.
Gray, M., R. McAfee, Jr., and C. L. Wolf, eds. 1974. *Glossary of geology.* Washington, DC: American Geological Institute.
Hall, S. A. 1977. Late Quaternary sedimentation and paleoecologic history of Chaco Canyon, New Mexico. *Bulletin of the Geological Society of America* 88:1593–1618.
Hevly, R. H. 1981. Pollen production, transport and preservation: Potentials and limitations in archaeological palynology. *Journal of Ethnobiology* 1:39–54.
Hill, J. N., and R. H. Hevly. 1968. Pollen at Broken K Pueblo: Some new interpretations. *American Antiquity* 33:200–10.
Iversen, J. 1941. Land occupation in Denmark's stone age. *Danmarks Geologiske Undersøgelse*, 2 Raekke, nr. 66.
Janssen, C. R. 1967. A postglacial pollen diagram from a small *Typha* swamp in northwestern Minnesota, interpreted from pollen indicators and surface samples. *Ecological Monographs* 37:145–72.
———. 1973. Local and regional pollen deposition. In *Quaternary plant ecology*, ed. H. J. B. Birks and R. G. West, 31–42. New York: John Wiley and Sons.
Jelinek, A. J. 1966. Correlation of archaeological and palynological data. *Science* 152:1507–09.
Kelso, G. 1970. Hogup Cave, Utah: Comparative pollen analysis of human coprolites and cave fill. In *Hogup Cave*, ed. C. M. Aitkens, 251–62. Salt Lake City: University of Utah, Anthropological Papers, no. 93.
King, F. B. 1977. An evaluation of the pollen contents of coprolites as environmental indicators. *Journal of the Arizona Academy of Science* 12:47–52.
King, F. B. and J. B. Johnson. 1977. Presettlement forest composition of the central Sangamon River basin, Illinois. *Transactions of the Illinois State Academy of Science* 70:153–63.
King, J. E. 1981. Late Quaternary vegetational history of Illinois. *Ecological Monographs* 51:43–62.
King, J. E., W. E. Klippel, and R. Duffield. 1975. Pollen preservation and archaeology in eastern North America. *American Antiquity* 40:180–90.
King, J. E., J. A. Lineback, and D. L. Gross. 1976. *Palynology and sedimentology of Holocene deposits in southern Lake Michigan.* Illinois State Geological Survey, Circular no. 496.
King, J. E., and W. H. Allen, Jr. 1977. A Holocene vegetation record from the Mississippi River valley, southeastern Missouri. *Quaternary Research* 8:307–23.
King, J. E., and T. R. Van Devender. 1977. Pollen analysis of fossil pack-rat middens from the Sonoran Desert. *Quaternary Research* 8:191–204.
Klippel, W. E., and J. Maddox. 1977. The early Archaic of Willow Branch. *Mid-Continental Journal of Archaeology* 2:99–130.

Laudermilk, J. D., and P. A. Munz. 1934. *Plants in the dung of Nothrotherium from Gypsum Cave, Nevada*. Washington, DC: Carnegie Institute Publication no. 453.

Leroi-Gourhan, A. 1975. The flowers found with Shanidar IV, a Neanderthal burial in Iraq. *Science* 190:562–64.

Long, A., R. M. Hansen, and P. S. Martin. 1974. The extinction of the Shasta ground sloth. *Bulletin of the Geological Society of America* 85:1843–48.

McAndrews, J. H. 1976. Fossil history of man's impact on the Canadian flora: An example from southern Ontario. *Supplement to the Canadian Botanical Association Bulletin* 9:1–6.

Maher, L. J., Jr. 1977. Palynological studies in the western arm of Lake Superior. *Quaternary Research* 7:14–44.

Martin, P. S. 1961. Pollen analysis of coprolites. In *A survey and excavation of caves in Hidalgo County, New Mexico*, ed. M. F. Laubert and J. R. Ambler, 101–04. Santa Fe: School of American Research, Monograph, no. 25.

———. 1963. *The last 10,000 years: A fossil pollen record of the American Southwest*. Tucson: Univ. of Arizona Press.

Martin, P. S., and J. Schoenwetter, 1960. Arizona's oldest cornfield. *Science* 132:33–34.

Martin, P. S., B. E. Sabels, and D. Shutler, Jr. 1961. Rampart Cave coprolites and ecology of the Shasta ground sloth. *American Journal of Science* 259:102–27.

Martin, P. S. and F. W. Sharrock. 1964. Pollen analysis of prehistoric human feces: A new approach to ethnobotany. *American Antiquity* 30:168–80.

Martin, P. S., and W. Byers. 1965. Pollen and archaeology at Wetherill Mesa. *American Antiquity* 31:122–35.

Mehringer, P. J., Jr. 1967. Pollen analysis of the Tule Springs site, Nevada. In *Pleistocene studies in southern Nevada*, ed. H. M. Wormington, and D. Ellis, 129–200. Carson City: Nevada State Museum Anthropological Papers, no. 13.

Mehringer, P. J., Jr., and C. V. Haynes, Jr. 1965. The pollen evidence for the environment of early man and extinct mammals at the Lehner Mammoth site, southeastern Arizona. *American Antiquity* 31:17–23.

Miller, J. A. 1973. *Quaternary history of the Sangamon River drainage system in central Illinois*. Springfield: Illinois State Museum Reports of Investigations, no. 27.

Pilcher, J. R., A. G. Smith, G. W. Pearson, and A. Crowder. 1971. Land clearance in the Irish Neolithic. *Science* 172:560–62.

Schoenwetter, J. 1962. The pollen analysis of eighteen archaeological sites in Arizona and New Mexico. *Fieldiana, Anthropology* 53:168–209.

———. 1966. A re-evaluation of the Navajo Reservoir pollen chronology. *El Palacio* 73:19–26.

———. 1970. Archaeological pollen studies of the Colorado Plateau. *American Antiquity* 35:35–48.

———. 1974a. Pollen analysis of human paleofeces from Upper Salts Cave. *Archeology of the Mammoth Cave area*, ed. P. J. Watson, 49–58. New York: Academic Press.

———. 1974b. Pollen records of Guila Naquitz Cave. *American Antiquity* 39:292–303.

Schoenwetter, J., and F. W. Eddy. 1964. *Alluvial and palynological reconstruction of environments, Navajo Reservoir District*. Santa Fe: Museum of New Mexico Press.

Schwegman, J. E. 1973. *Comprehensive plan for the Illinois Nature Preserves System. Part 2: The Natural Divisions of Illinois*. Rockford, IL: Illinois Nature Preserves Commission.

Shay, C. T. 1971. *The Itasca bison kill site: an ecological analysis.* St. Paul: Minnesota Historical Society.

Smith, B. D., ed. 1978. *Mississippian settlement systems.* New York: Academic Press.

Strahler, A. N. 1957. Quantitative analysis of watershed geomorphology. *American Geophysical Union Transactions* 38:913−20.

Tschudy, R. H. 1969. Relationship of palynomorphs to sedimentation. In *Aspects of palynology,* ed. R. H. Tschudy and R. A. Scott, 79−96. New York: John Wiley and Sons.

Van Devender, T. R., and J. E. King. 1971. Late Pleistocene vegetational records in western Arizona. *Journal of the Arizona Academy of Science* 6:240−44.

Van Zant, K. L. 1976. Late- and postglacial vegetational history of northern Iowa. Ph. D. thesis, Univ. of Iowa, Iowa City.

Vestal, A. G. 1939. Why the Illinois settlers chose forest lands. *Transactions of the Illinois State Academy of Science* 32:85−87.

Voss, J. 1937. Comparative study of bogs on Cary and Tazewell drift in Illinois. *Ecology* 18:119−35.

Webb, T., III, and R. A. Bryson. 1972. Late- and postglacial climatic change in the northern Midwest, U.S.A.: Quantitative estimates derived from fossil pollen spectra by multivariate statistical analysis. *Quaternary Research* 2:70−115.

Wendorf, F., ed. 1961. *Paleoecology of the Llano Estacado.* Santa Fe: Museum of New Mexico Press.

Wendorf, F., and J. J. Hester. 1975. Late Pleistocene environments of the southern high plains. Taos, NM: Fort Burgwin Research Center, Publication no. 9.

Willman, H. B., and J. C. Frye. 1970. *Pleistocene stratigraphy of Illinois.* Illinois State Geological Survey, Bulletin no. 94.

Winsor, R. A. 1975. Artificial drainage of east central Illinois. Ph. D. thesis, University of Illinois, Urbana.

Wood, W. R., and R. B. McMillan, eds. 1976. *Prehistoric man and his environments: A case study in the Ozark Highland.* New York: Academic Press.

Wright, H. E., Jr. 1968. History of the Prairie Peninsula. In *The Quaternary of Illinois,* ed. R. E. Bergstrom, 78−88. Urbana: Univ. of Illinois College of Agriculture, Special Publication no. 14.

———. 1976. The environmental setting for plant domestication in the Near East. *Science* 194:385−89.

———. 1977. Environmental change and the origin of agriculture in the Old and New Worlds. In *Origins of agriculture,* ed. C. A. Reed, 281−318. The Hague: Mouton.

Wright, H. E., Jr., T. C. Winter, and H. L. Patten. 1963. Two pollen diagrams from southeastern Minnesota: Problems in regional late-glacial and postglacial vegetational history. *Bulletin of the Geological Society of America* 74:1371−96.

Zubrow, E. B. 1971. Carrying capacity and dynamic equilibrium in the prehistoric Southwest. *American Antiquity* 36:127−38.

7

Theoretical and Practical Considerations in the Analysis of Anthrosols

ROBERT C. EIDT

ABSTRACT

Anthrosols are soils whose features have been altered by human activities, primarily through settlement and agricultural practices. The degree of potential alteration relates to both intensity and duration of settlement. Plant nutrients, organic matter, oxidation-reduction characteristics, pH values, and exotic contaminants are chemical soil factors reflecting human activities that can be quantified. Two physical changes that may be apparent in anthrosols are soil movement (excavation and deposition) and gross characteristics (color, texture, structure, and density).

To analyze anthrosols in an archaeologically useful manner, analyses should be made of nearby natural-soil profiles or from similar but known use areas for comparison. Field studies, plus a knowledge of soil-forming processes and soil-classification schemes, are necessary to obtain such profiles. Chemical analyses designed specifically for anthrosols—particularly phosphate analysis—can be applied both qualitatively in the field and quantitatively in the laboratory through a fractionation procedure.

The kinds of soil changes induced by humans reflect the nature of their activities. Plow farmers alter soils in different ways from those where similar numbers of people engaged only in hunting, or only in fishing, trading, livestock grazing, or industry. Anthrosols may be subdivided into *anthropogenic soils*, which have been intentionally altered, and *anthropic soils*, those changed unintentionally. In learning how to interpret the effects of different human activities on soil change, it is important to understand first how these changes occur. In principle, human-induced soil alterations can be brought about either by physical or by chemical processes. Since physical processes are always associated with chemical change, it is often difficult to separate the two, especially after time diminishes obvious physical evidence. Aside from physical changes caused by accelerated erosion or intentional soil transport, chemically induced changes represent the more dominant, longer lasting processes.

CHEMICAL SOIL CHANGES

Soils may be chemically affected in unimportant ways by human activities, but for practical purposes, we are interested in alterations significant enough to affect any of the following five soil factors: (1) macro- or micro- plant nutrients; (2) organic matter; (3) oxidation-reduction characteristics; (4) pH values; and (5) exotic (usually industrial) contaminants. These five important chemical factors of soil change can be measured and compared in any small laboratory with relatively inexpensive equipment.

MACRO- AND MICRONUTRIENTS

The terms *macro-* and *micronutrients* are most obviously relevant in agricultural situations. However, human activities even in nonfarming settlements alter the levels of the plant nutrients. The same chemicals are associated with both rural and urban settlements, varying from those in the body wastes, burials, or garbage disposals of primitive dwelling and work areas to those associated with the lawn fertilization, chemical treatment of streets, or industrial contamination found in modern cities.

Macronutrients are represented by the familiar N-P-K symbols—for nitrogen, phosphorus, and potassium—found on commercial fertilizer bags and by the elements calcium (Ca), magnesium (Mg), and sulfur (S). Because these six elements are used in relatively large quantities by all plants, vegetation removal by human activities depletes their concentration in soils. The first three nutrients are required in greater amounts by plants and may be restored after harvests by green manuring of soils or mixing them with garbage, dung, or commercial fertilizer. Of the other three, calcium and magnesium, which can raise the pH of acid soils, are worked into the soil as crushed limestone or dolomite. On calcareous soils, sulfur, added in the elemental form or as ammonium sulfate, lowers pH to more productive values. Manure, another common additive, contains magnesium, and ordinary commercial-grade superphosphate consists of some 13 percent sulfur. Depletion of macronutrients occurs where farming activities cause accelerated erosion (a physical change) or where harvesting is practiced without fertilization or crop rotation. Even primitive farmers learned to rotate land or even move settlement sites as local soils wore out.

Nonagricultural activities also affect the macronutrient levels in soil. The more direct and indiscriminate burial practices of some early peoples increased these soil constituents throughout residential areas. Cook and Heizer (1965) have reported that substantial annual additions of nitrogen, phosphorus, and calcium enter the soil from food waste, defecation, and urination in settlements and along paths and roads. Similarly, liquid and solid animal waste cause spot increases of these elements in residential areas, along trails, or in fields. Heidenreich and Navratil (1973) have recently found that soil magnesium increases where firewood has been burned. Copper sulfate and zinc are other enriching residues of wood ash (Goodyear, 1971). Natural nitrogen, potassium, and calcium, as well as the human-caused presence of these elements, are subject to quick loss by exposure to leaching, vegetation removal, and topsoil extraction. Phosphorus and potassium are especially sensitive to adsorption by plants; these elements may be completely eradicated from an area during herding activities (Allaway, 1975).

Eight elements constitute a group of micronutrients required in only small amounts for plant growth: iron, manganese, zinc, copper, boron, molybdenum, cobalt, and chlorine. The last joined the list in the 1960s, and the others have been generally recognized only since the 1950s. The list will expand as more is learned about plant and animal growth requirements and human effects on the soil.

Research has led to the understanding of the sensitivity of organisms (both plants and animals) to the narrow range of micronutrient amounts essential to proper growth. Too little (deficiency) or too much (toxicity) of any given element may produce harmful effects. Copper presents an interesting example of these effects: its deficiency in some soils in Australia and Sweden precluded using the land for grazing until recent times, when it became understood that conditions could be corrected by careful copper fertilization of pastures (Allaway, 1975).

Some research has been done on defining representative amounts of these nutrients. Table 7.1 indicates values for humid and arid region soils. It should be kept in mind when using the table that sandy and acid soils generally have lower macronutrient levels and that average micronutrient values are depressed in strongly alkaline, organic, and sandy soils.

ORGANIC MATTER

Soil organic matter, consisting mainly of compounds of carbon and nitrogen, exists in a raw form (10 percent of the total) and in a decayed form called humus (90 percent). Organic matter normally makes up less than 5 percent by volume of the soil but is one of the anthrosol factors most rapidly influenced by human activity. It plays a major role in the development of fertility, structural stability, and microorganism reaction in soils.

Studies of land use and crop yield show that both the organic-matter content and inorganic-nutrient levels of soil must be maintained in order to sustain the original vegetative productivity. This need is emphasized in warm climates, where greater intensities of weathering and leaching create a low level of soil mineral resources and where fertility is concentrated in a narrow band of organic material within a few centimeters of the surface (Gourou, 1961).

Table 7.1. Representative Amounts of Macro- and Micronutrients in Midlatitude Mineral Surface Soils

Element	Normal Range (in %)	Range in Humid Regions (in %)	Range in Arid Regions (in %)
N	0.02–0.50	0.15	0.12
P	0.01–0.20	0.04	0.07
K	0.17–3.30	1.70	2.00
Ca	0.07–3.60	0.40	1.00
Mg	0.12–1.50	0.30	0.60
S	0.01–0.20	0.04	0.08
	(in ppm)		
Fe	5,000 –50,000	[unavailable]	[unavailable]
Mn	200 –10,000		
Cl	10 – 1,000		
Zn	10 – 250		
Cu	5 – 150		
B	5 – 150		
Co	1 – 50		
Mb	0.2– 5		

SOURCE: Brady, 1974:23–24.

The principal natural sources of organic material are plant and animal remains. As these are decomposed by organisms in the soil, they mix with mineral matter to lend a darker color to the soil, which induces the familiar horizonation found in the upper level of soils (discussed below). As decay continues, organic compounds help dissolve soluble nutrients and minerals containing silicon, iron, calcium, and aluminum. One of the special functions of organic-matter decay is the release of chelating agents, such as oxalates, acetates, citrates, and tartrates, that can form organic iron, calcium, and aluminum compounds. Because these elements are used to form organic compounds, less calcium, iron, and aluminum are present to sorb (fix) phosphate. The small amount of phosphorus that escapes as soluble phosphate is taken up by plant roots or is leached downward along with other nutrients, where they either pass out of the system with drainage waters or become added to the lower soil horizons.

The natural formation of horizons is characterized in the upper portions of soil profiles by different degrees of organic-matter decay, or *humification*. Humus mixes with inorganic soil material unless extreme conditions, such as the low pH and temperatures of northern coniferous areas or the high rainfall and temperatures of the tropics, prevent decay or accumulation. The particles of completely decayed organic matter are so small (less than 200 μm) that they normally remain in colloidal suspension. Their negative surface charge attracts and holds nutrient cations from the soil solution and later releases them to living plant roots. Moreover, humic gums act as binding agents around which soil particles gather to form loose aggregates known as peds. Peds vary in size from a few millimeters to a few centimeters and produce first-class soil structure by forming spaces or voids that promote good drainage, root penetration, aeration, and temperature regulation.

The basis for good horizonation and soil structure is the so-called carbon cycle (fig. 7.1). This cycle is an open system involving inputs of carbon and nitrogen and other materials from the above-surface biomass and from the atmosphere, as well as from enzymes, microbes, and decaying matter just below the surface. It is different from the macro- and micronutrient cycle, which depends on inorganic rock components (the parent material of soil) for its basic renewal.

In nature, the carbon cycle may be interrupted if soil humus breaks down rapidly when organic compounds are oxidized. Carbon is converted to CO_2, which escapes as a gas, and remaining components enter the soil solution in ionized form. Organic phosphates, for example, revert to inorganic phosphates fixed by iron-, aluminum-, and calcium-retaining components. Minute quantities of the mineralized phosphate are ionized and become part of the "available" phosphorus required by plants for normal growth. Some of this soluble phosphorus escapes from the system during leaching, but even all the available phosphorus is only a small part of the total organic and inorganic phosphate fixed in the soil (fig. 7.2). The principle is the same for other nutrients, such as nitrogen, potassium, calcium, and manganese.

Since the carbon cycle is unstable, its delicate balance is easily disturbed by human activities, such as plowing. Annual organic-matter losses from mid-latitude plow farming are given in table 7.2. The practice of continuous harvesting, even by primitive methods such as those involving land rotation, depletes

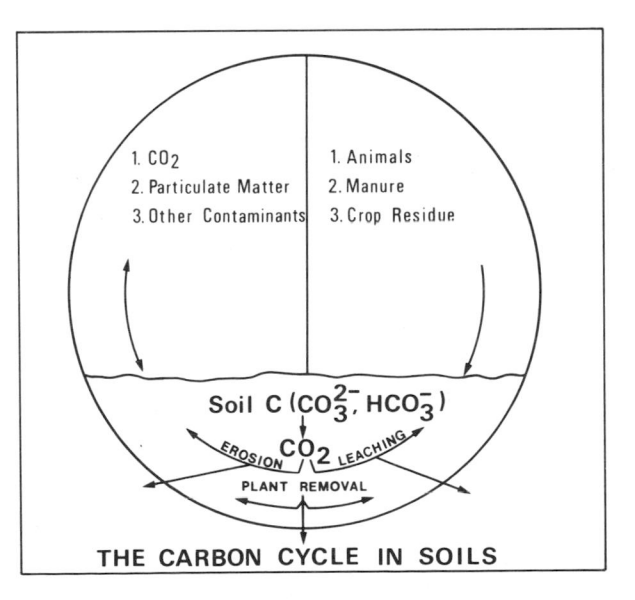

Fig. 7.1. Diagram illustrating the sources and removal of carbon in soils.

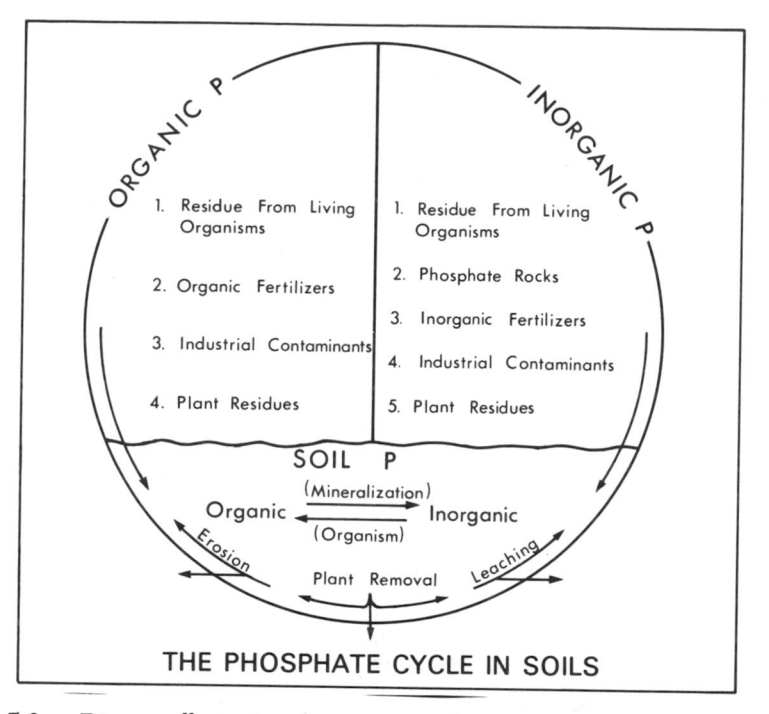

Fig. 7.2. Diagram illustrating the sources and removal of phosphorus in soils.

the organic matter in soil much more rapidly than it depletes inorganic plant nutrients. Young (1976) has reported that in cultivated tropical soils the ratio of organic-matter to plant-nutrient loss may reach 5:1. In the absence of organic-material renewal, the sun and wind dry soils so that upward movement of solutions takes place in rainless periods of as little as one day. Evaporation of colloidal sesquioxides produces a duripan near the surface as these compounds dry out and cement the soil particles.

Table 7.2. Annual Gains and Losses of Organic Matter in Soils

Crop System	Gain (in %)	Loss (in %)
Grasses	0.25	
Alfalfa	1.00 (in 3 yr.)	
Clover-timothy	1.25	
Common clovers	2.00	
Sweet clover	2.50	
Soybeans as hay-seed		0.5
Small grains		1.0
Most cultivated crops		2.0

SOURCE: Donahue, Schickluna, and Robertson, 1971:202.

Incorporation of mulches, manure, compost, trash, and chemical fertilizers and the planting of grass slow the loss of organic matter by direct addition and by causing increased growth of new root systems that decay in the ground. In practice, there is usually either an inexorable degradation of natural soil humus caused by human activities or a marked improvement in the overall condition—facts which make possible the detection of anthrosols by comparison of organic-matter values around settlements with those of unaffected natural soils (table 7.3).

OXIDATION AND REDUCTION

Oxidation and reduction occur in soil solutions because of the chemical interaction of microbes and enzymes in the presence of soil constituents and plant root systems. The extent to which oxidation or reduction takes place in a soil has a marked effect on the characteristics of the soil. Oxidation-reduction processes involve particularly nitrogen, iron, manganese, sulfur, and oxygen and vary with human activities, such as irrigation, drainage, and compaction. Measurement of oxidation and reduction effects can be highly useful in evaluating anthrosols.

When oxidation occurs, the resulting products are nearly always soluble. The example below represents atmospheric oxidation of iron pyrite in soil:

$$4FeS_2 + 15O_2 + 2H_2O \rightarrow 2Fe_2(SO_4)_3 + 2H_2SO_4.$$

During reduction, iron and aluminum substitute for oxygen and accept electrons produced by plant roots. A typical example is:

$$Fe(OH)_3 + e^- \rightarrow Fe^{2+} + 3OH^-.$$

(yields gray color)

Scheffer et al. (1976) have summarized the quantitative relationship between the redox potential (Eh) and the activity (concentration) of participating chemicals as follows:

$$Eh = E + \left(\frac{RT}{nF}\right)\ln\frac{Ox}{Red}, \tag{1}$$

Table 7.3. Soil Fertility Decline of Loamy Midwestern Soils with Length of Cultivation Period (from H. Jenny as shown in Millar and Turk, 1949)

Percent Relative Nitrogen Content	Years of Cultivation
88	5
78	15
74	25
69	35
67	45
65	55
63	65

where E = standard potential (constant), R = constant (8.315 joules/degree/mole), T = absolute temperature, n = number of electrons transferred, F = Faraday constant (96,500 coulombs), Ox = activity of oxidized phase, and, Red = activity of reduced phase. Combining constants and converting to base-10 logarithms yields:

$$\text{Eh} = E + \left(\frac{0.059}{n}\right)\log\frac{Ox}{Red}. \tag{2}$$

It is clear from equation (2) that Eh is a measure of the oxidation-reduction activity ratio (Ox/Red). The greater the oxidation, the higher the value of Eh. Conversely, the greater the reduction, the lower the value of the redox potential. If the activity of the chemicals involved is equal, then the redox potential corresponds to that of the standard (Eh = E). In other words, the value E is the numerical expression for the oxidation strength of the oxidized phase or the reduction strength of the reduced phase of a redox pair. The oxidized phase of a redox pair having a higher E is a better oxidizing agent than the oxidized phase of a redox pair having a lower E. These principles are basic to understanding the chemistry of soil reactions, both in the field and in the laboratory.

If E is measured for the commonly found redox pair Fe^{2+} and $Fe(OH)_3$, then

$$2Fe^{2+} + \tfrac{1}{2}O_2 + H_2O \rightarrow 2Fe(OH)_3$$

and E = +1.058V (measured at pH = O and at redox equilibrium). When this value is substituted in the Nernst-derived equation (2),

$$\text{Eh} = 1.058 - 0.177\text{pH} - 0.059\log Fe^{2+}. \tag{3}$$

Equation (3) indicates a relationship between redox potential and pH. It can be seen that the stability of given iron (or other) ions depends on both Eh and pH values. Jeffery (1960) and Greene (1963) have shown that the solution concentration of active Fe^{3+} ions increases by about 10^3 per unit decrease in pH. Thus, from equation (3), we can see that the lower the Fe^{2+} value and the lower the pH, the higher the value of Eh—that is, the greater the oxidation activity. Conversely, the redox potential declines as pH values rise.

Eh measurements vary considerably even within the same soil and, along with pH values, can change within hours after a heavy rain. The amount and duration of change can often be correlated with color alterations in the soil. Different iron oxides vary in color from gray, yellow, and brown to red, but the color of all soils are not determined by iron compounds. Thus, color examination alone does not tell the exact condition of soils or when discoloration occurred. Since the redox potential represents an index of the reduction-oxidation state, it gives a more accurate picture of the drainage changes in soils than do the subjective discoloration or mottling tests usually performed.

The higher the organic-matter content, the more responsive to change are

Eh values. Thus, Ah horizons, or the upper horizons with humus, have a more volatile redox potential—that is, one which lowers rapidly as soil moisture increases, or vice versa. Below the A horizons are B horizons, which have less organic matter and much more stable Eh values. Rodrigo (1963) notes that this difference may cause the development of podzols, since Fe^{2+} converted by reduction of Fe^{3+} in the upper horizon becomes transported in solution to a lower level, where it is oxidized and reprecipitated as Fe^{3+}. This is a significant part of the explanation for podzolization of northern European soils, where historical land clearing has altered the original Eh.

Eh values for some typical soils can be given for comparative purposes. In well-drained soils, the redox potential ranges from +400 millivolts (mv) in a clay to +450 mv in a silt loam, +550 mv in a peaty soil, +650 mv in a podzol, and approximately +750 mv in a chernozem. However, when soils are affected by everyday living in dwelling areas (compaction) or by vegetation removal, plowing, irrigation, and other human activities, redox values reflect the new conditions. In a compacted Iraqi tell, for example, Eh averages about +380 mv; it is +460 mv in lower, surrounding soils. Flooding has an even more pronounced effect. Pore spaces normally occupied by oxygen are filled with water, and any trapped gas is used within several hours to two days for the oxidation of organic matter. Anaerobic microorganisms then appear in growing numbers and begin to use organic matter as electron donors. This causes reduction of NO_3^- to N_2, of N_2 to NH_4^+, Mn^{4+} to Mn^{2+}, Fe^{3+} to Fe^{2+}, SO_4^{2-} to H_2S, HCO_3^- to CH_4, and H^+ to H_2. Within less than two weeks the Eh decreases to about −100 mv (Scheffer et al., 1976).

A temporarily flooded organic soil has an Eh of about −450 mv. Ultimately, the Eh of a waterlogged organic soil may reach as low as −700 mv (Bunting, 1967), that of an inorganic soil, about −200 mv. Volobuev (1964) has found that the Eh of solonchak meadow soils dropped from 420−460 mv before irrigation to 280−320 mv one day after watering. Since Eh values are regarded as an index of the intensity of reduction, they are measured under conditions approximating saturation.

It may be noted that H_2S usually forms where organic matter can provide sulfur. Fortunately, H_2S is short-lived as long as iron is present because it immediately helps precipitate iron sulfides (FeS_2) and iron sulfate [$Fe_2(SO_4)_3$]. Iron is so common in soils that the redox potential cannot normally decline sufficiently for production of excess H_2S, a substance toxic to nearly all life forms in soil. In addition to Fe^{2+}, soluble and potentially toxic manganese and aluminum ions are released under continuous reducing conditions—a feature that characterizes some buried soils. Presence of aluminum ions under these circumstances lends a greenish blue cast to the soil and may, like the gray of reduced iron, tend to mask normal color changes between previously formed A and B horizons. In such cases, other methods of investigating the evolution of horizons must be employed (Limbrey, 1975).

pH

Under normal soil conditions, electron exchange plays a major role in determining soil acidity. Water, which ionizes as H^+ and OH^-, provides a source of H^+ ions to the soil. A common soil reaction with a similar effect results when CO_2 mixes with water:

$$CO_2 + H_2O \rightarrow \quad H_2CO_3 \rightarrow H^+ + HCO_3^-.$$

carbonic
acid

All of these H^+-producing reactions are followed by the adsorption of the H^+ ions by minute, silica-clay colloids known as *micelles*. In the case of carbonic acid, the formation of soluble bicarbonates occurs:

$$2H^+ + 2HCO_3^- + M^{2+} \boxed{\begin{array}{c} - \\ \text{soil} \\ \text{micelle} \\ - \end{array}} \rightarrow \begin{array}{c} H^+ \\ H^+ \end{array} \boxed{\begin{array}{c} - \\ \text{soil} \\ \text{micelle} \\ - \end{array}} + M(HCO_3)_2.$$

In this reaction, the H^+ ions displace metallic cations (M), such as Ca^{2+}, held by the soil micelle, and the bicarbonates are carried away in solution, leaving the soil with an excess of H^+ ions (more acid). Chemical reactions that create an excess of hydroxyl ions produce basic soil conditions. A common example follows, where the OH^- concentration is greater than the H^+ concentration:

$$2Na^+ + CO_3 + 2H_2O \rightleftharpoons 2Na^+ + H_2CO_3$$

$$Na^+ \boxed{\begin{array}{c} - \\ \text{micelle} \\ - \end{array}} + H_2O \rightarrow \begin{array}{c} H^+ \\ H^+ \end{array} \boxed{\begin{array}{c} - \\ \text{micelle} \\ - \end{array}} + Na^+ + OH^-.$$

Soil pH is influenced not only by the H^+ ions present, but also by Al^{3+} ions. Aluminum, one of the most abundant soil elements, reacts as in the following simplified set of equations:

$$Al^+ \boxed{\begin{array}{c} - \\ \text{micelle} \\ - \end{array}} \rightleftharpoons Al^{3+}$$

$$Al^{3+} + H_2O \rightarrow [Al(H_2O)_5OH]^{2+} + H^{3+}$$

$$[Al(H_2O)_5OH]^{2+} + H_2O \rightarrow [Al(H_2O)_4(OH)_2]^+ + H^{3+}.$$

Jackson (1963) has termed the hydrated aluminum ion *hexaaluminohydronium* and believes that it exists as a stable ion in acidic solutions. If OH^- ions are added, the aluminum ions react by showing a tendency to polymerize. Polymeric formations occur in the interlayers of clays, such as montmorillonite and vermiculite. When complete layers are formed by these polymers, the term *chlorite* is applied (Jackson, 1963).

Although exchangeable hydrogen is considered the principal source of acidity, when the pH drops below 6 the aluminum in clay sheets becomes unstable and combines with OH^- ions from water. The process releases H^+ ions for adsorption and contributes further to soil acidity. The hydroloysis can be depicted as follows:

$$Al^{3+} \boxed{\text{– micelle –}} + 3H_2O \rightarrow \begin{matrix} H^+ \\ H^- \\ H^+ \end{matrix} \boxed{\text{– micelle –}} + Al(OH)_3.$$
$$\text{insoluble}$$
$$\text{gibbsite}$$

Aluminum is almost as important to soil acidity as hydrogen. However, no distinction between the two is made when pH is measured, since both are sources of H^+.

Soil pH normally has a tendency toward stability even though individual field values may vary within a few inches of each other. With the removal of ions from solution, the reserve ions on micellar surfaces tend to re-create the original conditions. This *buffering capacity* varies from soil to soil and depends on the type of clay minerals and organic matter present. Usually, the greater the reserve of exchangeable ions present in soil, the greater its buffering capacity. Consequently, the higher the percentages of clays and organic components in the soil, the slower will be the degree of pH alteration by human activities.

Marked alteration of soil pH changes chemical conditions, causing new plant colonizers to appear and alter the composition of the vegetation cover. High-pH halophytic plant types of arid zones are thus contrasted with those found in the same niche of more acidic, humid environments. The number of plant varieties present may indicate local pH variation. Since pH is closely related to the availability (solubility) of certain important nutrients, such as phosphorus, whole vegetation associations may be explained by these relationships (Adams and Walker, 1975). Certainly they are significant when it comes to choice of crops and soil preparation and establishment of appropriate agricultural systems.

The effects of human activities on soil pH vary with land use. Land clearing, for example, exposes soil formations to greater alterations in the heat budget and to leaching. When well-drained, cleared land is thoroughly moistened either by irrigation or by nature, plant nutrients such as calcium, magnesium, and potassium are displaced from exchange sites on clay micelles by preferential adsorption of H^+ ions. The process is accompanied by reaction between released nutrient cations and H_2CO_3 in the soil to form soluble bicarbonates. The result is leaching of nutrients—often by as much as are removed when plants are harvested—and lowering of the soil pH as H^+ ions accumulate. A pH decline of over one unit is not uncommon in these situations, given sufficient time.

Two other phenomena associated with land clearing significantly affect pH. One is cultivation. Long-term, properly managed cultivation is known for its tendency to raise pH in acid soils to about 6.5. The opposite happens in nonacid soils where annual rainfall above 600 mm lowers pH to below neutral. Improper cultivation can destroy soil structure, impede drainage, and cause unfavorable pH changes, as demonstrated by Laws and Evans (1949). For example, plowing to the same depth over many years results in the formation of a plow pan—an indurated layer at the bottom of the plow cut. Drainage is arrested at this point and a perched water table may keep soils overly moist in the plant-root zone. Eh is lowered as reduction begins and pH rises. Improper use of heavy machinery or trampling by livestock can create a traffic pan at the surface of soils that already have a platy or weak structure. Puddling then takes place. When rainwater stands for as long as a day or two, denitrification and other problems associated with reduction arise and may be followed by long-lasting or permanent pH changes.

The second phenomenon associated with land clearing is the change in vegetation cover. In northern Europe, centuries of deforestation and subsequent grazing have been followed by invasion of a heather vegetation (*Calluna* sp.), such as in the Lüneburger Heide area, West Germany, which has brought on an acid-soil condition where none existed before (Scheffer et al., 1976). The planting of food and feed crops whose removal depletes the quantity of soluble nutrients likewise lowers pH. In Wisconsin soils, Walsh (1972) found that significant acidification occurred when legume top growth was harvested because of removal of as much as 200 pounds per acre of calcium and magnesium. Finally, the planting of certain crops, such as pine trees, may ultimately produce low pH because of the organic acids formed during decomposition of the foliage.

Artificial soil drainage also causes changes in pH; drainage of peat bogs increases pH up to almost two units. Fertilization is another activity that can produce changes in pH. Fertilizer that contains or produces NH_4 acidifies soil by creating strong acids; for example,

$$(NH_4)_2SO_4 + 4O_2 \rightarrow 2HNO_3 + H_2SO_4 + 2H_2O$$

and

$$NH_4NO_3 + 2O_2 \rightarrow 2HNO_3 + H_2O.$$

Although soil acidification by fertilization was recognized as early as the 1920s (Pierre, 1928), the trend has increased with use of more and more nitrogen-type fertilizers. Since phosphate fertilizer has also increased in usage, it might be expected that the $H_2PO_4^-$ and H^+ ions released would acidify the soil. However, the phosphate immediately reacts with aluminum and iron to form insoluble compounds. In the process, liberated hydroxyl ions neutralize the H^+ formed and no long-term acidification results, unless the fertilizer contains nitrogen.

Land reclamation may be associated with drainage, as in the case of the polders of northern Europe. Desalinization chemicals containing calcium and ammonia are spread over the exposed marine sediments to remove sodium (by replacement) until root crops can be raised. After considerable time, during which organic matter builds up the soil structure, crop diversification with less salt-tolerant species follows. Near small settlements where population pressure has forced adaptation to swamps or areas of flooding, land has also been reclaimed by construction of platforms, mounds, and planting ridges whose pH contrasts with the rest of the area. This practice was typical in many parts of pre-Columbian Latin America (Denevan, 1970; Parsons and Bowan, 1966; Eidt, 1959).

In arid climates, when drainage of groundwater, irrigation water, or a combination of the two results in the presence of capillary water in the soil, salinization and/or alkalinization occurs with an accompanying rise in pH. The pH level is significant here because above 8.5 the flocculation caused by free salts is inhibited, and the structure of the soil ultimately collapses because peds become dispersed. If sufficient quantities of the clay particles are then lost by eluviation and an impermeable B horizon forms, the soil may be permanently ruined. Such soil destruction is marked by bare patches in which crops will not grow and which spread rapidly. Large-scale examples visible from the air are those in the Imperial Valley of California and in the Indus and Upper Ganges regions of Pakistan and India.

Soil pH changes of smaller magnitude than those discussed above occur during both dry periods and seasonal variations in temperature. Decreases in soil pH also accompany the drying of samples in the laboratory—indicating a need to refrigerate samples awaiting analysis (the best temperature appears to be about 4°C). The pH of normal soils tends to decrease in summer, probably from acids created by greater rainfall and its leaching effects and by interactions between plant roots and microorganisms. In winter, the pH rises slightly as low temperatures inhibit these activities. By contrast, Ballantyne (1963) has shown that several years of abnormally high precipitation can cause a relative rise in pH.

In an interesting example of interpretation of naturally induced pH changes, Zeuner has explained three periods of loess deposition in Wallertheim, Germany, where pH variations at certain levels led to dating of artifacts. During cold, windy periods near the edge of glacial activity, the first layer of loess covered Mousterian implements. Periglacial conditions then ceased, and a subsequent warm period provided sufficient rainfall to dissolve calcareous minerals leached from the upper part of the loess. At a later time, more loess covered the site and buried the first soil. The upper part of the second covering was also decalcified by rains, and the process was repeated a third time. The three periods have been correlated with the three cold maxima of the Würm glaciation, leading Zeuner to believe that the underlying implements were made by man before the onset of the last major glacial period (Pyddoke, 1961).

Alterations in pH play an important role in determining the availability of nutrients in the soil. Seasonal variations of pH with dry-wet cycles influence the amount of available nutrient material during these environmental cycles. Therefore, chemical tests based on available nutrients such as are sometimes used to indicate human activities (phosphate tests) may not give reliable results. In the seasonal variation caused by melting snow and ice, for example, intermittent oxidation periods assist a rise in soil pH, which opens a phosphate window for plant uptake. However, when pH drops below 6, aluminum ions become active and remove excess phosphorus:

$$Al + H_2PO_4 + 2H_2O \rightarrow 2H + Al(OH)_2H_2PO_4.$$
$$\text{"fixed" phosphate}$$

Among the micronutrients, iron, magnanese, and copper exist in more than one valence state. Higher pH values accompany the oxidation of these micronutrients. The condition creates a decline in available (soluble) forms. On the other hand, the hydrous oxide of Fe^{3+} precipitates at a pH of about 3. Low pH values found in poorly drained soils release both iron and manganese in toxic amounts, whereas deficiencies may occur at high pH values.

From these examples of pH effects, it is evident that human-induced pH changes may lead us to expect various types of chemical alterations in the soil. Interpretation of the pH changes should be made only after taking natural changes into account.

EXOTIC CONTAMINANTS

One of the important ways human activities alter soils chemically is by addition of trace amounts of metals and hydrocarbons. These chemicals often depress certain nutrients to the point of deficiency and/or augment others to the point of toxicity. In either case the results are lower crop yields or the accumulation in crops of poisonous chemicals that enter the food cycle. The major effects date from the industrial revolution and have increased rapidly since the turn of the present century. Within the last 50 years, per capita demands for copper, lead, and zinc have tripled (Warren and Delavault, 1971) and demand for organic chemicals used in catalytic petroleum cracking have risen tenfold, from a world production of 0.3 billion tons to an estimated 3.0 billion.

The most significant trace metals along with their representative background amounts in American and world soils are given in table 7.4.

Bertine and Goldberg (1971) believe coal combustion to be an important source of contamination by heavy metals. Supporting evidence is presented by Klein and Russell (1973) in their study of fallout around a coal-burning power plant. Soil samples were taken from established wooded areas with good drainage near Holland, Michigan, both downwind of the plant and in unaffected areas. Analytical results by atomic adsorption showed significant soil contamination in line with prevailing winds, as indicated in table 7.5. Soils were affected to a depth of 2 cm over an area of 115 square miles.

Mercury increases in soils have also been traced to coal burning (Joensuu, 1971). The appearance and persistence of metallic ions depends on many factors, including chemical and physical structure of colloids, valence, ionic radius (size) of the metal in question, and flocculating effects (Lisk, 1972). Little is known about the effects of given pH and Eh values and the conditions of fixation versus availability of ions. The lower the background amount of a particular chemical,

Table 7.4. Average Contents of Trace Elements in American and Global Soils

Element	Average American Content (in ppm)	Range of Global Contents (in ppm)
Mn	560	200 – 3,000
Ni	20	5 – 500
Zn	54	10 – 300
Cu	25	2 – 100
Co	10	1 – 40
Pb	20	2 – 100
Mo	—	0.2 – 5
Cd	—	<1

SOURCE: Warren, 1973:11.

Table 7.5. Trace Metals in Soils near a Coal-Burning Power Plant

Metal	Content in Background Soil (in ppm)	Content in Enriched Soil (in ppm)
Ni	2.4	4.0
Zn	26.3	35.0
Cu	2.8	4.6
Co	2.3	4.6
Cd	0.55	1.46

SOURCE: Reprinted by permission from Klein and Russell. Copyright 1973, American Chemical Society.

the more modest the increase required to bring about an alteration in the soil composition, if not in its evolution as well. The different chemical and pollution levels result in changes in land-use capabilities, which in turn create more complicated chemical relationships between soils and their use by humans. Some of these new relationships have not existed long enough for their soil prints to be easily identifiable but will one day be relevant to analysis of abandoned settlements.

Documentation of a settlement's abandonment due to soil contamination is difficult to establish, but the case of the village of Shipham, Somerset, in Great Britain, offers insight into the process in historic times. Land around Shipham is hillocky and is locally called "gruffy ground." The hillocks, it has been learned by the Department of the Environment, contain toxic levels of cadmium over twice as high (100 ppm) as those causing major health problems in Japan; in some places, cadmium levels approach 1,000 ppm. The contamination dates from the early industrial revolution, and it is probable that abandonment of the village will be proposed (MacLeod, 1979).

PHYSICAL SOIL CHANGES

The physical soil changes introduced by human activities may be divided for analytical convenience into two broad groups: those involving soil movement and those resulting in altered color, bulk density, soil structure, drainage, and texture.

The most common types of soil movement are associated with construction of structures and preparation of associated work areas or fields. The settlement proper grows in height as rubble accumulates. The city of London, for example, rests today on 4−5 m of man-made materials, including ashes and bricks (Sherlock, 1931). In the Middle East, tells were built up where mud walls had disintegrated after abandonment and resettlement. Soils in these tells are finer and deeper and have a more uniform structure and a grayer color that those of the immediate surroundings. Quite often they will have little or no horizonation because of the aridity of the area and the constant mixing that was carried out under foot. In more humid areas, whole settlements were sometimes originally placed on artificial platforms, and surviving settlements of this sort are still elevated more than a meter or two above the surroundings. This is the case in German, Dutch, and Belgian coastal areas, where the well-known *Worft* settlements occur (Mayhew, 1973), and in swampy zones of South America (Laeyendecker-Roosenburg, 1966). Houses were crowded together on these platforms, the soils of which became different from their surroundings through mixing and accumulation of rubble and drainage changes. Buried natural soils exist at the bases of these settlement sequences.

In fields around early farm settlements, accumulation of soil often occurred at right angles to the direction of plowing on sloping land. Soil creep caused the downslope buildups that in England became known as *lynchets*. In cases of rather extreme slope, terraces or retaining walls were built to hold soil brought in to level the land for farming. The ancient Greeks, Incas, and peoples of Southeast Asia are known for this type of construction. Lynchets and terraces can cover buried soils that may reveal earlier phases of cultivation. In more level regions, such as northern Europe, plowing in medieval times without a reversible moldboard caused a piling up of soil at the centers of fields, so that kilometer-long S-shaped plow ridges developed (Baker and Butlin, 1973). Some of these features were clearly preserved by forests planted after farming had been abandoned. Evans (1978) notes in a recent contribution that linear bank-and-ditch earthworks, called "ranch boundaries," characterize some Celtic fields, just as field walls characterize the farms of Bronze Age Highland Britain.

In poorly drained parts of Latin America, artificial fields were constructed by pre-Columbian farmers, who may have raised fish stocks and traveled by canoe in the flooded areas between fields (Parsons and Bowen, 1966). The fields themselves have different shapes, many of which have been described by Denevan (1970). Eidt (1977, 1981, in press) has classified them as platforms (many rows of crops), mounds (2−4 rows), and planting ridges (one row of crops), depending on size and use. Even today, when they have been so worn down as to be hard to see from the ground, their soils are more friable, better structured, and often darker than those around them.

Sometimes early farmers dug pits in arid lands so that rainwater would collect in them for use in crop production. This was a common practice in parts of northern Mexico and Peru. Topsoil washed into these pits and maintained healthy yields.

Occasionally, productive soils such as mor or peat formations were almost entirely removed and replaced by stable-bedding mixtures of heather and grass, sods, forest litter, manure, garbage, seaweed, and calcareous sea sand. Centuries of this process produced layers nearly a meter in depth in Ireland and northern Europe. These so-called *plaggen* (German for "to cut sod") soils are uniformly richer, browner, and deeper than surrounding soils. The various depths of plaggen have been used to reconstruct the reclamation history of old settlements in Europe (Pape, 1970). A settlement stratigraphy allowing relative settlement age to be determined is possible owing to the fact that the earliest reclamation probably occurred on the best podsols (moderpodsols), followed by reclamation of less favored types, such as xeropodsols, hydropodsols, and gleys.

Occupational debris, such as middens, represent other artificial features whose materials influence surrounding soils or may develop soils of their own. Generally, prehistoric middens are garbage heaps consisting of shell-food remains, such as those of the coasts of Denmark, California, or New Zealand. Examples of modern middens are Mount Trashmore, in Dupage County in the Chicago area—a garbage heap now 130 feet high; Mount Downey, California, a large dung heap built from dairy-cattle waste material southeast of Los Angeles; and the well-known Freshkill Landfill on Staten Island, where the dumping of 10,000 tons of garbage daily has leveled and raised an extensive area. Mine tailings, another type of artificial debris, also make their mark on the soil landscape; like prehistoric middens, they are not only physical but also chemical artifacts that leave behind various macro- and micronutrients, trace metals, hydrocarbons, and even excessive radiation levels, such as those in some soils at Denver, Colorado (from turn-of-the century radium tailings).

Ditches, borrow pits, and cache pits are further examples of physical features that alter color traits or produce new conditions, such as iron pans. These have been amply described by Limbrey (1975) and Pyddoke (1961). Effects of irrigation ditches have been analyzed by Fernea (1970), who points out that the portion of an irrigation ditch nearest the offtake receives more than double the silt that is dropped off at the tail. Gibson (1974) remarks on the change in soil quality along irrigation canals stemming from siltation. The coarser sediments precipitated with initial slowing of water produce more permeable soils than those that develop farther downstream. Gibson notes that variation in soil quality characterizes irrigated farmlands in ancient Mesopotamia because canal and ditch banks obstruct surface rainwater flow for varying lengths of time. Following water clogging, changes in both redox potential and salinization ensue due to salts derived from the river waters. After a time, land becomes useless and must be abandoned; fallow periods of more than a hundred years are required for significant recuperation. Such desertification of land is typical in the Fertile Crescent, the loess region of northern China, north Africa, the western United

States and the Soviet Union, due to overgrazing, improper irrigation techniques, and poor farming practices in marginally humid areas.

Burning of vegetation has altered chemical and physical traits of soils. Pyddoke (1961) states that low red heat (800°C) will convert yellow-brown limonite soil to red hematite. Since the process is irreversible, local reddening of soil may provide evidence of human occupation. Change in texture, indicated by the presence of ash fines, and reduced percolation rates, which accelerates flooding, are recognized effects. Rapp (1978) speculates that the pink cast of ancient mud bricks used in Greece reflects firing to 500°C, possibly from an ordinary house fire.

A dramatic example of soil erosion is the famous Dust Bowl case of the 1930s. At that time, the agricultural system was not in harmony with physical conditions of the region, and winds began to remove topsoil in amounts so great that dust clouds several miles high over parts of Texas, New Mexico, Colorado, Oklahoma, and Kansas carried away hundreds of millions of tons of soil. Some of the material was dumped on farms in drifts 25 feet high (Lockeretz, 1978). Years later, soil from this region was identified in places 500 miles away (Donahue, Schickluna, and Robertson, 1971). It is sometimes possible to note humic-horizon loss in such cases by comparing open fields with areas not used for crops.

Another type of soil alteration, less visible except in its effects, is a structural change measured by degree of compaction. Compaction is stated in terms of *bulk density*, the weight in grams of a thoroughly dry soil sample divided by its volume in cubic centimeters. This volume includes both solid and gaseous parts of the soil. Thus, bulk density is high for soils with coarse particles—sand, for example—and lower for fine-textured soils, because of their greater pore space. The effects of human activities, such as the use of cropping machinery, are so great on soil that crumb or granular structures can be compacted to platy structures with high bulk density in as little as two years. Compaction also characterizes dwelling floors, human and animal paths, railroad beds, roadways, and runways. The bulk densities of compacted settlement soils, unlike those of normal soils, can surpass 2 g/cm³. The degree of compaction should be determined by comparison with unaffected soils in the same area.

Soil *texture* refers to the size of the soil's component particles and is broadly classified as sandy, silty, or clayey. Since particle size in mineral soils is not easy to change, texture is normally a stable physical factor in spite of human activities. However, like soil color, texture can change if enough material containing a new component of different size, such as humus, is added for a long time or in large quantities to a soil. Moreover, burning can clog pore spaces with ash, the effects of which may alter drainage traits and erosion rates. Mining refuse can likewise bring about textural alterations.

One of the most significant aspects of soil texture is its control over structure, or the grouping of soil particles into aggregates known as *peds*. Because of this control, various types of aggregation are normally encountered from horizon to horizon. Types of structure (Clarke, 1971) are platy, prismatic, blocky, and granular, which is best for farming. The creation and stabilization of granular soil structure is the object of liming, cultivating, adding organic matter, and drainage. Marked differences in soil structure are associated with different land uses. Soil used for pasture grasses has a better structure than the same soil under constant cultivation, for example. Briggs (1977) reports old pasture soils with an average structural stability of 70−80 percent, compared with a value of 5−10 percent for newly cultivated grassland, and states that a hundred years may be required for recuperation of lost stability following cultivation.

Whatever the land use, the structural properties mentioned above can be measured and compared. Often enough, observed differences between background soils and those with artificially altered features assist the investigator in identifying anthrosols. The more similar the structure of native background soils and the suspected anthrosols, the lower is the risk that differences are due to normal differences in soil types.

BACKGROUND SOIL ANALYSIS

The analysis of anthrosols requires that background soils, unaffected by human activities, be investigated in the field and in the laboratory. Although it is impossible in a chapter of this length to present a detailed examination of these two facets of investigation, an approach can be outlined that will suffice for initiating and organizing research. The approach, derived from the modern literature and from practical experience, stresses initial avoidance of soil-identification schemes that are suitable primarily for advanced taxonomists and that require expansive laboratory equipment. Inherent in this approach is a recommendation that both field and laboratory research be carried out by the investigator and not delegated to outside parties for analysis.

FIELD STUDIES

By far the simplest and most effective system for recognizing, defining, and classifying soils is to dig a profile pit from surface to (if possible) parent rock; the pit should face south in the northern hemisphere so that good lighting can assist Munsell-color-chart interpretation and photography. This is followed by description and analysis of the major soil horizons, designated A, B, and C, as outlined by Dukuchaev and his Russian colleagues starting about 50 years ago. The technique is still undergoing modification and refinements, some of the major ones stemming from work by Kubiëna (1953), Hole and Hironaka (1960), and the Soil Survey Staff of the United States Department of Agriculture (1975) and reviewed in Buel, Hole, and McCracken (1980) and others.

The essence of the horizonation system is to define world soil types, beginning with the most marked variations in soil character as observed in the soil profile pit. This global type, or great soil group, is designated by a noun, such as *chernozem, solonchak,* or *braunerde.* The next lower category, the subtype, possesses a descriptive adjective, such as *degraded* chernozem or *podzolized* braunerde. In recent years, the use of prefixes has become more common: *para-*braunerde, *iron*podzol. Following subtype, more detailed changes are noted as soil varieties: *mull* pararendzina. Finally, a soil-series term maintains the standard method for identifying local soil-forming factors by use of local terms, as in *Fargo* chernozem. It should be noted that the entire system rests on careful identification and analysis of soil horizonation. Horizonation, in short, depicts the kind and intensity of chemical and physical transformations through time that can most easily be identified in the field. Since horizonation represents a synoptic view of the interactions of all the pertinent soil-forming processes, a brief review of these processes is considered essential.

SOIL-FORMING PROCESSES

The soil-forming processes are described in detail in Scheffer et al. (1976), Limbrey (1975), and Butzer (1971). Horizonation processes are both physical and chemical in nature. Natural physical processes are characterized by mixing of soil particles. Mixing occurs in many ways, such as by the root action of plants and the feeding, growth, elimination, and mineralization of soil fauna. Effects of these forms of mixing are noted at depths of more than one meter in grassland regions. Sometimes, life forms such as ants and termites bring soils from comparable depths to the surface. Coarse material from above may fall into holes created by these insects, bringing about further mixing.

Cycles of wetting and drying and heating and cooling (freeze and thaw) promote soil mixing through expansion and contraction processes. Wide shrinkage cracks in the soil, formed after drying and freezing, permit filling by gravity fall to depths of more than a meter. In wet-dry savannas, this process is called *tirsification* or self-mulching. Freeze-thaw mixing in cold areas is called *cryoturbation*: fine eolean surface materials are brought below while stones work their way upward. The intensity of cryoturbation depends on the number of freeze-thaw periods (Williams, 1973; Zeuner, 1959; Hole, 1961). Human disturbance of soils by cultivating promotes mixing, often of a very uniform type. Accelerated erosion contributes to mixing as well.

All of these processes, coupled with the effects of precipitation, contribute to the downward movement of fine materials or colloids. This movement, sometimes called *lessivage*, occurs by dispersal (peptization). Dispersal works most effectively at pH values between 4 and 7 where the stability of soil structure is weakest. Transported brown, clay-humus materials, as well as oxides and hydroxides, are deposited at depths where peptization ceases or evaporation begins. Deposition of a sufficient amount of fine materials can cause impeded drainage. Soils characterized by this form of mixing in the B horizon are called para-braunerde forms. The horizon is known as a textural B horizon because of the translocated clays; it is always associated with a brown color, hence the terms *browning* or *braunification* are sometimes used to describe the process. Chemical processes of change may also cause browning. Oxidation of iron compounds—such as goethite, $\alpha FeO(OH)$—produces browning and, if the color becomes slightly reddish, rubifaction; if hematite is present, definite reddening (ferrugination) takes place.

Chemical processes also bring about other changes, such as podzolization, for which an acidic condition and the presence of organic matter in the A horizon are prerequisites. Water-soluble organic compounds, such as phenols and chelates, form complexes with iron and aluminum. When these are transported to lower horizons, the higher pH brings about precipitation (Bloomfield, 1953), causing a whitish, or bleached (washed-out), upper *eluvial* horizon. The lower depositional, or *illuvial*, horizon has a lower sesquioxide ratio in the clay fraction compared with the eluvial horizon. After many years, the process results in an increase in the dithionite-soluble iron and occluded phosphate in the illuvial horizon, which may become cemented with black organic compounds and is then called ortstein.

Waterlogging produces horizonation due to reduction of Fe^{3+} and Mn^{4+} compounds. The dissolved compounds enter the soil solution and leave behind a gray horizon that has no brown or red colors. Only when the water evaporates and air enters the soil are the Fe^{2+} and Mn^{2+} compounds reoxidized. The iron oxides formed are goethite ($\alpha FeO(OH)$, brown) and lepidocrocite ($\gamma FeO(OH)$, orange). Where drainage is poor, uneven oxidation intensities produce discolored spots, or mottles. The process receives the name *gleization*, from the term *gley*, meaning gray, waterlogged soil. With better drainage, large pore spaces and rifts in the soils become filled with concretions or "tonged" mottles, and the term *pseudogleization* describes the phenomenon.

In tropical and subtropical climates with heavy precipitation and high temperatures, removal of dissolved organic and inorganic compounds leaves behind a relatively intensified concentration of iron and aluminum in the upper layers. Partial leaching of silica and even of iron and aluminum oxides occurs. The iron oxides may be deposited around other soil particles or they may form concretions. Breakdown of clays to form kaolinite, goethite, and gibbsite reflects the extent of chemical weathering. This process is called ferralitization. When the Fe- and Al-enriched horizon case hardens upon exposure to air, laterite forms. Alternating wet and dry warm climates favor the process of laterization.

Poorly drained sites sometimes form collecting areas for organic materials preserved under anaerobic conditions. Peat and muck soils develop as the materials accumulate and decompose. Under permanent water, deep humus forms soils known as *dy* (peat muck), *gyttja* (mineral-rich gyttjas are gray; humus-rich, black or brown-black), and *sapropel* (stinking slime). The general process is called paludification.

Aird, semiarid, and humid coastal regions are subject to the process of salinization, especially in poorly drained soils of high clay content. Sulfate, chloride, and carbonate ions predominate, but when sodium accumulates at the expense of other ions, the term *alkalization* is used to describe the process. If the water table remains relatively high and stable, sodium salts may form crusts on the surface by capillary action, and a soil type named *solonchak* emerges. If a drop in the water table finally permits drainage, further degradation produces solonetz soils that develop leached-clay horizons above the so-called natric (salt) horizon.

These fundamental soil-forming processes are responsible for creating the various distinctions among A, B, and C horizons, to which we now return in detail.

A Horizons

A horizons are those containing the greatest concentrations of organic matter and living organisms in inorganic soils. They may be distinguished by several sub-horizons that can easily be recognized in the field.

Ah: *h* for humus near the surface
Ab: *b* for brown, indicating a degraded, formerly black horizon, as in degraded chernozem
Ae: *e* for eluvial, indicating an ash-colored layer in podzolic soils and podzols
Al: *l* for lessivage, indicating transported clay as found in parabraunerde soils (argillic horizon)
Ap: *p* for plow, indicating an A horizon mixed by plowing to an average depth of about 15 cm
Ag: *g* for gley conditions

On top of the A horizon there may be several kinds of humus. Common designations for these follow.

O_L: *L* for litter, consisting of original vegetative material, essentially unaltered, but possibly discolored
O_F: *F* for fermentation, consisting of partially decomposed vegetative matter
O_H: *H* for humus, meaning completely decomposed vegetative matter

Soils with all three humus types and a poor mixing of the mineral segment with the humus (O_H) segment have a typically low pH (3–4) and a carbon/nitrogen (C/N) ratio of about 20 (moder). Under favorable conditions of climate and soil, humus is mixed with the mineral components, so the three subdivisions above are not discernible (mull). The C/N ratio is about 10 and pH is approximately neutral. Where plant remains are difficult to decompose, pH drops to 3 and the C/N ratio rises to approximately 40 (mor).

Transitional horizons are designated as A1, A2, and A3 and depend on minor changes within the main formation. If problems occur in separating horizons, the symbols A/B, or A/C (where no B is present) may be used. Occasionally, incipient soils are encountered in which there is no A horizon. Such soils are already inhabited by organisms producing humus in the well-aerated root zone nearest the surface. This characteristic layer of raw soils bears the name (A), pronounced the "A-bracket horizon."

B Horizons

B horizons lie between A and C horizons and contain residual or transported iron oxides; colors here are brown, shades of orange, or red. They may be distinguished by several subhorizons, as listed below.

Bv: *v* indicates weathering (with browning not derived from illuviation), as in braunerde soils (from the German *Verwitterung*)
Bs: *s* for enriched sesquioxides
Bh: *h* for enriched humus
Bsh, Bhs: Enriched in both sesquioxides and humus, one or the other dominating
Bt: *t* for clay (from the German, *Ton*) that has been translocated to the B horizon from above
Bg: *g* for gley

Transitional B horizons are designated B1, B2, and B3, depending on the need for identifying slight changes in physical traits, such as color and texture. Where change to the C horizon is indefinite, B/C is used. Where a B horizon exists because of deep chemical weathering without translocation of upper-layer materials, the designation is B-bracket, or (B). Where enrichment and fixing of alluvial material develop in the surface area, such as in the crusts of some humid soils or in deserts, the symbol B/A applies. B/A horizons having only weak humus contents are designated B/(A).

C Horizons

C horizons represent unweathered or little-altered parent material lying under the solum. Subhorizon designations follow.

Cv: *v* indicates slight weathering
Cn: *n* from new, referring to fresh, unweathered parent rock
Cc: *c* for CaCO$_3$ enrichment

In recent years, several other horizons have been added to the above lists (Semmel, 1977).

Go: G for gley or groundwater soil; *o* for oxidation horizon at the zone of mixing between water table and air as table rises and falls seasonally; mottling present
Gr: *r* for reduction horizon with standing water and low oxygen content
Sd: S for perched water horizon in pseudogley soils; *d* for impermeable (from the German *dicht*); mottling present
Sw: *w* for water in perched water horizons with gray colors
K horizons: Any subsurface horizon whose morphology is determined by carbonate content (50 percent by volume or more)
R horizons: Consolidated bedrock under soil

Principal Soil Types

Soils whose characteristics are derived from an interweaving of similar physical and chemical processes yielding the same results belong to the same general class or type. Outlined below are the major soil types with some indentifying characteristics, partially modified from outlines in Kubiëna (1953), Semmel (1977), Scheffer et al. (1976), and others.

TERRESTRIAL SOILS

(A)-C Soil Profile

(A)-C soils are raw soils in the beginning stages of development. Only a hint (if any) of humus horizon exists, and the soil traits derive from the parent material.

1. Syrosem: From the Russian name for raw soil in temperate climates; includes
   a. a lithosol (largely solid parent material) with subdivisions based on type of rock (granite syrosem, limestone syrosem) and transitional forms (such as rendzina syrosem, ranker syrosem)
   b. a regosol (largely loose parent material) with similarly based subdivisions (loess syrosem, dune syrosem) and transitional forms (such as pararendzina syrosem)
2. Råmark: From the Swedish word for raw soil in cold regions; subdivisions determined by cryoturbation traits
3. Yerma: From the Spanish word for "desert"; raw soil with primarily physical weathering traits; subdivisions according to parent material (such as dust, sand, or stone yerma)

A-C Soil Profiles

A-C soils are soils that have been weathered more than raw soils have been, so that a clear Ah horizon is present; they range from soils without much depth (numbers 1–3 below) to deep soils (numbers 4–6). All have Ah-C profiles.

1. Ranker: From the German *Rank*, for "slope"; thin soils developed from syrosem, råmark, or yerma; classified according to parent rock (basalt ranker), humus form (moder ranker), and transitional types (podzol ranker)
2. Rendzina: From the Polish word for scraping sound made during plowing; deeper soil types formed from limestone or gypsum; subdivisions according to humus (mull rendzina, moder rendzina), parent material (dolomite rendzina), stage of development (browned rendzina), and transitional type (terra fusca rendzina)
3. Pararendzina: Similar to rendzina, but with less than 50 percent $CaCO_3$ in parent material
4. Tirs: African name for self-mulching soils with high montmorillonite content; build microrelief from swelling and shrinking (gilgai, from the Australian word for tiny water hole); called grumosol, vertisol (in the United States), regur (in India)
5. Pelosol: From the Greek, *pelos*, for "clay"; similar to tirs soils but without marked self-mulching traits
6. Chernozem: Russian for "black earth"; deep Ah horizon with mull humus, characteristic of steppe climates

A-B-C *Soil Profiles*

A-B-C soils represent more completely developed soils with complex horizonation. They are usually developed from Ah-C soils.

1. Braunerde:
Ah-Bv-C soil profile with browning and associated relative increases in clay content following decalcification; subdivision according to nutrient content, pH value, transitional type (ranker braunerde, pseudogley braunerde)

2. Parabraunerde:
Ah-Al-Bt-C profile showing clay translocation Bt; subdivisions according to identifying traits (banded parabraunerde, base-poor parabraunerde)

3. Terrae calcis:
Ah-Bv-C profiles formed over limestone bedrock; subdivided into terra fusca (dark brown from influence of goethite) and terra rosa (red from presence of hematite); high-clay-content remnants of weathered bedrock present

4. Podzol:
O_L-O_F-O_H-Ah-Ae-Bh-Bs-C profile; Russian name for ash-colored soil (Ae) formed under conditions of acidic upper layers, ample precipitation, raw humus-building vegetation; characterized by whitish Ae horizon and translocated humus (Bh); may be classified as iron podzol (Bs predominates) or humus podzol (Bh predominates); Bs may form base for ortstein

5. Latosol:
Intensely weathered, deep A-B-C soil generally found in the tropics; few unweathered remnant minerals remain; Fe and Al oxides dominate and kaolinite clay forms; Latosols have good aggregate (ped) stability because of Fe flocculation; Fe-Al concretions are common, and if distributed horizontally may form basis for laterite crust; crusts favored near upper groundwater level in alternating wet-dry conditions

6. Plastosol:
Similar to latosol soils but having poor ped stability and a tendency toward compaction

HYDROMORPHIC SOILS

A-B-C *Soils*

1. Pseudogley:
Ahg-B-C profile; formed under alternating seasonal conditions of waterlogging after heavy precipitation followed by periods of drying; subdivided according to transitions (braunerde pseudogley, pelosol pseudogley)

2. Stagnogley:
Aheg-B-C profile; formed under very long seasonal waterlogging; subdivided according to humus forms (typical stagnogley, torf stagnogley)

A-G *Soils*

1. Gley:
Ah-Go-Gr profile; formed under influence of groundwater within 1.5 m of surface; subdivisions according to special traits (carbonate gley, oxigley, pelogley) and transitional forms (braunerde gley, podzol gley)

2. Wet gley: AhGo-Gr profile; formed in areas where groundwater is within 20 cm of surface
3. Peat gley: Ah-Gr profile; Ah is 15–30 percent of organic matter
4. Moor gley: (Ah)-Gr; gley overlain with up to 30 cm of torf; sometimes classified as T-(Ah)-Gr (*T* for torf)
5. Sapropel: Ah-G; fetid slime under strong reducing conditions; H_2S odors

(A)-C Soils

1. Rambla: (A)-C profile in groundwater soils near streams; raw alluvial soils with no humus horizon distinguishable; also called warp soils

A-C Soils

1. Dy: Swedish name for muddy acid soils with humus gels at bottom of brown waters
2. Gyttja: Swedish name for gray, gray-brown, or blackish wet soils rich in nutrients and organisms; muddy, inky smell, under water

Further discussion of detailed field descriptions of soils and of various taxonomic systems may be found in Olson (1981), Clarke (1971), and Buringh (1979).

ANTHROSOL ANALYSIS

Anthrosols sometimes can be detected quite easily by horizonation. In the case of a well-developed anthrosol, such as that in a tell, the soil formation is markedly different from that of the surrounding area. However, human activities often produce significant compositional and evolutionary alterations in soils that can best be detected by chemical means. It is therefore essential in thorough work to turn to laboratory analysis.

Two main groups of chemicals enter soils as a result of human activities: chemicals whose quantities are relatively large and may be measured in percentage changes; and those measured in parts per billion or relatively few parts per million, known as trace amounts. Of the chemicals that enter the soil in largest amounts during normal activities around settlements, Cook and Heizer (1965) place possible annual increments of elements in the range of approximately 0.1–0.4 percent (of the amount already present in the soil) for calcium, 0.7–6.7 percent for nitrogen, and 0.5–10 percent for phosphorus. If these values are representative of the potential annual macro- and micronutrient increments around settlements, it might be assumed that their abundance would provide information about the type, intensity, distribution, and duration of human activities. However, this assumption is complicated by the removal, in varying periods of time, of significant amounts of these additives by leaching, oxidation, and reduction. Therefore, their analysis must be undertaken with great attention to local variations in physical-chemical conditions. Of all the nutrient elements, only phosphorus (as phosphate) suffers practically no loss during natural chemical processes. Fortunately, phosphorus is associated with nearly every human activity because of its role as a basic ingredient of the DNA molecule and its presence in fertilizers, detergents, pesticides, oils, foodstuffs, medicines, body waste, garbage, and other substances. The bonding energy of phosphorus with calcium, iron, and aluminum is so great that phosphate has almost no vertical or horizontal migratory tendencies. It is therefore the most reliable of the macro- and micronutrients that can be investigated in anthrosol analysis.

The question might be asked whether trace elements can indicate the type and distribution of human activity. Unfortunately, most trace-element abundances only reveal information about human activities that have occurred since the industrial revolution. The chemical revolution, which accounts for the most significant introduction of these materials into the soil, began in earnest and in a widespread fashion only after the relatively recent development of petroleum refining. Finally, the cost of analyzing samples for trace metals and organic products has been prohibitive. New methods, such as LPLC (low-pressure liquid chromatography), are being developed for the detection and quantification of organic residues from human activities, but further research is needed before data on trace substances will be useful in understanding more fully an archaeological site.

Returning to phosphate analysis, two additional comments can be made. First, because of soil-fixing characteristics, settlement-related phosphate is present in reasonably stable forms over the entire time period of settlement activity. This advantage and the stability of phosphates have long caused hope that they could be utilized to reveal ancient human activities where no physical evidence survives. A step closer to analyzing type of land use came from research starting in the 1950s that produced a simple and quick field test for identifying settlement-related phosphate in terms of intensity and areal distribution (Eidt, 1973, 1977). The method utilizes an extracting solution of 35 ml $5N$ hydrochloric acid mixed with 5 g ammonium molybdate, dissolved in 100 ml cold water. A reducing reagent is prepared by dissolving 0.5 g ascorbic acid in 100 ml water. Two drops of the first reagent are placed on a 50 mg soil sample resting on a small, phosphate-free filter paper in order to extract the settlement phosphate. Thirty seconds later, two drops of the second reagent are applied. A blue ring forms when phosphate is present, the color traits and proportions indicating different amounts of phosphate in the sample. All readings must be made after a given time interval, usually two minutes. This is an effective ring-chromatography field test now used in many parts of the world.

Concurrent with development of the ring test, another test was gradually produced involving a separation of phosphate into three main types by a procedure known as fractionation (Chang and Jackson, 1957). This test is a nonoverlapping separation based on differences in solubilities of the three types and must be performed in the laboratory. Fraction I is easily extractable phosphate, fraction II is the tightly bound or occluded iron and aluminum phosphates, and fraction III is apatite and other tightly bound calcium phosphates. All fractions are determined quantitatively in parts per million and then expressed as a percentage of the total inorganic phosphorus. Continued experimental work has indicated that these percentages are diagnostic for broad categories of land use (Eidt, 1977) and has led to the proposal that if sufficient data can be assembled, percentages for soil samples from sites whose history is unknown can be compared with those for known land-use groupings and matched for similarities. Theoretically, phosphate patterns should be different in areas put to different land uses, such as crop-production and industrial areas or grazing and residential lands. Initial research has borne out this theory, as may be seen from table 7.6,

Table 7.6. Correlations of Inorganic Soil Phosphate Fractionation Percentages with Various Kinds of Land Use

Source of Soil (sample depth = approx. 10 cm)	Fractionation Percentages			Total PO_4^{3-}-P (in ppm)	Land-Use Types
	I	II	III		
1. Stavanger, Norway	84	7	9	1,256	Mixed-vegetable cultivation
2. Stavanger, Norway	87	6	7	1,298	Mixed-vegetable cultivation
3. Wahlstedt, Germany	83	6	11	368	Mixed-vegetable cultivation
4. Hamburg, Germany	82	10	8	78	Mixed-vegetable cultivation
5. Lüneburg, Germany	80	15	5	399	Mixed-vegetable cultivation
6. Bogotá, Colombia	85	11	4	206	Mixed-vegetable cultivation
7. Lüneburg, Germany	44	49	7	43	Forest (pine)
8. Bayreuth, Germany	48	48	4	178	Forest (pine-deciduous)
9. Southern Hamburg, Germany	53	38	9	274	Forest (beech-oak)
10. Southern Wisconsin*	50	37	13	418	Forest (oak)
11. Southeastern Wisconsin*	34	54	12	315	Forest (maple, basswood)
12. Northeastern Argentina	38	28	34	2,324	Residential (abandoned)
13. Varanasi, India	34	34	32	3,237	Residential (abandoned)
14. Southeastern Wisconsin	38	24	37	844	Residential (abandoned)
15. Southeastern Wisconsin	37	30	33	836	Residential (abandoned)
16. Southern Wisconsin	40	20[†]	40	1,393	Residential (modern)

Source: Values for no. 10 calculated from data in Hsu and Jackson, 1960. All other soils analyzed in Soils Laboratory, University of Wisconsin—Milwaukee, by R. C. Eidt, K. M. McBride, and/or D. Meyer.

*Note that the more acidic vegetation types (pine, mixed pine) have different effects on phosphate fractions than the more basic types (maple, basswood). This is apparent even in areas with similar soil types, as may be seen from nos. 10 and 11.

[†]Note that in sample no. 16 the occluded phosphate (fraction II) is lower, reflecting the more recent age of the settlement (occlusion is time-dependent).

which shows correlations between type distribution, and concentration patterns of the three phosphate fractional percentages and certain land uses.

As the statistical evidence accumulates, it appears that the diagnostic nature of phosphate fractional percentages is being confirmed within certain recognizable constraints. It is known, for example, that extremely ionic environments have skewed fractional values that relate to pH disturbance. Methods of adjusting such values for matching with other soils are being developed. Since most settlement soils fall within "normal" pH values, normal fractionation methods are usually adequate.

Until now one disadvantage of the fractionation method has been the time required for the quantitative analysis. Without expensive automated equipment, about 10 days are needed to process a batch of 10–15 soil samples— roughly the same time it takes for ^{14}C sample analysis. The time required involves several shelf days of chemical oxidation. Another disadvantage has been complexity of laboratory technique. However, the latest version is inexpensive and lends itself well to the small laboratory.

The complete method of phosphate analysis, consisting of a rapid ring test in the field and rapid laboratory fractionation, offers the advantage of using only 1 g soil samples, which means that the landscape need not be disturbed. Regardless of whether it is employed with conventional methods of archaeology, it offers a valuable addition to the search for more accurate location and interpretation of anthrosols. Finally, it holds promise of pointing toward new directions for accurate and systematic classification of anthrosols.

Soil phosphate fractionation should be carried out in conjunction with studies of organic matter, Eh, and pH. Special procedures appropriate to anthrosol analysis are summarized in the appendix.

Acknowledgments

The author wishes to thank the National Science Foundation (Grant SOC 77–24524) and the Graduate School and the Center for Latin America of the University of Wisconsin—Milwaukee for financial assistance during research and writing of this manuscript.

APPENDIX: NEW METHOD FOR DETERMINING SOIL PHOSPHATE FRACTIONATION

This new soil phosphate fractionation method is much simpler than standard methods now in use, but equally accurate and reliable. It has been developed under NSF Grant SOC 77-24524.

FRACTION I

NaOH/NaCl−Extractable Phosphate

Shake 1 g air-dried, lightly ground, sieved (500 μm) soil with 40 ml 0.1N NaOH/1N NaCl for 12 hours. Centrifuge for 30 min at 2,500 rpm. Pipette 2 ml aliquot into a 25-ml volumetric flask. Add 6 ml H_2O (distilled). Neutralize by adding drops of 6N HCl or 6N NaOH until mixture is clear. Add 2 ml standard Murphy−Riley solution. After 15 min read at 882.5 nm on spectrophotometer against blank prepared in the same way except with NaOH/NaCl extractant instead of soil aliquot and against standard prepared in the same way except with NaOH/NaCl extractant (2 ml), standard P solution (2 ml), H_2O (6 ml), and Murphy−Riley solution (2 ml). NOTE: If original soil aliquots are discolored dark by organic matter, color must be cleared by taking 10 ml soil aliquot, adding 10 ml XAD−2 solution, centrifuging 10 min at 1,500 rpm and filtering with No. 2 Whatman paper. Then proceed as with original aliquot.

CaCo₃-sorbed−Extractable Phosphate

Remove NaOH/NaCl solution remaining from procedure outlined above from soil-sample centrifuge tube by suction. Discard liquid. Wash soil twice by adding 25 ml 1N NaCl each time. Shake and centrifuge 30 min at 2,500 rpm. Discard wash. Add 50 ml sodium citrate−sodium bicarbonate solution and place tube in 82°C water bath for 30 min. Stir every 3 min. Centrifuge 15 min at 2,500 rpm. Transfer supernatant to 100 ml volumetric flask by suction. Add 1 drop 3N $FeCl_3$ to flasks. Fill to volume with H_2O. For colorimetry, pipette 2 ml aliquot into 25-ml volumetric flask. Add 4 ml H_2O, 2 ml molybdate solution, and 6 ml $SnCl_2$. Shake. After 10 min read against blank at 725 nm. Use blank prepared with 2 ml extractant (instead of soil aliquot), 4 ml H_2O, 2 ml molybdate, and 6 ml $SnCl_2$. Standard has 2 ml extractant, 2 ml standard solution, 2 ml H_2O, 2 ml molybdate solution, and 6 ml $SnCl_2$. NOTE: If original soil aliquots are discolored (amber or dark brown), clear by following procedure noted above. PO_4^{3-}-P is calculated in ppm and added to NaOH/NaCl-phosphate above.

FRACTION II

Na-citrate−, Na-bicarbonate−, Na-dithionite−Extractable Phosphate

Wash soil remnant from step above with 25 ml 1N NaCl by vortexing and centrifuging for 30 min at 2,500 rpm. Add 50 ml 0.22N Na-citrate/0.11N Na-bicarbonate after discarding wash by suction. Add 50 ml of extractant to a clean test tube. Place both in 82°C water bath. Stir every 3 min until 15 min pass. Add 1 g Na-dithionite and heat 15 min more. Remove, cool, centrifuge for 15 min at 2,500 rpm. Transfer supernatant to 100-ml volumetric flask by suction. Add 25 ml NaCl to soil, then vortex and centrifuge for 10 min at 2,500 rpm. Remove liquid and add to 100-ml flask. Add 25 ml NaCl to blank flask. Add 1 drop 3N $FeCl_3$ to each 100-ml flask. Cover flask mouths with filter-paper cones and allow to stand 8 days for oxidation of dithionite. Take colorimetric reading as for second part of fraction-I procedure explained above.

FRACTION III

HCl-Extractable Phosphate

Wash sediment remaining in centrifuge tube in 25 ml 1N NaCl. Vortex, centrifuge for 30 min at 2,500 rpm, and discard wash by suction. Add 40 ml 1N HCl to sediment, vortex, shake for 4 hours at 40 rpm on 24-inch wheel, or equivalent. Centrifuge for 15 min at 2,500 rpm. Pipette 2 ml aliquot into 25-ml volumetric flask. Add 14 ml H$_2$O. Neutralize as in fraction-I procedure, first part. Add 4 ml Murphy–Riley solution. Prepare standard with 2 ml 1N HCl, 14 ml H$_2$O, and 4 ml Murphy–Riley solution. Neutralize. Read after 15 min in spectrophotometer set at 882.5 nm.

Spectrophotometric values are first computed in parts per million for each fraction; then total phosphate is calculated by addition of the three fractional values. Computations are by standard formula for colorimetric conversion:

$$ppm = Q_s(C_s) \times \frac{Q_e}{Q_a} \times \frac{A_u}{A_s},$$

where Q_s = quantity of standard in ml, C_s = concentration of standard in μg/ml, Q_e = quantity of extracting solution in ml, Q_a = quantity of aliquot unknown in ml, A_u = absorbance value of unknown, and A_s = absorbance value of standard. The percentage of each fraction is then calculated and the three percentages compared, as in table 7.6. NOTE: Stock and standard solutions are prepared as in R. C. Eidt, *Field and Laboratory Analysis of Anthrosols* (Milwaukee, 1977).

ANALYSIS OF ORGANIC MATTER

Organic matter can be analyzed by various methods, one of the most useful being that devised by Walkley and Black (1934). Their method is based on the fact that chromic acid oxidizes organic matter equivalent to 77 percent of the total carbon present. Soil organic matter contains approximately 58 percent carbon. Using these facts, a close estimate of the content of organic matter can be made, although the presence of large amounts of reduction agents, such as chlorides (see below) and iron, may lead to high results.

Place 1 g dry soil (0.1 g if peat is being used; less than 2 g if nonorganic soils with 1 percent organic matter are being used) in a 500-ml Erlenmeyer flask. Pipette 20 ml 1N K$_2$Cr$_2$O$_7$ into the flask and swirl. Add 20 ml concentrated H$_2$SO$_4$ (to eliminate interference from soil chlorides if there are any). Allow to stand 30 min. Add 200 ml H$_2$O. Add ml 85% H$_3$PO$_4$, 0.2 g NaF, and 30 drops indicator (prepared by dissolving 3.71g 1,10-phenanthroline monohydrate and 1.74 g FeSO$_4$·7H$_2$O in 250 ml water). Back-titrate to a burgundy color with 0.5 N ferrous solution (prepared by dissolving 196.1g Fe(NH$_4$)$_2$(SO$_4$)$_2$·6H$_2$O in 800 ml H$_2$O containing 20 ml concentrated H$_2$SO$_4$ and diluting to 1 liter; this solution should be standardized daily against the dichromate solution). Prepare a blank using same procedure without soil.

Milliequivalents of oxidizable material (Ox) per gram of soil may be calculated as follows:

$$\text{meq Ox/g} = \frac{(\text{ml Fe}^{2+} \text{ for blank} - \text{ml Fe}^{2+} \text{ for sample}) \times N \text{ of Fe}^{2+} \text{ solution}}{\text{weight of soil in g}}.$$

Percent of carbon is computed as follows:

$$\%C = \text{meq OM/g} \times \frac{12}{4} \times \frac{1}{1000} \text{ g/meq} \times 0.77^{-1} \times 100,$$

where 12 is the atomic weight and 4 is the valence of carbon.

The basic reaction equation is

$$C_2H_{12}O_6(OM) + K_2Cr_{12}O_7 \rightarrow 6CO_2 + Cr_2(SO_4)_3 + K_2SO_4 + 6H_2O.$$

Percent of organic matter is calculated by using the formula:

$$\%C = 0.58\% \text{ OM, or, } \%OM = \frac{\%C}{0.58}.$$

A useful rule of thumb in estimating organic-matter content of the soil plow zone is to double the amount of organic carbon determined in the laboratory. Since organic matter varies through the different soil horizons, it is diagnostic of the environmental conditions that cause the differences. If the percentage of organic matter exceeds 20, the soil is no longer a mineral soil and is classed as an organic soil.

Eh DETERMINATION

Eh values may be determined on the same air-dried soil samples used for pH measurements. The sample should be swirled after the pH reading, then allowed to stand 3 hours. Eh is then measured with a platinum probe inserted in place of the normal pH probe. The meter is set in the millivolt mode during Eh measurements. If millivolts cannot be read directly, the pH mode is used and the reading converted with the equation:

$$E_0 = 59(7-R),$$

where R is the scale reading with the platinum electrode and E_0 is in millivolts.

Because of variations of Eh with pH, values obtained are often adjusted to a pH of 7 by adding 59 mv (at 25°C) for each soil pH unit above 7, or by subtracting 59 mv for each soil pH unit below 7:

$$pH - 7\text{-adjusted Eh} = E_0 + 59(pH - 7).$$

Since the reference electrode is normally a calomel half-cell with a voltage registering 245 mv (at 25°C) below that measured against the hydrogen electrode of standard tests, the difference in voltage must be added to the reading. Eh may be calculated from the following equation:

$$\text{corrected Eh} = (E_0 + 245) + 59(pH - 7),$$

where pH is read from the sample tested.

Care should be taken to clean the platinum probe with water between measurements. Occasionally it may be necessary to clean the platinum in 6N HCl. The probe should be washed thoroughly afterward in water and wiped with soft tissue.

A review of Eh soil theory appears in Black (1968) and in other items listed in the reference list of this chapter.

pH DETERMINATION

A special method of pH determination is used for anthrosol analysis. The common technique of mixing relatively large soil samples with water has the disadvantages that much more soil is needed than is sometimes convenient to ship from archaeological sites to the laboratory and that the pH values vary for different soil-solution ratios. Measurement of soil pH in $0.01M$ CaCl$_2$ approaches an ideal answer to such problems, and, although preparation of the stock solution is somewhat lengthy, the results are worth the effort. In addition to the possibility of using minute (1 g or less) soil samples and varying soil-solution ratios, the so-called liquid junction potential is minimized. The latter has to do with the discrepancy in readings between the solution and the soil body itself.

Place 2 g lightly ground, air-dried samples in a 50-ml beaker. Add 5 ml $0.01M$ CaCl$_2$. Swirl solutions and allow them to settle for 30 min prior to reading. Place the pH probe at the interface between settled soil and soil solution, or simply in the liquid portion.

The stock solution of $3.6M$ CaCl$_2$ is prepared by weighing 263 g CaCl$_2 \cdot 2H_2O$ in a preweighed 600-ml beaker. Dissolve with water. Do not exceed 450 ml total solution. Transfer all liquid to 500 ml volumetric flask. If solution is cloudy, filter with No. 40 Whatman filter paper.

To standardize the stock CaCl$_2$ solution, titrate with $0.1N$ AgNO$_3$. (Prepare by dissolving 8.49 g AgNO$_3$ in 100-ml beaker. Pour into 500-ml flask and bring to volume.) Pipette 20 ml stock solution into 1-liter volumetric flask and bring to volume. Pipette 25 ml of this diluted CaCl$_2$ solution into a 150-ml beaker and add 1.0 ml of 5.0% K$_2$CrO$_4$ indicator. (Prepare indicator by dissolving 5.0 g K$_2$CrO$_4$ with exactly 95 ml water.) Place beaker on stirring apparatus at moderate speed. Pour AgNO$_3$ solution into a 50-ml burette, record starting level, titrate by slowly releasing AgNO$_3$ into the diluted CaCl$_2$/K$_2$CrO$_4$ solution. A temporary reddish color will appear gradually until it becomes permanent at the end point. Compute normality of dilute CaCl$_2$ as follows:

$$N \text{ AgNO}_3 \times \text{quantity AgNO}_3 = N \text{ CaCl}_2 \times \text{quantity CaCl}_2.$$

Next, calculate molarity of stock CaCl$_2$ by using known molarity of the diluted CaCl$_2$:

$$\text{volume stock CaCl}_2 \times M \text{ stock CaCl}_2 =$$
$$\text{volume diluted CaCl}_2 \times M \text{ diluted CaCl}_2.$$

If the molarity of the stock solution is in the range of several hundredths molar, no other adjustments are necessary. If not, dilute stock solution or add concentrated CaCl$_2$ (slightly higher than 3.6 molar) as required. Repeat titration to determine actual molarity of adjusted stock solution.

To prepare $0.01M$ CaCl$_2$ for pH determination, dilute the appropriate amount of stock solution, based on actual molarity, in a 2-liter flask. Adjust pH of this solution to $5.0-6.5$ by adding a small amount of Ca(OH)$_2$ or HCl. Check electrical conductivity. Value should be 2.32 ± 0.08 mmho/cm at 25°C.

REFERENCES

Adams, J. A., T. W. Walker. 1975. Some properties of a chronotoposequence of soils from granite in New Zealand: 2. Forms and amounts of phosphorus. *Geoderma* 13: 41–51.

Allaway, W. H. 1975. *The effect of soils and fertilizers on human and animal nutrition*. Washington, D.C: U.S.D.A. Agricultural Information Bulletin no. 378.

Baker, A. R. H., and R. A. Butlin, eds. 1973. *Studies of field systems in the British Isles*. Cambridge: Cambridge Univ. Press.

Ballantyne, A. K. 1963. Recent accumulation of salts in the soils of southeastern Saskatchewan. *Canadian Journal of Soil Science* 43:52–58.

Bertine, K. K., and E. D. Goldberg. 1971. Fossil fuel combustion of the major sedimentary cycle. *Science* 173:233–35.

Birkeland, P. W. 1974. *Pedology, weathering, and geomorphological research*. New York: Oxford University press.

Black, C. A. 1968. *Soil Plant Relationships*. 2d ed. New York: J. Wiley.

Bloomfield, C. 1953. A study of podzolization, Parts I and II, *Journal of Soil Science* 4:5–23.

Brady, N. C. 1974. *The Nature and Properties of Soils*. 8th ed. New York: Macmillan.

Briggs, D. 1977. *Soils*. London: Butterworth.

Buel, S. W., F. D. Hole, and R. J. McCracken. 1980. *Soil genesis and classification*. Ames: Univ. of Iowa Press.

Bunting, B. T. 1967. *The geography of soil*. London: Hutchinson.

Buringh, P. 1979. *Introduction to the study of soils in tropical and subtropical regions*. 3d ed. Wageningen: Center for Agricultural Publishing and Documentation.

Butzer, K. W. 1971. *Environment and archaeology*. 2d ed. Chicago: Aldine.

Chang, S. C., and M. C. Jackson. 1957. Fractionation of soil phosphorus. *Soil Science* 80:133–44.

Clarke, G. R. 1971. *The study of soil in the field*. 5th ed. Oxford: Oxford Univ. Press.

Coleman, R. 1966. The importance of sulfur as a plant nutrient in world crop production. *Soil Science* 101:230–39.

Connor, J. J., and H. T. Schacklette. 1975. *Background Geochemistry of some rocks, soils, plants, and vegetables in the coterminous United States*. Geological Survey Professional Paper 574-F. Washington, D.C.

Cook, S. F., and R. F. Heizer. 1965. *Studies on the chemical analysis of archaeological sites*. Berkeley: Univ. of California Publications in Anthropology no. 2.

Denevan, W. M. 1970. Aboriginal drained-field cultivation in the Americas. *Science* 169:647–54.

Donahue, R. L., J. C. Schickluna, and L. S. Robertson. 1971. *Soils: An introduction to soils and plant growth*. 3d ed. Englewood Cliffs, NJ: Prentice-Hall.

Eidt, R. C. 1959. Aboriginal Chibcha settlement in Colombia. *Annals of the Association of American Geographers* 49:374–92.

———. 1973. A rapid chemical field test for archaeological site surveying. *American Antiquity* 38:206–10.

———. 1977. Detection and examination of anthrosols by phosphate analysis. *Science* 197:1327–33.

———. 1981. Rural society and land use change in the Highland Basins of Colombia. *Latin American Studies* 3:25–45.

———. In press. *Advances in abandoned settlement analysis: Application to prehistoric anthrosols in Colombia, South America*, University of Wisconsin—Milwaukee, Center for Latin America.

Evans, J. G. 1978. *An introduction to environmental archaeology*. Ithaca, NY: Cornell Univ. Press.

Fernea, R. A. 1970. *Shaykh and Effendi: Changing patterns of authority among the El Shabana of southern Iraq*. Cambridge, MA: Harvard Univ. Press.

Gibson, M. 1974. Violation of fallow and engineered disaster in Mesopotamian civilization. In *Irrigation's impact on society*, ed. T. E. Downing and M. Gibson, 7–19. Tucson: Univ. of Arizona Press.

Goodyear, F. H. 1971. *Archaeological site science*. London, Academic Press.

Gourou, P. 1961. *The tropical world: Its social and economic conditions and its future status*. London: Longmans, Green.

Greene, H. 1963. Prospects in soil science. *Journal of Soil Science* 14:1–11.

Heidenreich, C. E., and S. Navratil. 1973. Soil analysis at the Robataille site: Part I. Determining the perimeter of the village. *Ontario Archaeology* 20:25–32.

Hole, F. D. 1961. A classification of pedoturbations and some other processes and factors of soil formation in relation to isotropism and anisotropism. *Soil Science* 91:375–77.

Hole, F. D., and M. Hironaka. 1960. An experiment in ordination of some soil profiles. *Soil Science Society of America, Proceedings* 24:309–12.

Hsu, P. H., and M. L. Jackson. 1960. Inorganic phosphate transformations by chemical weathering in soils as influenced by pH. *Soil Science* 90:16–20.

Jackson, M. L. 1963. Aluminum bonding in soils: A unifying principle in soil science. *Soil Science Society of America, Proceedings* 27:1–10.

Jeffery, J. W. O. 1960. Iron and the Eh of waterlogged soils with particular reference to paddy. *Journal of Soil Science* 11:140–48.

Joensuu, O. I. 1971. Fossil fuels as a source of mercury pollution. *Science* 172:1027–28.

Klein, D. H., and D. Russell. 1973. Heavy metals: Fallout around a power plant. *Environmental Science and Technology* 7:357–58.

Kubiëna, W. L. 1953. *The soils of Europe*. Madrid: Consejo Superior de Investigaciones Científicas.

Laeyendecker-Roosenburg, D. M. 1966. A palynological investigation of some archaeologically interesting sections in northwestern Surinam. *Leidse Geologische Medeldelingen* 38:31–36.

Laws, W. D., and D. D. Evans. 1949. The effects of long-time cultivation on some physical and chemical properties of two rendzina soils. *Soil Science Society of America, Proceedings* 14:15–19.

Limbrey, S. 1975. *Soil science and archaeology*. London: Academic Press.

Lisk, D. J. 1972. Trace metals in soils, plants, and animals. *Advances in Agronomy* 24:267–325.

Lockeretz, W. 1978. The lessons of the Dust Bowl. *American Scientist* 66:560–69.

MacLeod, A. 1979. Warning of "poison" in the ground fails to shake English villagers. *Christian Science Monitor*, Jan. 25, p. 8.

Mayhew, A. 1973. *Rural settlement and farming in Germany*. London: B. T. Batsford.

Millar, C. E., and L. M. Turk. 1949. *Fundamentals of Soil Science*. New York: J. Wiley, 237.

Olson, G. W. 1981. *Soils and the environment*. New York: Chapman and Hall.

Pape, J. C. 1970. Plaggen soils in the Netherlands. *Geoderma* 4:229–54.

Parsons, J. J., and W. J. Bowan. 1966. Ancient ridged fields of the San Jorge River floodplain, Colombia. *Geographical Review* 56:317–43.

Pierre, W. H. 1928. Nitrogenous fertilizers and soil acidity. *Journal of the American Society of Agronomy* 20:254–69.

Pyddoke, E. 1961. *Stratification for the archaeologist*. London: Phoenix.

Rapp, G. R. Jr., 1978. Lithological studies. In *Excavations at Nichoria in southwest Greece*, ed. G. R. Rapp, Jr., and S. E. Aschenbrenner, 225–33. Minneapolis: Univ. of Minnesota Press.

Rodrigo, D. M. 1963. Redox potential—with special reference to rice culture. *Tropical Agriculturist* 119:85–100.

Scheffer, F., P. Schachtschabel, H.-P. Blume, K. H. Hartge, and U. Schwertmann. 1976. *Lehrbuch der Bodenkunde*. 9th ed. Stuttgart: Enke.

Semmel, A. 1977. *Grundzüge der Bodengeographie*. Stuttgart: B. G. Teubner.

Sherlock, R. L. 1931. *Man's influence on the Earth*. London: Butterworth.

Soil Survey Staff. 1975. *Soil taxonomy*. Washington, DC: U.S.D.A. Agricultural Handbook no. 436.

Volobuev, V. R. 1964. *Ecology of soils*. Trans. A. Gourevich. Jerusalem: S. Monson.

Walkley, A., and I. A. Black. 1934. An examination of the Degtjareff method for determining soil organic matter, and a proposed modification of the chromic acid titration method. *Soil Science* 37:29–38.

Walsh, L. M. 1972. *Profitable management of Wisconsin soils*. Madison: American Printing and Publishing.

Warren, H. V. 1973. Some trace element concentrations in various environments. In *Environmental Medicine*, ed. G. M. Howe and J. A. Loraine. London: in press, 9–24.

Warren, H. V., and R. E. Delavault. 1971. Variations in the copper, zinc, lead, and molybdenum contents of some vegetables and their supporting soil. In *Environmental Geochemistry*, ed. H. L. Cannon and H. C. Hopps. Boulder, CO: Geological Society of America Memoir 123:97–108.

Williams, R. B. G. 1973. Frost and the works of man. *Antiquity* 47:19–31.

Young, A. 1976. *Tropical soils and soil survey*. Cambridge, England: Cambridge Univ. Press.

Zeuner, F. E. 1959. *The Pleistocene period: Its climate, chronology and faunal successions*. London: Hutchinson.

8

Geophysical Surveying of Archaeological Sites

JOHN W. WEYMOUTH and ROBERT HUGGINS

ABSTRACT

In this chapter the two most commonly applied geophysical methods for assessing archaeological sites are described in detail. Magnetic surveying responds to contrasts in the magnetic properties of soils, which can be brought about by, among other causes, human activities such as burning, by humic decomposition, by compaction, and by the introduction of structures. Resistivity surveying responds to differences in the electrical conductivity of soils, which can be brought about by, among other causes, alterations of the natural soil profile through the construction of mounds or ditches, by compaction, or by the introduction of structures. Magnetic surveying is independent of soil moisture but will not respond if a structure is composed of the same magnetic material as the surrounding soil. Resistivity surveying is strongly dependent on soil moisture and on the contrast in porosity between a structure and its surrounding matrix.

Magnetic surveying is faster and easier to interpret but cannot easily be carried out near interfering magnetic sources, such as modern buildings or power lines. Resistivity surveying is slower and somewhat more difficult to interpret but is free from the interference of nearby buildings and power lines.

Like all remote sensing methods, those described here are nondestructive and are considerably more economical than test excavations, if they are properly conducted. A geophysical survey carefully coordinated with an archaeological program can provide valuable information for the planning and execution of that program, as the examples given for both techniques demonstrate.

The last 25 years have seen many applications of geophysical survey techniques in archaeological site surveying. The methods usually emerged from small-scale field experiments by physicists and geophysicists; they were then applied primarily by European archaeologists. In the last decade, however, there has been a growing awareness among American archaeologists of the value of geophysical methods in site location and mapping.

The various geophysical techniques for gathering information about subsurface features all depend, in one way or another, on differences in electric, magnetic, or elastic (seismic) properties of rocks and sediments. The techniques may be classified as passive or active. In the first category, existing force fields are measured directly without instrumentally generated signals, and the results are interpreted in terms of subsurface features perturbing the field. Magnetic, thermal, and gravity measurements fall in this category. In the second, or active, category, instrumentally generated signals pass through the subsurface and are then detected and recorded. Seismic techniques, electromagnetic techniques (including use of the simple metal detector, the pulsed-induction metal detector, and the soil conductivity meter), earth resistivity measurements, and the recently developed ground-penetrating radar are all active devices. This chapter will deal with one technique in each category: magnetic and earth resistivity surveying. For summaries of most geophysical methods see Aitken (1974) or Tite (1972).

MAGNETIC SURVEYING

SUMMARY OF METHOD

Magnetic surveying (or prospecting), as practiced on archaeological sites, consists of measuring the magnitude of the earth's magnetic field at each point on a grid established over the site. Variations in the magnetic properties of the subsurface material (sediments, rocks, or artificial materials such as brick) can produce an observable variation (anomaly) in the measured magnetic field. Anomalies may be caused by artificial structures such as walls, ditches, foundations, fire hearths, pits, or even an area of more intensive habitation. The task of interpretation—to separate the results of human activity from geological variations in subsurface materials—is guided by knowledge of the physics of soil magnetization and by the manipulating and displaying of the data in various ways so as to reveal significant patterns. At any site, successful application of the method depends on the magnetic properties of the local subsurface, the extent and nature of the human activity, the burial depths of artificial and natural features, and, finally, the care taken in field measurement and analysis.

HISTORICAL DEVELOPMENT

For several decades geophysicists have used magnetic surveying in the search for minerals, but until 20 years ago the instruments available were not sufficiently sensitive for archaeological applications. The development of the proton magnetometer provided just the needed sensitivity. Belshe (1957) seems to have been the first to experiment with a proton magnetometer in an archaeological context. He was followed by Aitken, Webster, and Rees (1958), and soon the Oxford University group under Aitken was obtaining results of archaeological usefulness. Subsequently, groups developed in Germany (Scollar, 1961), Italy (Lerici, 1961; Linington, 1964), and France (Hesse, 1962). The literature in this field is now fairly extensive. Reports of technical aspects, until recently, have been concentrated in four journals: *Archaeometry*, *Prospezioni Archeologiche*, *Revue d'archéometrie* and *Archaeo-Physika*. The literature on applications is scattered throughout a number of geophysical and archaeological journals; extensive references may be found in Aitken (1974) and Tite (1972).

In the United States, an early practitioner of magnetic surveying was the MASCA group at the University Museum, University of Pennsylvania, under Rainey (Rainey and Ralph, 1966) and Ralph (1964; Ralph, Morrison, and O'Brien, 1968). Their applications for the most part were in the Old World, as was the survey work by Rapp and Henrickson (in McDonald and Rapp, 1972). To our knowledge the first recorded application on this continent was at Angel Mounds by Black and Johnston (1962). Subsequent applications in this hemisphere are documented in Ezell et al. (1965), Morrison, Clewlow, and Heizer (1970), Breiner and Coe (1972), Arnold (1974), Nashold (1977), and von Frese (1978). The MASCA group has conducted some magnetic surveys in the eastern United States (Bevan, 1975).

The present writers recently started magnetic surveying on sites in the Central Plains (Weymouth, 1976) and, in conjunction with the Midwest Archaeological Center (National Park Service) and other agencies, have surveyed or analyzed data from surveys covering approximately 33 hectares in about 10 states and a few sites outside the hemisphere (Weymouth and Nickel, 1977; Weymouth, 1979; Bleed et al., 1980).

In the last few years several groups and individuals throughout the United States have started using magnetic surveying on sites. In one case a group from Michigan State University has covered a fairly large area (Mason, 1981).

THEORY

The magnetic field at any point on the earth can be defined, for our purposes, as the direction taken by a compass needle freely suspended there. The direction can be specified in terms of *declination*, the angle between true north and the horizontal component of the earth's field, and *inclination* (or *dip*), the angle between horizontal and the direction of the total field. The field strength or magnitude is proportional to the maximum torque exerted on the compass needle by the field. In this chapter the unit of magnetic field strength used is the *gamma* (1 gamma is also equal to 1 nanotesla, the SI unit). The earth's magnetic field in the United States varies from roughly 49,500 to 59,500 gamma, and its inclination varies from 60° to 75° below the horizontal.

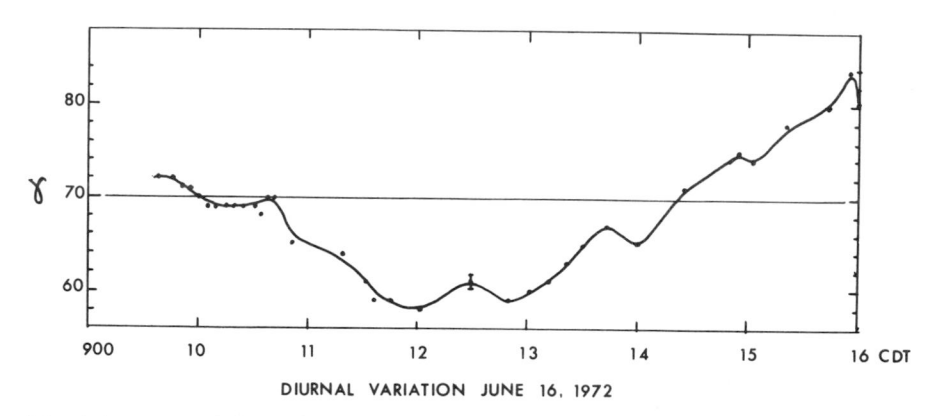

Fig. 8.1. Typical diurnal variation recorded by a stationary proton magnetometer.

Another aspect of the magnetic field with which we must be concerned is its variation with time. In a regular diurnal variation the magnitude decreases during the middle of the day by approximately 20 or 30 gamma from higher morning and evening values (fig. 8.1). During magnetic storms larger variations occur over time periods of a few hours to days.

Magnetic Properties of Soils

In the presence of a magnetic field, material such as soils, rocks, and ferrous objects can become magnetized. Such a magnetization is said to be *induced*. In addition to induced magnetization, which vanishes when the applied field is removed, some materials exhibit *remanent* magnetization, magnetization that persists in the absence of an applied field. Baked clays and some rocks retain a thermoremanent magnetization after being heated to several hundred degrees centigrade and then cooled in a magnetic field. Remanent magnetization can also arise from chemical change or from the settling of small particles in a magnetic field (see pp. 240–43).

A range of time responses exists between the extremes of permanent magnetization and the very rapid component of induced magnetization. Because the time response depends on particle sizes in the soil, parts of soils can become magnetized very rapidly while other parts change their magnetization very slowly. (This phenomenon of *viscous magnetization* is explained in detail on pp. 243, 245.)

The compounds in soils which are important in causing magnetization are hematite ($\alpha-Fe_2O_3$), magnetite (Fe_3O_4) and maghemite ($\gamma-Fe_2O_3$). The latter two compounds are much more strongly magnetic than the first, their saturation magnetization being approximately 200 times that of hematite. Since soils contain a few to several percent iron oxides, these compounds and their conversion from one form to another are the significant factors of soil magnetization. Two measures of the response of a material to magnetization are its *magnetic susceptibility*, which is the ratio of magnetization (dipole strength per unit volume) to magnetic field strength, and its *specific susceptibility* (dipole strength per unit mass per unit field). This latter quantity is measured in emu per gram. (For definitions of units see Aitken, 1974:140.)

Typical values for specific susceptibility are as follows (all in units of 10^{-6} emu/g).

Limestone, some unbaked clays	10
Subsoils	50–100
Topsoils	100–1,000
Heated soils, fired clays	1,000–2,500

Natural and anthropogenic activities can cause a conversion from hematite to magnetite or maghemite, thus resulting in a greater susceptibility. Such conversion processes, explored by Le Borgne (1955, 1960), occur when hematite is reduced to magnetite either during heating, as in a hearth, or during anaerobic decomposition in humic soil. Consequently, topsoils usually have a higher susceptibility than subsoils and hearths, while burned houses and trash pits may be even more magnetic. For a complete discussion on various aspects of soil magnetization relating to magnetic surveying of archaeological sites, see Graham and Scollar (1976).

Measurement of the Magnetic Field

The three instruments primarily used for archaeological surveys are the proton free-precession magnetometer, the fluxgate magnetometer, and the cesium or rubidium magnetometer. The proton magnetometer is the least expensive and by far the most widely used. Its sensor consists of a coil surrounded by a hydrogen-rich liquid (water or kerosene). A polarization current through the coil creates in the liquid a magnetic field many times more intense than the earth's. This field partially polarizes the hydrogen's nuclear protons, which are spinning magnetic dipoles. The polarization current is then quenched and the protons precess (gyrate) in the field of the earth. For the few seconds that the protons precess coherently, a voltage is induced in the coil. This voltage is amplified, the frequency measured, and the results displayed in gammas. In older commercial units the total cycle time may be 7 seconds, but this can be shortened to 3 or 4 seconds with no loss of sensitivity; the normal sensitivity is 1 gamma, which can be increased to 0.25 gamma in the usual portable models. Commercial units are now available with cycle times of 1.5 seconds and sensitivities to 0.1 gamma.

The fluxgate magnetometer measures the component of the vector field along the axis of a coil and is thus strongly direction-dependent. This disadvantage can be overcome by combining two instruments as a *gradiometer*, which is direction-independent. A gradiometer measures the magnetic gradient or difference in magnetic field at the two sensors. The requirement for parallel alignment of the detector units is stringent; therefore, the cost of a differential fluxgate magnetometer is high.

In cesium or rubidium magnetometers atomic electrons of these two elements (in a vaporized state) replace the nuclear protons of the proton magnetometer; otherwise the operational principles are comparable. Both fluxgate and cesium or rubidium magnetometers have a high sensitivity and also provide continuous measurements of the field, but their costs are greater than that of a proton magnetometer.

Anomalies Produced by Local Features

An isolated magnetic feature (termed a *source*) whose dimensions are small compared with the sensor distance produces the simplest, so-called dipole anomaly. A *normal dipole anomaly* results from induced magnetization (such as in a small pit), where polarization will be in the same direction as the earth's field. However, if the magnetization is permanent (such as in a piece of iron, a burned rock, or a fire hearth), the polarization may be in a different direction than that of the earth's field and the resulting anomaly is termed a *nonnormal dipole anomaly*.

The total magnetic field in the neighborhood of a normal dipole anomaly is the combination (vector sum) of the uniform, downward-pointing field of the earth and a weak, dipole field from the source feature. The profile of magnetic values measured along a south-north line is represented in figure 8.2. Three characteristics of this anomaly type may be noted:

1. The maximum intensity of the magnetic profile is displaced to the south of the source by about one-third the source–sensor distance.

2. The full width of the profile at half maximum is about equal to the source–sensor distance.

3. The negative region, due to the source dipole field opposing the earth's field, is about 10 percent of the maximum intensity. As in the first characteristic, this is true at midlatitudes.

Deviations of a magnetic anomaly from these characteristics imply a nonnormal dipole—that is, a source with permanent magnetization. Long, narrow pieces of iron or burned rocks in particular produce anomaly profiles with large deviations from those of normal dipole anomalies. Thus, if the minimum is not north of the maximum, or if the minimum deviates appreciably in magnitude from 10 percent of the maximum, one can conclude that the source is not a feature resulting from induced magnetization.

The magnitude of the anomaly depends strongly on the source–sensor distance, decreasing proportionally with the inverse of the cube of that distance. The magnitude also depends on source volume, V, and magnetic susceptibility contrast, k. For further discussion see Breiner (1973), Tite (1972), Aitken (1974), or Weymouth (1976).

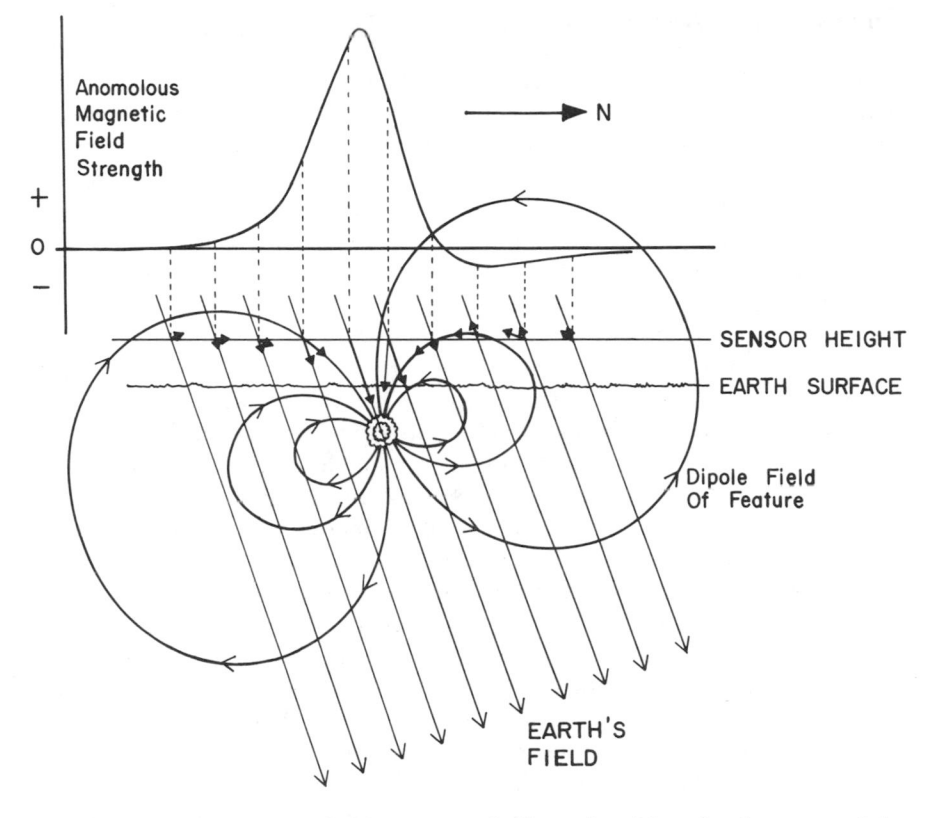

Fig. 8.2. Combination of a dipole magnetic field produced by a local source and the magnetic field of the earth. The plot is that of the magnitude of the sum of the two fields as measured by a magnetometer passing over the source.

A filled ditch oriented east-west is like a row of parallel dipoles, with a negative anomaly north of the maximum. For other shapes—ditches, plates, etc.—see Aitken and Alldred (1964), Breiner (1973) or Linington (1972).

APPLICATION OF THE METHOD

The method to be used in surveying a site depends on the information sought. If one wants information on possible linear features, such as ditches or walls of known orientation, or if the location of small features (such as fire hearths) in buildings of presumed location are known approximately, then one or a few magnetometer traverses may suffice. If surveying is done over a short period of time, it is not necessary to use a second or reference magnetometer; plotting profiles of such traverses may be sufficient to reveal the information desired.

If simple traverses do not suffice, then the problem is to seek patterns of anomalies in a two-dimensional mapping of the magnetic field over the site.

Mapping is accomplished by measuring the field on a grid of points over the site. When more than a few minutes are needed to survey the area, some method must be used to correct for the temporal variations of the earth's field. Without such corrections, the resulting magnetic map will be distorted and spurious anomalies will appear, particularly along traverse rows. The basic idea behind all corrections is the assumption that the earth's field changes in time simultaneously everywhere over a region considerably larger than the site being mapped. To carry out the corrections, several approaches can be used.

If only one magnetometer is available, it alone can be used to correct for temporal variations. The operator simply intersperses repeated readings at a single reference station between groups of grid readings—for instance, after each row of points. The reference readings are then plotted against time and a curve drawn through them. In this way reference values corresponding in time to each grid value are estimated and subtracted from the grid values. This procedure, however, can still result in incomplete corrections and spurious anomalies, particularly linear anomalies along traverse lines.

To survey a site properly, particularly if it is more than a few meters in area, two magnetometers in either of two modes must be employed. In the differential mode, one magnetometer is kept at a fixed reference point while the other measures values at the grid points. Because the instruments are operated simultaneously, any difference between the two readings represents the total field magnitude, corrected for time variation, at that point. In the gradiometer mode, two sensors are kept a fixed distance apart—1 m, for instance—in a vertical array. Both magnetometers are again operated simultaneously, and the difference in readings is recorded. These values are *gradient magnitudes*, the vertical spatial change of field magnitude. As in the differential mode, the temporal variations are canceled, but in addition any long-range trends affecting both sensors equally are canceled, whereas local anomalies (which will more strongly affect the lower sensor) will be recorded. The writers prefer the differential mode for any extensive survey.

Before starting the field survey of an extensive archaeological site, all visible geological and archaeological features must be examined. Also, the nature, size, and depth of features, as well as the possible existence of iron artifacts, should be ascertained, as this information will shape the field strategy and aid in the final interpretation of the data. Those conducting the survey should work closely with the archaeologists, so both parties share an understanding of the archaeology of the site and the geophysical method.

Some evaluation of the survey's possible success may be obtained by measuring the magnetic susceptibility of typical soil samples from the site. These samples should be taken at representative stratigraphic levels, as well as inside and outside any archaeological features of the site. By measuring the susceptibility of the samples before and after heating in a reducing atmosphere, it is possible to evaluate expected anomaly sizes and to get some information on the extent of anthropogenic hematite-magnetite conversion.

The size of the survey grid unit is determined by the size of the features expected. Since the number of magnetic field measurements will be proportional to the square of the grid spacing, this choice is a compromise between detail sought and time available. Generally speaking, the grid spacing should be comparable with, or somewhat smaller than, the linear dimensions of expected anomalies. Features of a meter or two in size near the ground surface can be surveyed with a 1-m grid. If the features are deeper, then the spatial extent of the anomalies will be larger (as well as weaker) and a larger grid spacing can be used. On historical sites, linear features such as walls or cellars can be picked up with a larger unit. One approach on large sites is to use a coarse grid (such as 2 m) and then concentrate with a 1- or ½-m grid in areas of greater magnetic activity.

In choosing sensor height above the surface, two considerations must be noted. First, since the width of an anomaly increases with the source–sensor distance, a greater sensor height will result in less resolution between anomalies from neighboring sources. An approximate rule is to have the source–sensor distance no greater than the intersource distances that one wishes to resolve. This suggests a source–sensor distance equivalent to or less than the grid spacing.

Second, the sensor height must be selected so as to reduce the relative contributions from surface noise arising from variations in surface-soil magnetization. The noise contribution relative to the signal will decrease with increasing sensor height. Probably the best compromise is to set sensor height at between 40 and 60 cm for a 1-m grid.

A necessity in all archaeological work is the location of some permanent reference points. This is no less true in the case of geophysical surveys. When surveying a site, particularly if it is large, the area should be subdivided into squares or blocks, each of which is treated as a unit. We have found that a block 20 grid units on a side is convenient. This area will have 21 × 21, or 441, grid points and can be surveyed in about 70 to 90 minutes, depending on site conditions. Including the time needed to lay out blocks and set up equipment, it is possible to survey three to five blocks in a day. If possible, these blocks should be oriented along magnetic north, since this orientation aids in the interpretation of anomalies. For a large site the blocks can be arranged in hectares, five blocks along an edge. The coordinates of grid points in each hectare can be designated I north, J east, where I and J run from 1 to 101. This is a convenient size array to be handled as a single matrix of values in computer programs.

OPERATIONAL DETAILS

The following discussion assumes survey operation in the differential mode, with one magnetometer sensor moved from grid point to grid point and one kept stationary for reference readings. We thus speak of moving sensor values, reference sensor values, and their differences.

Our technique for locating the grid points within a block is as follows. Stakes are placed at the four corners of the block and two ropes marked with grid units are placed, one at each end of the block, from stake to stake. Another rope marks the grid points along a traverse row from end rope to end rope. To avoid confusion and aid systematic data recording, all traverses should be conducted in the same direction, either south to north or west to east.

A systematic form for recording data in the field should be carefully planned. The writers use data sheets calling for the following information: site, block, data, grid unit, and names of persons operating the magnetometers. In addition, the height of the sensor and the sensitivity of magnetometers should be noted. Each sheet has space for four rows with 21 grid points and columns for the moving-sensor value, stationary-sensor value, and their difference, the last in case it is desirable to hand-calculate these values. The time of day is recorded at the start of each row. This is desirable if problems of analysis arise and one wishes to reconstruct events at the time; recorded data can also then be correlated with magnetic storm information.

Errors can arise in the hand-recording of data, so it is preferable to machine-record data with a data-collecting circuit connected to the magnetometer. Such a system involves expense and design time but considerably improves reliability.

Before taking measurements, it is important to evaluate the local magnetic environment. Vehicles can produce a shift of roughly 1 gamma at 30 m. Such disturbances are acceptable only if they remain stationary throughout the survey. Larger and closer amounts of iron, if stationary and not too large, can be treated by mathematical filtering techniques applied to the data. Power lines, moving trains, and other nonstationary sources must be avoided.

The person holding the moving sensor must be carefully checked before starting. Steel shoe tips, belt buckles, bank cards with magnetic strips, and even eyelets in hats can cause trouble. Repeated readings should be taken with the person assuming several positions relative to the sensor. Variations in readings should be random and not greater than one or two times the "least count" of the instrument (the smallest possible change in value displayed).

Even if the magnetic field is absolutely stationary in time, a random scatter or noise in reading values of about ±0.5 times the least count will be observed. When two instruments are recorded at the same time and the difference taken, this random scatter becomes ±0.7 times the least count. In actual practice, the reproducibility of difference values (taken by repeated readings at the same grid point, repositioning the sensor each time and calculating the standard deviation) is more like three to five times the least count. This variation in the difference values is a measure of the least amount of noise to be expected. Anomalies smaller than this variation may be lost in this noise unless they are observed on several grid points.

The reference magnetometer should be close enough to the grid points so that communication between the operators is not impaired but not so close that the two sensors interact. Two or 3 m is sufficient. It is important not to locate the reference sensor in a strong magnetic gradient. In such a gradient the instrument can lose sensitivity and give erratic readings. The location should be checked for this possibility by taking repeated readings, with the sensor moved slightly between readings. Variations no greater than random noise indicate lack of a gradient.

One final and very important field procedure is the measuring of "common rows." Relocating the reference magnetometer between recording two blocks will result in a constant shift in the difference values between the two blocks. This may be corrected by measuring one row common to the two blocks twice, once with each block. The average shift in the differences on the common row is then the correction factor that is added to all the values of one of the blocks to bring them to the same "level" as the other block. If the standard deviation in the distribution of shifts in differences on the common row is greater than 3 to 5 times the least count of the magnetometers, then the correction factor may not be reliable. This can arise by careless positioning of the sensor or by passing over a strong gradient due to an anomaly on the common row. In this latter case small and unavoidable shifts in sensor positions can cause large shifts in recorded values. Either a subset of values on the common row that avoids the anomaly should be used or other common rows should be measured. These problems are particularly troublesome with matching blocks recorded on several days on a large site.

INTERPRETATION

The first requirement in the analysis of magnetic data is to produce a matrix of magnetic field values corrected for diurnal variations to be used in all subsequent mapping and profiling. Since the number of values from a site can be quite large, computer processing becomes desirable or even necessary. The following discussion assumes the use of a computer.

Usually, results are presented in various ways so that individual anomalies and patterns of anomalies can be identified. The simplest treatment is to plot profiles along traverse rows, this is useful in seeking linear features of known orientation.

For any more complicated situation it is necessary to examine the areal (two-dimensional) patterns of anomalies. Various forms of magnetic contour maps are generated in which the magnetic field strength is treated as a "height" on a map of the site. Although simple contour maps can be hand-drawn by recording the magnetic field values on a grid of points and drawing lines of equal magnetic field magnitude through the grid of values, this process can be very tedious and is not likely to be often repeated if the map needs to be redrawn with a different scale or contour interval. Thus it becomes necessary to produce contour maps by a computer.

The most convenient types of computer contour maps are those produced by a line printer. Such maps are quickly generated and easily interpreted because of the visual advantage obtained through use of the shading variations of print characters. They have the disadvantage of relatively low resolution determined by the print character size. In the simplest map type the magnetic field values at the grid points are sorted according to predetermined intervals (usually no more than 10) and print characters of different shades are printed at the grid points according to these intervals. The result is effective for sites of more than one hectare.

The map next in complexity is based on interpolation between grid points and prints several characters per point. A commercial program package that makes such maps with a wide range of options is SYMAP (1975). A simpler program to obtain some of these same results is described by Davis (1973).

The traditional type of line contour map can be programmed to output on a plotter. Such a map can have much higher resolution but lacks the immediate visual impact of a shaded map. Color contour-mapping equipment is also available, but it is expensive.

EXAMPLES OF MAGNETOMETER SURVEYS

The Knife River Indian Villages National Historic Site

The objective of this National Park Service project is to assess and develop into a national historic site a collection of Mandan, Hidatsa, and other Native American village sites grouped at the confluence of the Knife and Missouri rivers north of Bismarck, North Dakota (Weymouth and Nickel, 1977). The Midwest Archaeological Center (MWAC) of the National Park Service has been gathering magnetic survey data at Knife River since 1976, with the present writers analyzing them. Two magnetometers in the differential mode have been used to collect data, which since 1978 have been recorded automatically with an electronic data logger.

One of these villages—Big Hidatsa (32ME12), occupied in the late eighteenth and early nineteenth centuries—covers several hectares and is marked today by many depressions left by earth lodges. These show up clearly on the magnetic record, partly due to the topography, but also due to differences in soil susceptibility inside and outside the lodges: some lodges are visible on the magnetic maps but are not visible in the topography. The most characteristic anomaly of these lodges is their central fire hearths, which lie at depths of 40 to 120 cm below the ground surface and produce anomalies of 20 to 50 gamma. In addition, anomalies associated with smaller interior or exterior fire hearths are visible. Midden areas between houses produce magnetic high regions. These correlations have been established by soil probes and by test excavations.

Figure 8.3 is a grid-point map of data taken over two seasons (1977 and 1978) at Big Hidatsa. Each print character corresponds to data on 1 grid point, with an interval of two grid points between different characters. The map illustrates a way in which one can organize a large survey. Each set of blocks obtained on one day has been balanced against other sets by using common rows. Each hectare is handled as a separate matrix in the computer processing.

Figure 8.4 is a topographic map of Sakakawea Village (32ME11), another village of the same time period at Knife River. It may be compared with figure 8.5, a SYMAP magnetic map of the same region with a 9-gamma interval (Weymouth and Nickel, 1977). Figure 8.6 is the southeastern part of this map with a contour interval of 5.6 gamma. The anomalous regions at N21−31, E80−89 is due to a house not visible on the topographic map. Two excavations in

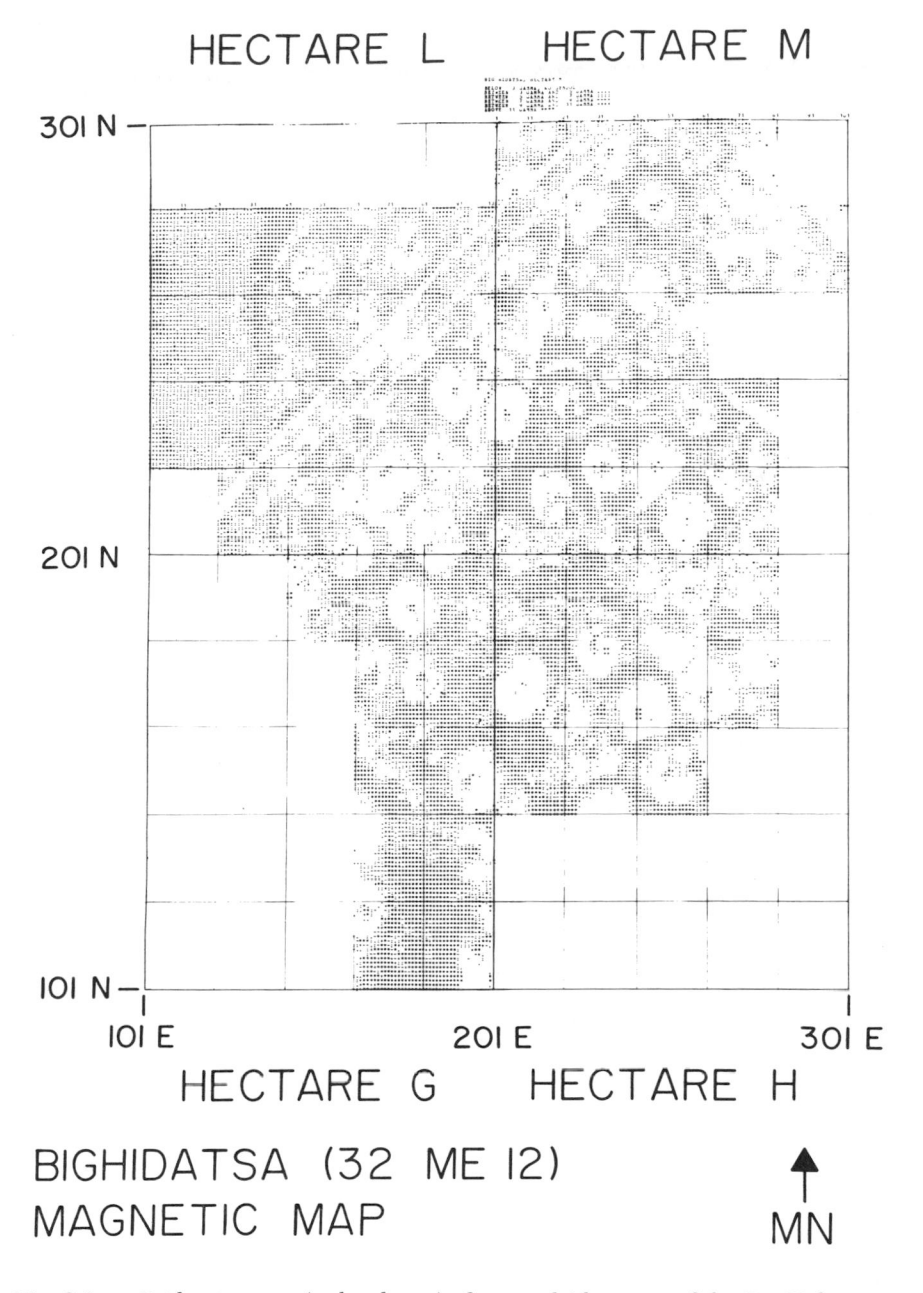

Fig. 8.3. Grid-point map (or level map) of parts of 4 hectares of the Big Hidatsa site (32ME12), Knife River Indian Villages National Historic Site. Each print character is at a grid point. Each increment in character darkness represents an increase in two gamma.

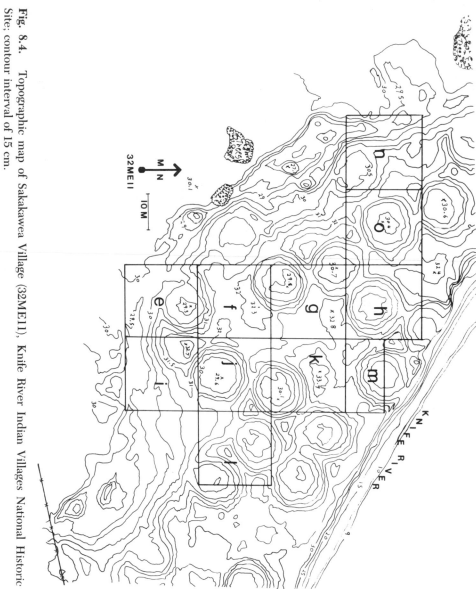

Fig. 8.4. Topographic map of Sakakawea Village (32ME11), Knife River Indian Villages National Historic Site; contour interval of 15 cm.

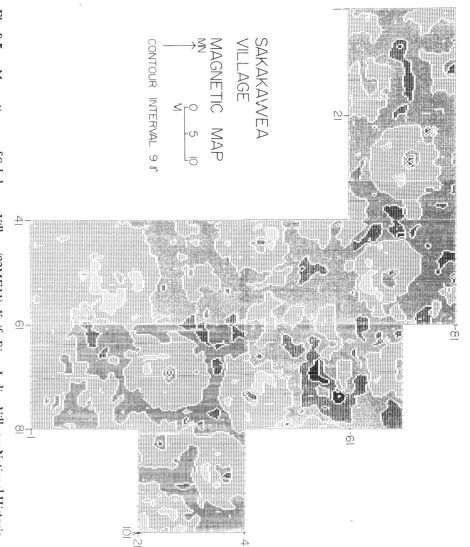

Fig. 8.5. Magnetic map of Sakakawea Village (32ME11), Knife River Indian Villages National Historic Site; contour interval of 9 gamma.

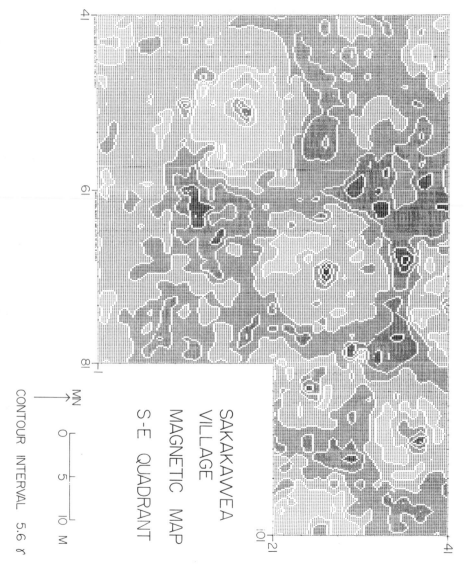

Fig. 8.6. Magnetic map of Sakakawea Village, southeast quadrant; contour interval of 5.6 gamma.

this area established that the anomaly at N36, E69 between the houses was caused by an exterior fire hearth about 60 cm in diameter and 40 cm below surface and that the smaller anomaly just to the south inside the house was caused by a group of burned rocks—a presumed sweat lodge.

Figure 8.7 is a magnetic map of house 6 in block O in the northwest quadrant, taken with the sensor at our usual height of 60 cm. Figure 8.8 is a map of the same area but based on readings taken with the sensor 120 cm high. Some of the smaller and sharper anomalies produced by near-surface sources disappear in

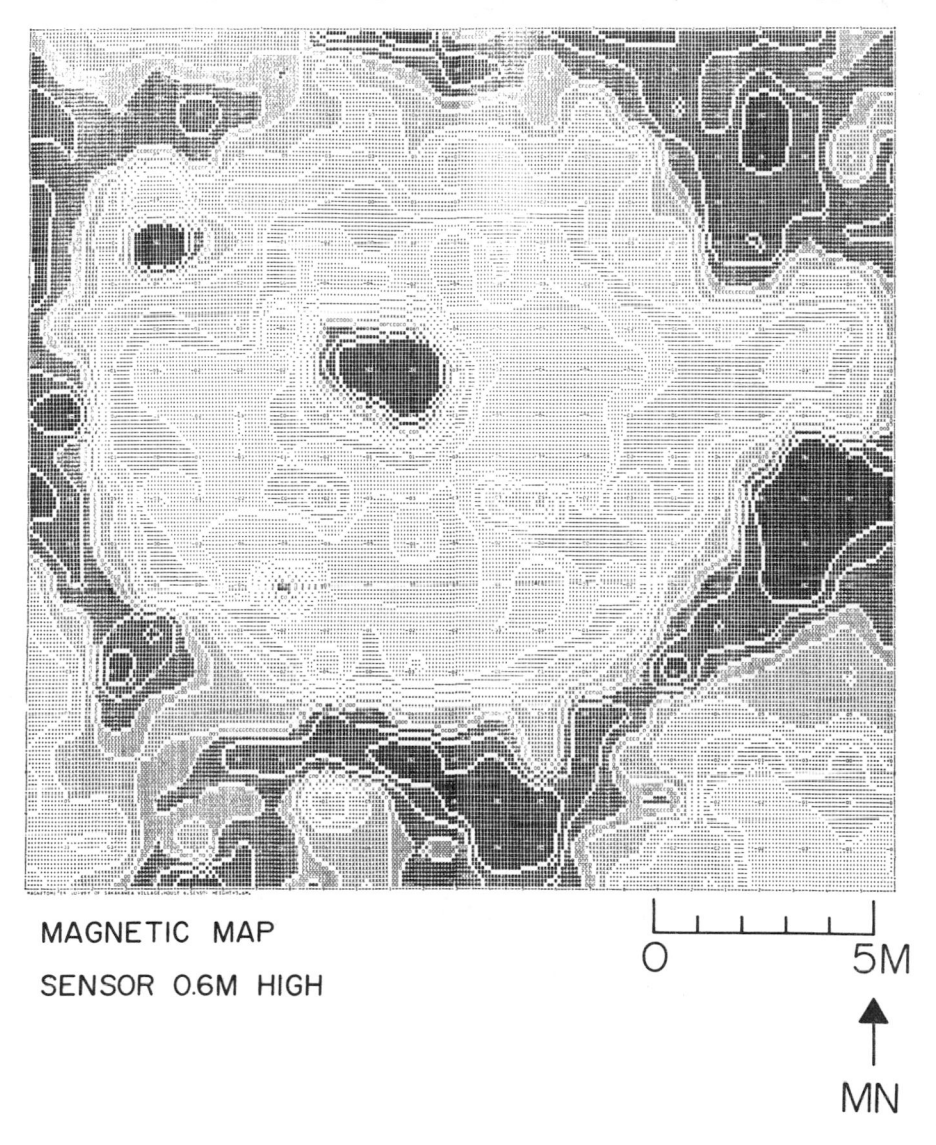

MAGNETIC MAP

SENSOR 0.6M HIGH

0   5M

MN

Fig. 8.7.   Magnetic map, house 6, Sakakawea Village; sensor height of 0.6 m, contour interval of 2 gamma (except at highest and lowest levels).

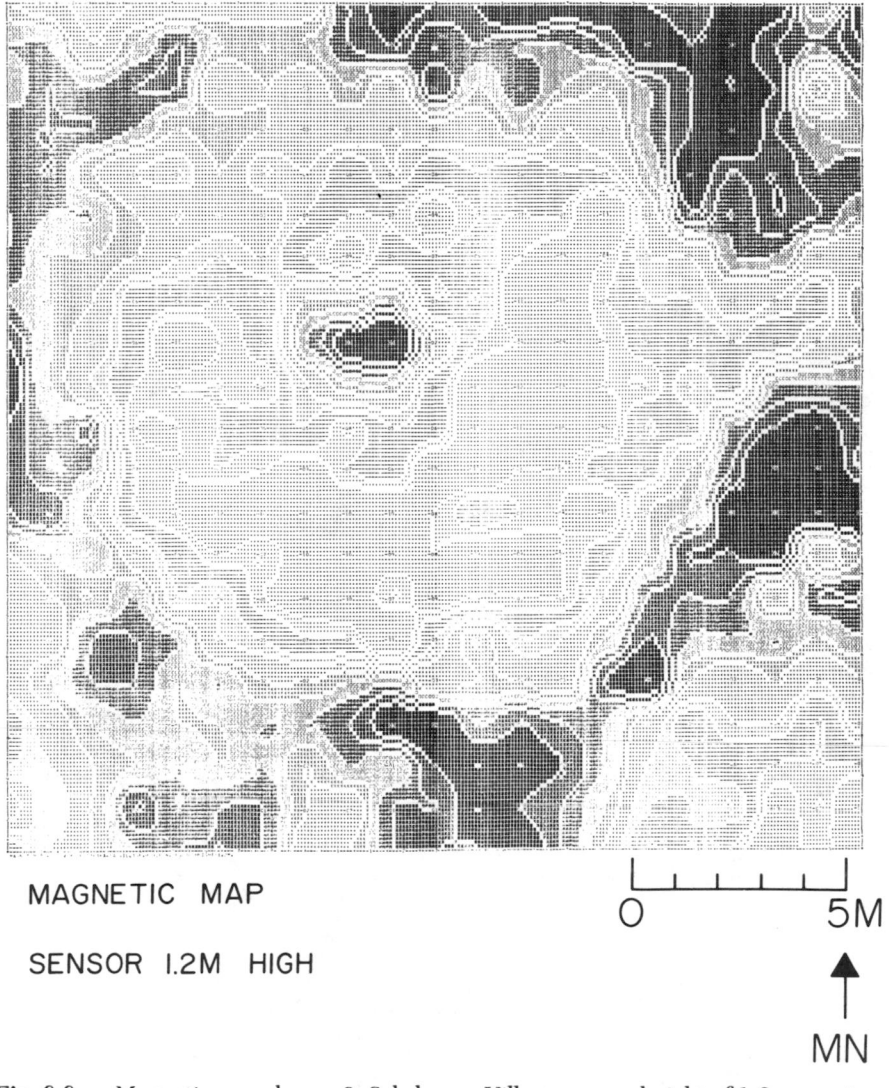

MAGNETIC MAP

SENSOR 1.2M HIGH

0   5M

MN

Fig. 8.8. Magnetic map, house 6, Sakakawea Village; sensor height of 1.2 m, contour interval of 1 gamma (except at highest and lowest levels).

the second map, while the fire-hearth anomaly produced by a larger and deeper source persists.

Fort Union National Historic Site

This site, a trading post in North Dakota from about 1830 to 1865, has undergone considerable excavation. The data for the survey were obtained by the MWAC and analyzed by the present writers.

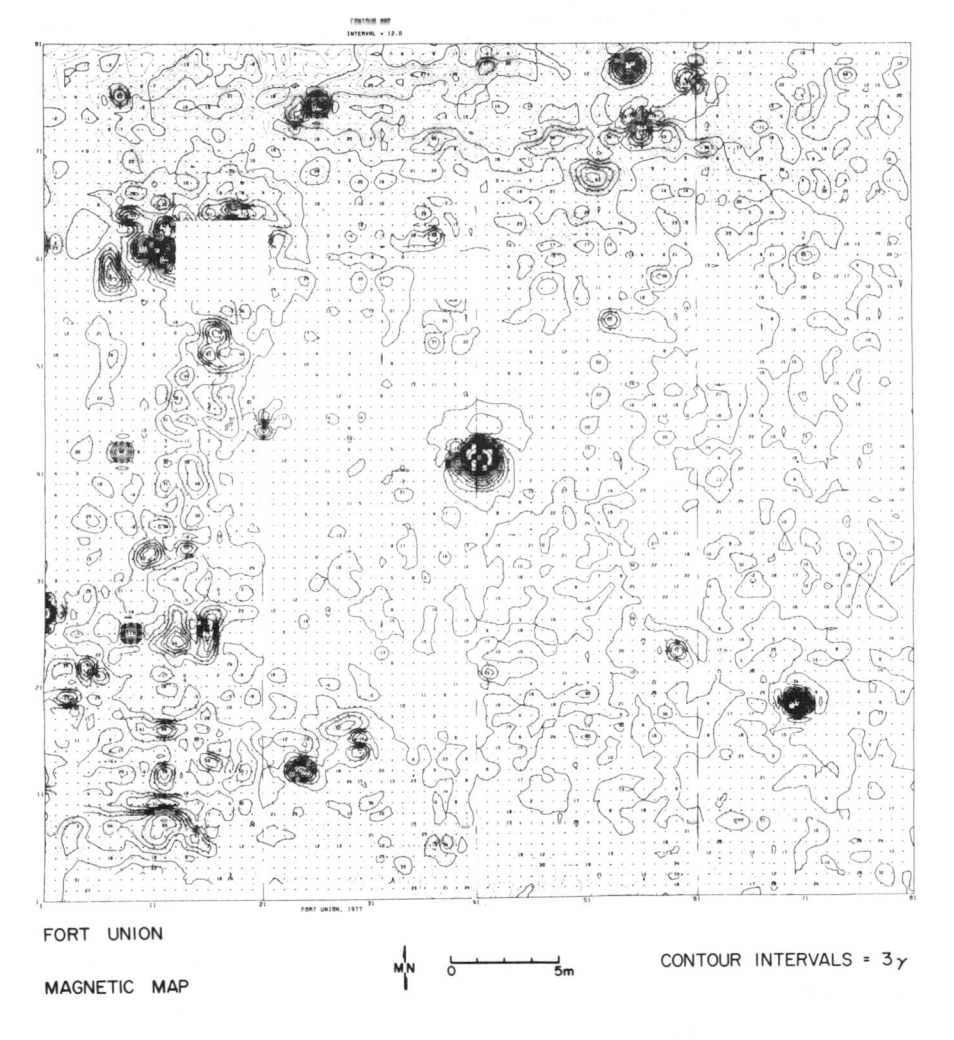

FORT UNION

MAGNETIC MAP

M↓N   0 ┣━━━━━━┫ 5m   CONTOUR INTERVALS = 3γ

Fig. 8.9. Magnetic line contour map, Fort Union National Historic Site; contour interval of 3 gamma.

Figure 8.9 is a line contour of the total survey area. A number of anomalies can be seen, particularly along the position of the western range of houses. Some of these are due to previous excavations and some arise from original wall foundations and fire hearths. The central area, a parade ground, is relatively free of anomalies except for the strong anomaly in the center due to a modern flagpole base. This survey illustrates two points. A strong anomaly at N19, E70.5 in the southeast quadrant of the map (see fig. 8.10) was actually located earlier by personnel from MWAC who ran a series of seven 20-m-long traverses using a 1-m grid unit and one magnetometer. On a map of these data (fig. 8.11) the large

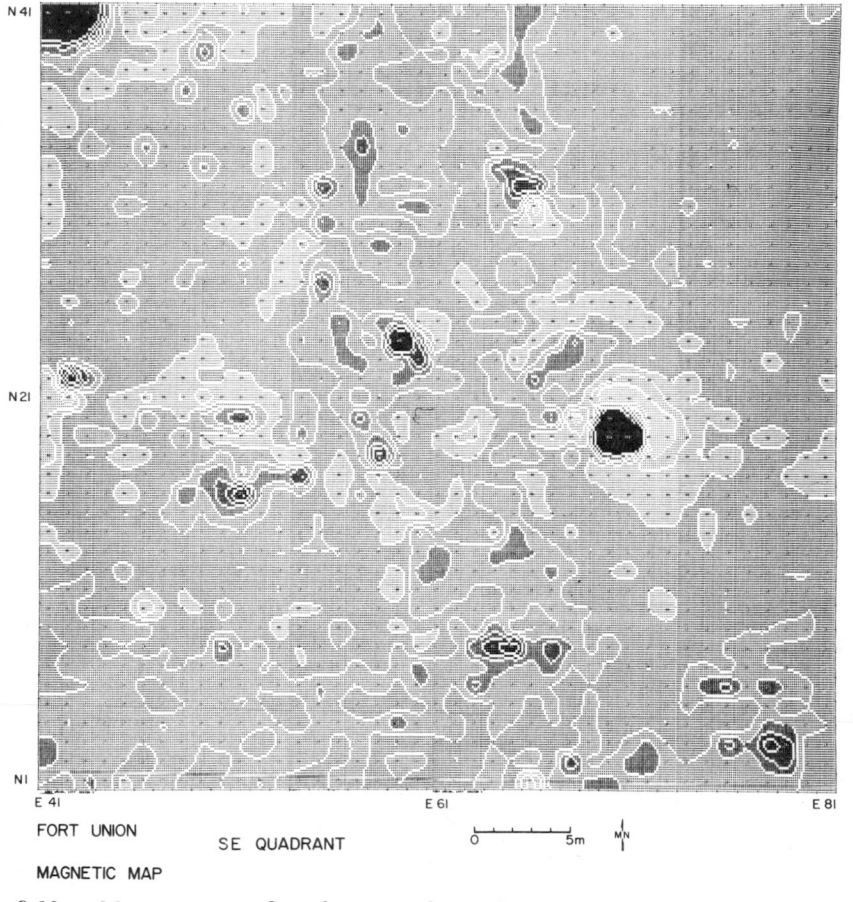

Fig. 8.10. Magnetic map of southeast quadrant of Fort Union National Historic Site; contour interval of 5 gamma.

anomaly and a number of smaller anomalies are reproduced. Such an operation is not usually successful on a larger area. Figure 8.12 is a magnetic line contour map resurveyed in part of the southwestern corner using a grid interval of 50 cm. Comparison with figure 8.9 shows several anomalies duplicated, as well as more detail.

Ward and Guerrier Trading Post

The Ward and Guerrier Trading Post is located adjacent to the Fort Laramie National Historic Site, Wyoming. As part of a program to construct a parking lot and visitors' center, it was necessary to determine the location of the trading post that was known to be in this vicinity around 1858. Excavations in 1963 had

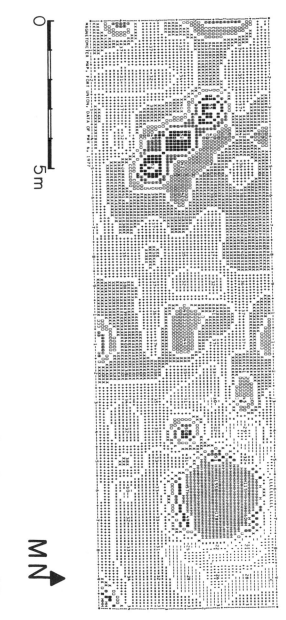

Fig. 8.11. Magnetic map produced from 7 east-west traverses in southeast quadrant of Fort Union National Historic Site; no diurnal correction, contour interval of 3.5 gamma.

0 ⊢━━━━━━━━━━━┥ 5 m

MN

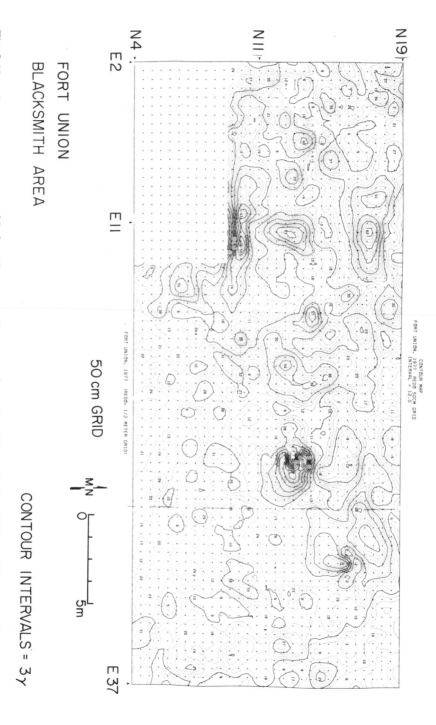

N19

N11

N4

E2   E11   E37

FORT UNION
BLACKSMITH AREA

50 cm GRID

CONTOUR INTERVALS = 3γ

FORT UNION, 1977 (REDD; 1/2 METER GRID)

CONTOUR MAP
FORT UNION, 1977 REDD 50CM GRID
INTERVAL = 12.0

M N
0   5m

Fig. 8.12. Line contour map over blacksmith area, southwest quadrant, Fort Union National Historic Site, using 50-cm grid unit; contour interval of 3 gamma.

revealed a small, square foundation (presumed to be a smithy), several bits and pieces of foundations, a row of postholes, and a sizable trash pit containing iron material that was not removed at that time. The data for the 1963 excavation, however, had been lost and only the approximate position of the site was known. In 1976 the MWAC opened up three excavation units and surveyed the area with two magnetometers, as shown in the magnetic map of figure 8.13 (Weymouth, 1979). Three observations from these excavations and from the survey led to the conclusion that the previous excavations had been found again and determined the position of the main structure of the trading post.

First, the excavation unit and survey information in the easternmost block established this to be the smithy area with the now rediscovered small foundation. Second, the large anomaly at N31, E31 was assumed to be the iron-filled trash pit. Third, a row of sharp anomalies (marked by the dotted line in fig. 8.13) was believed to mark a wall of the main building of the post. The length of this line is approximately 100 feet, which agrees with what is known of the post. This interpretation is further supported by a row of post molds at the end of the line uncovered in the 1963 excavation and indicated in the excavation trench at about N21, E5. Figure 8.14, a 3-D view of the results, again shows the line of anomalies, the strong anomaly due to the trash pit, and the group of anomalies in the smithy area.

The Dolores Archaeological Program

The final example is drawn from the Dolores Archaeological Program (DAP), a long-term study of the archaeological resources in the Dolores River valley, southwest Colorado, before its inundation by dam construction. As part of the resource assessment a large number of sites, each encompassing two to ten blocks, have undergone magnetic surveying.

Because most surveys in this program have been followed by excavations or at least blading of the plow zone, it has been possible to obtain rapid checks on the archaeological significance of magnetic anomalies. This has improved our ability to interpret data and make predictions. Following the 1979 season, a study was made of the correlation between predictions made from the magnetic maps and the results of the various excavations. The pit structures and other large architectural features were of sufficient volume and extent to produce anomalies over several meters that could be readily detected; of 26 such "high-priority" anomalies, excavation revealed that 23 were caused by cultural features.

We show as an example of the larger anomalies results from site 5MT2193, surveyed in 1978. Figure 8.15, a magnetic map of this site, shows in the western half two large anomalies that were presumed due to two pit-house structures. These anomalies extend over several meters but have a maximum of only about 15 gamma. Use of quarter-gamma sensitivity here was justified, but a 2-m grid spacing could have been used if only this type of anomaly was being sought.

Upon excavation, the sources of the anomalies were shown to be pit-house structures originally about 1.5 m deep that subsequently were filled in with

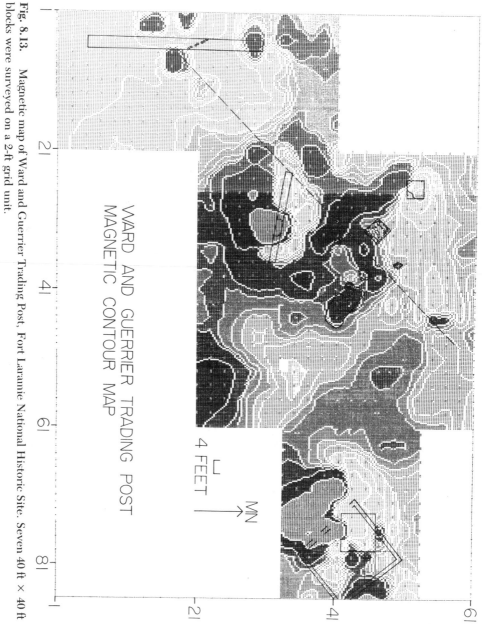

WARD AND GUERRIER TRADING POST
MAGNETIC CONTOUR MAP

MN

4 FEET

Fig. 8.13. Magnetic map of Ward and Guerrier Trading Post, Fort Laramie National Historic Site. Seven 40 ft × 40 ft blocks were surveyed on a 2-ft grid unit.

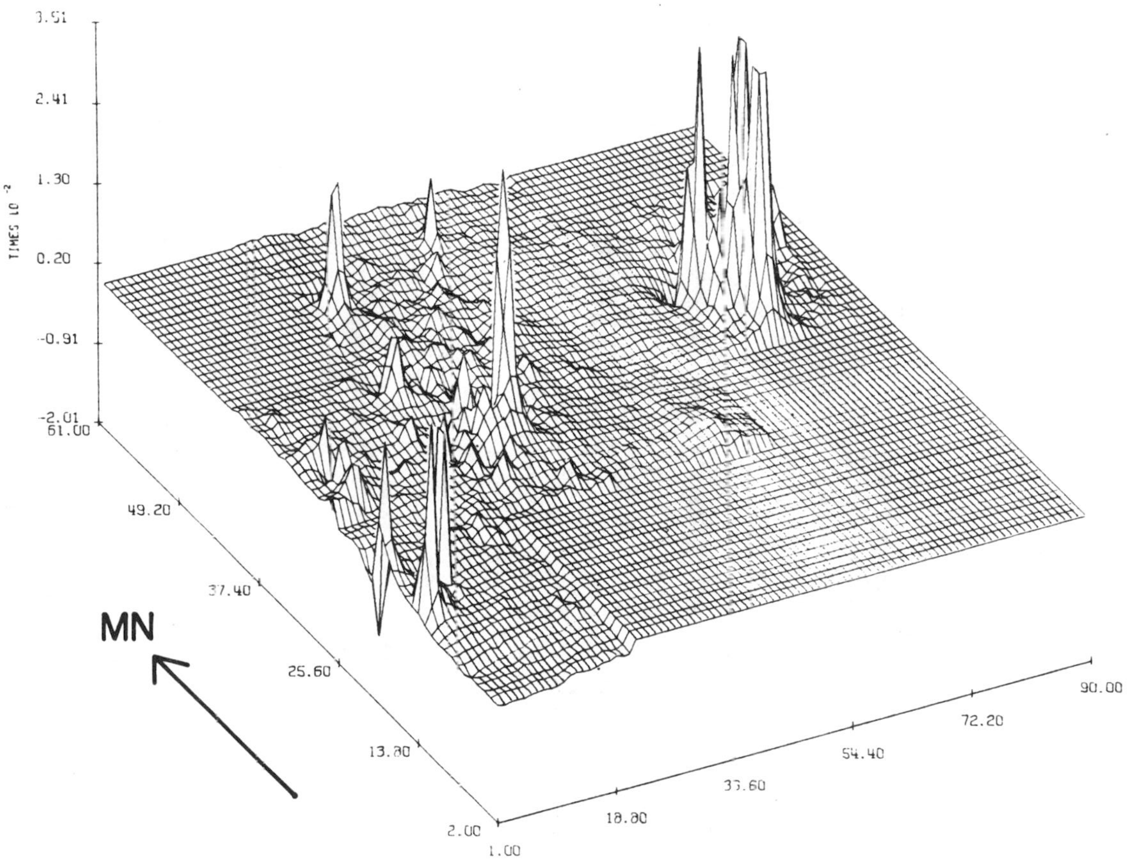

Fig. 8.14. Three-dimensional representation of magnetic values, Ward and Guerrier Trading Post, Fort Laramie National Historic Site.

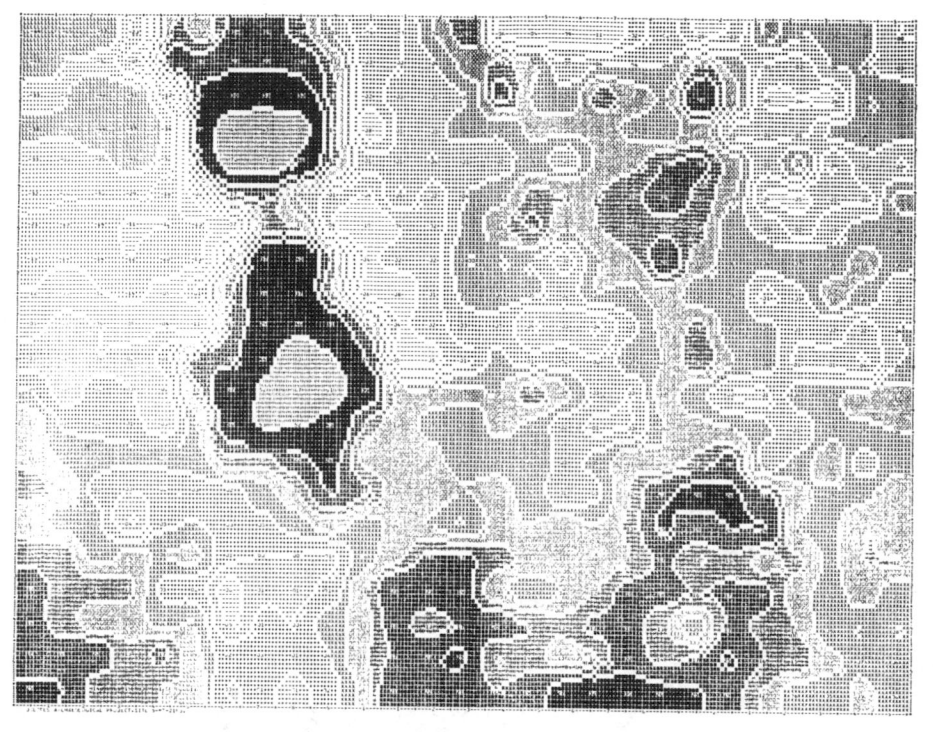

DOLORES ARCHAEOLOGICAL PROJECT

SITE 5MT2193

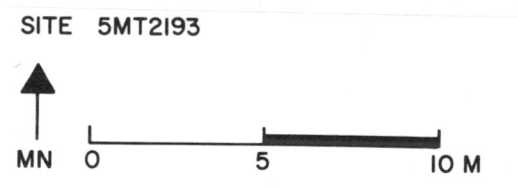

MN   0   5   10 M

Fig. 8.15.   Magnetic map of site 5MT2193, Dolores Archaeological Program, contour interval 1 gamma in midrange.

burned-roof fall and soil. The shapes of the anomalies followed the shapes of the sources fairly faithfully, including the air vent in the southern structure. As a further test of the source-anomaly relation a model calculation was made, representing the southern pit house by a box with a hearth on the floor and a sphere for the air vent to the south. This model is represented in figure 8.16, along with a three-dimensional representation of the magnetic results. The susceptibility values had to be multiplied by four in order to obtain a magnetic map (figure 8.17) comparable with the original map, which was plotted in quarter-gamma units.

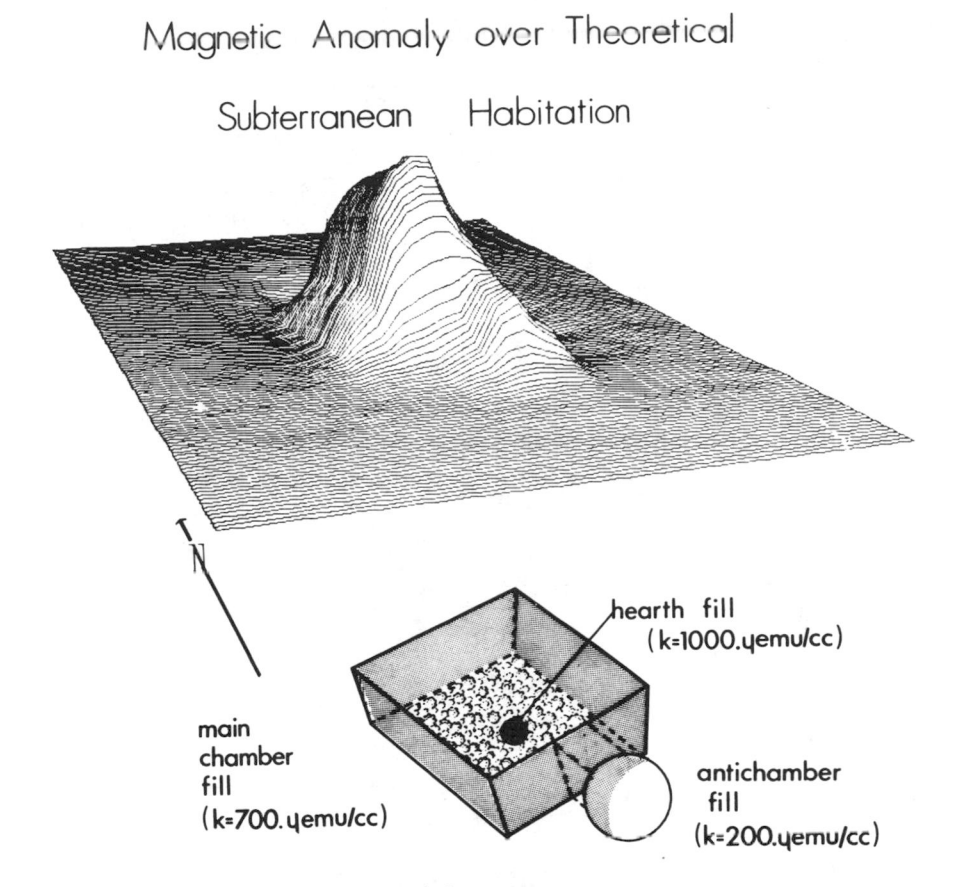

Fig. 8.16. Shape of feature and magnetic susceptibility values used to generate a model calculation simulating a magnetic anomaly produced by a pit-house. Upper half of figure shows three-dimensional view of magnetic values produced by calculation.

RESISTIVITY SURVEYING

SUMMARY OF METHOD

Resistivity surveying on archaeological sites indicates spatial differences in sediment moisture: the presence of features, architecture, activity areas, and other archaeological remains can be detected if the amount of moisture they retain is different from that retained by surrounding sediment. Location of these anomalies or contrasts involves careful measurement of the sediment resistivity at

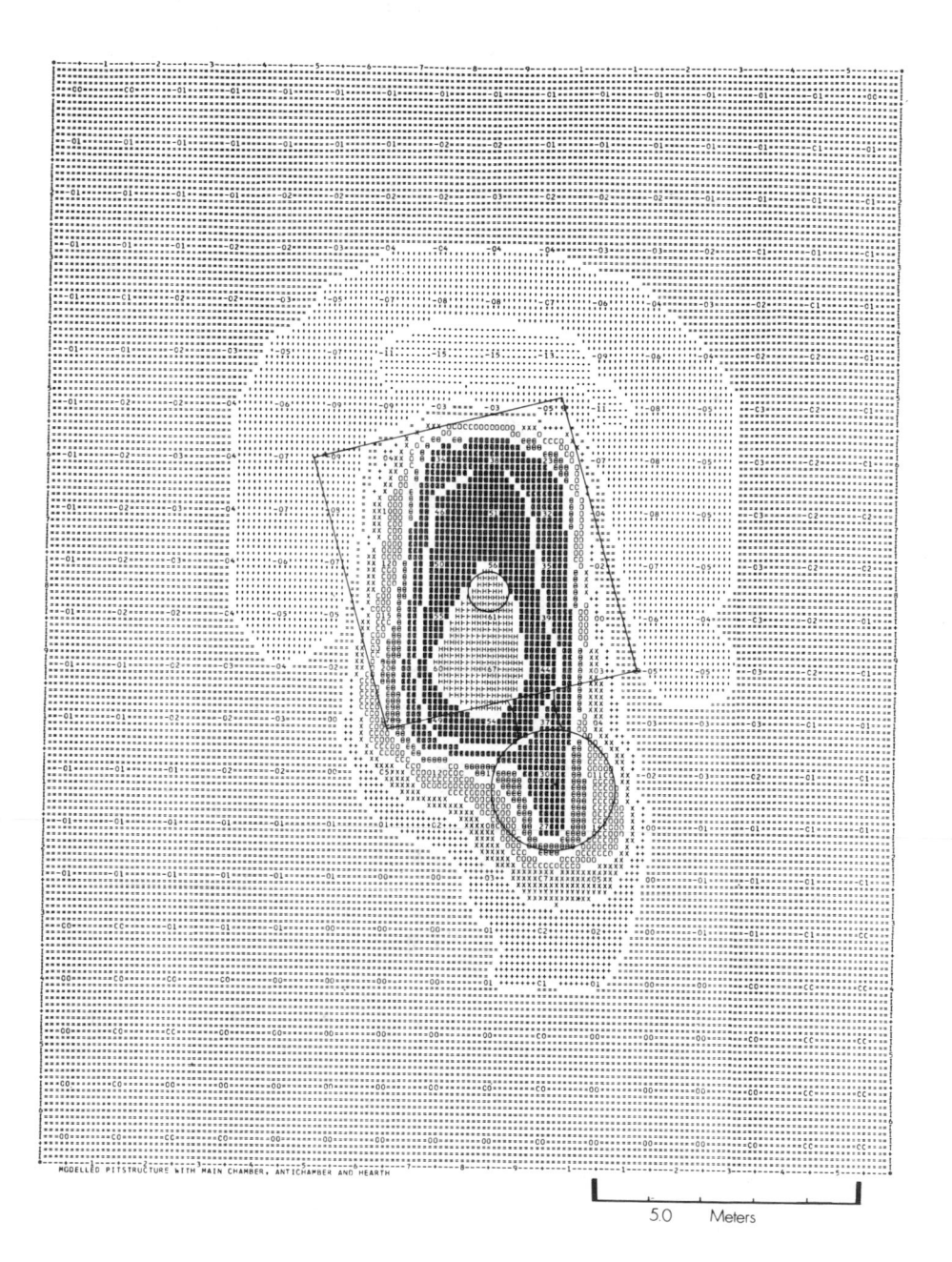

MODELLED PITSTRUCTURE WITH MAIN CHAMBER, ANTICHAMBER AND HEARTH

5.0   Meters

Fig. 8.17. Magnetic map produced by model calculation simulating pit-house anomaly. Outline shows the position of the feature below the surface.

discrete points on the surface along traverses or on a grid of points. The collected data are usually displayed as profiles or as electrical-resistivity contour maps. Because of the amount of data involved, a computer is an invaluable assistant. The data are then computer-enhanced (if necessary) and interpreted, and areas of potential archaeological interest are located on the basis of a clear understanding of the principles discussed in the following sections.

HISTORICAL DEVELOPMENT

Measuring the resistivity of the subsurface has long been used as a method for exploring geologic structure. A practical measuring technique was first introduced by Wenner (1915) and significant improvements were made by Schlumberger (1920). Resistivity measurements are still widely applied in geophysical exploration for mineral deposits and gravel beds. An early archaeological use of the method was in 1947 by H. Lundberg (De Terra, 1947:162−64), to locate fossil human remains at Tepexpán, Mexico. Since that time resistivity has been utilized successfully by Atkinson (1952), Clark (1969, 1975), Carr (1977), Leith, Schneider, and Carr (1976), and Ginzburg and Levanon (1977), among others, in a variety of archaeological contexts. For an exhaustive treatment of the historical development of the technique, refer to Van Nostrand and Cook (1966).

THEORY

In order successfully to conduct and interpret a resistivity survey, a grasp of basic electrical theory is necessary, beginning with the nomenclature. Electric *current* is defined as the rate of flow of charge passing through a cross section of a conducting medium for a specific length of time. To cause charge to flow, a *voltage* (also known as *potential difference*, a measure of the energy used to move the charges) must be applied. When a voltage is applied and a current flows, a *resistance* is encountered to the movement of the charge, which is dependent on the characteristics of the medium in which the charges are moving. These three physical quantities are related by Ohm's law,

$$\text{resistance} = \frac{\text{voltage}}{\text{current}}, \text{ or}$$

$$R = \frac{V}{I}. \tag{1}$$

Resistance is measured on Ohms (Ω), voltage in volts (V), and current in amperes (A). In a conductor of length L and cross section area S the voltage difference per unit length can be thought of as the moving force, the current as the quantity that is moved, and the resistance as the opposition encountered by moving the current. From Ohm's law we can develop the concept of *resistivity* by incorporating into equation (1) the geometry of the medium. Resistivity is a more useful quantity than resistance in the examination of an archaeological site since it is specific to the medium and independent of the geometry of the material being surveyed. Resistivity (ρ) is defined as

$$\rho = \frac{V/L}{J}, \tag{2}$$

where V/L is the change of voltage with distance in the direction of current flow and J is the current density in the medium in which charge is flowing. The basic unit of resistivity is the ohm-meter or ohm-centimeter ($1\ \Omega\text{-m} = 100\ \Omega\text{-cm}$). If a specified current is flowing in a known geometrical shape, we can deduce the resistivity of the material, providing the voltage difference is known. The inverse of resistivity ($1/\rho$) is known as the *conductivity*, although in the discussion following, we will consider only resistivity. More complete discussions of these concepts are available in most basic physics texts, such as Resnick and Halliday (1966).

Resitivity Measurements

The concept of subsurface resistivity measurements can be illustrated in an actual field situation. Current is induced in the ground by inserting in the earth two metal probes that are connected to a battery. In this idealized case, distribution of voltage and current in a uniform earth is well understood (a model is shown in fig. 8.18). Also shown are current- and voltage-measuring devices to indicate both the amount of charge flowing between the current probes and the voltage in the area of interest between the two potential or voltage probes.

By calculating the volume affected by the flow of current, we can derive an expression for the average resistivity within the measuring probes:

$$\rho = 2\pi a V/I, \tag{3}$$

which is easily calculated because the distance between the probes *(a)* is known and the current *(I)* and voltage *(V)* are measured quantities. The resistivity is correct only for this particular probe configuration (or probe array), as other probe geometries change the volume of earth affected by the current flow. In archaeological applications, the prime concern is not the absolute value of the resistivity at any one point but the change between readings.

We can alter this simple model to illustrate how the current and voltage are affected by some inhomogeneity in the uniform earth—for instance, a trash-filled pit. The voltage and current deviate from the normal pattern and the resistivity measurement of the earth between the two voltage probes changes (fig. 8.19).

If these measurements are continued by moving all four probes from grid point to grid point, we can generate a series of readings indicating the lateral variations in electrical resistivity to a depth approximately equal to the separation between the voltage-measuring probes. By increasing the spacing between the probes for any given survey, one can examine a greater volume (and therefore depth) of material.

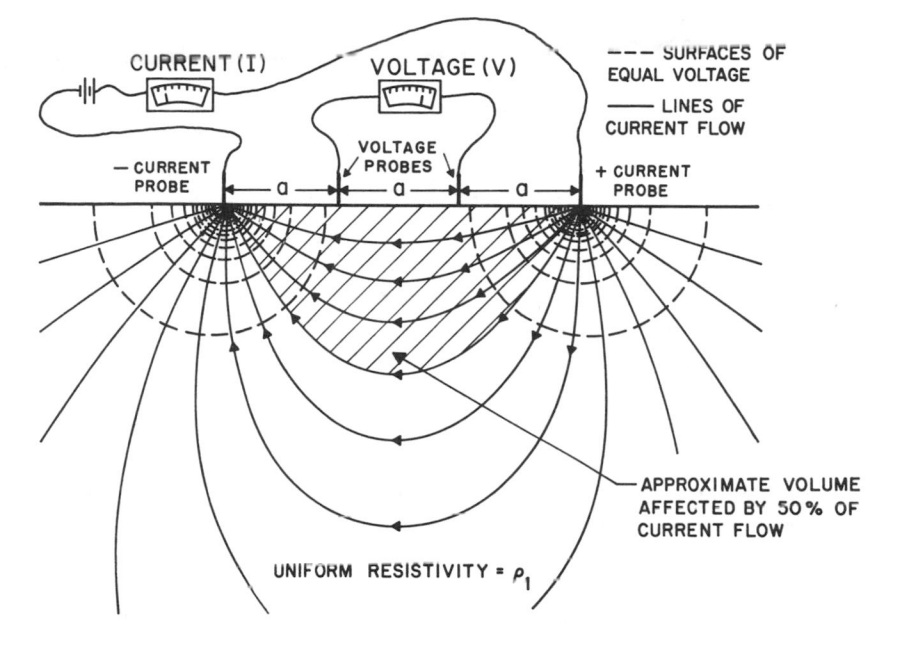

Fig. 8.18.   Distribution of current and voltage in a homogeneous earth.

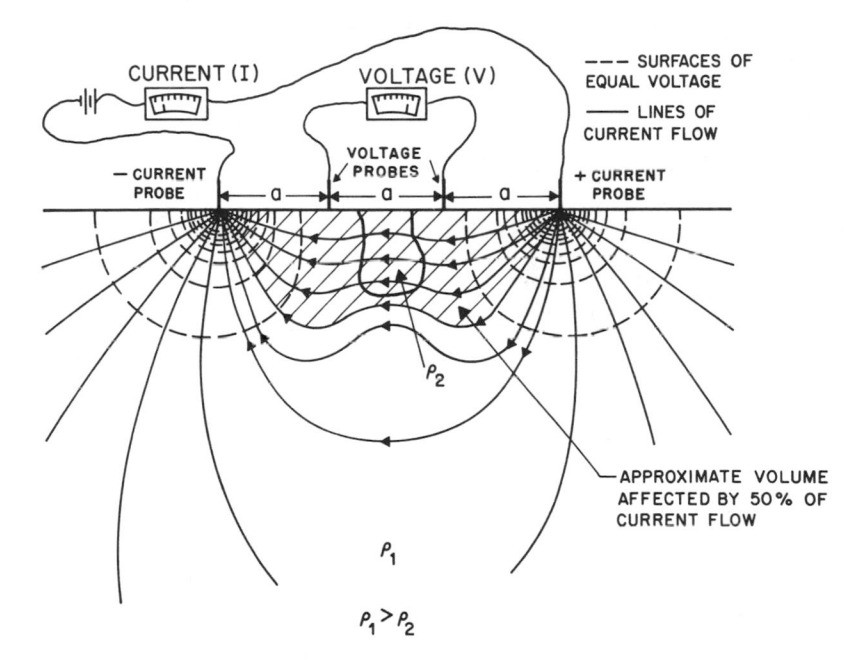

Fig. 8.19.   Distribution of current and voltage in the presence of a feature with a lower resistivity than that of the surrounding medium.

ELECTRICAL PROPERTIES OF SOIL SEDIMENTS

In the following paragraphs only sediments on which soil profiles have developed are considered, although resistivity measurements may be made in other sediment types (sand, till, mud) as well. The conduction of current in soils is largely an electrolytic phenomenon—that is, moisture in a soil containing free charged particles is responsible for the current flow. The resistance to currents flowing in all soil types depends directly upon the following variables:

1. *Soil moisture content*, which at archaeologically significant depths is usually generated by rainfall, with occasional contributions from areas having high water tables or from nearby streams. In general, soils receiving little rainfall have a high average resistivity and conduct electricity poorly. Seasonal variation in the total amount of rainfall also affects the resistivity (Al Chalabi and Rees, 1962; Clark, 1975). The amount of water that the soil can contain is determined by soil porosity, which exhibits wide spatial variation according to soil type, shape of the constituent grains, and amount of compaction.

2. *Permeability;* although a soil might have a high water content, current cannot flow unless connections exist between its interstitial pores.

3. *Ion content;* the ions responsible for conduction in the soil come from dissolved salts, such as calcium and sodium carbonates. They may be derived from a variety of cultural and noncultural sources: from the soil itself, underlying geologic strata, rainwater, modern agricultural fertilizers, or compounds generated by cultural processes. Figure 8.20 illustrates the pronounced effect of the addition of small amounts of dissolved salts on soil resistivity (Tagg, 1964).

4. *Temperature* affects resistivity, particularly when freezing of the groundwater takes place. Fortunately, most field surveys can be performed when the temperature is above 0°C, where daily variations in temperature are not sufficient to affect the resistivity in an archaeological context.

SOIL RESISTIVITY

The variables outlined above show wide spatial variation depending on climatic, geologic, and edaphic conditions. Consequently, the resistivity of different archaeological sites changes dramatically as well. Typical values of resistivity for different soils are given in the following table (Tagg, 1964).

Type of Soil	Resistivity (Ohm-cm)	
Loams	500–	5,000
Clays	800–	5,000
Clay sand and gravel mixture	4,000–	25,000
Sand and gravel	6,000–	10,000
Slates, shales, sandstone, etc.	1,000–	50,000
Crystalline rocks	20,000–	1,000,000

Similarly, wide variation in resistivity can be encountered on a small scale on an individual site.

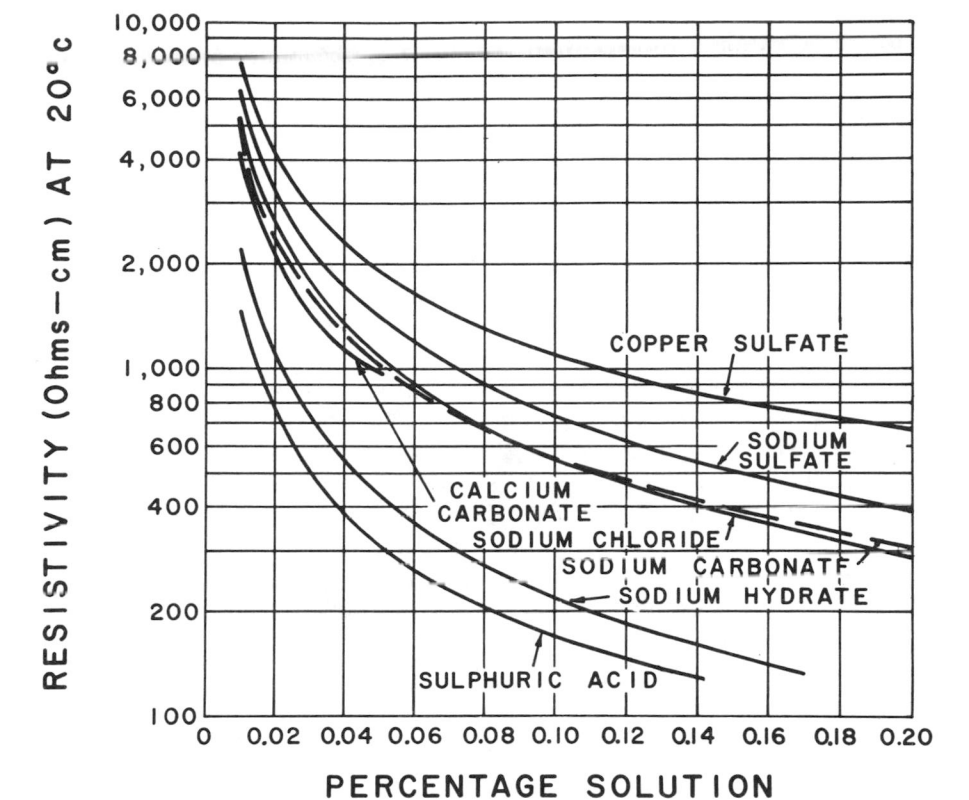

Fig. 8.20.   Typical resistivity curves of solutions (Tagg, 1964).

Differences in soil moisture, dissolved salts, and like factors also are responsible for producing the *culturally* formed resistivity contrasts that are detected at archaeological sites. Linear features, such as fortification ditches a few meters long, are the most susceptible to detection when the surrounding medium and climatic conditions allow a change in their moisture content. Stone alignments and foundations may exhibit detectable resistivity contrasts because of their marked difference in water retention. House floors or other compacted activity areas are visible occasionally due either to decreased porosity or to moisture that has accumulated on their surfaces. Midden areas, which are often high in soluble ions and have a larger volume of interstitial-pore space, can show distinct resistivity contrasts. The resistivity contrast of filled pits, although more subtle than the other, more extensive features, are sometimes discernible.

Although the phenomena responsible for conduction of current in soils and archaeological sites are understood fairly well, one cannot easily predict which archaeological features will be detectable by resistivity surveying or whether the soil noise will confuse or mask cultural resistivity contrasts. Since the state of the physical remains depends on environment and cultural history, a feature which is easily located by resistivity surveying in one area may be imperceptible in another.

INSTRUMENTATION AND FIELD PROCEDURES

Probes

The probes used for the transfer of current or measurement of voltage usually consist of mild steel rods pointed on one end, with suitable hand holds on the other to facilitate insertion in the soil. Ideally, the probes should act as point sources, but since they must be inserted a finite distance into the ground this is not the case. The actual value of resistivity will change if the insertion depth is varied for one or all of the probes, so steps must be taken to ensure uniformity. This can be accomplished by the use of an adjustable ring on the probe that will arrest insertion at the desired depth. Field conditions usually dictate what is reasonable, although to maintain adequate precision in determining the resistivity the insertion depth should be less than 20 percent of the distance between the nearest adjacent probes (Aitken, 1974; Van Nostrand and Cook, 1966). For a 1-m spacing, the writers have found that insertion to 5–10 cm provides adequate contact in all but the driest soils.

Care must also be taken in situating the probes relative to one another. For example, when using small spacings on the order of 30 cm in a colinear array such as the Wenner or double dipole, errors in the measured resistivity can be as large as 15 percent for the misplacement of a probe by 3 cm (Aitken, 1974). Fortunately, archaeological applications seldom require spacings smaller than 50 cm, as this seems to be the minimum size of detectable features. At 50 cm between probes, one should attempt a positional accuracy of ± 1 cm.

Probe Configurations

For simplicity, only the four-in-line probe configuration, commonly known as the Wenner array, was shown in the previous discussions. In practice, a variety of arrays are effective in an archaeological context, the choice among them depending on the terrain, the size of expected features, and the experience or familiarity of the operator with one array or another. Figure 8.21 shows a plan layout of the more common types. Descriptions of each array follow:

i. The Wenner array is the most often used and has several advantages. It produces the largest percentage change over most lateral soil variations. Most instruments are set up to accommodate this array. With the use of a fifth probe and a rotary switch, the last probe can be deactivated and "leap-frogged" to the front of the array while the instrument operator takes a reading using the other four probes. This allows each additional reading to be taken with the insertion of only one probe instead of moving the entire array. The main disadvantage of the Wenner array is that it tends to produce subsidiary peaking in the data, typified by large excursions in the readings before and after a resistivity contrast occurs.

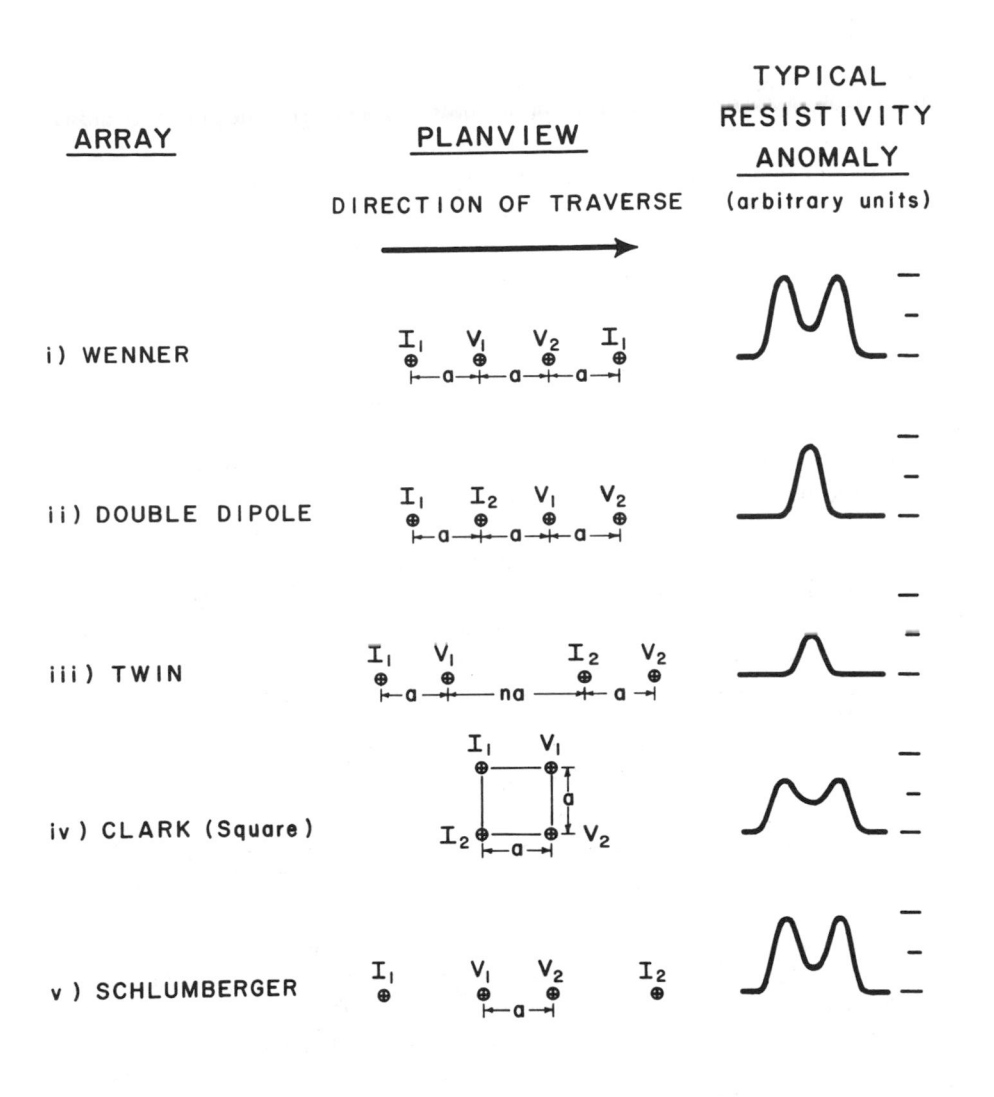

Fig. 8.21. Diagram of probe configurations (arrays) commonly used in archaeology with a typical anomaly for each due to an idealized feature.

This gives rise to M- and W-shaped anomalies, as illustrated in figure 8.21i.

ii. The double-dipole array is an attempt to reduce the subsidiary peaking that occurs in the Wenner array while retaining similar sensitivity and mobility. As shown in figure 8.21ii, only a single peak is encountered over a feature of interest, making interpretation of profiles and gridded data less complex (Clark, 1975). A five-probe system can be used in a fashion similar to the Wenner system, thus speeding data collection. The apparent disadvantage of the double-dipole array is the increased fall-off with depth, which means that deeper features contribute less to the readings.

iii. The twin array (fig. 8.21iii) devised by Aspinall and Lynam (1970) used one fixed current-voltage pair (I_1 and V_1) that remains fixed throughout the survey while the other pair (I_2 and V_2) is used in the measuring procedure. The motivation for this alteration is economy: the roving pair can be mechanically linked together, resulting in the least amount of actual probe movement of any of the configurations. The twin array also has the interpretational advantage of producing single-peak anomalies. Its main disadvantage is reduced sensitivity, producing the least percentage change from background of any of the other arrays. However, in regions where resistivity contrasts are large the twin array is probably the ideal choice. It should be noted that the value of n should always be at least 30 or confusing effects will occur.

iv. The square array designed by Clark (1968) was an attempt to reduce difficulty in the field and in interpretation (fig. 8.21iv). Clark configured the probes in such a way that they can be used as the legs of a table while the instrument and other recording equipment rest on top. This seems like a compact and suitable solution in many instances. Its main disadvantage is the fixed distance between the probes, usually at a maximum of 1 m, with the effect that the array can be used only in the detection of features that are less than 1 m in depth.

v. The Schlumberger array (fig. 8.21v) is more commonly used in large-scale geophysical applications but has been used successfully in an archaeological context (Rees and Wright, 1969). Its sensitivity is comparable to that of the Wenner array, but subsidiary peaking is again a problem. The distances between I_1 and V_1 or I_2 and V_2 should always be much larger than a.

Choice of the correct array for any given field situation is not a straightforward procedure. Despite the problems with peaking, the present writers generally use the Wenner array, sometimes verifying a particular anomaly with the double-dipole or the more expeditious twin array, both of which yield single peaks.

Instrumentation

In the direct-current cases presented up to this point, a simple ammeter and voltmeter would be adequate to measure the current and voltage needed to calculate the resistivity. In actuality, several problems arise with this method. First, small, chemically derived voltages develop between the probes and the ground, causing measurement errors. Second, probes gradually become polarized during current flow because build-up of charge near the probes alters the measured resistance. Finally, measuring probes can detect small, naturally occurring currents flowing directly through the earth, which introduce discrepancies. All of these error-producing effects can be minimized by using an alternating current with a judiciously chosen frequency, typically between 10 and 500 Hz. Most instruments available today are designed in this manner and many are calibrated to read resistivity directly for a specific array.

Field Procedures

Lateral variations in soil resistivity are measured by using linear traverses or a regularly spaced grid, the choice depending primarily on required resolution and economy. If the expected archaeological features are ditches or palisades, or features measuring several meters on a side and lacking interior detail, then traverses are likely the best choice. However, if features on the order of a meter or two in diameter are sought, then more definition can be obtained by measuring with a gridded regular interval. A locational system similar to that described for magnetometer surveying is adequate, providing a nonconductive material, such as dry cloth tapes or ropes, is used.

Establishing the electrical properties of the site before systematic surveying aids in choosing the array and spacing. If the location of any suitable feature is known, a trial traverse can be run using either the Wenner array or the double-dipole or both. The array spacing for this test and the later complete survey should be contrived to allow current flow to encompass as much of the feature as possible. The depth measured by the array is about equal to the distance between the voltage probes. In other words, a volume roughly equal to a hemisphere of diameter a is measured. It is permissible to have the probe spacing *at most* equal to the width of the feature, providing its width and depth are roughly the same. A higher degree of confidence in the existence of an anomaly is obtained if three or more readings can be taken over the feature. If its width is greater than its depth, the array spacing should be reduced accordingly. For example, if some horizontal feature, such as a house floor, is encountered at a depth of 30 cm with a width of 4 m and exhibiting a resistivity contrast of 5:1, the following anomalies (expressed in arbitrary units depending on soils) would be observed for various probe spacings.

Probe Separation (m)	Strength of Anomaly (arbitrary units)	Percentage of Maximum Possible Anomaly
0.3	3.75	94%
0.5	3.40	85%
1.0	2.70	67.5%
1.5	1.80	45%

The test traverse should be extended 10 or 15 m on either side of a prospective feature to ascertain ground-noise conditions. It is often advantageous to traverse a nearby area devoid of cultural features to determine what natural resistivity variation one might encounter. If the overall noise level appears high compared with the anomaly produced by the feature, it is probably best to continue using the double-dipole or Wenner array; if significant change is detected, one might choose the Clark or the twin array to speed data collection. The choice of the array is usually made on the basis of the operator's experience and familiarity with an array and the affinity of the instrument used to one probe configuration or the other. When little is known about the geometry of the features on a site, a probe spacing of 1 m using the Wenner or the double-dipole array is a judicious choice.

If traverses are used for field measurement, they should extend well away from the region of archaeological interest so that additional background information can be determined. If a survey takes place over an extended time and periodic rainfall occurs, steps should be taken to adjust the traverses relative to one another to correct for the increased soil moisture. This can be accomplished by running a common traverse before and after a rainfall and using the average difference as a constant for adjusting subsequent measurements.

Where the ground is level, easily penetrated, and clear of excessive vegetation, the writers have taken between five and seven readings per minute or, considering the repositioning of measuring ropes involved, 150 to 190 readings per hour. When a 1-m grid spacing is used, an area of about $1,000-1,200$ m^2 can be covered in a day, although such large surveys are arduous. Hard ground slows the work, as the probes must be pounded to ensure proper contact.

Field notes should be maintained with observations on variations in soil type, ease or difficulty in probe insertion, change in the density of vegetation, and noticeable topographic features. These factors affect the resistivity and can often be of valuable assistance during interpretation. A log of the rainfall for the period of the fieldwork is also a valuable aid, since large amounts of rain can drastically affect the magnitude and even the polarity of resistivity anomalies (Clark, 1975; Al Chalabi and Rees, 1962).

Several simple field techniques can be used to distinguish more readily the resistivity anomalies due to background variation in areas where noise from surrounding soil is high. When the validity of a reading is suspect, the traverse should be rerun for verification. With a simple alteration, instruments can quickly read both the double-dipole and Wenner arrays at a single station without repositioning the probes. In agricultural regions where cultivation has truncated features, the confusing effects of disturbed topsoil can be reduced and the relative contribution from features increased by using the Barnes Layer Method (Barnes, 1952, 1954; Carr, 1977, 1982). By running two or three traverses over the area of interest and using increased probe spacings for each traverse, sets of data are generated that encompass greater soil depths. These data sets can be treated to reduce contributions from specific layers to enhance resistivity contrasts from cultural features or to reduce noise from topsoils.

INTERPRETATION

Once data are collected, they are processed and displayed in a manner that will enhance the resistivity contrasts. Little alteration of data is needed before preliminary display, since many instruments produce readings that are already converted to resistivity and some compensate for the array being used. The only other preliminary processing that might be necessary is the addition of an empirical climatic factor, used to compensate for rainfall during a lengthy survey.

Because individual traverses are utilized more often in resistivity than magnetic surveying, profiling the data tends to be a more common display technique. Profiles can be plotted readily by hand and on occasion should be done in the field, in order to check noise levels or the feasibility of a particular array or spacing. For large surveys, where an appreciable amount of data is collected, a computer is a necessity.

Where data have been collected on a grid system, contour plotting yields better resolution and more information. In these instances computer maps such as SYMAP or line contours can be used.

Because the signal-to-noise ratio is difficult to quantify in resistivity surveying, in archaeological contexts the data are interpreted qualitatively. Essentially two types of noise are involved: *correlated noise*, caused by the contributions from natural soil variation, and *uncorrelated noise*, the sum of instrument variation, difference in probe spacing and depth, and the occasional poor contact of a probe with the ground. Both sources of noise are sufficient to mask contrasts from archaeological features, so steps should be taken to recognize and to minimize their occurrence.

Correlated noise from natural soil variations can be as large or larger than the signal and in these instances can be distinguished only when the noise signature has a different shape that that of the signal. Traverse data from an area lacking buried cultural remains as the best preliminary clue to recognizing correlated noise. If no specific correlated-noise anomalies are recognizable from natural soil variation, then the standard deviation of all the readings on the test traverse provides a reasonable background level above which to look for cultural anomalies.

Uncorrelated noise levels can be minimized by field procedures. If a measurement on a traverse is greater than the preceding one by a predetermined amount (usually the standard deviation of the test traverse), then the probe contacts should be checked and the reading retaken. Recognition of both correlated and uncorrelated noise can be aided by the use of alternate probe configurations.

Figure 8.21 shows typical anomalies for features similar to pits or ditches. Anomalies are similar in shape for most other archaeological features of that size, but the responses will vary in magnitude. The feature in figure 8.21 has a higher resistivity than the surrounding media such that $\rho_2/\rho_1 > 1$. The anomalies would be inverted if the resistivity of the feature were lower (i.e., $\rho_2/\rho_1 < 1$). Resistivity contrasts between larger features may be manifest as an average change in the measurements, rather than as an individual, symmetrical anomaly. If the resistivity contrast is large, subsidiary peaking may be present at the boundary. Models of features have been used to aid in interpretation, but the mathematical calculation for the expected anomalies quickly becomes formidable, even for simple geometric shapes (Cook and Van Nostrand, 1954; Grant and West, 1965; Telford et al., 1976).

EXAMPLES

The following examples are all drawn from a survey at the Knife River Indian Village National Historical Site in the vicinity of Sakakawea Village (32ME11). The survey was undertaken in cooperation with the Midwest Archaeological Center, National Park Service, to assess the potential responses of experimental features and to examine what cultural areas were amenable to detection using resistivity.

Sterile Region

On the periphery of the village an imitation "cache pit" was excavated and refilled with moistened earth. Profiles across this anomaly using the Wenner and the twin array are shown in figure 8.22. The Wenner array produces the classic W-shaped anomaly, whereas the twin array produces a narrower, single peak. The magnitude of the twin anomaly is uncharacteristically large; it would normally be quite a bit less than the magnitude of the Wenner anomaly.

Suspected Palisade

To test for the existence of a fortification ditch inferred by the presence of a

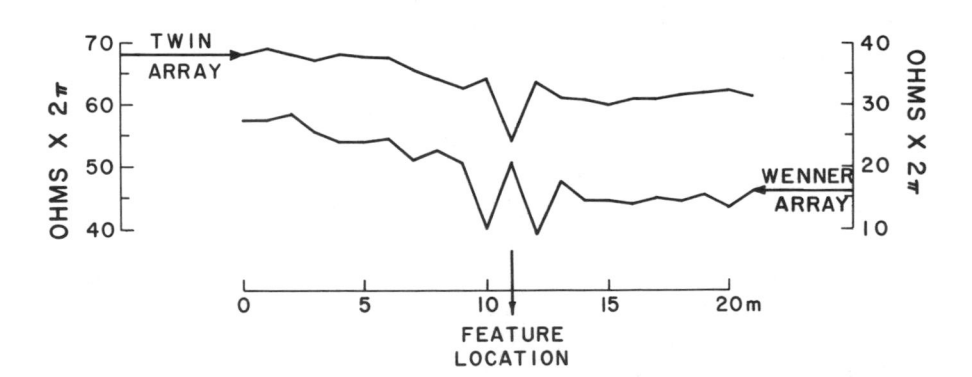

Fig. 8.22. Resistivity anomalies over earth-filled pit produced by the Wenner and twin arrays.

visible difference in soil in the river bank where the site is partially truncated, several radial traverses were run from the center of the village across the surrounding depression. The first traverse, which is located north of the area of figure 8.4, is shown in figure 8.23. When a marked area of low resistivity was encountered using probes spaced 1.0 m apart, the traverse was rerun with a spacing of 1.5 m, which shows a reduced contrast. From this we can infer that the source is of limited depth, probably no more than 1 m. Because the other traverses revealed no anomalous responses in this region, it is likely that the source of this anomaly is localized, indicating no detectable fortification ditch.

House 6

Using a sample spacing of 1 m, we collected two matrices of data over house 6 (block O, fig. 8.4) by running traverses first eat to west and then north to south. This method provides more information on the geometry of features and assures more accuracy in each reading. In figure 8.24 the average of the two profiles is displayed, a technique that reduces contributions from subsidiary peaking for features of certain geometries and reflects some of the larger trends on the site. In the center of the map is an oblong region of lower resistivity corresponding to the center of the depression (see fig. 8.4). We can infer a region of higher moisture and likely a reduced amount of compaction from this data. Surrounding this central low region is a ring of higher resistivity that appears to lie well within the depression. This area is apparently more compact, thus having a reduced moisture content and a higher resistivity. A particularly notable portion of the high ring is a crescent-shaped anomaly on the southeast edge. Wilson (1934) mentions that livestock were kept in the interior of lodges and penned close to the walls. It is conceivable that this high anomaly defines the boundaries of a horse corral or other area of excessive traffic.

The midden region surrounding the depression has mostly a low resistivity due to the reduced compaction and higher ion content. It is interesting to

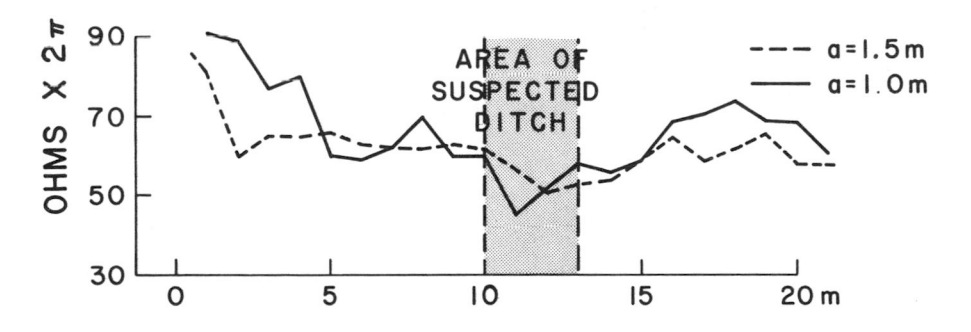

Fig. 8.23. Resistivity profiles across a suspected ditch using the Wenner array; with spacing set at $a = 1.0$ m and $a = 1.5$ m.

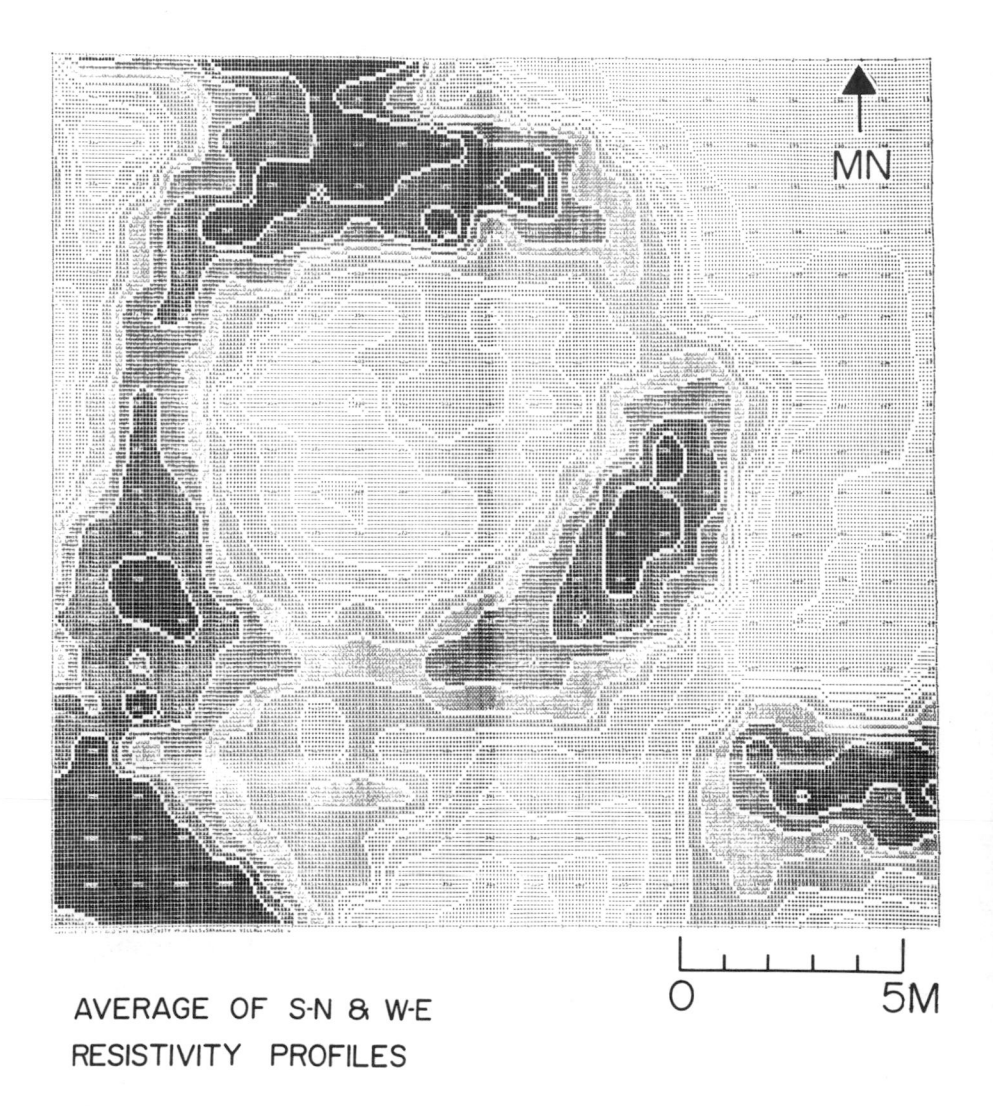

AVERAGE OF S-N & W-E
RESISTIVITY PROFILES

0   5M

Fig. 8.24. SYMAP contour map of the average of north-south and east-west profiles over house 6, Sakakawea Village.

compare figure 8.24 with figure 8.7, which shows the magnetic field over this same house. The general outline of the lodge is present in both, but each method shows distinctly different properties of the same area.

Acknowledgments

We wish to express our thanks to the following agencies, whose contracts have permitted us to develop our techniques and gain valuable experience: the Tulsa

District U.S. Corps of Engineers, the Dolores Archaeological Program (Bureau of Reclamation), the Oklahoma Archaeological Survey, and the Arkansas Archaeological Survey. In particular we wish to thank the Midwest Archaeological Center, National Park Service, and F. A. Calabrese and R. Nickel of that agency for long and continued support, encouragement, and collaboration.

REFERENCES

Aitken, M. J. 1974. *Physics and archaeology*. Oxford: Clarendon Press.

Aitken, M. J., G. Webster, and A. Rees. 1958. Magnetic prospecting. *Antiquity* 32: 270–71.

Aitken, M. J., and J. C. Alldred. 1964. A simulator-trainer for magnetic prospection. *Archaeometry* 7:28–35.

Al Chalabi, M. M., and A. I. Rees. 1962. An experiment on the effect of rainfall on electrical resistivity anomalies in the near surface. *Bonner Jahrbucher* 162: 266–71.

Arnold, J. B. 1974. A magnetometer survey of the nineteenth century steamboat Black Cloud. *Bulletin of the Texas Archaeological Society* 45:225–30.

Aspinall, A., and J. T. Lynam. 1970. An induced polarization instrument for detection of near surface features. *Prospezioni Archeologiche* 5:67–75.

Atkinson, R. J. C. 1952. Méthodes électriques de prospection en archéologie. In *La découverte de passé*, ed. A. Laming, 59–70. Paris: Picard.

Barnes, H. E. 1952. *Mapping and subsurface exploration for engineering purposes*. Michigan State Highway Department Research Board Bulletin, 65.

———. 1954. Electrical subsurface exploration simplified. *Roads and Streets* 97:81–84.

Belshé, J. C. 1957. Recent magnetic investigations at Cambridge University. *Advances in Physics* 6:192–93.

Bevan, B. 1975. A magnetic survey at Les Forges du Saint-Maurice. *MASCA Newsletter* 11:1.

Black, G. A., and R. B. Johnston. 1962. A test of magnetometry as an aid to archaeology. *American Antiquity* 28:199–205.

Bleed, P., M. Yoshizaki, W. Hurley, and J. W. Weymouth. 1980. A Preliminary Report on the 1978 Excavation at the Yagi Site, Japan. Technical Report No. 80–14. Division of Archaeological Research. University of Nebraska.

Breiner, S. 1973. *Applications manual for portable magnetometers*. Sunnyvale, CA: Geometrics.

Breiner, S., and M. D. Coe. 1972. Magnetic exploration of the Olmec civilization. *American Scientist* 60:566–75.

Carr, C. 1977. A new role and analytical design for the use of resistivity surveying in archaeology. *Mid-continental Journal of Archaeology* 2:161–93.

———. 1982. *Handbook on soil resistivity surveying*. Evanston, IL: Center for American Archeology Press.

Clark, A. J. 1968. A square array for resistivity surveying. *Prospezioni Archeologiche* 3:111–14.

———. 1969. Resistivity surveying. In *Science in archaeology*, 2d ed., ed. D. Brothwell and E. S. Higgs, 695–708. London: Thames and Hudson.

———. 1975. Archaeological prospecting: A progress report. *Journal of Archaeological Science* 2:297–314.

Cook, K. L., and R. G. Van Nostrand. 1954. Interpretation of resistivity data over filled sinks. *Geophysics* 19:761−70.

Davis, J. C. 1973. *Statistics and data analysis in geology.* New York: John Wiley and Sons.

De Terra, H. 1947. A preliminary note on the discovery of fossil man at Tepexpán in the Valley of Mexico. *American Antiquity* 13:40−44.

Ezell, P., J. R. Moriarity, J. D. Mudie, and A. I. Rees. 1965. Magnetic prospecting in southern California. *American Antiquity* 31:112−13.

Ginzburg, A., and A. Levanon. 1977. Direct current resistivity measurements in archaeology. *Geoexploration* 15:47−56.

Graham, I. D. G., and I. Scollar. 1976. Limitation on magnetic prospection in archaeology imposed by soil properties. *Archaeo-Physika* 6:1−125.

Grant, F. S., and G. F. West. 1965. *Interpretation theory in applied geophysics.* New York: McGraw-Hill.

Hesse, A. 1962. Geophysical prospecting for archaeology in France. *Archaeometry* 5:123−25.

Le Borgne, E. 1955. Susceptibilité magnétique anormale du sol superficiel. *Annales de géophysique* 11:399−419.

———. 1960. Influence du feu sur les propriétés magnétiques du sol. *Annales de géophysique* 16:159−95.

Leith, C. J., K. A. Schneider, and C. Carr. 1976. Geophysical investigation of archaeological sites. *Bulletin of the International Association of Engineering Geology* 14:123−28.

Lerici, C. M. 1961. Archaeological survey with the proton magnetometer in Italy. *Archaeometry* 4:76−82.

Linington, R. E. 1964. The use of simplified anomalies in magnetic surveying. *Archaeometry* 7:3−13.

———. 1972. A summary of simple theory applicable to magnetic prospecting in archaeology. *Prospezioni Archeologiche* 7/8:9−60.

Mason, R. 1981. Large-scale archaeomagnetic surveys of the Barton and Vinton Townsites. Paper delivered at the Fourteenth Annual Meeting, Society for Historical Archaeology (New Orleans).

McDonald, W. A., and G. Rapp, Jr., eds. 1972. *The Minnesota Messenia expedition.* Minneapolis: Univ. of Minnesota Press.

Morrison, F., C. W. Clewlow, and R. F. Heizer. 1970. Magnetometer survey of the La Venta Pyramid. *Contributions of the University of California Archaeological Research Facility* 8:1−20.

Nashold, B. W. 1977. An archaeological magnetic survey at Cahokia. M.A. thesis, University of Illinois at Chicago Circle.

Rainey, F., and E. K. Ralph. 1966. Archaeology and its new technology. *Science* 153:1481−91.

Ralph, E. K. 1964. Comparison of a proton and a rubidium magnetometer for archaeological prospecting. *Archaeometry* 7:20−27.

Ralph, E. K., F. Morrison, and D. P. O'Brien. 1968. Archaeological surveying utilizing a high-sensitivity difference magnetometer. *Geoexploration* 6:109−22.

Rees, A. I., and A. E. Wright. 1969. Resistivity surveys at Barnsley Park. *Prospezioni Archeologiche* 4:121−24.

Resnick, R., and D. Halliday. 1966. *Physics.* New York: John Wiley and Sons.

Schlumberger, C. 1920. Analysis of electrical prospecting of the subsoil. *Académie de sciences (Paris), Comptes rendus* 170:519–21.

Scollar, I. 1961. Magnetic prospecting in the Rhineland. *Archaeometry* 4:74–75.

———. 1965. A contribution to magnetic prospecting in archaeology. *Beihefte der Bonner Jahrbucher* 15:21–92.

SYMAP. 1975. *Synagraphic computer mapping.* Cambridge, MA: Harvard University, Laboratory for Computer Graphics and Spatial Analysis.

Tagg, G. F. 1964. *Earth resistances.* London: Pitman.

Telford, W. M., L. P. Geldart, R. E. Sheriff, and D. A. Keys. 1976. *Applied geophysics.* Cambridge, England: Cambridge Univ. Press.

Tite, M. S. 1972. *Methods of physical examination in archaeology.* New York: Academic Press.

Van Nostrand, R. G., and K. L. Cook. 1966. *Interpretation of resistivity data.* U. S. Geological Survey, Professional Paper no. 499.

von Frese, R. R. B. 1978. Magnetic exploration of historical archaeological sites as exemplified by a survey at Ft. Ouiatenon (12T9). M.S. thesis, Purdue University.

Wenner, F. 1915. A method of measuring earth resistivity. *Bulletin, U.S. Bureau of Standards 12, Scientific Paper 258,* pp. 469–78.

Weymouth, J. W. 1976. A magnetic survey of the Walth Bay site, Midwest Archaeological Center (NPS). Occasional Studies in Anthropology, no. 3.

———. 1979. Magnetic surveying of archaeological sites. In *Proceedings of the First Conference on Scientific Research in the National Parks,* vol. 2, 941–47.

Weymouth, J. W., and R. Nickel. 1977. A magnetometer survey of the Knife River Indian villages. *Plains Anthropologist* 22:104–18.

Wilson, G. L. 1934. The Hidatsa earth lodge. *Anthropological Papers of the American Museum of Natural History* 23:340–420.

9

Archaeomagnetism

D. H. TARLING

ABSTRACT

Many materials in an archaeological site contain particles that were magnetized at a specific time when the materials were heated, chemically changed, or deposited. This magnetization can be used for relative and absolute dating, with a potential accuracy of some 25 years for materials formed in these ways during the last few thousand years. However, many archaeological artifacts have yet to be studied and the accuracy of absolute dating depends on the establishment of a precise chronology for areas of the Earth's surface some thousands of kilometers in extent. The study of the magnetic properties of archaeological materials may also be of value in sourcing (provenancing), reconstructing artifacts, and evaluating the nature of past technologies—although such other applications are only now being investigated.

Most archaeological and geological materials contain magnetic particles, if only as very minor impurities, which means that many substances of archaeological interest are potentially available for magnetic study. Because the geomagnetic field gradually changes in direction and intensity, materials that became magnetized at a specific time can be dated *absolutely*, by comparison with established records of geomagnetic change, or *relatively*, by comparison with other archaeological material. A material's magnetic properties will also vary because of differences in composition; these differences can be used to distinguish the original sources of lithic materials such as obsidian. Magnetic parameters can be useful in determining, for example, the orientation of metal casts or coins when cooled and thus produce evidence bearing on the technology of their manufacture. In these and other archaeological applications, the fundamental difficulty lies in finding materials that acquired their *primary* magnetization at a specific time and then isolating this magnetization from any later *secondary* magnetizations that may have been acquired.

Absolute dating by archaeomagnetism presents further problems, such as obtaining accurate field orientations of samples and establishing the reliability of the chronological record against which comparison is to be made. If these can be solved, the principles of archaeomagnetism are straightforward (Aitken, 1970, 1974; Tarling, 1971, 1975), and the technique has the advantages of being cheaper than many other scientific dating methods, usually nondestructive, and applicable to materials that may not be datable by other methods.

In this chapter the ways in which archaeological materials may become magnetized are described, followed by a brief outline of the methods of collection, measurement, and analysis of these materials. Progress in establishing good chronological records of past geomagnetic variations will then be discussed prior to indicating some of the ways in which archaeomagnetic observations can be related to other archaeological studies.

THE MAGNETIZATION OF ARCHAEOLOGICAL MATERIALS

Figure 9.1 illustrates some materials in an archaeological site that may be suitable for archaeomagnetic study. Several other materials, such as hearths, metal casts, well fillings, and geological sediments, may also prove usable.

A material containing magnetic particles can become magnetized in the direction of the Earth's magnetic field (a) when it is cooled after heating, (b) when it undergoes chemical changes affecting the size of the magnetic grains, or (c) if it is composed of previously magnetized particles redeposited by water or wind. For particles of the common iron oxides—magnetite and hematite—with diameters in the range of 10^{-3} to 10^{-5} cm (fig. 9.2), this primary magnetic alignment with the geomagnetic field becomes blocked within them for millions or even billions of years. However, smaller or larger particles are less magnetically stable. As their primary alignment is gradually lost they can acquire a fourth generic magnetization type—viscous—while lying in the changing geomagnetic field.

THERMAL MAGNETIZATION

When a material is heated to temperatures over 700°C, thermal agitation prevents any alignment of the magnetization of its magnetic grains so that the material, as a whole, is nonmagnetic. In cooling, the magnetization associated with the electron spin in individual magnetic grains becomes linked by atomic exchange forces, and their direction of magnetization becomes aligned with any existing magnetic field. Under normal archaeological circumstances, the existing magnetic field will have been that of the Earth. As cooling continues, this orientation becomes "frozen" into the material so that at ambient temperatures

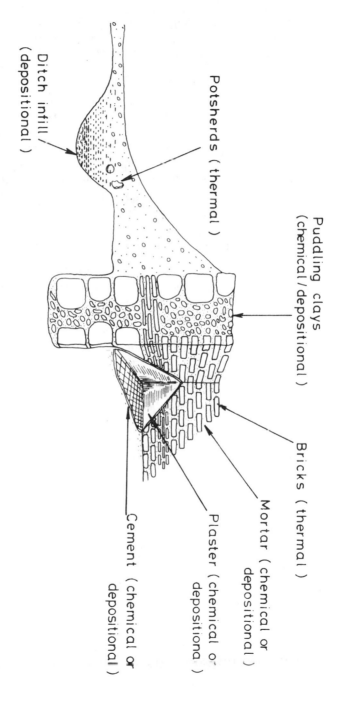

Puddling clays
(chemical/depositional)

Potsherds (thermal)

Ditch infill
(depositional)

Bricks (thermal)

Mortar (chemical or
depositional)

Plaster (chemical or
depositional)

Cement (chemical or
depositional)

Fig. 9.1. Some materials in an archaeological site that may be suitable for archaeomagnetic study. This diagram demonstrates only certain features and it is clear that many other materials, such as hearths, metal casts, well infill, and geological materials, may also be suitable for archaeomagnetic study.

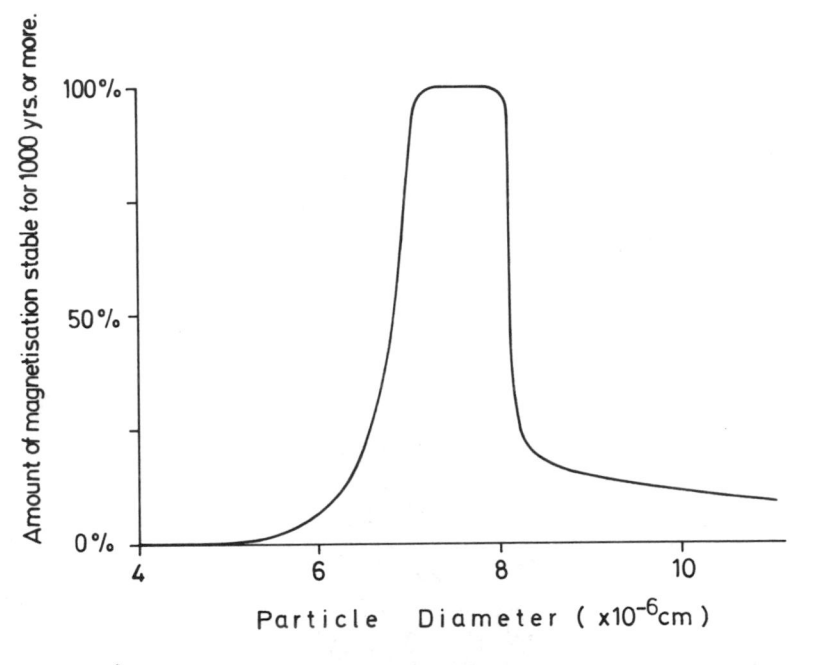

Fig. 9.2. The stability of magnetization for different grain sizes of magnetite. Particles of titanomagnetite, e.g., magnetite (Fe_3O_4), less than 600 Å in diameter do not carry a stable magnetization, whereas those larger than 850 Å carry only a small amount of stable magnetization. Particles of hematite have a high stability over an even wider range of grain sizes. Most materials have a range of grain sizes and therefore include both stable and unstable components of magnetization.

the high-temperature magnetization acquired by the stable-sized particles will last until the material decomposes. Particles larger or smaller than those in the stable grain-size range (fig. 9.2) also become aligned at this time. Gradually, however, they lose their original alignment and very slowly become magnetized along any new direction of the Earth's magnetic field, with those closest to the stable range being the least free to change direction.

Any artifactual material that has ever been heated is likely to possess a magnetization that, if isolated from the contribution of unstably magnetized particles, can be attributed to the time of heating. Most archaeomagnetic studies have been carried out on such materials: kilns (fig. 9.3), hearths, sherds, tiles, bricks, etc. Other materials known to have often acquired a thermal remanence include coins, metallic casts and their clay cores, obsidian, ash layers, slag heaps, and so forth. Most of these have not yet been subjected to rigorous examination, but preliminary studies have shown that they all appear to be potentially useful in archaeomagnetic studies. Although materials that have acquired a remanence by heating are generally the best for archaeomagnetic study, they may be so strongly magnetized as to cause local distortions of the geomagnetic field and

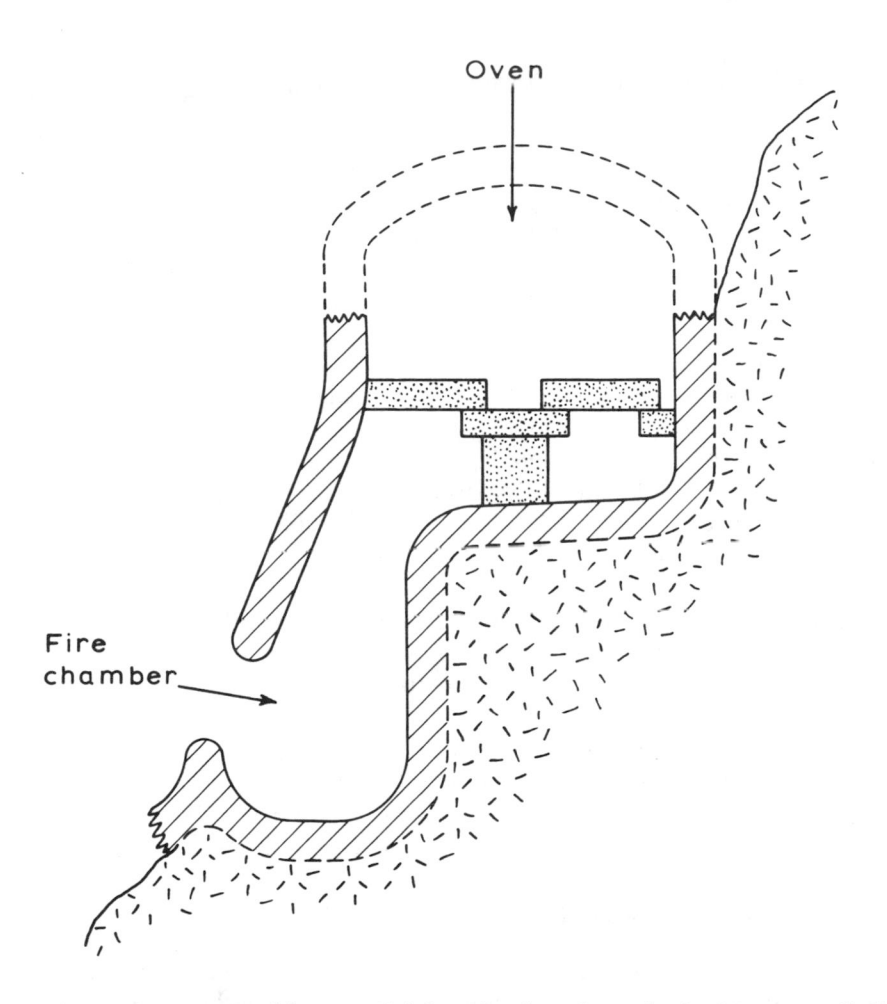

Fig. 9.3. Cross-sectional diagram of a kiln. Kilns, hearths, and so forth are generally the most suitable sites for standard archaeomagnetic study, as their composition usually includes clays containing magnetic particles that have been heated well above their Curie temperatures. One problem for large kilns is that the magnetization of the kiln itself can affect the local geomagnetic field, making magnetic compass orientation suspect, and another is that the magnetization of the walls may have somewhat distorted the field (possibly 2–3° in direction and 2–3 percent in intensity) in which the inner parts of the kiln were cooling.

thus affect magnetic compasses used to orient them at the archaeological site. The materials themselves may also be magnetically anisotropic and hence distort the field in which they cooled (Harold, 1960; Weaver, 1962; Rogers, Fox, and Aitken, 1979). However, such refractions are generally less than 3° in direction and a few percent in intensity and, with care, should be largely removed during partial demagnetization (discussed below).

CHEMICAL MAGNETIZATION

If only extremely fine grained particles are present in a material, their direction of magnetization is unstable and readily follows any changes of the Earth's magnetic field. But chemical changes can occur that result in these particles growing into the magnetically stable size range, and then the orientation of any field affecting them as they enter this range becomes locked within the material for an indefinite time. Further growth can expand the grains out of the stably magnetized range, but there are generally sufficient grains of just submicroscopic size for a detectable stable magnetization to be retained. Thus, archaeological materials that have been subjected to a specific chemical change may have a remanence associated with that event. Few studies of such materials have yet been made and most chemical remanence encountered has been acquired during weathering over protracted periods of time. If the magnetization acquired this way is stable, then it is extremely difficult or impossible to isolate it from any other generic type. It is generally advisable, at this stage, to avoid routine sampling of materials that show signs of long-term weathering.

Materials such as puddled clays, cements, and mortars may contain a chemical remanence associated with the time that they set. Present research, however, suggests that this magnetization is essentially *detrital* rather than chemical in origin—that is, the smallest magnetic particles rotate into alignment with the Earth's field and then become locked in this position as the material sets. Mortar, in any case, is often difficult to analyze with success, as it commonly contains ground-up pottery. If not finely comminuted, the pieces of pottery may cause the samples to be inhomogeneously magnetized and obscure the magnetization associated with cementation.

DEPOSITIONAL (DETRITAL) MAGNETIZATION

Thermal and chemical magnetizations both arise from the alignment of electron spins by internal atomic forces within the particles themselves and not by the actual, physical alignment of the particles. However, when already magnetized particles are deposited by wind or (more commonly) water, they act as microscopic, suspended compass needles and are themselves aligned by the Earth's magnetic field. While these particles are in suspension within air or water, they are aligned by the Earth's magnetic field. Upon deposition, however, some of this alignment is usually lost, so that while the horizontal component (declination) is still oriented to magnetic north, the magnetic angle relative to the horizontal plane (inclination) is usually less than the actual geomagnetic field value (Griffiths et al., 1960). Where deposition is on a sloping surface, a further decrease in inclination may occur, although, again, declination is largely unaffected. In subaqueous environments most of this inclination misalignment is recovered while the particles are at the water-sediment interface (fig. 9.4a); the particles are then surrounded by water and so are free to rotate into alignment with the Earth's field (Irving and Major, 1964). Noel's work (1980) indicates that the remanence in most unconsolidated subaqueous sediments probably arises from such postdepositional rotations and probably accounts for the general absence of inclination errors in such sediments. By implication, then, sediments in which the finer particles are rapidly isolated from interstitial fluids are likely to be the best carriers of a remanence acquired at a specific time. For example, some ditch sediments carry a specific magnetization related to the time of the ditch infill (fig. 9.4b), whereas some lake and pond sediments apparently acquire their remanence over a longer period of time and so are less reliable indicators of the ambient geomagnetic field during accumulation.

Because of uncertainty about the way in which sediments have acquired their remanence, most archaeomagnetic studies have avoided their use. Uncertainty still exists about many of these factors, and sediments are clearly less-reliable indicators of past magnetic-field parameters than are fired materials; nevertheless, good results have been obtained from a range of unconsolidated lake sediments (Mackareth, 1971; Thompson, 1973; Creer et al., 1975). Studies of pond and cave sediments (Creer and Kopper, 1974), ditch infills, and the like suggest that they too may give reliable archaeomagnetic dates. It must be emphasized, however, that the processes of magnetization in sediments are complex and only poorly understood; such materials are subject to physicochemical changes in areas of changing water-table level, percolating groundwaters, and so forth. Such uncertainties preclude the use of sediments for archaeomagnetic-intensity studies at the present stage, but many sedimentary environments are potentially suitable for directional study and future research seems likely to allow improved assessment of such environments.

Consolidated sediments probably carry a magnetization of postdepositional origin, either supplemented or dominated by chemical remanence associated with the disintegration of the individual detrital particles or by the cementation of mineral-rich fluids. In an archaeological context, such chemical processes are insignificant in sediments except under unusual circumstances. This type of remanence could be associated with the magnetization of mortars, cements, puddling clays, and so forth.

Viscous (Time-dependent) Magnetization

As most archaeological materials contain iron-oxide impurities with a wide range of grain sizes, they will normally exhibit both stable and unstable magnetizations. The magnetization of the smallest and largest of the unstably magnetized particles will be free to follow the direction of the geomagnetic field, but particles in the middle size range will have magnetizations that only gradually follow changes of the field. When a range of particle sizes is present, as in most archaeological materials, the nonstable grains will have acquired secondary magnetizations over several years or even several thousand years while stably

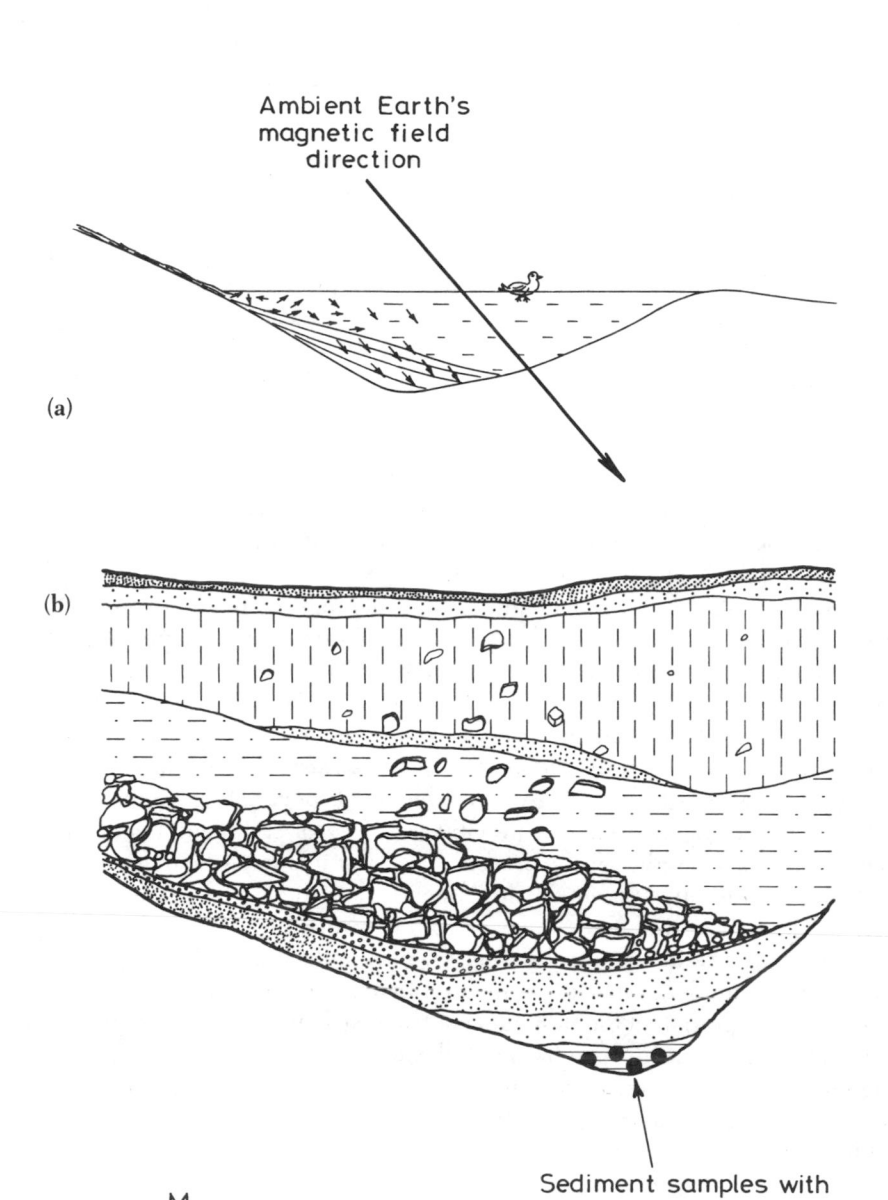

Ambient Earth's
magnetic field
direction

(a)

(b)

M.

0   1

Sediment samples with
detrital magnetisation

Fig. 9.4. (a) As sediment grains are washed into a pond, lake, or ditch, they become aligned during deposition, but this may be distorted by the effect of currents and the unevenness of the bottom surface. The alignment is regained immediately after deposition as the magnetic grains rotate back into the direction of the ambient field. (b) Even ditch-infill materials have been found suitable for archaeomagnetic study, although preliminary studies indicate that the *very first* sediments trapped in the ditch are generally the most effective carriers of the original magnetization (after A. J. Clark, pers. com., 1980).

magnetized particles will still carry the primary thermal, chemical, or detrital remanence. The rate at which these viscous magnetizations are naturally acquired can be measured in the laboratory and used to determine how long that material has been lying in its present position in the geomagnetic field (discussed below). Generally, it is essential to remove these unstable components in order to define and use the original primary magnetization. As viscous magnetization is by definition unstable, it can be randomized by moderate heating and then cooling in zero magnetic field (fig. 9.5a). The thermal vibrations during heating agitate the electrons in the unstable particle sizes sufficiently for them to lose any alignment. Thus, when cooled in zero external magnetic field, their magnetizations become frozen in random positions and cancel each other. Though also agitated by heating, the electron spins in the stable particles remain unable to move out of their original alignment until reheated to the higher temperature at which they were originally acquired; after cooling they will retain this alignment. Similar electron agitation can also be induced by placing the samples in low, alternating magnetic fields, which again randomize the less-stable magnetizations but leave the original magnetization untouched (fig. 9.5b). At each incremental demagnetization, the direction of remanence associated with this secondary magnetization changes, while the primary magnetization will remain constant. At some stage during the demagnetization procedure, such high temperatures or alternating magnetic fields are applied that the stable remanence itself starts to become randomized. Under normal conditions, therefore, the demagnetization procedure results in the initial removal of the viscous remanence acquired over the last few decades or centuries, after which the primary stable remanence is isolated; the direction then shows little or no change until at much higher fields or temperatures it begins to fluctuate randomly.

The stable direction persists over a range of incremented demagnetization treatments, but this range differs from one sample to the next due to variations in the grain size of the magnetic particles, different amounts of more-stable secondary magnetization, etc. It is, therefore, necessary to examine all samples over a range of demagnetization treatments in order to determine the direction of the primary content. Where the samples are similar, such as those from the same level at a site, then detailed demagnetization of a few samples is usually adequate to define the optimum treatment by which one can isolate stable magnetic components in the other samples.

Although partial demagnetization effectively reduces or removes the time-dependent viscous magnetization, it is not effective against secondary magnetizations that have been acquired either thermally or chemically. For example, refired materials will normally show a stable remanence associated with the last firing. If, however, this firing was at a lower temperature than the original heating and was below the Curie point of the minerals, then both thermal components can be determined. For example, a pot may have been fired initially at temperatures above 675°C and, for some reason, reheated to about 400°C. In this case the magnetic properties obtained at the original, higher temperature

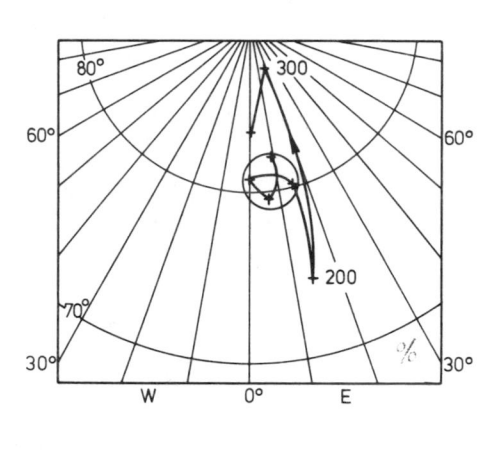

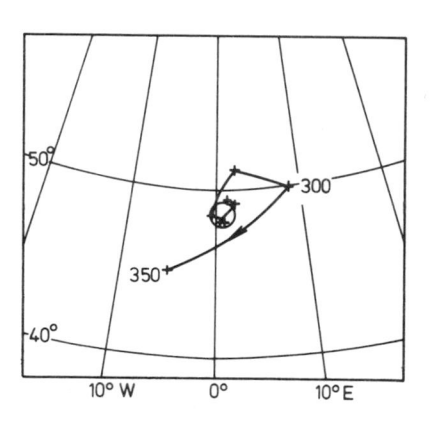

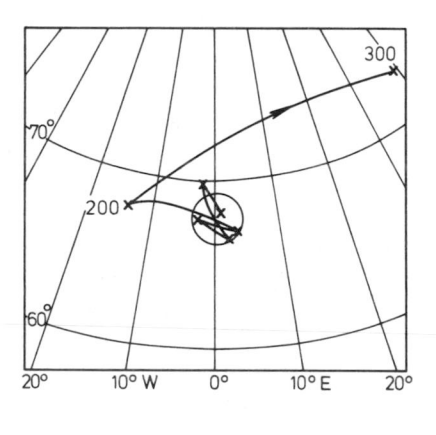

Fig. 9.5. Examples of demagnetization of some archaeological materials. The change in direction during (a) thermal demagnetization of an Israeli Byzantine hearth sample, (b) alternating-magnetic-field demagnetization of a Greek kiln sample, and (c) a late eighteenth-century pond unconsolidated sediment. The precisions (95 percent probability) of the mean directions are, respectively, 1.6°, 0.7°, and 1.5°.

can be retrieved by heating the pot to 410–420°C, just above the later temperature. In most archaeological situations, chemical changes occurring during preservation are more difficult to remove and may obliterate any preexisting stable magnetization.

The determination of the intensity of the past field (Thellier and Thellier, 1959) is somewhat more tedious than defining the direction of magnetization associated with the stable component. The intensity of the ancient external magnetic field—assumed to be the Earth's—is directly related to the intensity of a sample's magnetization acquired by contemporary thermal or chemical processes. Thus, the stable remanence intensity now observed needs to be compared directly with the intensity acquired by the same sample subjected to the same process in a known external field:

$$\text{ancient EFI} = \frac{\text{natural SRI}}{\text{laboratory SRI}} \times \text{laboratory EFI},$$

where EFI is the external field intensity and SRI is stable remanence intensity. If no viscous remanence is present, this equation can be solved by measuring the total natural stable remanence, then heating the material above its Curie temperature and measuring its SRI after it has cooled down to room temperature in a laboratory-controlled external magnetic field. As viscous remanence is generally present, the comparison must normally be made between the intensity of natural thermal remanence acquired over a range of temperatures and that of a remanence acquired during cooling over the same range of temperatures in a known external magnetic field.

At the moment, this repetition of the magnetization process in the laboratory is practicable with adequate precision only for objects that have been fired (fig. 9.6). Comparison of the stable remanence isolated during thermal demagnetization with the laboratory-induced magnetization over the same range of demagnetization treatments is clearly not possible if chemical changes occur in the magnetic minerals during the heating and cooling cycles. Conveniently, many archaeological materials (including potsherds) tend to be chemically inert during this process, but where such chemical changes are detected, the experiment must be abandoned if precise determinations of past intensities are required (Barbetti et al., 1977). In material in which chemical changes may occur during the thermal processing, purely magnetic techniques may be applied (Shaw, 1974): the magnetization acquired by materials subjected simultaneous-

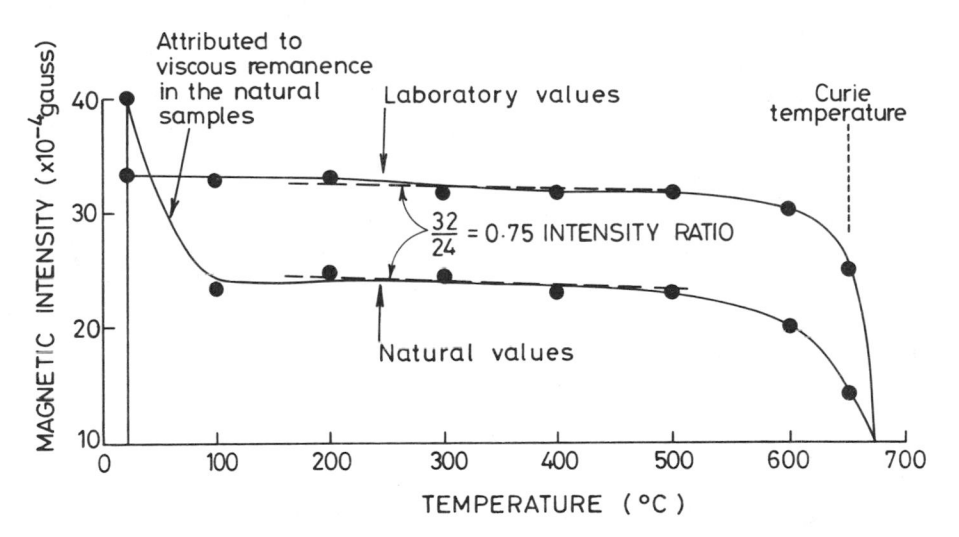

Fig. 9.6. The determination of intensity of the past geomagnetic field. Comparison of the demagnetization of the original, natural magnetization of a baked sample with that of the magnetization given to the same sample after cooling it from above its Curie temperature in a known magnetic field. The ratio of the two intensities of stable remanence (0.75) indicates that the strength of the past geomagnetic field was 75 percent of that used in the laboratory.

ly to direct and alternating magnetic fields (anhysteretic remanent magnetization) is almost identical to a magnetization acquired thermally. At present, however, this technique is uncertainly reliable, at least for the precision required in archaeomagnetic intensity investigations.

Recent experiments (Games, 1977) indicate that comparison of the stable intensity of magnetization of puddled clays with that acquired when the clays are repuddled in a known magnetic field can also be used as a method for determining the past geomagnetic-field intensity. Although still in an early stage, there is a clear potential for developing such comparative techniques and eventually extending them to chemically magnetized materials and sediments deposited under other conditions.

POTENTIAL ACCURACY OF ARCHAEOMAGNETIC DATING

The Earth's magnetic field is presently changing its direction and intensity at different rates in different areas, and archaeomagnetic studies have shown that these rates of change also vary with time. Thus, different accuracies in defining the primary magnetization of an archaeological sample are required at different times for precise dating. If the present rate of change in Britain is taken as typical (fig. 9.7), then a total error of $\pm 1°$ in the determination of the direction of the earth's magnetic field corresponds to an error in age assessment of ± 5 years, and an error of ± 1 percent in the determination of field intensity corresponds to an error of ± 20 years. Such accuracies are well within the range of most magnetometers (Collinson, 1975) when analyzing standard sample volumes of many archaeological materials (cylinders 2.1 cm high $\times$ 2.5 cm in diameter, or 5.3 cm high $\times$ 5.5 cm in diameter; or cubes of 2.1 cm or 4.5 cm on a side). Similar but slightly less accuracy may be obtained for most irregular shapes and sizes using special magnetometer designs (Molyneux, 1971), but it is often difficult to demagnetize such samples properly because conventional demagnetization instruments are not designed for them. The precision of archaeomagnetic dating is not, therefore, limited by instrumental accuracy of measurement, but by the difficulties of (1) adequately isolating the primary magnetization; (2) orienting the samples prior to measurement; and (3) allowing for inhomogeneous and anisotropic magnetic properties.

Partial demagnetization by heating or applying an alternating magnetic field (fig. 9.5) is generally very effective in removing low-stability components, but in some samples the small total amount of stable magnetization may be obscured both by random components left after demagnetizing the lower stability fraction and by instrumental noise. In such cases, instrumental noise can be reduced by determining the stable magnetization over a range of demagnetization levels, and some other errors can be averaged out by detailed sampling. Thus, total error values are presently of the order of $\pm 3°$ in direction and ± 5 percent in intensity.

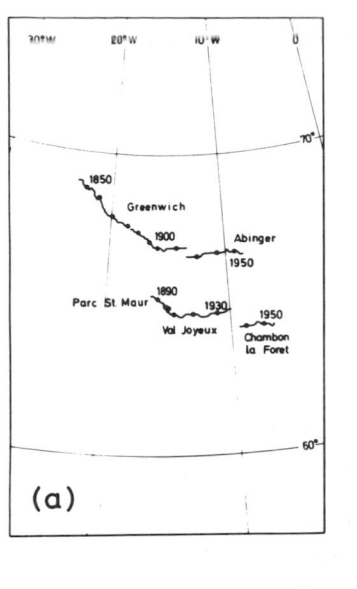

(a)

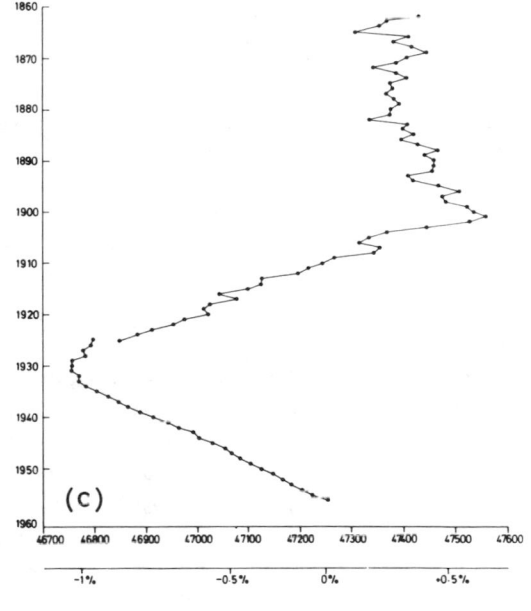

Fig. 9.7. Observatory records of the geomagnetic field near London and Paris. Annual means for Greenwich (1862—1925), Abinger (1925—1956), Parc St. Maur (1883—1900), Val Joyeux (1901—1936), and Chambon la Foret (1936— 1957) for (a) the declination and inclination, (b) the latitude and longitude of the corresponding geocentric magnetic dipole, and (c) the total intensity of the geomagnetic field.

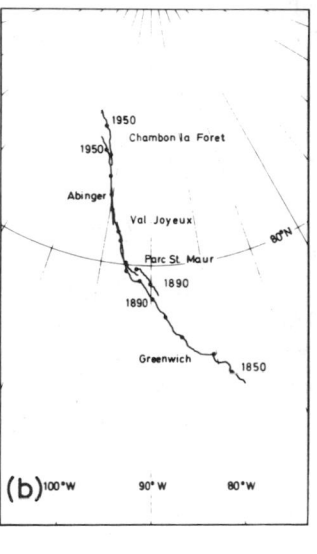

(b)

COLLECTION AND ORIENTATION OF SAMPLES

Standard-sized samples can be drilled from consolidated materials such as kiln walls, vitrified layers, etc., using portable drills (fig. 9.8a); alternatively, samples measuring 12 × 9 × 6 cm can be hand-collected from which standard cubes or cylinders are cut in the laboratory. Nonmagnetic standard-sized holders, such as plastic tubs (2.1 cm high × 2.4 cm in diameter, or 5.3 cm high × 5.5 cm in diameter) or boxes (2.1 cm or 4.5 cm wide), can be pushed into less consolidated materials such as pond sediments with little disturbance.

Friable materials are more difficult to sample. One system is to carve out a pillar with nonmagnetic tools over which standard sample holders can be placed (fig. 9.8b). The pillar is secured within the holder by filling any gaps with gypsum, polyurethane foam, or other nonmagnetic substances. When set, the orientation of the pillar is determined; it is then removed inside its holder, and the ends of the holder are sealed.

Where the material to be sampled is consolidated, the simplest method is to attach flat discs to the sample (fig. 9.8c), using nonmagnetic adhesives. The surface of the attached discs are then oriented, and the sample is removed and placed within standard-sized holders for later measurement.

Archaeomagnetic dating by means of declination and inclination requires that the samples be individually oriented before removal. The top surface of the tub or cube holding a material, while still in situ, is oriented both relative to horizontal and relative to north, using a compass (fig. 9.8a). When samples are being collected from strongly magnetized localities such as kilns, where the materials themselves may distort the local geomagnetic field, a magnetic compass should be avoided. A simple sun-compass consists of a vertical pin, the shadow of which is measured relative to an arrow on the horizontal upper surface of the sample holder. If the time is noted (within 2 minutes) and the sampling latitude is known, the direction of the arrow can be determined from an ephemeris, or by calculation, to within 1° of true north. Even materials such as potsherds and metal casts, which are no longer in their original positions, may carry some indications that can be used to reestablish that orientation. Wheelmarks, rocking edges, glaze runs, and the like can be used as an estimate of the original horizontal position of the potter's wheel; if the pots were stacked systematically in the kiln, as is sometimes the case for glazed ware, then such marks can be an indication of horizontal within the kiln. Under such circumstances, the direction of the remanence can be measured relative to the horizontal or vertical indicators and thus indicate the inclination of the synchronous geomagnetic field.

Dating by means of intensity alone does not require orientation, but it is often worth determining orientation, as it enables the directions of the magnetic-intensity components to be determined. Such data may indicate how long the samples have been in their present position, or they may show that certain components removed during partial demagnetization do actually correspond with the expected viscous magnetic direction.

In view of the importance of accuracy in orientation, particular care is essential to ensure that structures being sampled have not tilted, subsided, or moved since their magnetization was acquired. Movements such as the "fall out" of walls can introduce systematic errors that may not be averaged out even if the entire structure is sampled. Naturally, it is also important to ensure that the samples remain in situ until the orientation procedure is complete and that no movement of the sample takes place after it is within the holder. With care, accuracy in orientation can usually be achieved within some ±1−2°, but it is generally necessary to collect several samples—six or more—from the same horizon. This ensures that sampling errors are, to some extent, averaged out in the determination of the mean direction of the previous geomagnetic field at that locality. Taking this many samples also allows a good statistical estimate to be made of the precision of the determination.

Major errors in estimates of the previous direction or intensity of the geomagnetic field may occur if the samples are unevenly magnetized: sediments may contain unevenly distributed pebbles or include a large number of potsherds carrying a strong thermal magnetization. Such magnetic inclusions will have a random orientation that may dominate the magnetization carried by cement or other secondarily deposited materials. One can sometimes reduce these errors by increasing the volume of the sample, but anomalous observations usually are better eliminated from any assessment of the parameters of the past geomagnetic field.

ABSOLUTE ARCHAEOMAGNETIC TIME SCALES

Records of the Earth's magnetic field obtained for the last 100−350 years show that *secular variations*—long-term changes in the field—are broadly similar to each other within regions of 250,000−500,000 km^2 and that their pattern shows a drift westward at a speed of some 0.2° of longitude per year (Bullard et al., 1950). This basic picture has been only slightly modified from more recent geomagnetic, paleomagnetic, and archaeomagnetic studies (Cox, Hillhouse, and Fuller, 1975) covering longer periods of time. Three time-varying properties of the Earth's magnetic field can be used for archaeological dating: its intensity and the directions of its declination (the horizontal angle from true north) and inclination (the vertical angle from horizontal). As these variables are virtually independent (there is some correlation between the intensity and inclination of the field), it is only rarely that exactly the same combination of all three will have been repeated in the past at any one locality.

Comparisons of the remanent magnetism between nearby archaeological sites, therefore, permits relative dating: samples with identical magnetic properties are almost certainly of the same age. It is not always possible (or practical) to determine all three parameters of a past magnetic field. Changes in geomagnetic intensity generally occur more slowly than directional changes and thus provide a coarser distinction of age, although there are particular time periods when the intensity of the field was varying more rapidly (Walton, 1979). Nevertheless, for most archaeological situations an identity of sample age, or lack of it, can be readily determined using magnetic methods.

In localities where a record of geomagnetic-field changes exists, then absolute dating is possible by comparison with the established record of as many magnetic parameters determined at a particular site as possible. Unfortunately, direct measurements of the Earth's field exist only for the last 400 years and only

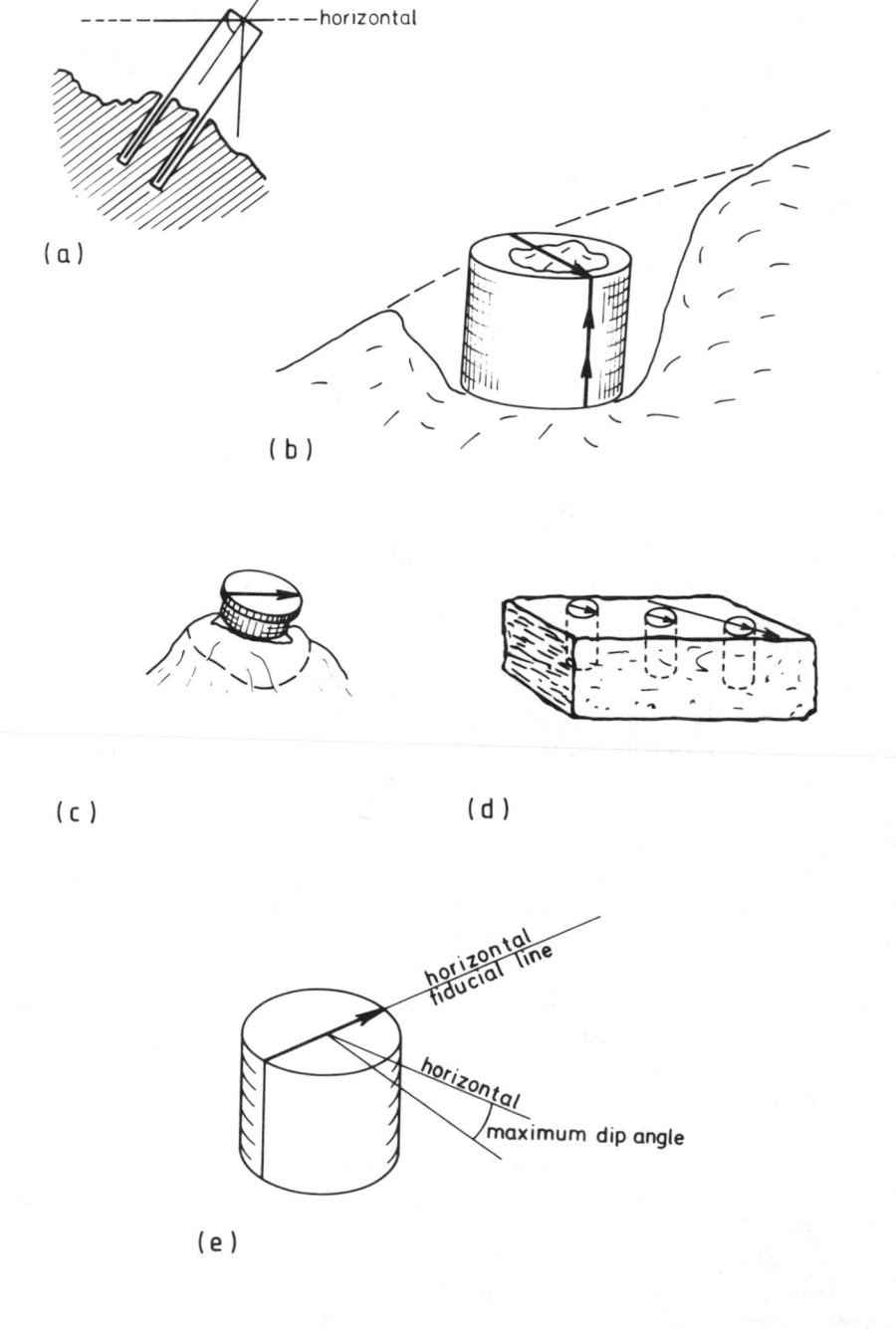

horizontal

(a)

(b)

(c)

(d)

horizontal
fiducial line

horizontal

maximum dip angle

(e)

in specific places such as London and Paris: most magnetic observatories have been recording field variations for less than 100 years. Correlations therefore can rarely be made with direct observations of the Earth's field. Our theoretical understanding of the origin and controls on the geomagnetic field is still completely inadequate to allow estimates of past secular-variation patterns. This means that past records can be built up only by means of magnetic studies of dated materials. The large scale of secular variation usually permits the combination of results from different localities to evaluate the secular change for large regions—for example, over northwestern Europe (fig. 9.9), the Mediterranean region, or the southwestern United States. However, such records will obviously incorporate any errors in the initial magnetic study and in the dating methods originally used to establish the age of the magnetization. Nonetheless, excellent directional records are becoming available for western Europe (Thellier, 1966, 1981; Aitken, 1970), the southwestern United States (Wolfman, pers. com., 1978), Japan (Hirooka, 1971), East Africa (Skinner, Illes, and Brock, 1975), and the Soviet Union (Rusakov and Zagniy, 1973). These records have been constructed largely from measurements of the direction of remanence of fired or burnt materials, particularly kilns and hearths; especially in Japan, studies of historically recorded lava flows have provided supplementary data. Some of the dates are from historic records, but most are based on stratigraphic considerations and ^{14}C dates.

Fig. 9.8. Methods of sampling and orienting archaeological-geological materials. Where good, solid materials exist, samples can be (a) drilled, leaving a core still attached to the outcrop. The direction of drilling, relative to true north, is then measured and marked on the core, and the angle of drilling from horizontal is also measured. The core can then be removed with the fiducial line as the direction of drilling. Specimens can then be cut from the core. (b) Partially consolidated materials can be excavated, leaving a pillar of material over which a plastic tube can be placed. The gaps between the inside of the tube and the pillar are packed with paper, magnetically clean gypsum, polyurethane foam, or other nonmagnetic substances. The orientation of the top of the tub is then measured. If the tub is horizontal, an arrow is placed on top in any direction and the direction of the arrow measured relative to true north. The tub and pillar can then be removed. If the tub is inclined, then a horizontal arrow is drawn through the diameter of the top of the tub; the direction of the arrow is determined and the maximum slope (at right angles to the arrow) is measured from horizontal (see also e, below). (c) Solid or partially solid materials can be sampled by gluing a plastic disk onto the material and then determining the direction and slope of the surface of the disk as was done for the tub. (d) Flat surfaces on solid materials can be oriented using the same procedure as for the top of the tub. The whole piece can then be removed and later cut into specimens. (e) If the material is soft, such as pond sediments, then a plastic tub can be forced into the material, ensuring that little disturbance takes place. The top surface of the tub can then be oriented as in (b), above.

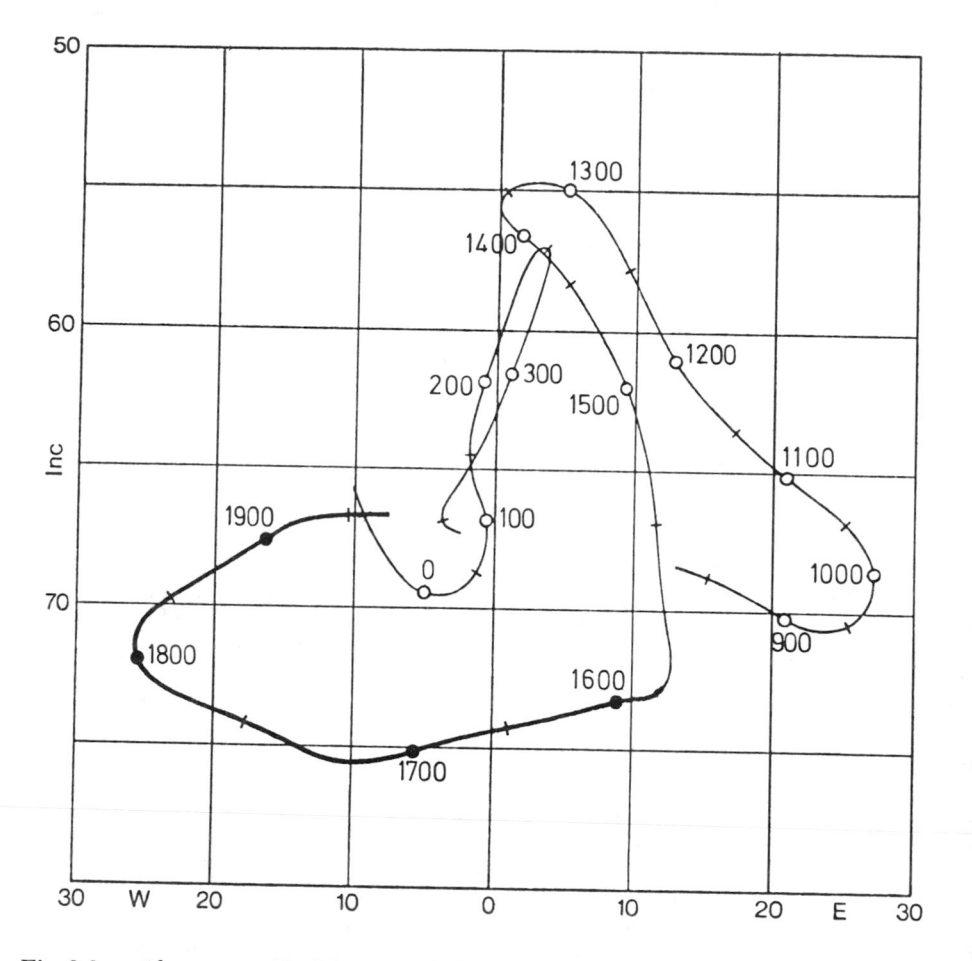

Fig. 9.9. Observatory (thick line) and archaeomagnetic (thin line) determination of the geomagnetic field in Britain, from the Iron Age to the present (courtesy of M. J. Noel).

Advances have recently been made in the measurement of long cores of lake and sea sediments (Molyneux and Thompson, 1973). These sediments, which have accumulated gradually over archaeological or longer periods of time, can provide a continuous record of geomagnetic-field changes that may be used for magnetic dating. Studies of sediments from Lake Windermere, England (Mackereth, 1971; Thompson, 1973) show that a swing in declination is recorded in them corresponding with the peak westward swing around A.D. 1820 that was observed in London. Declination swings have been determined in older sediments, and the speed of collection and measurement is leading to an accumulation of records from many parts of the world. Sedimentary records for changes in inclination are generally much poorer, however. Oscillations in the inclination record are poorly defined and may be controlled mainly by sedimentation factors.

The sedimentary record is further complicated by the variability of deposition rates and the occasional interspersal of levels of deposition and levels of erosion. These facts notwithstanding, it is generally necessary to assume a constant rate of sedimentation between different levels that have been dated by other means (particularly ^{14}C), and even these dates are often suspect. Where the rate and type of sedimentation is seasonally controlled, such as in the varved glacial and postglacial lake clays of Scandinavia and North America, it is possible to count backwards from the present for some 14,000–15,000 years or more to produce a chronology much more reliable than that of other dating methods (Tauber, 1970; Fromm, 1970). Uncertainties still exist, particularly in correlating between different varve localities. The use of a magnetic parameter in such correlations should improve the reliability of the chronology, and work in progress (Noel, 1975; Noel and Tarling, 1975) should provide an annual record of geomagnetic-field changes in Scandinavia and probably in North America. Such records will be subject to variability due to sedimentation factors, although these are likely to have little effect during quiet winter deposition. Magnetic analyses of varve sediments and fired materials will need to be integrated for full monitoring of all three parameters of the geomagnetic field.

As secular variations are essentially regional in extent, it is necessary to construct separate records for each subcontinental area. Although usually dominated by local changes, each variation also incorporates changes in the Earth's field that are worldwide in nature. The most spectacular of these occurs when the entire polarity of the Earth's field changes (Cox, 1975) so that the north magnetic pole is located within some 20–30° of the south geographical rotation pole. A complete polarity reversal most recently occurred some 730,000 years ago; it and still older reversals are mainly of interest for hominid and paleolithic studies (fig. 9.10). Dating such polarity reverses is still difficult, although comparison of dates from various parts of the world using different methods means that they are becoming more precisely dated. Observation of the polarity of magnetism, for example, can often be diagnostic when the approximate age is known. It is also significant that these major changes in the nature of the geomagnetic field coincide with, and possibly cause, changes in climatic patterns of possible archaeological importance.

Less spectacular but of possibly more archaeological importance are apparently major changes in the geomagnetic field that create marker horizons allowing intercontinental calibration. During the last 700,000 years or so of the Brunhes Normal epoch, five major worldwide "excursions" of the geomagnetic field have so far been delineated (fig. 9.10), and others undoubtedly exist. Major changes in either declination or inclination during such excursions imply that the virtual magnetic-pole position, relative to the site being examined, deviates significantly from its usual range and may even cross the equator. So far these

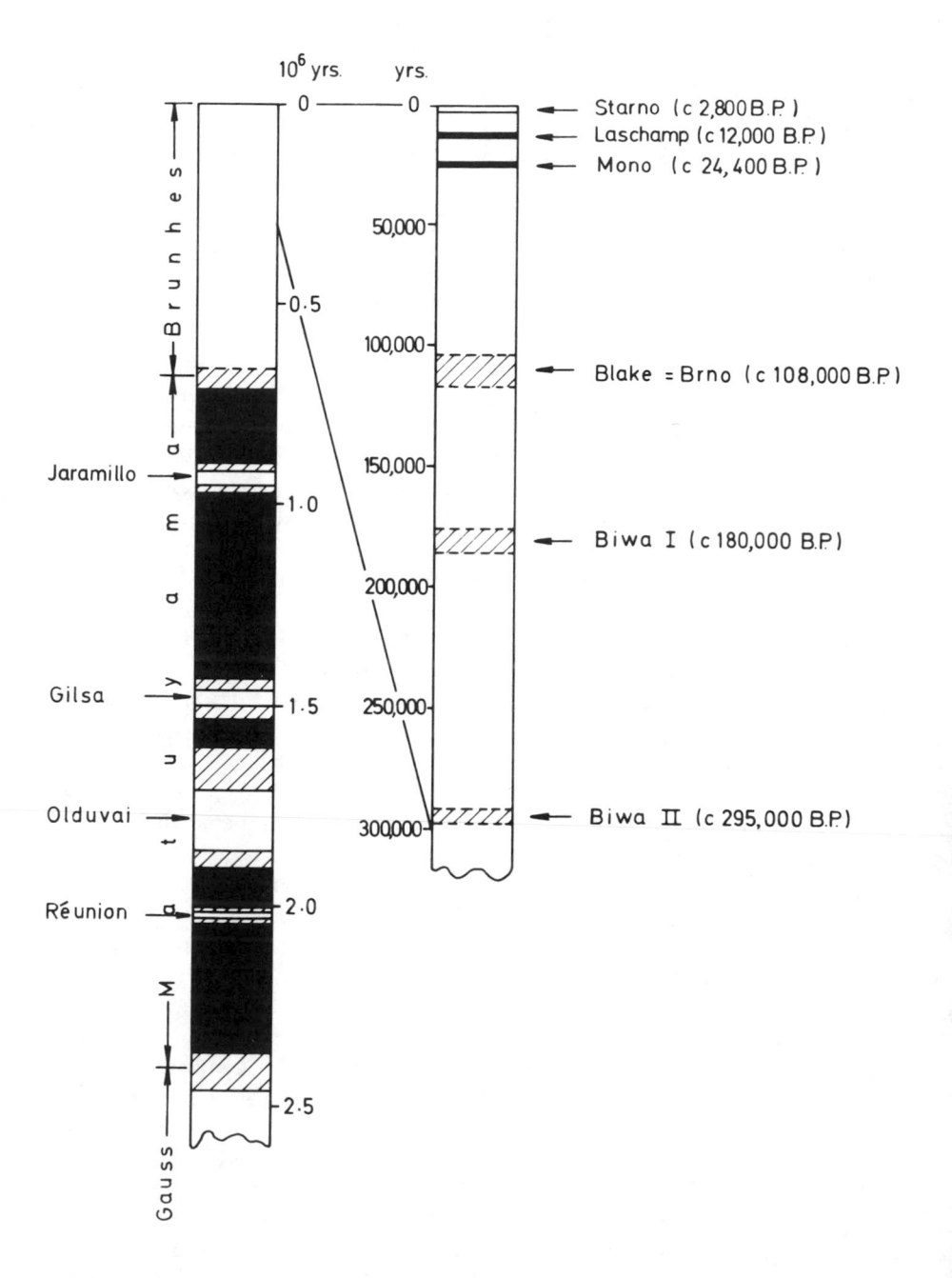

Fig. 9.10.   Geomagnetic polarity reversals and events during the last 2.5 million years.

excursions have not been examined in sufficient detail to be accurately character-ized, although they appear to last for a few hundred years, with changes in intensity possibly preceding changes in direction. Most excursions (Noel and Tarling, 1975) have been detected in sedimentary sequences, so it is not clear if there are significant synchronous changes in intensity of the geomagnetic field, although this seems probable. The Starno event, for example, has been identi-fied in postglacial sediments in Sweden and may correlate with old anomalous observations of pottery of a similar age in Greece, but the reality and extent of this event have yet to be conclusively established.

All methods so far described depend on isolating the stable component of magnetization by eliminating the time-dependent magnetization. It is also pos-sible to measure the amount of time-dependent magnetization, estimate the rate at which it was acquired, and thus estimate how long the material had been lying in the orientation in which it was found. The amount of time-dependent magne-tization can be obtained by observing the field decay rate during partial de-magnetization, but an estimate of the rate at which it was acquired is difficult. In the laboratory, the rate of acquisition of magnetization by a sample over a few days or weeks can be estimated, especially at elevated temperatures; such rates can then be extrapolated to determine the time during which the natural viscous remanence was acquired. Heller and Markert (1973) used this method to obtain moderately consistent dates for three strongly magnetized stones from Hadrian's Wall, northern England, but the uncertainties were somewhat large for any one determination and several fundamental values are necessarily assumptions:—for example, an average intensity of the earth's field has to be estimated for all methods. Although it seems unlikely that this technique can be used as a routine magnetic dating method—at least until measurement techniques and theoretical understanding improve—it is certainly practical for coarse dating, such as distin-guishing fakes from genuine artifacts by determining whether the material has been in situ for only a few decades or for considerably longer.

Finally, magnetic dating of buildings or parts of buildings is possible if the buildings were oriented at the time of their construction by magnetic compass (Searle, 1974). Between the eleventh and seventeenth centuries in western Europe, it was thought that magnetic north was constant and related to attraction by the north star and ultimately to the pole of the Universe. Buildings such as churches and features such as graves constructed during this period were often oriented magnetically and, therefore, reflect a synchroneity with a previous magnetic declination and not true geographic north. As the early descriptions of the compass in western Europe were of fairly advanced forms of the device, it is of considerable interest to determine exactly when magnetic orientation was first used, thus indicating the introduction of the compass. The study of alignments of structures of religious significance, in particular, could be used either for dating by relation to an established record or for helping to establish a record if the age of the building is known; it could also provide technological information on the evolution of surveying techniques.

OTHER APPLICATIONS

Almost as important as dating is the application of magnetic-analysis techniques to other archaeological studies. The behavior of the magnetization of pottery or metals can, for example, be used to determine the temperature of later heating. In a similar application, the temperature at deposition of volcanic ash from a mid-second-millennium B.C. eruption of Thera (which is thought to have contributed to the decline of the Minoan civilization) can be shown to have been low (Tarling, 1978); it could not have constituted a *nuée ardente* eruption, such as that responsible for the destruction of Pompeii in A.D. 72 or of St. Pierre in 1902.

The magnetization of component sherds can assist in reconstructing the original vessel or in estimating its shape from a few noncontinguous sherds (Burnham and Tarling, 1975). The magnetization of metal casts of known age can be used to estimate the angle at which the cast was made by comparing its magnetic inclination with the past inclination of the Earth's magnetic field (fig. 9.11a). For example, a sample of baked clay from the interior of an Indian Buddha has been used to show that the statue was originally cast upside down while facing east. Similarly, the few Roman coins so far examined suggest that these coins were all hot-struck with the emperor's head on the lower die (fig. 9.11b).

A further application of archaeomagnetism is in the magnetic sourcing of lithic raw materials. A preliminary study of obsidian artifacts and chippings and their outcrops in the Mediterranean region (McDougall, 1978) has shown that most known sources have characteristic magnetic parameters, which indicates that the derivation of such artifacts, and hence trade routes, can readily be determined (see fig. 9.12). The study indicated that the precision of magnetic sourcing is comparable to that of neutron-activation techniques (which measure trace-element abundances) and that such magnetic sourcing can be carried out very rapidly (a few minutes per sample), nondestructively, and very cheaply. Obsidians are relatively simple to source, and it seems probable that other materials, both natural and artificial, may also be suitable for sourcing based on their magnetic parameters.

The recent development of techniques for the rapid determination of the magnetic anisotropy of rocks (Rathore, 1975), even when the anisotropy is extremely weak, is likely to be of considerable value in determining the petrofabric of sediments within an archaeological context. The petrofabric and hence the magnetic fabric of such sediments strongly reflects their mode of accumulation and subsequent deformation by compaction, groundwater motions, tectonic disturbances, etc. Although such studies have so far been carried out only on an experimental basis, they have a potential application in archaeological investigations concerned with the origin and history of site sediments. Similarly, the magnetic susceptibility of sediments may be of major use in the correlation of sedimentary sequences (Thompson, 1973), for example, between different cores of lake sediments or between sediment samples within an archaeological site.

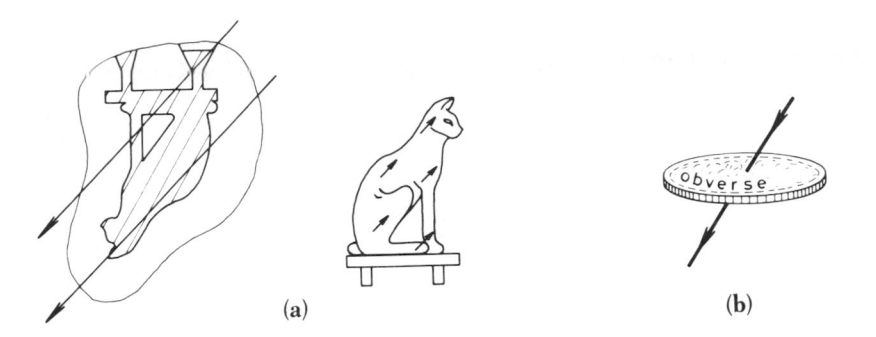

Fig. 9.11. The orientation of some archaeological objects, determined through use of their magnetization. (a) Casts acquire a magnetization as they cooled; this allows their position, during casting, to be determined from their magnetization. (b) Coins appear to acquire a thermal magnetization when struck. In the few Roman coins so far examined, the magnetic-inclination vector passes out of the coin through the emperor's head, unless the coin had been later silvered, indicating (i) that the lower die always carried the Emperor's head and (ii) that the silvering process was carried out with the coins on their edges.

There is some tentative evidence for relationships between the strength of the geomagnetic field and the production of ^{14}C (Bucha, 1970), climatic change (Kawai, 1972; Wollin et al., 1973) and even health (Malin and Srivastava, 1979). These relationships are, to say the least, speculative, but it is obviously important that such investigations be pursued in view of the possible significance of such factors to the development of cultures. Also, our capability to predict future geomagnetic-field changes is itself partially being improved from an increasing understanding of geomagnetic-field behavior and from restraints placed on theoretical models by archaeomagnetic observations.

CONCLUSION

The *potential accuracy* of archaeomagnetic dating is high, probably better than ±25 years for most materials of the last 2,000−3,000 years. At the moment, the *precision* is somewhat less due to "noise" in laboratory methods for isolating the stable magnetization and particularly to the accuracy of collection and orientation of samples in the field. The potential degree of precision is mainly dependent on the construction of an accurate chronology against which reference can be made, and thus on the precision of those independent dating techniques used in its construction. Although quite good scales are being built up by painstaking studies of well-documented materials, it will be several decades before adequate coverage will be available for the world. Meanwhile, good approximations are becoming available from the study of sediment sequences, although these pro-

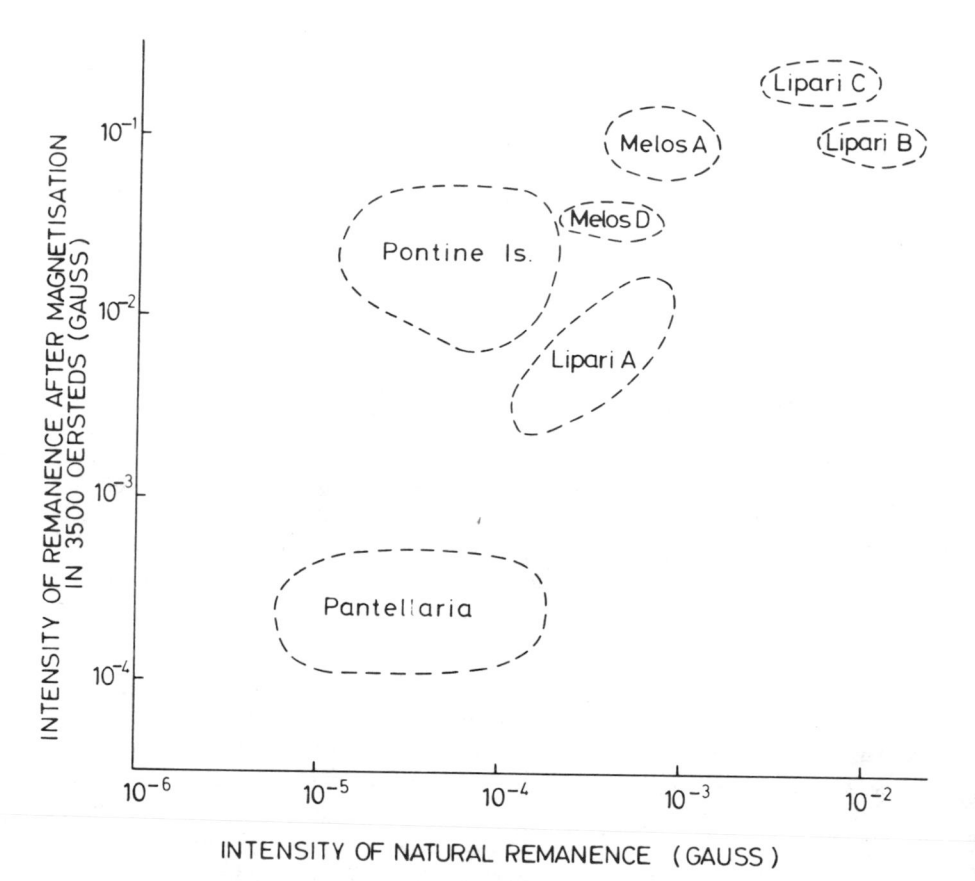

Fig. 9.12. Obsidian sourcing using magnetic properties. Determination of the original intensity of magnetization and that of the magnetization acquired by small chips of obsidian placed in strong magnetic fields allows the unique assignment of the source of the obsidian artifacts in most Mediterranean obsidians so far investigated (after McDougall, 1978).

vide little reliable information on field-intensity variations and are somewhat poor indicators of variations in inclination. The technique does have the major advantages that it is cheap and usually nondestructive and that it can often be applied to a variety of materials that are not always suitable for other methods of absolute dating.

Archaeomagnetic studies are still in their infancy, but they appear to have a potential application for absolute dating on both short and long time scales. At the moment, fired materials are by far the most suitable objects for such dating, although ongoing studies indicate that a wide variety of materials may in fact be suitable. In addition to dating, however, the magnetic properties themselves can provide information valuable in solving archaeological conservation and sourcing problems and data on the deposition and modification of sedimentary sequences, in particular for Quaternary geological and archaeological studies.

REFERENCES

Aitken, M. J. 1970. Dating by archaeomagnetic and thermoluminescent methods. *Philosophical Transactions of the Royal Society, London* A269:77–88.

————. 1974. *Physics and archaeology*. Oxford: Clarendon Press.

Barbetti, M. F., M. W. McElhinny, D. J. Edwards, and P. W. Schmidt. 1977. Weathering processes in baked sediments and their effects on archaeomagnetic field-intensity measurements. *Physics of the Earth and Planetary Interiors* 13: 346–54.

Bucha, V. 1970. Influence of the Earth's magnetic field on radio-carbon dating. In *Radiocarbon variations and absolute chronology*, ed. O. Olsson, 501–51. New York: Wiley Interscience.

Bullard, E. C., C. Freedman, H. Gellman, and J. Nixon. 1950. The westward drift of the Earth's magnetic field. *Philosophical Transactions of the Royal Society, London* A243:67–92.

Burnham, R. J. P., and D. H. Tarling. 1975. Magnetization of sherds as an assistance to the reconstruction of pottery vessels. *Studies in Conservation* 20:152–57.

Collinson, D. W. 1975. Instruments and techniques in palaeomagnetism and rock magnetism. *Review of Geophysics and Space Physics* 13:659–86.

Cox, A. 1975. The frequency of geomagnetic reversals and the symmetry of the non-dipole field. *Reviews of Geophysics and Space Physics* 13:35–52.

Cox, A., J. Hillhouse, and M. Fuller. 1975. Palaeomagnetic records of polarity transitions, excursions, and secular variation. *Review of Geophysics and Space Physics* 13:185–89.

Creer, K. M., and J. S. Kopper. 1974. Palaeomagnetic dating of cave paintings in Tito Bustillo Cave, Asturias, Spain. *Science* 186:348–50.

Creer, K. M., L. Molyneux, J. P. Vernet, and J. J. Wagner. 1975. Palaeomagnetic dating of 1-meter cores of sediment from Lake Geneva. *Earth and Planetary Science Letters* 28:127–32.

Fromm, E. 1970. An estimation of errors in the Swedish varve chronology. In *Radiocarbon variations and absolute chronology*, ed. O. Olsson, 163–72. New York: Wiley Interscience.

Games, K. 1977. The magnitude of the palaeomagnetic field: A new, nonthermal non-detrital method of using sun-dried bricks. *Geophysical Journal of the Royal Astronomical Society* 48:315–30.

Griffiths, D. H., R. F. King, A. I. Rees, and A. E. Wright. 1960. The remanent magnetization of varved clays from Sweden. *Monthy Notices of the Royal Astronomical Society, Geophysical Supplement* 7:103–14.

Harold, M. R. 1960. Magnetic dating: Kiln wall fall-out. *Archaeometry* 3:45–47.

Heller, F., and H. Markert. 1973. The age of viscous remanent magnetization of Hadrian's Wall (Northern England). *Geophysical Journal of the Royal Astronomical Society* 31:395–406.

Hirooka, K. 1971. Archaeomagnetic study for the past 2000 years in southwest Japan. *Memoirs of the Faculty of Science, Kyoto University, Series of Geology and Mineralogy* 38:167–207.

Irving, E., and A. Major. 1964. Post-depositional remanent magnetization in a synthetic sediment. *Sedimentology* 3:135–43.

Kawai, N. 1972. The magnetic control on the climate in the geologic time. *Proceedings of the Japan Academy* 48:687–89.

McDougall, J. 1978. An analytic study of obsidian from Europe and the Near East by examination of magnetic properties. M.Sc. thesis, Bradford Univ.

Mackereth, F. J. H. 1971. On the variations of the direction of the horizontal component of remanence magnetization in lake sediments. *Earth and Planetary Science Letters* 12:332–38.

Malin, S. R. C., and B. J. Srivastava. 1979. Correlation between heart attacks and magnetic activity. *Nature* 277:646–48.

Molyneux, L. 1971. A complete result magnetometer for measuring the magnetization of rocks. *Geophysical Journal of the Royal Astronomical Society* 24:429–33.

Molyneux, L., and R. Thompson. 1973. Rapid measurement of the magnetic susceptibility of long cores of sediment. *Geophysical Journal of the Royal Astronomical Society* 32:479–81.

Noel, M. 1975. The palaeomagnetism of varved clays from Blekinge, southern Sweden. *Geologiska Foereningens I Stockholm Forhandlingar* 97:357–67.

Noel, M. 1980. Surface tension phenomena in the mangetization of sediments. *Geophysical Journal of the Royal Astronomical Society* 62:15–25.

Noel, M., and D. H. Tarling. 1975. The Laschamp geomagnetic 'event.' *Nature* 253:705–07.

Rathore, J. S. 1975. The magnetic fabric of rocks. Ph.D. diss., Univ. of Newcastle-upon-Tyne.

Rogers, J., M. M. W. Fox, and M. J. Aitken. 1979. Magnetic anisotropy in ancient pottery. *Nature* 277:644–46.

Rusakov, O. M., and G. F. Zagniy. 1973. Archaeomagnetic secular variation study in the Ukraine and Moldavia. *Archaeometry* 15:153–58.

Searle, S. 1974. The Church points the way. *New Scientist* 61:10–13.

Shaw, J. 1974. A new method of determining the magnitude of the palaeomagnetic field. *Geophysical Journal of the Royal Astronomical Society* 39:133–41.

Skinner, N. J., W. Illes, and A. Brock. 1975. The recent secular variation of declination and inclination in Kenya. *Earth and Planetary Science Letters* 25:338–46.

Tarling, D. H. 1971. *The principles and applications of palaeomagnetism*. London: Chapman and Hall.

———. 1975. Archaeomagnetism: The dating of archaeological materials by their magnetic properties. *World Archaeology* 7:185–97.

———. 1978. Magnetic studies of the Santorini tephra deposits. In *Thera and the Aegean world*, vol. 2, ed. C. Doumas, 195–210. London: Thera and the Aegean World.

Tauber, H. 1970. The Scandinavian varve chronology and C14 dating. In *Radiocarbon variations and absolute chronology*, ed. O. Olsson, 173–96. New York: Wiley Interscience.

Thellier, E. 1966. Le champ magnétique terrestre fossile. *Nucleus* 7:1–35.

———. 1981. Sur la direction du champ magnétique terrestre, en France, durant les deux derniers millenaires. *Physics of the Earth and Planetary Interiors* 24: 89–132.

Thellier, E., and O. Thellier. 1959. Sur l'intensité du champ terrestre dans la passé historique et géologique. *Annales de géophysique* 15:285:376.

Thompson, R. 1973. Palaeolimnology and palaeomagnetism. *Nature* 242:182–84.

Walton, D. 1979. Geomagnetic intensity in Athens between 2000 B.C. and A.D. 400. *Nature* 277:643–44.

Weaver, G. H. 1962. Archaeomagnetic measurements on the second Boston experimental kiln. *Archaeometry* 5:93–107.

Wollin, G., G. Kuhla, D. B. Ericson, W. B. F. Ryan, and J. Wollin. 1973. Magnetic intensity and climatic change. 1925–1970. *Nature* 242:34–37.

10

Tephrochronology and Its Application to Archaeology

VIRGINIA STEEN-MCINTYRE

ABSTRACT

Tephra *are fragments of solid material—crystal, glass, rock—ejected into the air by a volcanic eruption. Tephrochronologists and tephrostratigraphers study deposits and samples of these fragments in order to characterize, correlate, and date them, thereby providing archaeologists working within the area of tephra fall with recognizable time lines to aid them in their research. Tephra from a single eruption may blanket an area of several thousand square kilometers and drastically alter the environment and patterns of human activity. Ash, where it is thick, effectively protects buried artifacts from disturbance and plunder. The fragments themselves may provide material for radiometric dating. This chapter describes tephra, presents examples of how its study has benefited archaeologists in Mexico and El Salvador, and explains how tephra samples are collected, examined, and dated.*

Volcanic eruptions are natural phenomena that bring fresh or rejuvenated magmatic material to the surface of the earth, whereupon the solidified fraction weathers to become part of the geochemical cycle. They are devastating to local ecosystems and awesome to human experience, with memories of them being preserved in myth, song, and legend.

A range of energies is released by volcanic eruptions. Some are small and produce primarily steam and gases, along with a minor amount of previously solidified rock fragments. In others the eruption is characterized mainly by an outpouring of lava from a vent or fissure. Still other eruptions are much more violent: instead of the magma issuing forth quietly as lava, pent-up gases cause it to explode or rush downslope as a glowing, incandescent cloud (*nuée ardente*) from breaches in the vent wall.

Fragments thrown into the air by an explosive volcanic eruption are called pyroclasts or *tephra*. This material is to be distinguished from *lava*, the molten rock that flows as a stream from rifts, fissures, or vents. The term *tephra* ("ashes") was introduced into the literature by the late Sigurdur Thorarinsson (1944, 1954), an Icelandic geologist who saw the need for a collective term for all airborne pyroclastic material regardless of size, shape, or composition of fragment. As it is used today, tephra refers both to air-fall and pyroclastic flow material (Thorarinsson, 1974).

Those who study tephra are called *tephrochronologists* or *tephrostratigraphers* (the names have been used interchangeably); they seek to characterize and date tephra from different eruptions so as to recognize it when it occurs in isolated exposures. Usually, the goal of such research is to provide limiting ages and dated marker horizons for sedimentary deposits associated with tephra layers. Archaeologists, pedologists, geomorphologists, engineers, and stratigraphers then may use these dated layers in their own research.

Tephrochronology is a new discipline. Many archaeologists are not familiar with it nor with the potential value of tephra layers as correlative marker horizons, source of radiometric dates, evidence for paleodisasters, or protective sediment covers for the preservation of artifacts. This chapter will describe tephra, present examples of how its study has benefited archaeologic research in the Americas, and show how tephra is collected, examined, and dated.

EARLY INVESTIGATIONS

Study of tephra components, layers, and soils has been conducted for decades in other countries, but in the Americas tephrochronology prior to 1965 was the concern of only a handful of people, most notably A. Swineford and colleagues, H. Williams, I. S. Allison, C. R. Keyes, and H. A. Powers in the United States; V. Auer in South America; W. H. Mathews and S. R. Capps in Canada and Alaska; and F. Mooser in Mexico. Since the 1960s, primarily through interest generated by the publications of R. E. Wilcox, G. A. Izett, J. A. Westgate, D. G. W. Smith, and their colleagues, research into problems involving New World tephrochronology has increased substantially. Over 300 citations for the period 1965–1973 are listed for this area alone in the *World Bibliography and Index of Quaternary Tephrochronology* (Westgate and Gold, 1974), which also lists 180 references under the heading "Archaeology."

Many excellent articles on tephra have appeared since 1973. Several which summarize current tephra research throughout much of the world have been collected in one volume (Self and Sparks, 1981). Others that would be of special interest to archaeologists, either because they describe tephra deposits that occur in archaeologic sites or deposits that have the potential for containing sites, include the following:

North America: Bloomfield and colleagues (1977a, 1977b); Davis (1978); Izett (1981); Kittleman (1973, 1979a, 1979b); Lambert (1979); Lambert and

Valastro (1976, in preparation); Lemke et al. (1975); Lerbekmo et al. (1975); Mehringer, Blinman, and Peterson (1977); Moody (1978); Mullineaux (1974), Mullineaux, Hyde, and Rubin (1975); Porter (1978); Sheridan and Updike (1975); Smith and Leeman (1982); Smith, Okazaki, and Knowles (1977a, 1977b); Souther (1977); Steen-McIntyre (1977b); Steen-McIntyre, Fryxell, and Malde (1981); Westgate (1977, 1982); Westgate and Briggs (1980); Westgate and Fulton (1975); Westgate and Evans (1978); Wood (1977).

Central America and South America: Drexler et al. (1980); Hahn, Rose, and Meyers (1979); Hart (1983); Hart and Steen-McIntyre (1983); Koch and McLean (1975); Riezebos (1978); Rose, Grant, and Easter (1979).

Mediterranean area: Barberi et al. (1978); several articles in Doumas (1978); Federman and Carey (1980); Keller et al. (1978); Lirer et al. (1973); Richardson and Ninkovich (1976); Thunell et al. (1979); Vitaliano and Vitaliano (1974); Watkins et al. (1978).

North Atlantic: Larsen and Thorarinsson (1978).

The Pacific: Blong (1982); Hodder and Wilson (1976); Hogg and McCraw (1983); Howorth and Rankin (1975); Ikawa-Smith (1978a, 1978b); Kohn (1970, 1979); Kohn and Topping (1978); Machida (1976); Machida and Arai (1978); McCraw (1975); Ninkovich et al. (1978); Vucetich and Howorth (1976).

Africa: Coppens et al. (1976); Hay (1976); Johanson and Edey (1981); Laury and Albritton (1975).

STATUS OF TEPHROCHRONOLOGY

In spite of the work that has been done, tephrochronology today remains an inexact discipline. The reason lies mainly in the inability of most workers to amass the analytical data required for unquestionable correlations between two tephra exposures. These can be made with complete confidence only if all the following criteria are met:

1. Stratigraphic, paleontologic, paleomagnetic, and radiometric age relations are compatible.
2. The essential characteristics of the glass shards and phenocrysts match. (Essential characteristics, described below, are those that remain constant throughout the tephra deposit, from volcanic source to far downwind.)
3. The tephra layer of interest has one or more characteristics by which it may be distinguished from any and all others that occur in the region (Wilcox and Izett, 1973).

In practice, it is a rare sample that satisfies all criteria—most satisfy only one or two. In addition, the last requirement presupposes that all other tephra units in the study area have been discovered and examined, a situation seldom encountered. Partly because of these problems, tephrochronology today is anything but the well-established discipline that some suppose. Therein, perhaps, lies its charm.

COMPOSITION OF TEPHRA

The main components of an average tephra layer are volcanic glass, heavy mineral phenocrysts (specific gravity > 2.86), lightweight phenocrysts (specific gravity < 2.86), and minor amounts of xenocrysts and xenoliths (fig. 10.1). *Phenocrysts* are mineral crystals that were growing within the liquid magma at the time of eruption. *Xenocrysts* and *xenoliths* are, respectively, crystal and rock fragments torn or melted from the walls of the magma chamber or vent. The proportion of these components varies from sample to sample and from layer to layer.

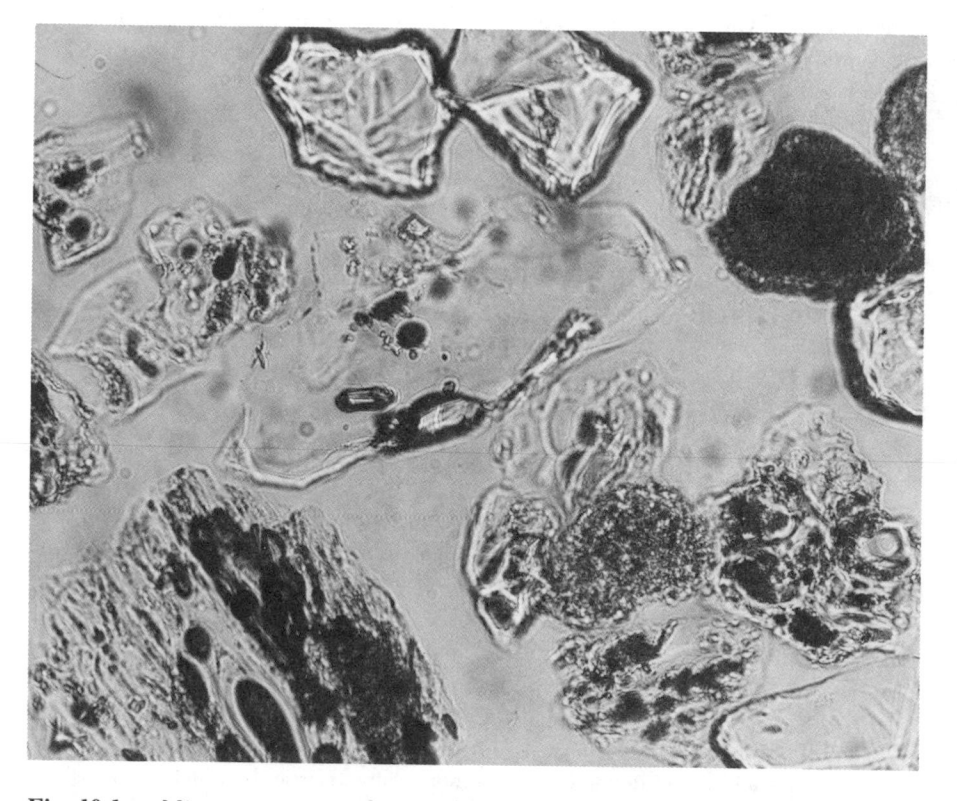

Fig. 10.1. Microscopic view of a sample of 1,700-year-old *tierra blanca joven (tbj)* tephra, collected in central El Salvador. The fragments are less than 0.25 mm in size and are mounted in an immersion oil with a refractive index (n) slightly less than that of the volcanic glass. The dark masses at upper right are either rock or organic matter. The bright, transparent objects with dark outline at upper center and to the right are crystal fragments. The other fragments are volcanic glass shards. One at the center is quite dense and clear; it contains a tiny phenocryst (microphenocryst). The rest of the shards are pumiceous, and many of the bubble cavities are filled with an opaque, claylike substance, the product of tropical weathering.

Heavy minerals are rich in the elements iron, magnesium, and calcium; lightweight minerals are rich in sodium, potassium, aluminum, and silica. Volcanic glass contains all these elements in varying amounts. Often the glass and crystals are intimately mixed in two frothy rock types, pumice and scoria. *Pumice* is light in color and relatively rich in silica (fig. 10.2). It is known as a *silicic* or *acidic* tephra. A large portion of its bulk may be composed of bubble cavities, or vesicles; as a result it is lightweight, often with a specific gravity less than that of water. Pumice readily soaks up and retains water and is often added to potting soils as a conditioner. *Scoria* is dark (usually shades of red, black, dark gray, or brown), rich in iron, and known as a *basic* tephra. It also is a vesicular rock, but in general the fragments are heavier than pumice. In warm, moist climates scoria weathers rapidly to form a fertile soil (fig. 10.3).

TEPHRA LAYERS AND TIME

A mantle of tephra from a single volcanic eruption is deposited over the landscape in an instant of geologic time. For major eruptions, this mantle can extend over an area the size of several states. Once the eruption that produced the tephra has been dated, the tephra mantle becomes an almost ideal chronologic marker horizon for archaeologic research—it can occur in virtually any type of sediment, is relatively easy to recognize (once one knows what to look for), and is

(a)   (b)

Fig. 10.2. Coarse fragments of (a) Mazama and (b) Glacier Peak pumice, collected from B3 soil horizons at two sites on the eastern slope of the Cascade Range, Pacific Northwest. Note the vesicular, almost spongy nature of Mazama tephra compared with that of Glacier Peak. The Glacier Peak fragments (ca. 11,250 years old) are considerably more rounded and stained than are the Mazama fragments (6,700 years old; Mehringer, Blinman, and Peterson, 1977). This in large part is due to the difference in their ages.

Fig. 10.3. Interior of a hill exposed in 1978 during highway construction southwest of the city of San Salvador, El Salvador (Hart and Steen-McIntyre, 1983, Fig. 2-3). A light-colored deposit of silicic tephra, the *tbj* tephra, is draped over a weathered core of more basic volcanic debris. According to Olson (1983), it is the deeply weathered basic material that formed the rich, fertile soil cultivated by the Late Preclassic Maya. By contrast, the soil that has developed on the younger tephra layers is much less productive.

independent of locally derived radiometric dates and relative dating methods. True, the mantle quickly develops "holes" as erosion strips the tephra fragments from more exposed parts of the landscape, but in protected areas, such as rockshelters or caves, and in areas of active deposition, like the banks of streams or rivers, remnants of the original tephra mantle are often preserved.

An intact remnant of tephra layer uncovered during excavation should be of keen interest to the archaeologist. It rests upon the land surface that existed at the time of the eruption. Artifacts that occur at the basal contact probably were in use at that time; and, if the deposit is thick enough, they have not been disturbed since. Artifacts found lower down in the stratigraphic section are older than the tephra unit; those above it, younger. Such a relationship, basic and simple as it is, can have surprisingly wide repercussions in archaeologic research, as will be shown below.

Reworked tephra layers can also be useful. Except for some special cases of intense mixing, reworked layers can be no older than the eruption that produced the ejecta. In many cases they are essentially the same age: relatively pure deposits of reworked tephra often represent an early readjustment of slopes and drainage patterns to the set of new conditions created when the mantle was first emplaced.

A tephra layer is a useful marker horizon within a single excavation. Its value increases severalfold when it is recognized in several excavations within a single complex site or in several widely dispersed sites. In such cases intrasite and intersite stratigraphic correlations can be made with confidence.

As with other scientific disciplines, tephrochronology possesses a set of axioms that are acknowledged either explicitly or implicitly by anyone working in this field:

1. Volcanic eruptions from any single eruptive source represent discrete points in time.
2. Within the area of tephra fall, fragments are deposited impartially upon all unshielded, nonvertical surfaces.
3. In most depositional environments, and in sedimentary units of Miocene or younger age, a given layer of silicic tephra will preserve an assortment of recognizable characteristics that can be correlated.
4. Remnants of primary (uneroded) tephra blankets or deposits of slightly reworked material are best preserved in environments of low potential energy (e.g., lowlands, river floodplains).
5. Wherever a remnant of primary tephra blanket is preserved, it mantles the land surface that existed just prior to the eruption.
6. A primary deposit of tephra may be dated radiometrically using mineral or organic components available at the collecting site. It also may be dated by various methods of correlation with an identical, dated deposit or with the eruption that produced the deposits.
7. Except for tephra that has been intimately mixed with older sediments by organisms, frost action, or other physical processes, the maximum limiting age for deposits of reworked tephra is determined by the age of the parent eruption.

Theoretically, within the area of tephra deposition, a tephrochronologist is limited only by his or her ability to find and recognize primary tephra horizons and reworked deposits. Actually, studies to date have been performed mostly on layers of silicic tephra that have been deposited during the past 20 or so million years, since early Miocene time. Tephra that either is older than this or more like scoria in composition tends to be highly weathered and is more difficult to analyze.

APPLICATIONS OF TEPHROCHRONOLOGY IN ARCHAEOLOGIC RESEARCH

TIERRA BLANCA JOVEN TEPHRA, CENTRAL AMERICA

A long-standing problem in the study of the cultures of Mesoamerica has been the origin of Classic Maya civilization. What caused it to flourish? Just prior to its beginning at approximately A.D. 300, new styles of artifacts suddenly appeared in the Maya Lowlands of Guatemala and Honduras. Mayanists have postulated that either new trade routes had opened up or that a substantial human migration had taken place. One of the possible sources for the new artifacts was the Southeast Maya, or Guatemalan-Salvadoran Highlands (Sheets, 1976).

Recent work has shown that, at about the same time, a massive volcanic eruption took place from vents within Lake Ilopango in central El Salvador, the *tierra blanca joven (tbj)* tephra eruption of A.D. 260 (Hart and Steen-McIntyre, 1983). The tephra blanket, which is still a half-meter thick 75 km from the source vent (fig. 10.4), is believed to extend as far north as central Mexico and may even have reached south Texas (fig. 10.5).

There is strong reason to believe that the *tbj* eruption and the tephra mantle it produced devastated much of western El Salvador and that survivors of the catastrophe moved northward into the Maya Lowlands, bringing with them their pottery and other implements, or at least ideas of how pottery should be made. Little or no Protoclassic/Early Classic and Middle Classic pottery (*ca.* A.D. 250–650) has been found in western El Salvador, although sherds of earlier and later ceramic types abound below and above the ash blanket (Beaudry, 1983). What Protoclassic/Early Classic materials occur, for example mammiform tetra-pod bowls and Usulután ware, often are found on the ancient ground surface directly below the tephra layer. Such types of pottery are also reported from several sites farther to the north (Sheets, 1976).

The *tbj* tephra has gone unrecognized in archaeologic sites north and west of El Salvador, although it undoubtedly occurs in many of them (fig. 10.5). It is expected that Mayanists, once alerted to the possible presence of this tephra, will search diligently for it in their excavations. By close examination of the ceramic sequences found both below and above this dated ash layer, they will be able to test independently a hypothesis set forth by Sheets in 1976: that the introduction of Protoclassic/Early Classic artifacts into the Maya Lowlands was caused by the influx of refugees fleeing the devastation produced by the eruption of the *tbj* tephra from Lake Ilopango, El Salvador. If his hypothesis is correct, the bulk of the Protoclassic/Early Classic artifacts will be found only *above* the *tbj* tephra in Lowland sites.

CERÉN TEPHRA, EL SALVADOR

The Cerén site, located in central El Salvador approximately 20 km northwest of San Salvador, was discovered during excavation for a building. The site, of

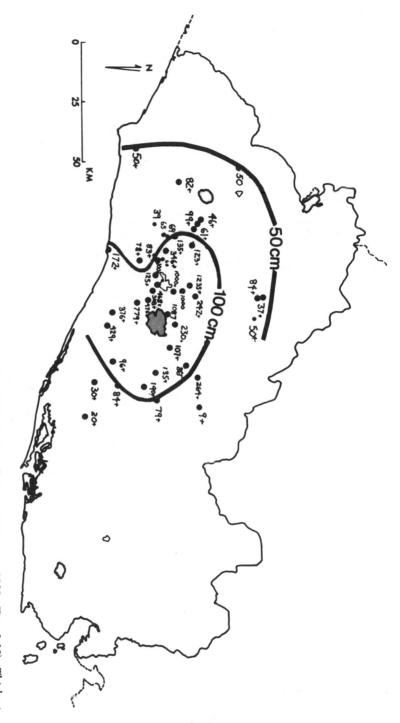

Fig. 10.4. Total thickness of the *tbj* tephra mantle as determined in 1978 (Hart and Steen-McIntyre, 1983, Fig. 2-10). Thickest deposits occur near Lake Ilopango, the irregular gray area around which the isopachous lines curve. The area outlined in black to the west of the lake is San Salvador.

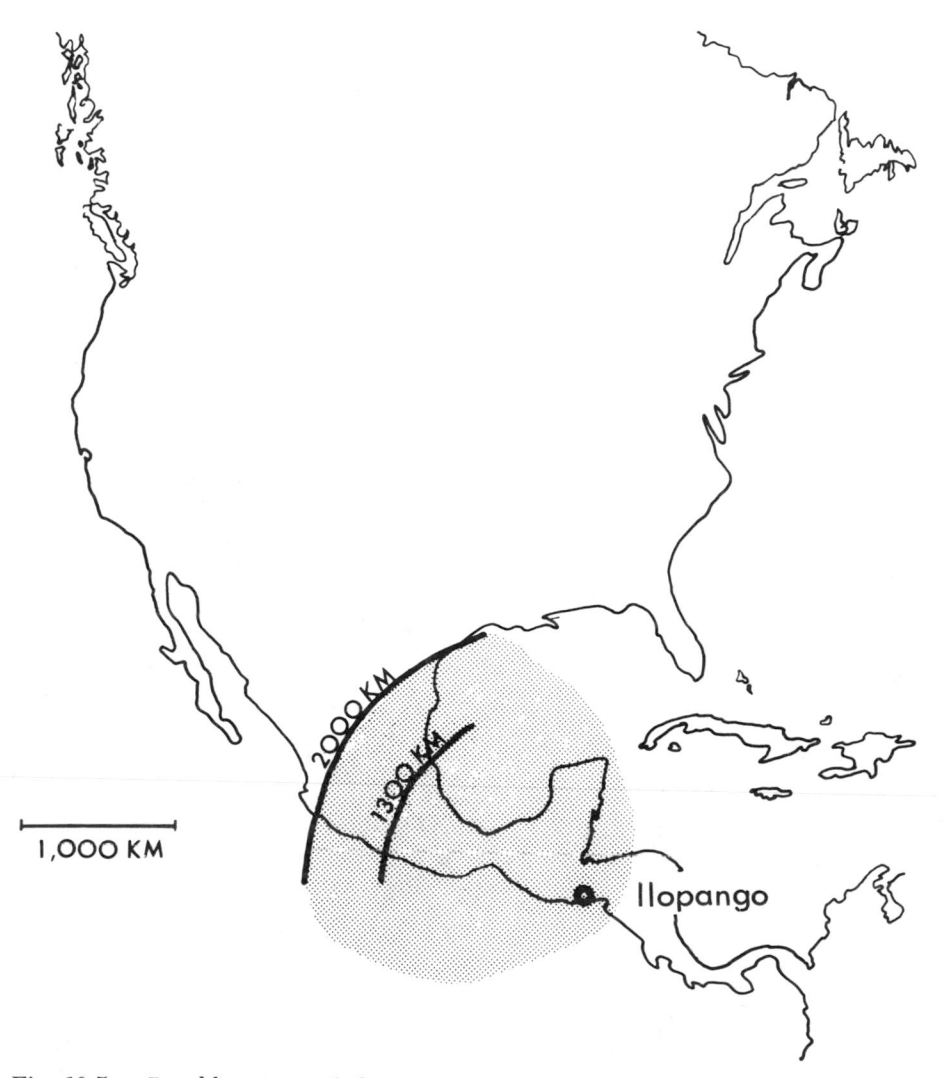

Fig. 10.5. Possible extent of *tbj* tephra in the northwest direction. The arcs are superposed distances of known tephra fall for two historic eruptions: La Soufrière, 1902, and Katmai, 1912 (Hart and Steen-McIntyre, 1983, Fig. 2-15). Fresh glass shards and glass-mantled phenocrysts, possibly *tbj* tephra, have been recovered from samples of "Maya clay" taken from lake-core sediments in the Petén, northern Guatemala (Steen-McIntyre, 1981a).

unknown extent, lies buried beneath 5 m of black ash and scoria bombs erupted from Laguna Caldera, a nearby cinder cone (Hart, 1983). Exposed during archaeologic excavations in 1978 were at least two Classic-period Maya domestic structures and an associated *milpa* (cornfield) (Zier, 1983). One house (fig. 10.6)

Fig. 10.6. Cerén site, El Salvador, during the spring of 1978. Exposed are the sectioned walls and floor of a Classic-Period Maya farmhouse. The structure was built upon reworked *tbj* tephra, and remnants of the original ash layer can still be seen (white layer at base). The house was subsequently buried by several meters of black cinders and ash, the Cerén tephra. In a nearby *milpa*, charred fragments of young corn plants were discovered still in the position of growth (Zier, 1983).

was sectioned by the bulldozer before it could be studied by archaeologists. Also removed and apparently lost at that time were several polychrome vessels and human bones.

Radiocarbon dates on organic material, including charred thatch from the roof of the house, give a mean date for the eruption of approximately A.D. 500. The site can be no older than A.D. 260, as it is built upon the *tbj* tephra of that age, described above.

The eruption that produced the Cerén tephra was a small one—the tephra blanket probably extends no more than 10 km from the vent. For local residents, however, the effects were devastating. Skeletal material removed by the bulldozer and flattened scoriaceous clasts embedded in the prehistoric ground surface (Zier, 1983) suggest that the eruption was quick, hot, and violent and that few in the immediate vicinity escaped.

As a result of the sudden deposition and excellent preservation of the Cerén tephra, archaeologists have a rare opportunity to excavate with confidence that the artifacts they uncover at the base of this unit—a blob of unworked clay, some paint pigment, a child's toy—actually were in use at the time of the eruption. Already one surprising fact has come to light: the abundance of polychrome pottery to be found in a simple farmhouse (Zier, 1983; Beaudry, 1983). Previously, well-made polychrome pottery was thought to have been a privilege of the ruling class.

HUEYATLACO TEPHRA, CENTRAL MEXICO

The Valsequillo Reservoir area, located south of the city of Puebla in central Mexico (fig. 10.7), has long been famous for its wealth of vertebrate Pleistocene fauna (Osborn, 1905). In the badlands around the reservoir, remains of camel, horse, bison, mastodon, mammoth, four-horned antelope, peccary, tapir, sloth, glyptodon, short-faced bear, dire wolf, and sabre-tooth cat have been found (Irwin-Williams, 1967; Kurten, 1967; Guenther, 1968; Guenther, Bunde, and Nobis, 1973). Associated with this fauna are well-made artifacts of flaked chert and worked bone (Armenta Camacho, 1978). In the early 1960s four sites were located where faunal remains and artifacts occurred together in situ (Irwin-Williams, 1967). No organic matter was preserved at these sites and the animal bones were heavily permineralized. As a result, the artifacts could not be dated by the ^{14}C method (Irwin-Williams, 1978). The bones were also unsuitable for amino-acid dating (Bada, pers. com., 1978). Uranium-series dating of bone associated with tools at two sites, El Horno and Hueyatlaco, gave ages that at the time were thought to be impossibly old, in the order of 200,000–300,000 years (Szabo, Malde, and Irwin-Williams, 1969).

Fortunately, the artifact-bearing beds exposed at Hueyatlaco were shown by direct tracing via a connecting trench to pass beneath a nearby bluff of sediments 8 m high (fig. 10.8; Steen-McIntyre, Fryxell, and Malde, 1981). The artifact-bearing layers are grouped into a lower set of beds with edge-retouched tools (fig. 10.9a) (projectile points, scrapers made on blades and flakes with prepared striking platform) and an upper set characterized by well-made, bifacially worked artifacts (fig. 10.9b) (projectile points, knives; percussion and pressure flaking; burins, scrapers, wedges, knives on flakes and blades; prepared striking platform [Irwin-Williams, 1978, fig. 3]).

Two tephra horizons occur within the bluff: the meter-thick Hueyatlaco ash (fig. 10.8), a fine-grained airfall ash, and the Tetela brown mud, a mudflow unit

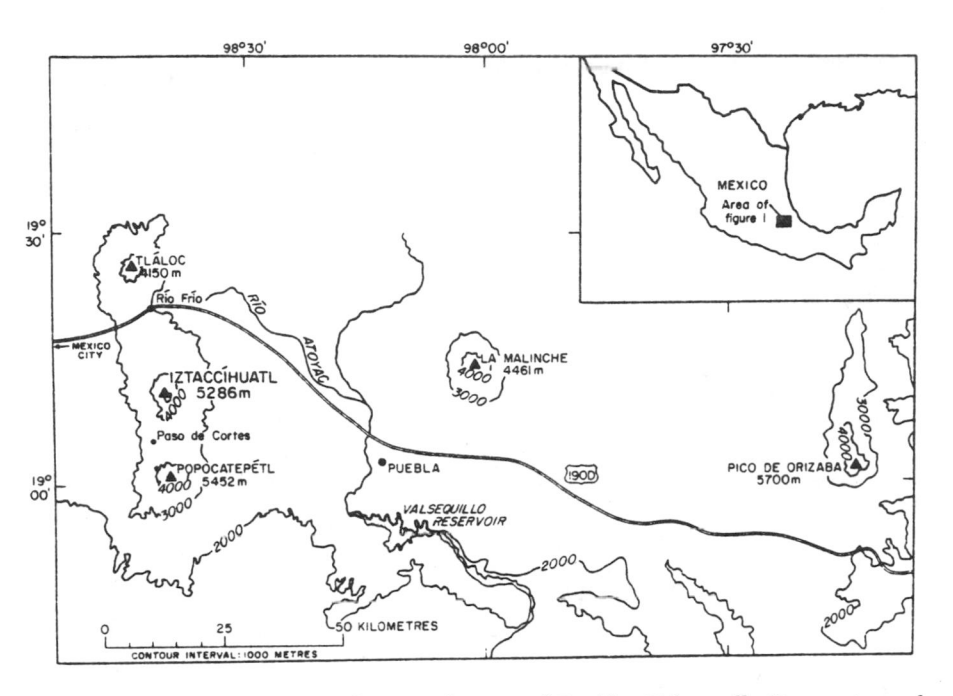

Fig. 10.7. Map of Mexico, showing the city of Puebla, Valsequillo Reservoir, and surrounding volcanoes. On the north shore of the reservoir, several sites have been located where Pleistocene fauna and well-made stone tools occur together in situ. Many tephra layers crop out in the bluffs surrounding the reservoir (Steen-McIntyre, Fryxell, and Malde, 1981, fig. 1).

with reworked chunks of pumice as large as cobbles sprinkled throughout it. A short way back from the bluff, the pumice-rice Tetela brown mud passes beneath another airfall-tephra layer, the Buena Vista lapilli. To complete the picture, another, unnamed tephra unit occurs in a small channel within the set of beds that contain bifacial tools.

The tephra layers exposed at Hueyatlaco were used to help date the sediments exposed at the site. In brief, our findings support those of Szabo—an age of 200,000–300,000 years. Fission-track dating of zircon phenocrysts (2 sigma) performed by C. W. Naeser of the U.S. Geological Survey gives ages of 370,000 ± 200,000 years for the Hueyatlaco ash and 600,000 ± 340,000 years for the Tetela brown mud pumice (Steen-McIntyre, Fryxell, and Malde, 1981). The Buena Vista lapilli has yet to be dated by this means: the unnamed tephra unit from within the upper artifact zone contains no zircon. Relative dating methods such as the extent of hydration of the volcanic-glass shards (tephra-hydration dating method), extent of etching and weathering of the heavy-mineral phenocrysts, and the depth of burial and subsequent dissection of the sediments support this great age.

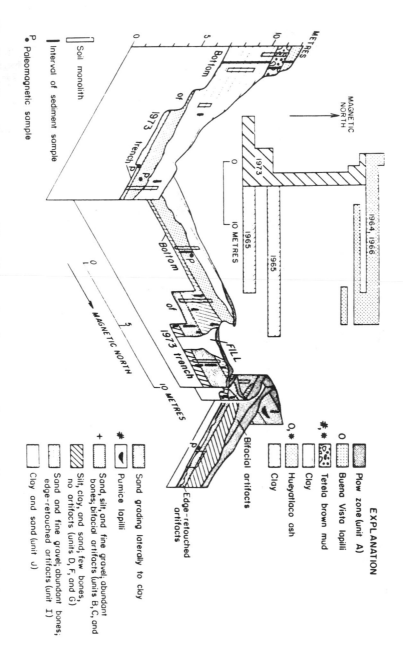

Fig. 10.8. Hueyatlaco excavation and fence diagram based on mapping done in profile in 1973 at a scale of 1 to 20 (Steen-McIntyre, Fryxell, and Malde, 1981, Fig. 6). The artifact-bearing beds underlie the Hueyatlaco ash, Tetela brown mud pumice, and Buena Vista lapilli. An unnamed tephra unit occurs in a small channel within beds that contain bifacial artifacts.

Fig. 10.9. Representative specimens from the artifact-bearing beds at Hueyatlaco: (a) is from the lower set of beds; (b) is from the upper set (Steen-McIntyre, Fryxell, and Malde, 1981, fig. 3). The artifacts have been fumed with ammonium chloride. Photographs by H. S. Rice, Washington State University, 1966.

Much work remains to be done on the Hueyatlaco site. One ton of sediment samples and stabilized stratigraphic sections taken from the trench walls in 1973 is waiting to be examined. The site is controversial to say the least (see Irwin Williams, 1969, 1978, 1981; Lorenzo, 1967; Malde and Steen-McIntyre, 1981), but the dating of the tephra layers and the sediments that contain them seems assured.

THE STUDY OF TEPHRA

In this section, we will see how tephra is examined and handled, first in the field, then in the laboratory. Emphasis will be placed on the petrographic examination of samples, although other methods of characterizing tephra will also be discussed. The role of the archaeologist in such a study is stressed.

SAMPLE COLLECTION

When a tephra layer of unknown age or provenance is exposed at an archaeologic site and more information about it is required, the ideal thing to do is to consult a tephrochronologist. He or she will make a regional survey to see what tephra layers occur in the area, collect samples from these layers, and make some attempt to date them, either radiometrically or, if there is no time or money for radiometric dates, on a relative basis. An effort will be made to choose samples from deposits of well-drained, medium-grained airfall tephra that preserve an upper depositional contact.

Lapilli are the medium-grained fragments (2−64 mm) of Fisher's classification (1961), as shown in table 10.1. Deposits of this size fraction are fine enough so that the stratification necessary for detailed sampling is quite apparent, yet the fragments themselves are large enough to be cleaned and handled individually to prevent contamination. Sample fragments are cleaned, crushed, and sieved, and the fine-sand fraction is given a general petrographic examination as described later in this section. Meanwhile, one or more tephra samples from the archaeologic site will also have been prepared and examined. By comparing properties, it may be possible to fit the tephra layer of interest into the regional tephrochronologic framework.

If the budget is small or time short, such a regional investigation, although valuable, is not usually possible. In the initial stages of a study, often only enough funds are available for an examination of the tephra exposed in the excavation trenches. Much can be learned from such a limited examination nonetheless. Fortunately, the work of collecting and preparing the samples can be done by the archaeologist or members of the crew at a considerable saving of time and money.

The first problem an archaeologist using tephrochronology as a research tool faces is to recognize a tephra layer when it occurs at a site. This can be difficult if one has never worked with the material before, especially if the layer is thin and stained with iron or organic matter. Coarse-grained tephra deposits, such as those covering the houses at the Cerén site, are no problem to spot; nor are thick layers of relatively young fine ash deposited in a semiarid environment (fig. 10.10). Near the distal edge of the tephra blanket, recognition is much more

Table 10.1.   Size of Naturally Fragmented Tephra Clasts and Weights and Volumes for Ideal Samples

Diameter of Coarsest Particle* (in mm)	Size Classification	Suggested Weight of Sample[†]	Approximate Volume of Sample[†]
Blocks and bombs (> 64)	Coarse-grained tephra	> 8 kg	—
Lapilli (64−2)	Medium-grained tephra	8−1 kg	—
Coarse ash ($2-\frac{1}{16}$)	Fine-grained tephra	500−125 g	500 cc (1 pint)
Fine ash ($< \frac{1}{16}$)	Fine-grained tephra	125 g	250 cc ($\frac{1}{2}$ pint)

*From Fisher, 1961.

[†]Modified from Krumbein and Pettijohn, 1938:32.

difficult. Here the tephra layer is very thin, fine-grained, and often stained (fig. 10.11). It can easily be mistaken for ash from a fire pit or for a layer of mineral salt, carbonate, gypsum, diatomite, phytoliths, or silt.

Layers that are not readily identifiable should be tested. If the material dissolves in water it is likely a mineral salt. If it fizzes and dissolves away in dilute hydrochloric acid (10% HCl) it is a form of calcite or dolomite. If it is inert in acid, but feels slippery, it probably is gypsum. If it feels gritty between fingers or teeth it may be diatomite, opal phytoliths, silt, or volcanic ash. Collect a small sample, rinse away the clay-size material, and examine the residue under an optical transmitted-light microscope. At 100× magnification, silt will be readily apparent and diatomite and plant opal will show regular structures that are organic in appearance; volcanic ash will consist of irregularly shaped glass shards, with or without bubble cavities, and fragments of mineral crystals (fig. 10.1; Hart and Steen-McIntyre, 1983, fig. 2.16).

Once a tephra layer is discovered in an excavation, it should be mapped as a separate unit and described in detail. This can be done at the time the trench

Fig. 10.10. Thick deposit of fine-grained Mazama ash (white layer) in rockfall at the entrance to Marmes rockshelter, southeastern Washington. Parent vent for the tephra is Mount Mazama (Crater Lake) in southwestern Oregon. The ash is approximately 6,700 years old (Mehringer, Blinman, and Peterson 1977). Exposed in the square pit beneath the layer of rock rubble are fine-grained floodplain sediments that include a thin layer of volcanic ash (not shown). These sediments contain partially permineralized fragments of human bone dated at between 11,000 and 13,000 years (Fryxell et al., 1968).

Fig. 10.11. Thin, tan layers of Mazama ash at Mummy Cave, Wyoming, near Yellow-stone National Park. It had been mistakenly identified as wood ash by the excavators (Wedel, Husted, and Moss, 1968). Its true nature could be determined only under the microscope.

profiles are drawn. Especially to be noted are the nature of the upper and lower contacts (table 10.2); color, using standard Munsell notation; and variation in particle size. The last can be accomplished best with a series of small soil sieves.

Next, the unit should be sampled. For thin layers, only one sample may be possible, and it may be no more than a few tenths of a gram in weight. For thick layers, several samples should be collected from a vertical cut. Table 10.1 gives ideal weights and volumes for tephra samples of various size ranges. In a pinch samples may be collected in doubled plastic bags, with an identification slip tied to the neck and/or placed between the two layers of plastic. For more permanent storage, cylindrical cartons of heavy cardboard in pint, quart, and gallon sizes work well.

From thick layers of tephra, samples should be collected from bottom to top in order to avoid contamination. The collecting pattern should be so devised as to include samples from each episode of eruption preserved in the deposit. These episodes may be recognized by such criteria as abrupt change in particle size or degree of sorting, distinctive parting layers, and significant unconformities. In addition, at least three samples (bottom, middle, top) should be collected from thick subunits so that later they may be checked for uniformity of characteristics.

Table 10.2. Nomenclature for the Boundary between Two Sedimentary Units or Soil Horizons

Property	Term	Description	
		Size Range	
		(in mm)	(in in.)
Distinctness	Very abrupt	< 1	< $\frac{1}{16}$
	Abrupt	< 2.5	< 1
	Clear	2.5–6.5	1–2.5
	Gradual	6.5–13	2.5–5
	Diffuse	> 13	> 5
Topography	Smooth	Nearly a plane	
	Wavy	Pockets with width > depth	
	Irregular	Pockets with width < depth	
	Broken	Discontinuous	

SOURCE: Modified from Soil Survey Staff, 1951:187–88.

Composite samples such as those recovered from a channel cut into the face of the exposure should be avoided; they tend to mask important petrographic and chemical changes within the sample interval.

Of special importance in sampling deposits of medium- and coarse-grained tephra are fine-grained parting layers of volcanic ash, especially when they rest unconformably upon older ejecta. Although they ordinarily are composed of fragments that were still suspended in the turbulent hot air while the bombs and lapilli fell to earth, occasionally they represent other large-scale explosions whose plumes were directed away from the collecting site, or still other eruptions from distant vents.

The bulk samples may then be sent to the tephrochronologist for preparation, examination, and possible dating. Not all the samples will be examined in detail, but it is always good to offer a choice from which to work. The archaeologist may choose, on the other hand, to have a crew do much of the initial sample-preparation work in the field laboratory. This is not only possible but is highly recommended. It can save the project time and money and will give all who work with the material an insight into the problems involved in the laboratory preparation of tephra. Full details of field and laboratory sample preparation are given in Steen-McIntyre (1977b).

ANALYTICAL METHODS

Most researchers involved in tephrochronology have entered the field from other disciplines, bringing their preferences and prejudices with them. As a result, several approaches to the characterization of tephra layers are available to the tephrochronologist. Those who are accustomed to working with long spans of time and with small amounts of fine-grained samples—for example, marine geologists—prefer chemical analysis of bulk samples as the best approach. They use methods such as instrumental neutron-activation analysis, atomic-absorption spectrophotometry, and X-ray fluorescence. Geochemists rely strongly on chemical analyses of tephra components, especially the glass shards and/or heavy-mineral phenocrysts. Still others, including archaeological geologists and those who study terrestrial sediments, commonly work with larger sample volumes. They tend to use a combination of petrographic and chemical methods to characterize and correlate tephra and they emphasize the phenocryst fraction as strongly as the volcanic glass. Significant papers concerned with these various methods are listed in Steen-McIntyre (1977b). Others include: Davis (1978); Drexler et al. (1980); Howorth and Rankin (1975); Kittleman (1979a, 1979b); Kohn (1970, 1979); Kohn and Topping (1978); Medlin, Suhr, and Bodkin (1969); Moody (1978); Richardson and Ninkovich (1976); Riezebos (1978); Theisen et al. (1968); Westgate (1977, 1982); and Westgate and Evans (1978).

At present the fields of tephrochronology and tephrostratigraphy are developing rapidly with promising new approaches to the problems of tephra correlation and characterization appearing each year. The more that is learned about the subject, however, the more it is evident that no single technique is entirely adequate for the study of tephra; rather, a combination of several techniques must be used (Wilcox and Izett, 1973).

SAMPLE DESCRIPTION

A tephrochronologist, in characterizing ejecta from a volcanic eruption, works with phenocrysts, glass shards, and sometimes the xenocrysts and xenoliths torn from the magma chamber, vent walls, plug, and cap rock. Of the many methods available for characterizing a tephra layer, one seems to be especially useful for tephra horizons uncovered at archaeologic sites, where the time interval between important tephra units can be very small, on the order of hundreds of years or less. It combines detailed study with the petrographic microscope and major-element analysis using the electron microprobe. Good examples of this approach are reported in Davis (1978); Moody (1978); Smith, Okazaki, and Knowles (1975); Westgate and Fulton (1975); and Westgate and Evans (1978). To these techniques I would add particle size analysis. The method is well suited for departments with limited research facilities. Other than a few specialized items, the equipment necessary is available in any college petrographic laboratory of moderate size. The electron-probe work may be subcontracted.

THE PETROGRAPHIC EXAMINATION

Two classes of characteristics—*essential* and *supplemental*—are available to the tephrochronologist who chooses to identify tephra by petrographic means (Wilcox and Izett, 1973). Essential characteristics reflect the physical and chemical nature of the parent magma at the time of eruption. Except for weathering and the passage of time, they are little affected by posteruptive processes. They include the crystallographic properties, morphology, internal structures, and chemical composition (as indicated by refractive index) of the phenocrysts and glass shards. Supplemental characteristics vary from sample to sample. They include such elements as size of the tephra clasts and components; the relative amount of phenocrysts, glass, and wall rock in the sample; and degree of weathering. Supplemental characteristics are strongly influenced by winnowing and differential settling of the components during transport from the source vent, by postdepositional reworking and alteration, and also by methods used in collecting and preparing the individual sample.

Both essential and supplemental characteristics are important and should be considered by the tephrochronologist during any petrographic examination. By reflecting conditions within the parent magma that may persist through several eruptions, essential characteristics are valuable in correlating tephra horizons with potential source vents. Supplemental characteristics, on the other hand, provide information about an individual episode of the eruption, the wind pattern that prevailed at the time, and the environment of deposition.

The petrographic examination of a tephra sample is conducted at two different levels: general and detailed. Most samples received from the archaeologist will be given a general examination; few will be chosen for detailed analysis because of the amount of time involved.

The General Examination

Information collected during the general petrographic examination is limited to what can be observed in the bulk sample, the reference slide, and in loose fragments in oil-immersion mounts, including data on essential and supplemental characteristics. Both a wide-field stereomicroscope and a polarizing microscope normally will be used. During this phase of the petrographic examination, glass shards are inspected for shape, transparency, inclusions, vesicularity, modal refractive index, and extent of hydration; the phenocrysts are inspected for type, relative percentage, inclusions, and color; the bubble cavities are examined for water. About four hours of microscope time per sample are needed for this work. Examples of the types of data that can be gathered during the general examination are shown in tables 10.3 and 10.4.

The Detailed Examination

Information from the detailed petrographic examination primarily concerns the tephra's essential characteristics. It is obtained by means of the polarizing microscope from individual mineral crystals mounted on the spindle stage and from counts of volcanic-glass shards in a series of refractive-index oils. In this phase of the examination, a refractive-index histogram of the volcanic glass is compiled from grain-count data on glass shards immersed in a series of high-dispersion refractive-index oils with n spaced at intervals of 0.002. Phenocrysts are mounted on the tips of wire spindles and examined in the spindle stage, a

Table 10.3. Information Gathered during the General Petrographic Examination

Source of Tephra (Age) (Sample numbers)	Volcanic Glass		Shard Type	Percent of Shards with Microphenocrysts	Type of Microphenocryst (Percent)					Percent of Heavy-Mineral Phenocrysts with Adhering Glass				
	Modal Refractive Index (at 25° C)	Range of Refractive Index			Pale green Amphibole	Orthopyroxene	Opaques	Apatite	Feldspar	Pale green Amphibole	Orthopyroxene	Clinopyroxene	Opaques	Apatite
St. Helens W (450 yr)														
77 (northwest lobe)	1.494	—	Pumiceous	5		5				58	29	5	7	1
74 (east lobe)	1.496	—	Pumiceous	22	1	18	3			27	53	9	9	2
Glacier Peak * (ca. 12,000 yr)														
78 (top upper unit)	1.503	—	Pumiceous	13	13	13				65	29	—	6	—
73 (base lower unit)	1.502	—	Pumiceous	18	18	18				36	43	—	21	—
Duncan Lake ** (ca. 34,000 yr)														
UA 539 (youngest)	1.505	1.501–1.507	Pumiceous	<9	3	<1	3	<1	<1			No data available		
UA 538	1.503	1.500–1.506	Pumiceous	<8	2	<1	3	1	<1					
UA 537 (oldest)	1.505	1.502–1.507	Pumiceous	<8	2	1	3	<1	<1					

*Samples supplied by R. Okazaki.

**Samples supplied by J. Westgate (see Westgate and Fulton, 1975, for more information).

Table 10.4. Hydration Data for Samples of St. Helens and Glacier Peak Tephra, Pacific Northwest

Source Vent	Name of Tephra Horizon (Position of sample)	Approximate Age (in yr B.P.)	Extent of Hydration Apparent Rind Thickness (in μm)	Extent of Hydration Actual Values (in μm)	Percent by Volume of H_2O in Vesicles of 100 Shards ≤ 0.1	≤ 1	≤ 5
St. Helens	W (Northeast lobe)	450	< 1	(all < 1)	100		
	W (East lobe)	450	1	(1, 1, 1, 1, 1)	97	3	
	Y (Southeast lobe)	3,400	1.5	(2, 1, 1, 2)	98	2	
	Y (North lobe)	3,500	1.5+	(1+, 1+, 2, 2)*	96	4	
			3.5	(4, 3, 4, 3)†			
	Set J (Main body)	8,000–12,000	5	(4, 5, 4, 6)	100		
Glacier Peak	(Upper unit)	12,000	5	(4, 6, 5, 5)	64	33	3
	(Lower unit)	12,000	5.5	(5, 6, 6, 5)	73	24	3
St. Helens	Set S (Near upper contact)	13,000	6	(6, 5, 6, 6)	95	5	
	Set M (Near basal contact)	18,000–20,000	6	(7, 5, 6, 5)	89	11	
	Set C (Near basal contact)	37,000?	Completely hydrated		87	13	

*Pumiceous shards, n = ca. 1.505.

†Dense shards, n ≤ 1.501.

SOURCE: Steen-McIntyre, 1981a. Samples provided by Rose Okazaki, Soils Department, Washington State University, Pullman.

handy, inexpensive device that often has been used in place of the universal stage. Individual crystals are examined for signs of zoning, sheathing, twinning, inclusions, optic sign, extinction angle (Z:C), lowest refractive index (n), and approximate optic angle (2V)—all well-known optical mineralogical properties. In some cases actual 2V is measured with the help of a stereo net. Eight hours of microscope time per sample should be considered the minimum amount necessary for this phase of the work. Figure 10.12 gives an example of the type of data that result.

For an expanded discussion of the general and detailed petrographic examination, see Steen-McIntyre (1977b:83–118).

ELECTRON-MICROPROBE ANALYSIS

Major-element analysis of tephra components using the electron microprobe has proved to be a powerful tool in tephra research, especially when it is used in conjunction with petrographic studies. It has two major advantages over other analytical techniques. First, single fragments rather than bulk samples are analyzed, thereby reducing dramatically the chance for contamination. Second,

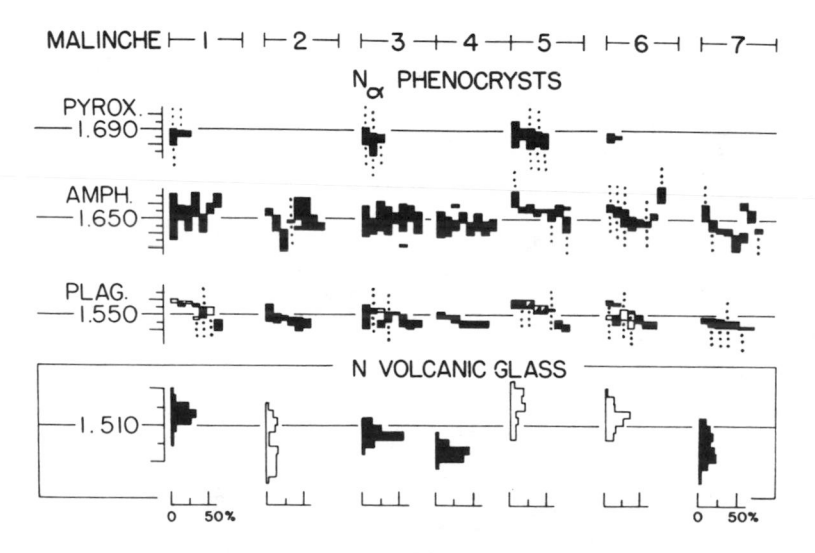

Fig. 10.12. Selected petrographic data for samples of coarse ash and lapilli, dated from greater than 25,000 years (1) to 8,000 years (7), collected on the western flank of La Malinche Volcano, Puebla, Mexico. Each vertical bar in a set refers to the refractive-index (n) range observed in one small crystal when viewed from the direction of lowest n. Volcanic glass in samples (1), (3), and (4) is completely hydrated; glass from sample (7) is incompletely hydrated. Glass from samples (2), (5), and (6) was hydrothermally altered and/or choked with microlites, and the extent of hydration of the shards could not be determined. For more information on these samples, see Steen-McIntyre (1977b: 114–17).

only very small amounts of material are needed, often no more than 100 grains.

The method of analyzing tephra with the electron microprobe was developed by D. G. W. Smith and colleagues at the University of Alberta (D. G. W. Smith and Westgate, 1969; Smith, Westgate, and Tomlinson, 1969). A modification of their procedure has been adopted by H. W. Smith and co-workers at Washington State University (H. W. Smith, Okazaki, and Knowles, 1977a, 1977b). Both require the services of an experienced operator to run the equipment.

Davis (1978), in his work on tephra layers of western Nevada, determined the percentage of eight elements in volcanic glass shards with the microprobe: silicon, aluminum, potassium, sodium, calcium, iron, magnesium, and titanium. H. W. Smith, Okazaki, and Knowles (1977a, 1977b) used only three: potassium, calcium, and iron. Differences in the relative proportion of these elements was found by D. G. W. Smith and Westgate (1969) to be most helpful in distinguishing among tephra layers that otherwise seemed identical.

Samples of cleaned, sieved volcanic glass for probe analysis are prepared in the same manner as for petrographic analysis. Although glass-rich samples are best to use (most operators require them), Davis found that with care he could obtain acceptable results with tephra that had not been given a density separation.

DATING

There are several ways to date tephra layers. Tephra samples and bracketing sediments may be tested for remanent paleomagnetism (Reynolds, 1975; Reynolds and Larson, 1972; Liddicoat et al., 1979, 1981). They may be dated indirectly by associated organic material (Fernald, 1962; Damon, 1968; Fulton, 1971), although the ^{14}C method should be used with caution when dating samples collected on the flanks of a volcano. Chatters, Crosby, and Engstrand (1969) sampled living plants near a fumarole (gas vent) that gave radiocarbon dates of several thousand years. Tephra layers may also be dated by associated fossil bone (Szabo, Malde, and Irwin-Williams, 1969; Steen-McIntyre, Fryxell, and Malde, 1981; Bada, Schroeder, and Carter, 1974; Bada and Masters-Helfman, 1975).

Radiometric ages of the tephra components themselves are obtained by the potassium-argon dating method on sanidine, leucite, or mica phenocrysts (Curtis, 1966; Damon, 1968; Dalrymple and Lanphere, 1969). Dates can also be derived from fission-track counts on zircon phenocrysts or volcanic-glass shards (Fleischer, 1979; Fleischer and Walker, 1975; Naeser, Izett, and Obradovich, 1980). Ages of samples often can be roughly estimated using the tephra-hydration dating method and by noting the extent of etching of the heavy-mineral phenocrysts (Steen-McIntyre, 1975, 1977a, 1977b, 1981a, 1981b). This section will outline techniques of fission-track dating, potassium-argon dating, tephra-hydration dating, and using the extent of etching of heavy-mineral phenocrysts to roughly estimate age.

FISSION-TRACK DATING

Fission-track dating is based on counting the tracks produced in a glass or crystal fragment by fast-moving atomic particles that result from radioactive decay. Two components of tephra most often dated by the fission-track method are zircon phenocrysts and volcanic glass. Dating of the glass has been limited to platy, nonvesicular shards to the exclusion of pumiceous fragments. For a description of the theory and technique, see the references listed above.

Glass shards to be examined for fission tracks must never be exposed to high heat, such as can occur on a hot tray or directly beneath a heat lamp. Heat anneals the fission tracks and the resulting date will be too young (Izett, pers. com., 1972).

To separate an adequate number of zircon phenocrysts for fission-track dating, it often becomes necessary to process a large volume of material. Zircons are scarce in most tephra units, rarely making up more than 1 percent of the heavy-mineral fraction. At least 50—100 grains are required for a good date, and the more the better. Here, panning is a good method for concentrating zircons (of specific gravity 4.6—4.7) from the bulk tephra sample. The dry concentrate is then treated with the heavy liquid methylene iodide (s.g. 3.3) to separate zircons from apatite (s.g. 3.2), and the zircon-rich fraction is passed several times through an isodynamic separator. At the last pass, provided all apatite and glass shards have been removed by previous treatment, only zircon phenocrysts should pass down the trough that collects minerals unaffected by the magnetic field.

The zircon fraction is next examined with a binocular microscope, and with a fine brush or needle the grains for analysis are hand-picked. To lessen the chance for contamination, only grains showing no signs of abrasion and/or having fragments of glass adhering to them are chosen.

POTASSIUM-ARGON DATING

Potassium-argon dating is a widely accepted method for dating very early archaeologic sites. For a description of the theory, see Dalrymple and Lanphere (1969). The following information was obtained from R. F. Marvin, of the U.S. Geological Survey (pers. com., 1976).

Tephra samples of early or middle Pleistocene age can be dated by the potassium-argon method using sanidine, biotite, or leucite phenocrysts, assuming such minerals are present in the sample and that they can be concentrated. In some instances, associated fresh, nonhydrated obsidian or fine-grained, fresh lavas (basalt, andesite, or felsite) could give usable ages, although the sample volume would have to be increased because of their small potassium content. When biotite, glass, or whole rock is used for analysis, however, much of what is measured is atmospheric argon; this will impair the measurement of radiometric argon and may produce a calculated potassium-argon age with a very large analytical error.

The feldspar sanidine is the best mineral for dating young volcanic material. By treating the phenocrysts with a dilute hydrogen flouride wash before the argon analysis, a considerable amount of adsorbed atmospheric argon is removed, thereby increasing the precision of the argon measurement and the calculated age. The treatment should also dissolve unwanted volcanic glass.

Concentration involves crushing the sample, sizing the material (0.15−0.25 mm is a convenient size range), using heavy liquids and, when available, a floor-model centrifuge for density separations, magnetic separation, and possibly some hand-picking for the final cleaning of the concentrate.

TEPHRA-HYDRATION DATING

Tephra-hydration dating is a method whereby the approximate age of a sample of rhyolitic or dacitic tephra can be obtained, provided enough is known about the chemical composition of the glass, shard morphology, vesicle density, and the environment of deposition and preservation. It utilizes the research of I. Friedman and colleagues on obsidian-hydration dating (Friedman and Smith, 1960; Friedman and Long, 1976, Ross and Smith, 1955; and cited references) and of E. Roedder and colleagues on the study of liquid water in pumice vesicles (Roedder and Smith, 1965; Roedder, 1972), but differs from both in the method of obtaining data. The method is explained in detail elsewhere (Steen-McIntyre, 1975; 1981a).

In the hydration of silicic glass shards, water penetrates the surface of fragments during weathering and diffuses into the interior across a well-defined front, raising the refractive index of the glass approximately 0.01 in the process. Following hydration of the glass itself, water continues to diffuse slowly into the vesicles by a process called *superhydration*, until finally even the vesicles are filled. The tephra-hydration dating method uses these time-dependent processes to approximate the age of silicic tephra 1,000 to 2,000,000 years old.

To apply the tephra-hydration dating method, it is necessary to note the proportions of hydrated to nonhydrated glass in individual shards of fine-sand size (0.25−0.125 mm), often by measuring the apparent thickness of the hydration rind, and to estimate average proportions of liquid water to water vapor in selected glass vesicles of 100 different fragments. These quantities represent, respectively, the extent of hydration and superhydration. The visual estimates are then compared with estimates for similar samples of dated tephra in order to obtain an approximate age for the undated sample. As a general rule, the smaller the proportion of nonhydrated glass and the greater the proportion of water in the vesicles, the older the sample.

Table 10.4 lists thicknesses of hydration rinds and extent of superhydration for ten samples of tephra from the Pacific Northwest. In these young samples, no vesicle contains more than 5 percent water, and most show considerably less.

Superhydration curves for eight dated samples of rhyolitic and dacitic tephra are given in figure 10.13. In general, the older the sample, the more the curve is

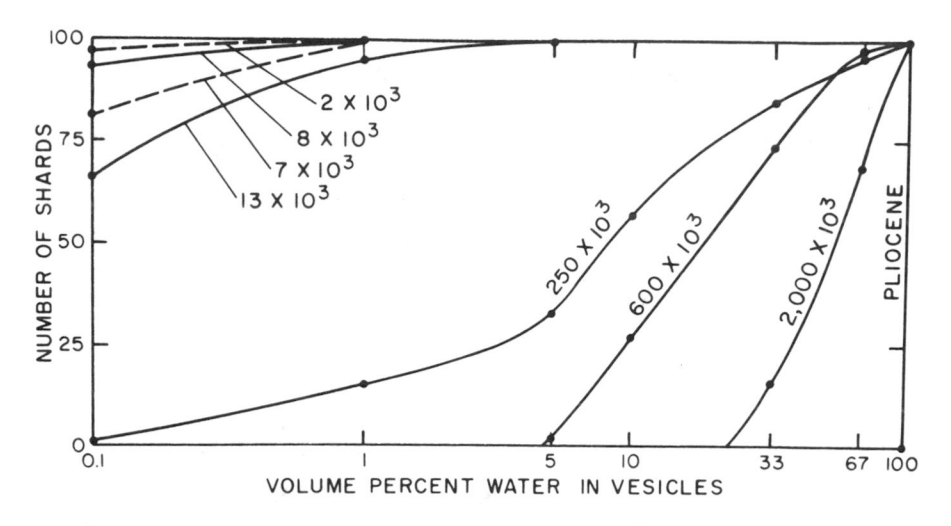

Fig. 10.13. Superhydration curves for naturally fragmented shards of pumiceous fine ash. Each curve plots the average volume of water in closed, spindle-shaped cavities 10–50 μm long that occur in 100 glass shards. The tephra samples are identified by their ages in thousands of years: Ilopango tephra, El Salvador, 2×10^3 years; Mazama ash, Washington State, 7×10^3 years; La Malinche ash, central Mexico, ca. 8×10^3 years; Glacier Peak ash, Montana, 13×10^3 years; Hueyatlaco ash, central Mexico, ca. 250×10^3 years; Pearlette type-O ash, Saskatchewan, 600×10^3 years; Pearlette type-B ash, Kansas, $2,000 \times 10^3$ years; Bidahochi ash, New Mexico, of Pliocene age. See Steen-McIntyre (1975), table 1 for more exact locations and actual dates.

displaced vertically and to the right of the graph. Note, however, that the 7,000-year-old glass (Mazama tephra) has more water in the vesicles than the 8,000-year glass (Malinche tephra). This is presumably due in part to different climatic conditions that exist at the collecting sites and in part to differences in the surfaces of the shards—Mazama tephra fragments are finely pumiceous and were collected in the hot, dry Columbia Basin of Washington, whereas Malinche fragments, relatively dense and coarsely vesicular, were collected near timberline high on the flanks of a glaciated volcano.

Recent work suggests that specific surface and glass chemistry are especially important factors in controlling the extent of hydration and superhydration for silicic tephra of late Quaternary age (Riezebos, 1978; Steen-McIntyre, 1981a). On the other hand, depositional environment plays an important role in the extent of superhydration for older tephra-glass samples, as suggested by the curves in figure 10.14. All four samples represented by the curves have been dated individually by radiometric means and correlated by various techniques with the 600,000-year-old Pearlette type-O tephra from Yellowstone Park. The difference in extent of superhydration for these four samples thus would appear to be related to differences in environment and underscores the necessity of

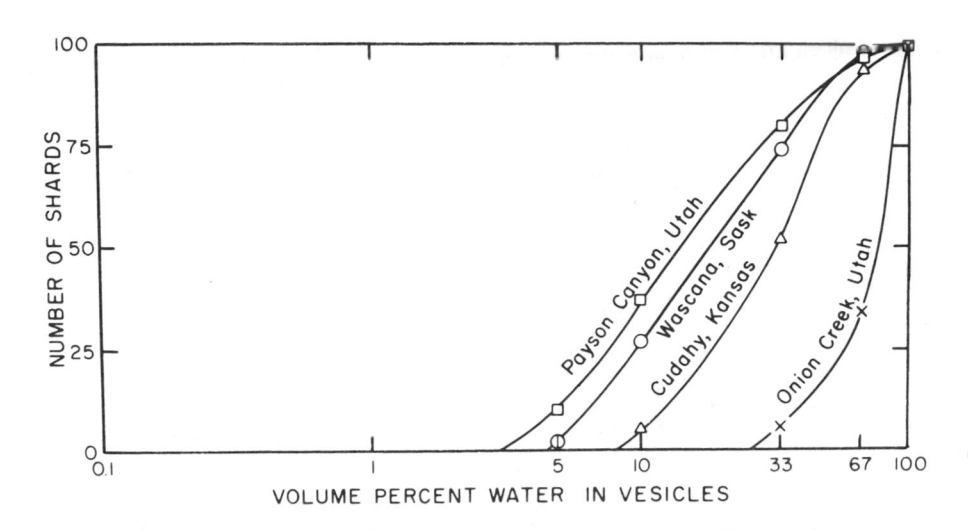

Fig. 10.14. Superhydration curves for dated samples of Pearlette type-O tephra (600,000 years old) collected from different environments. The Onion Creek location is in a hot, dry semidesert with scrub juniper as the dominant tree type (G. Richmond, pers. com., 1975). Environmental conditions at the other sample localities are not exactly known, but the Saskatchewan sample at least would come from a cooler and moister environment than that at the Onion Creek locality. For more detailed information on the samples, see Steen-McIntyre (1975), table 1.

taking this factor into account when comparing curves for dated and undated samples.

ETCHING OF PHENOCRYSTS

Intrastratal solution is the chemical solution that takes place within a sedimentary bed after deposition (Pettijohn, 1957:674). It produces in heavy minerals a characteristic etched appearance that I have referred to as "picket-fence" structure (Steen-McIntyre, Fryxell, and Malde, 1981), which has also been called "hacksaw" or "cockscomb" structure. Not all minerals in a sample are equally affected by intrastratal solution; they seem instead to be selectively etched, with the phenocrysts of basic magmas being more readily attacked (Goldich, 1938).

Several workers have made effective use of the selective solution of heavy minerals to obtain relative ages for deposits (Steen-McIntyre, 1977b, 1981a). Others point out that intrastratal solution depends not only on time but also on grain size of the sedimentary unit (Blatt and Sutherland, 1969), on weathering environment (Hay and Jones, 1972; Hay, pers. com., 1975), and on Eh, pH, and the hydraulic gradient of the groundwater (Walker, 1967).

Heavy-mineral phenocrysts from within pumiceous tephra clasts are ideal specimens to observe for signs for etching. Because they are contained within larger fragments, there is no danger of contamination from older, more highly etched minerals of the same species. In addition, the glassy envelope that surrounds the crystal protects it from the immediate effects of weathering (Hay and Jones, 1972). This is an especially valuable consideration for tephra samples collected in the tropics, where chemical weathering is both rapid and severe.

The extent of etching of phenocrysts seems to depend on climate, sedimentary environment, mineral size and species, the nature of the glass mantle, the chemical composition of the minerals, glass, and circulating fluids, and time. On the Island of Hawaii, phenocrysts from within clasts of basaltic Pahala tephra (10,000–17,000 years old) have begun to alter, whereas those from younger beds have not (Hay and Jones, 1972). On Saint Vincent in the West Indies, andesitic tephra deposits approximately 20,000 years old show decomposed lapilli fragments and slightly etched hypersthene crystals (Hay, 1959; pers. com., 1975). In temperate central Mexico, etching of hypersthene phenocrysts from dacitic tephra deposits is rare, and then only incipient in units 22,000–24,000 years old and entirely lacking in younger units (P. W. Lambert, pers. com., 1973). For

Fig. 10.15.   Etched heavy-mineral phenocrysts from pumiceous lapilli fragments collected in an unnamed tephra deposit within the upper artifact zone, Hueyatlaco archaeologic site, Puebla, Mexico. The tephra unit is associated with radiometric dates of approximately 250,000 years (Steen-McIntyre, Fryxell, and Malde, 1981). Phenocrysts of hypersthene are all highly etched. A slightly etched crystal of augite occurs near the base of the photo, and a fresh-looking, unetched hornblende crystal is seen at the left. Photomicrograph by R. B. Taylor, U.S. Geological Survey.

dacitic tephra fragments approximately 250,000 years old and collected in the same general area in Mexico (Hueyatlaco site, unnamed tephra layer from the upper artifact zone), hypersthene is severely etched, augite is slightly etched, and the hornblende still looks fresh (fig. 10.15).

CONCLUSIONS

Tephrochronology can be of great value in archaeological research. A tephra layer from a dated major eruption forms an almost ideal marker horizon—easy to recognize, widespread, cosmopolitan in its associations. Where thick, it preserves untouched the artifacts it buries; when old, it can be dated radiometrically using the tephra components themselves. To archaeologists, it represents a reference plane in both time and space to which artifacts and interesting features, separated perhaps by thousands of kilometers, may be compared.

REFERENCES

Armenta Camacho, J. 1978. *Vestigios de labor humana en huesos de animales extintos de Valsequillo, Puebla, Mexico.* Puebla, Mexico: Publicaciones del Consejo Editorial del Gobierno del Estado de Puebla.

Bada, J. L., and P. Masters-Helfman. 1975. Amino acid racemization dating of fossil bones. *World Archaeology* 7:160–73.

Bada, J. L., R. A. Schroeder and G. F. Carter. 1974. New evidence for the antiquity of man in North America deduced from aspartic acid racemization. *Science* 184: 791–93.

Barberi, F., F. Innocenti, L. Lirer, R. Munno, T. Pescatore and R. Santacroce. 1978. The Campanian Ignimbrite: A major prehistoric eruption in the Neapolitan area (Italy). *Bulletin volcanologique* 44:1–22.

Beaudry, M. 1983. The ceramics of the Zapotitán Valley. In *Archaeology and Volcanism in Central America: The Zapotitán Valley of El Salvador,* ed. P.D. Sheets, 161–90. Austin, TX: Univ. of Texas Press.

Blatt, H., and B. Sutherland. 1969. Intrastratal solution of non-opaque heavy minerals in shales. *Journal of Sedimentary Petrology* 39:591–600.

Blong, R. J. 1982. *The time of darkness.* Canberra: Australian National Univ. Press.

Bloomfield, K., G. Sanchez Rubio, and L. Wilson. 1977. Plinian eruptions of Nevado de Toluca volcano, Mexico. *Sonderdruck aus der geologischen rumdnschau* 66: 120–46.

Bloomfield, K., and S. Valastro, Jr. 1977. *Late Quaternary tephrochronology of Nevado de Toluca volcano, Central Mexico.* Overseas Geology and Mineral Resources, no. 46: London: Institute of Geological Sciences.

Brunnacker, K. 1967. Die sedimente der Crvena Stijena: Posban otisak iz glasnika zemaljskog muzeja, Bosne i Hercegovine, Sveska. *Arheologia, Sarajevo* 21–22: 31–65.

Chatters, R. M., J. W. Crosby, III, and L. G. Engstrand. 1969. *Fumerole gaseous emanations: Their influence on Carbon-14 dates.* College of Engineering Circular no. 32. Pullman, WA: Washington State Univ.

Coppens, Y., F. C. Howell, G. L. Isaac, and R. E. F. Leakey, eds. 1976. *Earliest man and environments in the Lake Rudolf Basin*. Pt. 1. Chicago: Univ. of Chicago Press.

Curtis, G. H. 1966. The problem of contamination in obtaining accurate dates for young geologic rocks. In *Potassium-argon dating*, ed. O. A. Schaeffer, and J. Zähringer, 151–62. New York: Springer-Verlag.

Dalrymple, G. B., and M. A. Lanphere. 1969. *Potassium-argon dating: Principles, techniques and applications to geochronology*. San Francisco: W. H. Freeman.

Damon, P. E. 1968. Radioactive dating of Quaternary tephra. In *Means of correlation of Quarternary successions*, ed. R. B. Morrison and H. E. Wright, Jr., 195–206. Salt Lake City: Univ. of Utah Press.

Davis, J. O. 1978. *Quarternary tephrochronology of the Lake Lahontan area, Nevada and California*. Nevada Archaeological Survey Research Paper no. 7. Reno: Univ. of Nevada.

Doumas, C., ed. 1978. *Thera and the Aegean world*, II. London: Thera and the Aegean World.

Drexler, J. W., W. I. Rose, Jr., R. S. J. Sparks, and M. T. Ledbetter. 1980. The Los Chocoyos Ash, Guatemala: A major stratigraphic marker in middle America and in three ocean basins. *Quaternary Research* 13:327–45.

Federman, A. N., and S. N. Carey. 1980. Electron microprobe correlation of tephra layers from eastern Mediterranean abyssal sediments and the Island of Santorini. *Quaternary Research* 13:160–71.

Fernald, H. T. 1962. Radiocarbon dates relating to a widespread volcanic ash deposit, eastern Alaska. *U.S. Geological Survey Professional Paper 450-B*, pp. B29–30.

Fisher, R. V. 1961. Proposed classification of volcaniclastic sediments and rocks. *Bulletin of the Geological Society of America* 72:1407–14.

Fleischer, R. L. 1979. Where do nuclear tracks lead? *American Scientist* 67:194–203.

Fleischer, R. L., and R. M. Walker. 1975. *Nuclear tracks in solids*. Berkeley: Univ. of California Press.

Friedman, I., and R. L. Smith. 1960. A new dating method using obsidian: Part I, The development of the method. *American Antiquity* 25:476–522.

Friedman, I., and W. Long. 1976. Hydration rate of obsidian. *Science* 191:347–52.

Fryxell, R., T. Bielicki, R. D. Daugherty, C. E. Gustafson, H. T. Irwin, and B. C. Keel. 1968. A human skeleton from sediments of Mid-Pinedale Age in southeastern Washington. *American Antiquity* 33:511–14.

Fulton, R. J. 1971. *Radiocarbon geochronology of southern British Columbia*. Canada Geological Survey Professional Paper 71–37.

Goldich, S. S. 1938. A study in rock weathering. *Journal of Geology* 46:17–58.

Guenther, E. W. 1968. Untersuchungen zur jungeiszeitlichen und nacheiszeitlichen geologischen und paläontologischen Geschichte. In *Das Mexiko-projekt der Deutschen forsungsgemeinschaft*. Vol. 1, *Berichte über begonnene und geplante arbeiten*, 33–37. Wiesbaden: Franz Steiner Verlag.

Guenther, E. W., H. Bunde, and G. Nobis. 1973. *Das Mexiko-projekt der Deutschen Forschungmeinschaft*. vol. 6, *Geologische und palaeontologische untersuchungen in Valsequillo bei Puebla (Mexiko)*. Wiesbaden: Franz Steiner Verlag.

Hahn, G. A., W. I. Rose, Jr., and T. Meyers. 1979. Geochemical correlation of genetically related rhyolitic ash-flow and air-fall ashes, central and western Guatemala and the equatorial Pacific. In *Ash flow tuffs*, ed. W. Elston and C. Chapin, 101–12. Geological Society of America Special Paper no. 180.

Hart, W. J. E. 1983. Classic to Postclassic tephra layers exposed in archaeological sites: Eastern Zapotitán Valley. In *Archaeology and Volcanism in Central America: The Zapotitán Valley of El Salvador*, ed. P. D. Sheets, 44–51. Austin, TX: Univ. of Texas Press.

Hart, W. J. E., and V. Steen-McIntyre. 1983. Tierra blanca joven tephra from the A.D. 260 eruption of Ilopango Caldera. In *Archaeology and Volcanism in Central America: The Zapotitán Valley of El Salvador*, ed. P. D. Sheets, 14–34. Austin, TX: Univ. of Texas Press.

Hay, R. L. 1959. Formation of the crystal-rich glowing avalanche deposits of St. Vincent, B.W.I. *Journal of Geology* 67:540–62.

———. 1976. *Geology of the Olduvai Gorge*. Berkeley: Univ. of California Press.

Hay, R. L., and B. F. Jones. 1972. Weathering of basaltic tephra on the Island of Hawaii. *Bulletin of the Geological Society of America* 83:317–32.

Hodder, A. P. W., and A. T. Wilson. 1976. Identification and correlation of thinly bedded tephra; the Tirau and Mairoa ashes. *New Zealand Journal of Geology and Geophysics* 19:663–82.

Hogg, A. G., and J. D. McCraw. 1983. Lake Quaternary tephras of Coromandel Peninsula, North Island, New Zealand: A mixed peralkaline and calcalkaline tephra sequence. *New Zealand Journal of Geology and Geophysics* 26:163–87.

Howorth, R., and P. C. Rankin. 1975. Multi-element characterization of glass shards from stratigraphically correlated rhyolitic tephra units. *Chemical Geology* 15:239–50.

Ikawa-Smith, F. 1978a. Chronological framework for the study of the Paleolithic in Japan. *Asian Perspectives* 19:61–90.

———. 1978b. Lithic assemblages from the early and middle upper Pleistocene formations in Japan. In *Early man in America from a circum-Pacific perspective*, ed. A. L. Bryan, 42–53. Edmonton, Alberta: Archaeological Researches International.

Irwin-Williams, C. 1967. Associations of early man with horse, camel and mastodon at Hueyatlaco, Valsequillo (Puebla, Mexico). In *Pleistocene extinctions: The search for a cause*, ed. P.S. Martin and H. H. Wright, Jr., 337–47. New Haven: Yale Univ. Press.

———. 1969. Comments on the association of archaeological materials and extinct fauna in the Valsequillo region, Puebla, Mexico. *American Antiquity* 34:82–83.

———. 1978. Summary of archaeological evidence from the Valsequillo region, Puebla, Mexico. In *Cultural continuity in Mesoamerica*, ed. D. L. Browman, 7–22. Chicago: Aldine.

———. 1981. Commentary on geologic evidence for age of deposits at Hueyatlaco archaeological site, Valsequillo, Mexico. *Quaternary Research* 16:258.

Izett, G. A. 1981. *Stratigraphic succession, isotopic ages, partial chemical analyses, and sources of certain silicic volcanic ash beds (4.0 to 0.1 m.y.) of the western United States*. Open-File Report 81-763, U.S. Geological Survey.

Johanson, D. C., and M. A. Edey. 1981. *Lucy: The beginnings of humankind*. New York: Simon and Schuster.

Keller, J., W. B. F. Ryan, D. Ninkovich, and R. Altherr. 1978. Explosive volcanic activity in the Mediterranean over the past 200,000 years as recorded in deep-sea sediments. *Bulletin of the Geological Society of America* 89:591–604.

Kittleman, L. R. 1973. Mineralogy, correlation, and grain-size distribution of Mazama tephra and other Post-glacial pyroclastic layers, Pacific Northwest. *Bulletin of the Geological Society of America* 84:2957–80.

————. 1979a. Geologic methods in studies of Quaternary tephra. In *Volcanic activity and human ecology*, ed. D. K. Grayson and P. D. Sheets, 49–82. New York: Academic Press.

————. 1979b. Tephra. *Scientific American* 241:160–77.

Koch, A. J., and H. McLean. 1975. Pleistocene tephra and ashflow deposits in the volcanic highlands of Guatemala. *Bulletin of the Geological Society of America* 86:529–41.

Kohn, B. P. 1970. Identification of New Zealand tephra-layers by emission spectrographic analysis of their titanomagnetites. *Lithos* 3:361–68.

————. 1979. Identification and significance of a late Pleistocene tephra in Canterbury District, South Island, New Zealand. *Quaternary Research* 11:78–92.

Kohn, B., and W. W. Topping. 1978. Time-space relationships between late Quaternary rhyolite and andesitic volcanism in the southern Taupo volcanic zone, New Zealand. *Bulletin of the Geological Society of America* 89:1265–71.

Krumbein, W. C., and F. J. Pettijohn. 1938. *Manual of sedimentary petrography*. New York: Appleton-Century-Crofts.

Kurtén, B. 1967. Präriewolf und Säbelzahntiger aus dem Pleistozön des Valsequillo, Mexiko. *Quartär* 18:173–78.

Lambert, P. W. 1979. Descripcióne preliminar de los estratos de tefra de Tlapacoya I. In *35,000 Años de historia de la Cuenca de Mexico*. Departamento de Prehistoria, Instituto Nacional de Antropología y historia, Mexico.

Lambert, P. W., and S. Valastro, Jr. 1976. Stratigraphy and age of upper Quaternary tephras on the northwest side of Popocatépetl Volcano, Mexico. American Quaternary Association, *Abstracts of the Fourth Biennial Meeting*, 143.

————. (in preparation). Stratigraphy and Age of Upper Quaternary Tephras on the Northwest side of Popocatépetl Volcano, Mexico. Departamento de Prehistoria, Instituto Nacional de Antropología y historia, Mexico.

Larsen, G., and S. Thorarinsson. 1978. H_4 and other acid Hekla tephra layers. Reprinted from *Jökull* 27:28–46.

Laury, R. L., and C. C. Albritton, Jr. 1975. Geology of Middle Stone Age archaeological sites in the main Ethiopian Rift Valley. *Bulletin of the Geological Society of America* 86:999–1011.

Lemke, R. W., M. R. Mudge, R. E. Wilcox, and H. A. Powers. 1975. Geologic setting of the Glacier Peak and Mazama ash-bed markers in West-Central Montana. *U.S. Geological Survey Bulletin* 1395-H:H1–H31.

Lerbekmo, J. F., J. A. Westgate, D. G. W. Smith, and G. H. Denton. 1975. New data on the character and history of the White River volcanic eruption, Alaska. In *Quaternary Studies*, ed. R. P. Suggate and M. M. Cresswell, 203–09. Wellington: Royal Society of New Zealand.

Liddicoat, J. C., R. S. Coe, P. W. Lambert, and S. Valastro, Jr. 1979. Paleomagnetic record in Late Pleistocene and Holocene dry lake deposits at Tlapacoya, Mexico. *Geophysical Journal of the Royal Astronomical Society* 59:367–78.

Liddicoat, J. C., R. S. Coe, P. W. Lambert, H. E. Malde, and V. Steen-McIntyre. 1981. Paleomagnetic investigation of Quaternary sediment at Tlapacoya, Mexico, and at Valsequillo, Puebla, Mexico. *Geofísica International* (Mexico) 20:249–62.

Lirer, L., T. Pescatore, B. Booth, and G. P. L. Walker. 1973. Two Plinian pumice-fall deposits from Somma-Vesuvius, Italy. *Bulletin of the Geological Society of America* 84:759–72.

Lorenzo, J. L. 1967. Sobre método arqueologico. México Instituto Nacional de Antropología y historia, bol 28.

McCraw, J. D. 1975. Quaternary airfall deposits of New Zealand. In *Quaternary Studies*, ed. R. P. Suggate and M. M. Cresswell, 35–44. Wellington: Bulletin of the Royal Society of New Zealand, no. 13.

Machida, H. 1976. Stratigraphy and chronology of late Quaternary marker-tephras in Japan. *Geographical Reports of Tokyo Metropolitan Univ. no. 11*, pp. 109–132.

Machida, H., and F. Arai. 1978. Akahoya ash—a Holocene widespread tephra erupted from the Kikai Caldera, southern Kyoshu, Japan. *Quaternary Research Japan* 17(3).

Malde, H. E., and V. Steen-McIntyre. 1981. Reply to comments by C. Irwin-Williams: Archaeologic site, Valsequillo, Mexico. *Quaternary Research* 16:418–21.

Medlin, J. H., N. H. Suhr, and J. B. Bodkin. 1969. Atomic absorption analysis of silicates using $LiBO_2$ fusion. *Atomic Absorption Newsletter* 8:25–29.

Mehringer, P. J., Jr., E. Blinman, and K. L. Peterson. 1977. Pollen influx and volcanic ash. *Science* 198:257–61.

Moody, U. L. 1978. Microstratigraphy, paleoecology and tephrochronology of the Lind Coulee site, central Washington. Ph.D. diss., Washington State Univ., Pullman, 237 pp. *Dissertation Abstracts*. Ann Arbor, MI: University Microfilms International, 7820112 (catalog no.)

Mullineaux, D. R. 1974. *Pumice and other pyroclastic deposits in Mount Rainier National Park, Washington*. U.S. Geological Survey Bulletin, no. 1326.

Mullineaux, D. R., J. H. Hyde, and M. Rubin. 1975. Widespread late Glacial and Postglacial tephra deposits from Mount St. Helens volcano, Washington. *Journal of Research, U.S. Geological Survey* 3:329–35.

Naeser, C. W., G. A. Izett, and J. D. Obradovich. 1980. *Fission track and K-Ar ages of natural glasses*. U.S. Geological Survey Bulletin, no. 1489.

Ninkovich, D., N. J. Shackleton, A. A. Abdel-Monem, J. D. Obradovich, and G. Trett. 1978. K-Ar age of the late Pleistocene eruption of Toba, north Sumatra. *Nature* 276:574–77.

Olson, G. 1983. An evaluation of soil properties and potentials in different volcanic deposits. In *Archaeology and Volcanism in Central America: The Zapotitán Valley of El Salvador*, ed. P. D. Sheets, 52–56. Austin, TX: Univ. of Texas Press.

Osborn, H. F. 1905. Recent vertebrate paleontology: Fossil mammals of Mexico. *Science* n.s. 21:931–32.

Pettijohn, F. J. 1957. *Sedimentary rocks*. 2d ed. New York: Harper and Brothers.

Porter, S. C. 1978. Glacier Peak tephra in the North Cascade Range, Washington: Stratigraphy, distribution and relationship to Late-Glacial events. *Quaternary Research* 10:30–41.

Reynolds, R. L. 1975. Paleomagnetism of the Yellowstone tuffs and their associated airfall ashes. Ph.D. thesis, Univ. of Colorado, Boulder.

Reynolds, R. L., and E. E. Larson. 1972. Paleomagnetism of Pearlette-like airfall ash in the western United States. *Geological Society of America, Abstracts with Programs* 4:405.

Richardson, D., and D. Ninkovich. 1976. Use of K_2O, Rb, Zr and Y versus SiO_2 in volcanic ash layers of the eastern Mediterranean to trace their source. *Bulletin of the Geological Society of America* 87:110–16.

Riezebos, P. A. 1978. Petrographic aspects of a sequence of Quaternary volcanic ashes from the Laguna de Fuquene area, Colombia, and their stratigraphic significance. *Quaternary Research* 10:401–24.

Roedder, E. 1972. *Composition of fluid inclusions*. U.S. Geological Survey Professional Paper 440–JJ.

Roedder, E., and R. L. Smith. 1965. Liquid water in pumice vesicles, a crude but useful dating method. Geological Society of America Special Paper 82. *Abstracts for 1964*, p. 164.

Rose, W. I., Jr., N. K. Grant, and J. Easter. 1979. Geochemistry of the Los Chocoyos Ash, Quezaltenango Valley, Guatemala. In *Ash flow tuffs*, ed. W. Elston and C. Chapin, 87–100. Geological Society of America Special Paper 180.

Ross, C. S., and R. L. Smith. 1955. Provenience of pyroclastic materials. *Bulletin of the Geological Society of America* 66:427–34.

Self, S., and R. S. J. Sparks (eds). 1981. *Tephra Studies, NATO Advanced Study Institutes* ser. C, vol. 75. Dordrecht, Holland: D. Reidel.

Sheets, P. D. 1976. *Ilopango volcano and the Maya Protoclassic*. University Museum: Research Records, no. 9. Carbondale: Southern Illinois University.

——— (ed.). 1983. *Archaeology and volcanism in Central America: The Zapotitán Valley of El Salvador*. Austin, TX: Univ. of Texas Press.

Sheridan, M. F., and R. G. Updike. 1975. Sugarloaf Mountain tephra: A Pleistocene rhyolitic deposit of base-surge origin in northern Arizona. *Bulletin of the Geological Society of America*. 86:571–81.

Smith, D. G. W., and J. A. Westgate. 1969. An electron probe technique for characterizing pyroclastic deposits. *Earth and Planetary Science Letters* 5:313–19.

Smith, D. G. W., J. A. Westgate, and M. C. Tomlinson. 1969. Characterization of pyroclastic units—a stratigraphic application of the microprobe. *Proceedings of the Fourth National Conference on Electron Microprobe Analysis*, paper 34.

Smith, D. R., and W. P. Leeman. 1982. Mineralogy and phase chemistry of Mount St. Helens tephra sets W and Y as keys to their identification. *Quaternary Research* 17:211–27.

Smith, H. W., R. Okazaki, and C. R. Knowles. 1975. Electron microprobe analysis as a test of the correlation of West Blacktail ash with Mount St. Helens pyroclastic layer T. *Northwest Science* 49:209–15.

———. 1977a. Electron microprobe analysis of glass shards from tephra assigned to Set W, Mount St. Helens, Washington. *Quaternary Research* 7:207–17.

———. 1977b. Electron microprobe data for tephra attributed to Glacier Peak, Washington. *Quaternary Research* 7:197–206.

Soil Survey Staff. 1951. *Soil survey manual*, supplemented in 1962. U.S. Department of Agriculture Handbook no. 18. Washington, DC: Government Printing Office.

Souther, J. G. 1977. Volcanism and tectonic environments in the Canadian Cordillera—a second look. In *Volcanic Regimes of Canada*. Geological Association of Canada. Special Paper 16, pp. 3–24.

Steen-McIntyre, V. 1975. Hydration and superhydration of tephra-glass—a potential tool for estimating the age of Holocene and Pleistocene ash beds. In *Quaternary Studies Bulletin 13*, ed. R. P. Suggate and M. M. Cresswell, 271–78. Wellington: Royal Society of New Zealand.

———. 1977a. Approximate dating of Quaternary tephra deposits using tephra glass hydration and etching of heavy mineral phenocrysts. *Geological Society of America, Abstracts with Programs* 9:1190–91.

———. 1977b. *A manual for tephrochronology : Collection, preparation, petrographic description and approximate dating of tephra (volcanic ash).* Idaho Springs, CO: published privately by the author.

———. 1981a. Approximate dating of tephra. In *Tephra studies*, ed. S. Self and R. S. J. Sparks, 49–64. NATO Advanced Study Institutes, series C, vol. 75. Dordrecht, Holland: D. Reidel.

———. 1981b. Tephrochronology and its application to problems in new-world archaeology. In *Tephra studies*, ed. S. Self and R. S. J. Sparks, 355–72. NATO Advanced Study Institutes, series C, vol. 75. Dordrecht, Holland: D. Reidel.

Steen-McIntyre, V., R. Fryxell, and H. E. Malde. 1981. Geologic evidence for age of deposits at Hueyatlaco archaeological site, Valsequillo, Mexico. *Quaternary Research* 16:1–17.

Szabo, B. J., H. E. Malde, and C. Irwin-Williams. 1969. Dilemma posed by uranium-series dates on archaeologically significant bones from Valsequillo, Puebla, Mexico. *Earth and Planetary Science Letters* 6:237–44.

Theisen, A. A., G. A. Borchardt, M. E. Harward, and R. A. Schmitt. 1968. Neutron activation for distinguishing Cascade Range pyroclastics. *Science* 161: 1009–11.

Thorarinsson, S. 1944. Tefrokronologiska studier på Island. *Geografiska Annaler* 26: 1–217.

———. 1954. The tephra fall from Hekla on March 29th, 1947. In *The eruption of Hekla, 1947–1948*, 1–68. *Nat. Islandica* 11(3).

———. 1974. The terms "tephra" and "tephrochronology." In *World bibliography and index of Quaternary tephrochronology*, ed. J. A. Westgate and C. M. Gold, xvii–xviii. Edmonton, Alberta: Univ. of Alberta.

Thunell, R., A. Federman, R. S. J. Sparks, and D. Williams. 1979. Age, origin, and volcanological significance of the Y-5 ash layer in the Mediterranean. *Quaternary Research* 12:241–53.

Vitaliano, C. J., and D. B. Vitaliano. 1974. Volcanic tephra on Crete. *American Journal of Archaeology* 78:19–24.

Vucetich, C. G., and R. Howorth. 1976. Late Pleistocene tephrostratigraphy in the Taupo District, New Zealand. *New Zealand Journal of Geology and Geophysics* 19: 51–69.

Walker, T. R. 1967. Formation of red beds in modern and ancient deserts. *Bulletin of the Geological Society of America* 78:353–68.

Watkins, N. D., R. S. J. Sparks, H. Sigurdsson, T. C. Huang, A. Federman, S. Carey, and D. Ninkovich. 1978. Volume and extent of the Minoan tephra from Santorini volcano: New evidence from deep-sea sediment cores. *Nature* 271:122–26.

Wedel, W. R., W. M. Husted, and J. H. Moss. 1968. Mummy cave: Prehistoric record from Rocky Mountains of Wyoming. *Science* 160:184–86.

Westgate, J. A. 1977. Identification and significance of late Holocene tephra from Otter Creek, southern British Columbia, and localities in west-central Alberta. *Canadian Journal of Earth Sciences* 14:2593–2600.

———. 1982. Discovery of a large magnitude, late Pleistocene volcanic eruption in Alaska. *Science* 218:789–90.

Westgate, J. A., and C. M. Gold, eds. 1974. *World bibliography and index of Quaternary tephrochronology.* Edmonton, Alberta: Univ. of Alberta.

Westgate, J. A., and R. J. Fulton. 1975. Tephrostratigraphy of Olympia interglacial sediments in south-central British Columbia, Canada. *Canadian Journal of Earth Sciences* 12:489–502.

Westgate, J. A., O. L. Hughes, N. D. Briggs, and V. N. Rampton. 1977. Tephra marker beds of late Cenozoic age in the Yukon territory. *Geological Society of America, Abstracts with Programs* 9:1222.

Westgate, J. A., and M. E. Evans. 1978. Compositional variability of Glacier Peak tephra and its stratigraphic significance. *Canadian Journal of Earth Sciences* 15:1554–67.

Westgate, J. A., and N. D. Briggs. 1980. Dating methods of Pleistocene deposits and their problems: V. Tephrochronology and fission-track dating. *Geoscience Canada* 7: 3–10.

Wilcox, R. E., and G. A. Izett. 1973. Criteria for the use of volcanic ash beds as time-stratigraphic markers. *Geological Society of America, Abstracts with Programs* 5:863.

Wood, S. H. 1977. Distribution, correlation and radiocarbon dating of late Holocene tephra, Mono and Inyo Craters, eastern California. *Bulletin of the Geological Society of America* 88:89–95.

Zier, C. 1983. The Cerén site: A Classic Period Maya residence and agricultural field in the Zapotitán Valley. In *Archaeology and Volcanism in Central America: The Zapotitán Valley of El Salvador*, ed. P. D. Sheets, 119–43. Austin, TX: Univ. of Texas Press.

11

A Successful Technique for the Radiocarbon Dating of Lime Mortar

ROBERT L. FOLK and SALVATORE VALASTRO, JR.

ABSTRACT

The writers have refined the method of radiocarbon dating of lime mortars by taking the following steps: (1) separating inert aggregate from live mortar by careful crushing and screening or by flotation; (2) using only the first fraction of CO_2 gas evolved, which comes only from the live mortar itself; (3) applying $\delta^{13}C$ and dendrochronological corrections. Examples of the success of this technique are described from Yugoslavia, Israel, and France for samples dating from the first century to the sixteenth century, with an accuracy of about 50 years, compared with the archaeological dates. Results are in agreement with radiocarbon dating of conventional charcoal.

Much radiocarbon dating relies upon analysis of samples of charred wood. This technique, however, is subject to a serious drawback—the date obtained from the charcoal indicates when that fragment of wood was still a living tree, not when it was used in the building. For example, let us say that a 150-year-old tree was cut down in A.D. 300 and used in the construction of a building. After the building was destroyed, the beams were reused to build a house in A.D. 600, and that house subsequently burned. Wood from the center of the tree is older than that near the bark; when wood burns, the outside of the log disappears first and the central core—the oldest wood—is more likely to be preserved as charcoal. Thus, the dates obtained on charcoal from this tree that was cut in A.D. 300 could well give a date of A.D. 150–200, when the tree was young, long before either structure was built. Lime-mortar dating, in contrast, gives the date of construction with much better precision.

The idea of dating lime mortar was conceived by G. Delibrias and J. Labeyrie (1965), who obtained excellent results with the technique in France. Dating was also attempted in England but without success (Baxter and Walton, 1970a, 1970b). To make it reliable and practical, we have refined the technique by (1) removing the mortar aggregate and dating only the "live" carbonate material, (2) using only the first fraction of CO_2 gas, which is evolved from the most reactive mortar, and (3) applying $\delta^{13}C$ and dendrochronological corrections (Folk, 1973; Valastro and Folk, 1974; Valastro, 1975; Folk and Valastro, 1976a, 1976b).

Lime mortar and plaster have been used for over 3,000 years (Gourdin and Kingery, 1975) and are made by the following process. Limestone ($CaCO_3$) is crushed and burned over a wood fire at over 1,000°C. The heat calcines the limestone, driving off the CO_2 and converting the limestone to CaO, the white powder known as quicklime (Clark, Bradley, and Azbe, 1940; White, 1939). When the material is ready to be used in construction, a variable amount of inert "aggregate" is added. Aggregate consists of sand or gravel, often obtained from a local river, and is used to add bulk and lessen shrinkage. Sometimes leftover broken bits of brick or marble are thrown in. Water is added to the quicklime-aggregate mixture, which is then used to affix building stones or to plaster walls. As mortar hardens it reabsorbs CO_2 from the atmosphere (White, 1939:367−77), converting back into cryptocrystalline calcite ($CaCO_3$) with a crystal size of 1 μm or less. Thus, on recrystallization during the hardening process, mortar absorbs a representative proportion of ^{14}C from the atmosphere and consequently can be used for dating the time of construction.

To use this technique, one must be sure that the mortar is indeed lime mortar, $CaCO_3$. Some mortar and plaster utilized in building is really plaster of paris (gypsum, $CaSO_4 \cdot 2H_2O$), which is useless for dating as it contains no carbon. Weak acid will cause true lime mortar to effervesce violently, whereas gypsum mortar will not have a visible reaction. True cement, calcium aluminosilicate, also lacks carbon and cannot be used for dating.

Three complicating factors deserve mention. If the limestone is improperly burned (at too low a temperature, or starting out with pieces that are too coarse), not all of it will be converted to CaO, and the residual pieces would yield dead carbon and date too old (Stuiver and Smith, 1965). Second, if the lime mortar was applied far inside a joint between two building stones, atmospheric CO_2 would have had difficulty penetrating and the mortar may not have set until some years after construction. However, this creates far less serious a problem than the inherent counting error of ^{14}C dating. Most mortar has apparently set by the first five or ten years (Baxter and Walton, 1970a:501), except in some very thick castle walls where soft mortar still occurs after 100 years (White, 1939:377).

A far more serious effect, which made attempts to date mortar in England fail, is the aggregate problem (Stuiver and Smith, 1965). If the aggregate used as filler in the mortar consists of quartz, chert, or feldspar sand, there is no problem, since all the carbon will come from the live mortar, and a correct date will be obtained. Luckily this was the situation in mortars successfully dated by Delibrias and Labeyrie (1965). However, if the aggregate contains sand or gravel-sized bits of reworked limestone, which do occur in some rivers draining terranes of carbonate rocks, or if the aggregate is made of leftover scraps of marble or limestone used in construction, then there is a serious problem. Unless special steps are taken, a large amount of dead carbon (that is, carbon so old that none of the radioactive isotope ^{14}C remains) from the aggregate will be present along with the live carbon (containing ^{14}C) from the mortar itself, and a date obtained on this mixture will be uselessly old. Previous workers have not attempted to separate the mortar from the aggregate and have simply crushed the bulk material. This elementary step of separating mortar from aggregate is the primary modification the writers have made to produce a workable technique.

SAMPLE PREPARATION AND PRETREATMENT

A simple technique is used for eliminating the dead carbon contributed by carbonate-rock chips in the aggregate (for more details see Valastro, 1975). The mortar sample is rinsed to remove dust and then is dried at approximately 100°C. For accurate dating, a piece of mortar about 2 kg in weight should be collected.

The dried mortar lumps are gently broken with a rubber pestle to approximately 1 cm dimensions in order to separate the binding mortar powder (which is soft and friable) from the aggregate. Care should be taken not to crush sand grains or pebbles used as aggregate but to make as clean a separation as possible between aggregate and live, white mortar powder by rubbing the sample with a soft rubber pestle. Aggregate grains are usually much harder than the mortar itself, so that separation is not a difficult problem but one that takes a little care and time. Next, the fine mortar powder is separated from the coarser aggregate by sieving through a series of 10-, 30-, 100-, 200-, and 230-mesh U.S. Standard sieves (respectively 2.0, 0.59, 0.149, 0.074, and 0.0625 mm). Any size screen can be used if, on inspection with a binocular microscope, it is evident that aggregate sand grains are not going down into the finer fractions. At any rate, the material that passes the screens should be inspected for presence of carbonate sand grains. This procedure is continued with freshly broken mortar lumps until approximately 300 g or more of powder passing through the 230-mesh screen (0.0625 mm) is accumulated.

A less laborious method of obtaining 300 g of mortar powder involves placing approximately 400 g of broken mortar fragments into a 2-liter Erlenmeyer flask. To the broken mortar is added enough water to make a supernatant liquid. This mixture is continuously agitated until a thick suspension of mortar powder is formed. The suspension is then passed through a 230-mesh screen and collected in a 3-liter beaker. The shaking is repeated in more water until very little suspension is formed and it is estimated that 300 g of the fine mortar power has been accumulated. After allowing the suspension to settle, the supernatant liquid is decanted from the beaker and the mortar sample is dried in an oven at approximately 100°C. The cohesive cake is weighed, then stored and kept dry in an aluminum-foil container until the sample is ready for chemical preparation.

CHEMICAL PREPARATION

It is desirable to have three full-sized gas samples for ^{14}C determination—that is, to have an amount of mortar powder that will generate three 6-liter volumes of CO_2 gas, which will eventually be converted into three 3-ml samples of benzene (C_6H_6) for counting. A 2-g portion of each piece of mortar is chemically analyzed to determine the amount of $CaCO_3$ present, and from this amount one can calculate the weight of mortar required to produce the three sufficient gas samples.

The ^{14}C analysis begins by weighing out the proper amount of mortar powder, which is put into a 3-liter round-bottomed flask with enough water added to cover the solids. The chemical train leading from the flask is evacuated of all the air present by means of a mechanical pump.

From a separatory funnel, altered to suit vacuum techniques and connected to the 3-liter round-bottomed flask by means of a 24−40 vacuum joint, a 1:3 HCl−H_2O acid solution is slowly added in a steady, drop-wise flow while the mortar sample in the flask is continuously agitated. Usually 155 ml of the acid solution is adequate to liberate enough CO_2 to form 3 ml of benzene, which is then utilized in the liquid scintillation-counting procedure.

Despite the effort to completely separate mortar from aggregate, some tiny pieces of dead limestone (from the aggregate or from unburned limestone pieces) still may be present in the sample. If all the CO_2 gas evolved from complete acidification of all the $CaCO_3$ present were collected, one would get contamination from these ^{14}C-depleted limestone fragments. Fortunately, the live mortar is very fine-grained, porous, and powdery, and it reacts very quickly in acid while the dead limestone is much harder, nonporous, and reacts more slowly. By taking the *first* gas evolved by the acidification, one gets CO_2 evolved from the more reactive live mortar while most of the dead carbonate comes off in the later stages of effervescence. Thus, in our technique the total amount of CO_2 evolved from each sample is divided arbitrarily into three portions. The first 6-liter volume of CO_2 to be evolved (designated the "first fraction") is the all-important one because it consists of the effervescence from the live mortar, which represents the actual ^{14}C concentration and gives an accurate date. The second and third 6-liter quantities of gas are more contaminated with dead carbon and in our samples have given spurious dates ranging from 200 to over 2,000 years older than the first fraction.

After preparing the CO_2 gas sample, the normal procedure for liquid scintillation techniques as practiced by the Radiocarbon Laboratory at the University of Texas at Austin is carried out as follows. The CO_2 sample is combined stoichiometrically with lithium metal to form lithium carbide (Li_2C_2), which is in turn hydrolized to form acetylene (C_2H_2). The acetylene is then trimerized, by means of a silicon vanadium catalyst, to benzene (C_6H_6). Benzene, the final product, is counted for residual beta radioactivity over a period of 24–48 hours, and the age of the sample is then calculated. Dates were corrected according to the dendrochronological calibration curve of Damon and colleagues (Damon, Long, and Gray, 1966; Damon, Long, and Wallich, 1972). The samples were also tentatively corrected for ^{13}C fractionation (Damon, Long, and Wallich, 1972) using an atmospheric CO_2 value of 7 per mil. All ^{14}C dates here were referenced to A.D. 1950 in earlier publications, but in order to bring them into congruence with the actual calendar we have added 23 or 27 years to each date, depending on the year in which the sample was analyzed (1973 or 1977).

RESULTS FROM STOBI

We have dated a series of samples from Stobi, a Hellenistic-Byzantine provincial capital in Yugoslavian Macedonia (fig. 11.1). Two large samples were collected in 1973 from the theater by E. M. Davis, one (Tx-1941) from the foundation of the analemma in the east parodos and the other (Tx-1942) from the south wall of the first radial corridor of the cavea, next to the west parodos. Archaeological evidence based on imported pottery points to an initial construction phase of the theater in the late first and second centuries A.D. (E. Gebhard, pers. com., 1974), but it is not yet clear whether one building episode is represented, with modification of the plan as the work proceeded, or whether there was more than one separate stage of construction.

We split each of the two mortar samples from the theater into three parts and prepared and dated each part separately. The three dates from Tx-1941 give an average date of A.D. 283 ± 40, and those from Tx-1942 give an average date of A.D. 273 ± 50. (Dated from reference year 1950, the dates would be A.D. 260 and 250, but since the analysis was done in 1973, we have moved the date up 23 years. All dates herein have been treated similarly.) These dates are in excellent statistical agreement with each other and represent contemporaneity in radiocarbon-dating terms. Since the archaeological evidence also is for contemporaneity, all six dates from the two samples can be averaged, giving a mean radiometric date of A.D. 278 ± 32 (corrected for $\delta^{13}C$ and dendrochronology).

Two further samples from the theater were dated in 1977. One sample (Tx-2488) from the porch of the scene building was radiometrically dated at A.D. 125 ± 80 and archaeologically dated at A.D. 110–130. Another sample (Tx-2489) was taken from an inner radial wall and dated A.D. 198 ± 69. On the average, the radiometric dates for the theater are about 50–100 years younger than the archaeological dates.

From the Episcopal basilica, sample Tx-1943 was obtained by Dr. Davis in 1973 from the socle foundation of the south wall of the principal construction stage. Architectural style, documentary evidence, and pottery dates this stage of construction close to A.D. 400. As with the samples from the theater, Tx-1943 was split into three parts that were prepared and counted independently of one

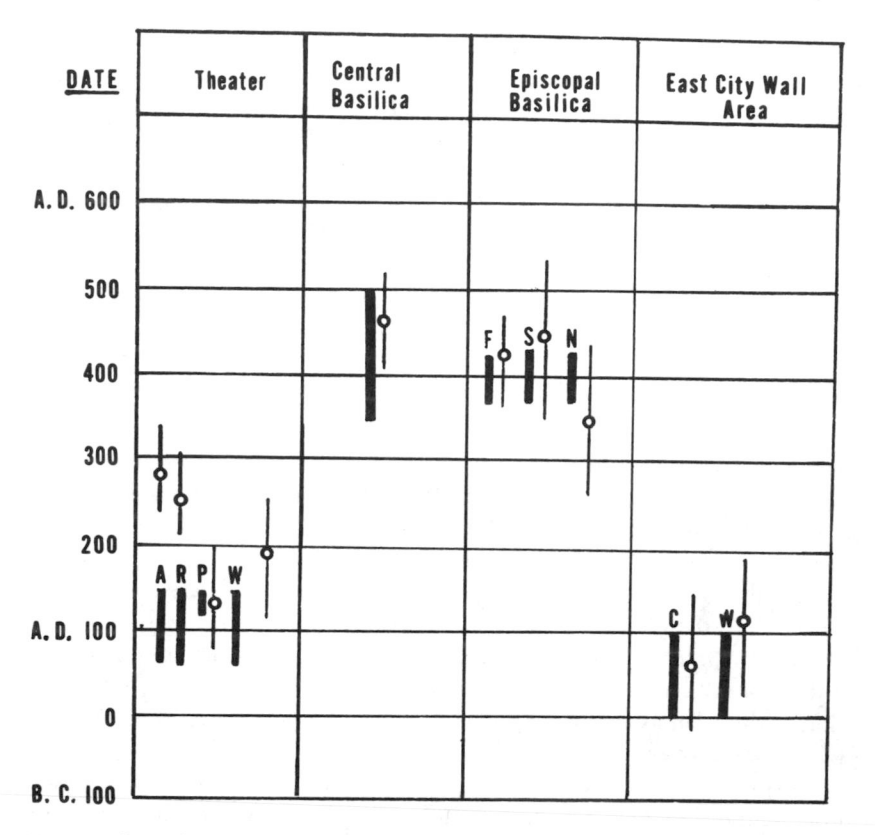

Fig. 11.1.   Chart showing comparison of estimates of archaeological age with radiocarbon dates of samples from Stobi, in Yugoslavian Macedonia. The archaeological estimate (half-life = 5730) is shown by the solid bar. The corresponding radiocarbon date is shown by the circle and the line immediately to the right of each archaeological date (the line gives standard deviation). Radiocarbon dates are corrected for dendrochronology and $\delta^{13}C$, with 23 or 27 years added to bring them into congruence with the actual calendar, depending upon whether the sample was run in 1973 or 1977. *Theater*: A, analemma; R, radial corridor; P, porch of scene building; W, radial wall. *Episcopal Basilica*; F, socle foundation; S, stylobate; N, north wall foundation. *East City Wall*, W, wall proper; C, Casa Romana.

another. The resulting average date is A.D. 302 ± 60. Charcoal dates on this same structure give ^{14}C dates of A.D. 340 and 290, agreeing with the mortar dates and testifying to the validity of the mortar-dating method (fig. 11.1). Correcting these dates for $\delta^{13}C$ gives an age of A.D. 400 ± 60.

Two later samples from the structure are dated as follows: southern stylobate foundation (Tx-2490), A.D. 370−535; foundation of northern wall (Tx-2490), A.D. 265−440. The average of all our radiocarbon dates thus agrees exactly with the archaeological age of about A.D. 400. One sample from the central basilica

(Tx-1944) was dated to A.D. 400–500, while its archaeological date is between A.D. 350–500.

A sample (Tx-2492) from the city wall, with an archaeological estimate of the very early Christian era, gave a radiocarbon date of 15 B.C. to A.D. 140. Another sample from part of the city wall that was incorporated into the Casa Romana (Tx-2494) dated radiometrically as A.D. 25–195. Both these confirm our earlier radiometric date on the Casa Romana of 50 B.C. to A.D. 70 (Tx-1940), which at that time was three centuries older than the supposed archaeological date (Folk and Valastro, 1976a:36). The conflict has now been resolved in favor of the radio-metric date.

RESULTS FROM OTHER AREAS

Tel Yin'am is a small archaeological site near Yavne'el, about 8 km south of Tiberias in the Galilee region of Israel, currently under excavation by Dr. Harold Liebowitz, Department of Oriental and African Languages, University of Texas (Liebowitz and Folk, 1980). In 1977 excavation revealed a very well preserved mikveh (Jewish ritual bath), consisting of steps leading down into a rectangular basin, all beautifully plastered. The plaster contained basalt and limestone fragments as aggregate. Pottery found on the mikveh floor (and one piece incorporated into the plaster itself) indicated a Roman-Byzantine date, according to Liebowitz. Two masses of plaster were dated at A.D. 505 ± 65 (Tx-3018) and A.D. 529 ± 65 (Tx-3019).

Another series of samples has been dated from a medieval church at Saint-Benigne near Dijon, France. Table 11.1 shows the results. The date of consecration of each particular section of the church is known from written records (see Malone, Valastro, and Varela, 1980). The radiocarbon dates were originally counted from 1950, but since the analysis was made in 1979, the dates have all been made later by 29 years (e.g., an original date of A.D. 1027 becomes A.D. 1056, the date in the table below).

Note that the radiocarbon dates appear to be about a half-century too young; nevertheless, the radiometric evidence has confirmed the several stages of construction of the building.

Table 11.1. Radiocarbon Dates on Mortar from a Church at Saint-Benigne, France

Archaeological Date	Radiocarbon Date
A.D. 535	554 ± 72
A.D. 800–900	929 (mean of two samples)
A.D. 1016	1055–1111 (range of 7 samples)
A.D. 1147	1186 (mean of 3 samples)
A.D. 1400–1600	1553 ± 53

CONCLUSIONS

Radiocarbon dating of mortar is a relatively new technique, full development of which has been retarded by problems of contamination by dead carbon. In this chapter we have shown that if care is taken to remove aggregate containing dead carbon and to use only the first evolution of CO_2 gas from the live mortar, promising results are obtained. It is much easier to collect chunks of mortar from an ancient building—which give the correct date of actual construction—than to search for bits of charcoal that give some vaguely relevant date indicating when the charcoal was part of a living tree.

The close clustering of the dates from split samples (and, in the case of the theater in Stobi, from multiple samples) is indicative of the precision of radiocarbon dating of mortar. Mortar dates also agree with charcoal dates in the one structure where a cross-check was obtained. Furthermore, the mortar dates fall close to the archaeological dates in the structures we have examined, even though these latter dates are subject to some fluctuation as new cultural discoveries are made. In the further development of the method many more samples from well-dated structures will, of course, be required. The patterns of date clusters may then be used to calibrate the radiocarbon dates against the known dates. At that time one may expect the radiocarbon dating of mortar to provide reliable dates for mortar construction in a building of unknown date.

If absolute dates through mortar dating should prove especially difficult to obtain, the technique may provide relative ages for differentiating building episodes not distinguishable through other lines of evidence. Our experience with the two mortar samples from the theater at Stobi, in which contemporaneity is attested to both by archaeological evidence and by radiocarbon dating, suggests this likelihood. Mortar dating may also add confirmation to stylistic dating. Thus, dating of mortar and plaster offers much in the way of detailed dissection of the anatomy of construction of ancient buildings. Successive stages of alterations in a building under construction are not always datable on stylistic or contextual grounds alone, and the application of mortar dating may well provide a valuable line of evidence for discerning such episodes.

Acknowledgments

We gratefully acknowledge Dr. James Wiseman, director of the Stobi project, for providing funds to visit Stobi, Yugoslavia, and Dr. E. Mott Davis for correction and consultation on the archaeological dates. L. S. Land (University of Texas) performed the $\delta^{13}C$ analyses, and Alejandra Varela assisted in the ^{14}C determinations. Connie Warren did the drafting, for which the University of Texas Research Institute provided funds.

REFERENCES

Baxter, M. S., and A. Walton. 1970a. Glasgow University radiocarbon measurements, III. *Radiocarbon* 12:496–502.

————. 1970b. Radiocarbon dating of mortars. *Nature* 225:937−38.

Clark, G. L., W. F. Bradley, and V. J. Azbe. 1940. Problems in lime burning: A new X-ray approach. *Industrial Engineering Chemistry* 32:972−76.

Damon, P., A. Long, and E. I. Wallich. 1972. Dendrochronologic calibration of the carbon-14 time scale. *Proceedings of the 8th International Conference on Radiocarbon Dating*, A28−43. Wellington, New Zealand.

Damon, P., A. Long, and D.C. Gray. 1966. Revised ^{14}C dates for the reign of Pharaoh Sesostris III. *Journal of Geophysical Research* 71:1055−63.

Delibrias, G., and J. Labeyrie. 1965. The dating of mortars by the carbon-14 method. *Proceedings of the 6th International Conference on Radiocarbon and Tritium Dating*, 344−47. Washington, DC.

Folk, R. L. 1973. Geologic contributions to archeology and dating techniques: Stobi, Yugoslavia. *Geological Society of America Annual Meeting, Abstracts*, 624.

Folk, R. L., and S. Valastro, Jr. 1976a. Radiocarbon dating of mortars at Stobi. In *Studies in the antiquities of Stobi*, vol. 2, ed. J. Wiseman, 29−41. Beograd: Naucno Delo.

————. 1976b. Successful dating of lime mortar by carbon-14. *Journal of Field Archeology* 3:203−08.

Gourdin, W. A., and W. D. Kingery. 1975. The beginnings of pyrotechnology: Neolithic and Egyptian lime plaster. *Journal of Field Archaeology* 2: 133−50.

Liebowitz, H., and R. L. Folk. 1980. Archaeological geology of Tel Yin'am, Galilee, Israel. *Journal of Field Archaeology* 7:23−42.

Malone, C., S. Valastro, Jr., and A. G. Varela. 1980. Carbon-14 chronology of mortar from excavations in the medieval cathedral of Saint-Benigne, Dijon, France. *Journal of Field Archaeology* 7:330−43.

Stuiver, M., and C. S. Smith. 1965. Radiocarbon dating of ancient mortar and plaster. In *Proceedings of the 6th International Conference on Radiocarbon and Tritium Dating*, 338−43. Washington, D.C.

Valastro, S. 1975. A new technique for the radiocarbon dating of mortar. M.A. thesis, Univ. of Texas.

Valastro, S., and R. L. Folk. 1974. New radiocarbon technique for dating of mortars. *Geological Society of America Annual Meeting, Abstracts*, 993−94.

White, A. H. 1939. *Engineering materials*. New York: McGraw-Hill.

12

New Approaches to Mineral Analysis of Ancient Ceramics

DIANA C. KAMILLI and ARTHUR STEINBERG

ABSTRACT

Many methods can be used to analyze the materials and firing histories of ancient ceramics; the approach presented here includes use of transmitted and reflected, unpolarized and polarized light, scanning electron microscope, electron microprobe, element density scanning, and photomicrograph analysis.

These methods were applied to a polished thin section and chip from a single 'Ubaid-style potsherd from Ur. The results showed the paste to contain quartz, calcic plagioclase, augite, chert, micas, magnetite, ilmenite, and chromite in a matrix of felsic remnants mixed with melt-produced mullite, hedenbergite, fayalite, and calcic plagioclase. The paint consists of relict refractory grains of zoned chromite and melt-produced pseudobrookite, hedenbergite, and fayalite. These assemblages suggest a firing temperature in excess of 1,000°C and reducing firing conditions. Similar data sets for several hundred other samples from the Middle East have helped to clarify patterns in ceramic provenance, transport, and development of ancient firing technology.

The important fact here, however, is that no one method could have produced these data. It is strongly suggested that several independent methods be used to analyze ancient ceramics; also that the analyst describe these methods clearly, present primary data, and note limitations of the approach. This would greatly help communication among geologists, archaeometrists, and archaeologists.

Ceramics, usually in the form of potsherds, are plentiful remnants of ancient societies, and archaeologists have spent much time investigating the evolution of ceramic forms, designs, and styles. Some researchers have taken a different route, however, and have studied the composition of the ceramics and the methods by which they were made, not only to establish provenance, but also to

313

determine the technology involved in manufacture. Some examples of papers by outstanding contributors to this field are Matson (1971), Noll, Holm, and Born (1975), Peacock (1970), Shepard (1954), Tite (1969), Tite and Maniatis (1975a, 1975b).

Researchers entering the field in the last ten years have often used a single analytical technique most familiar or available to them. The resulting proliferation of highly specialized methods has led to some problems in communication, not only between analysts and archaeologists but also among analysts. In addition, the analyst does not always consider which method will best answer the cultural problem being addressed. In the long run, the best approach is probably to consider the archaeological problem, list the methods that will produce the required data, then choose two or more overlapping techniques on the basis of familiarity and availability of equipment.

The purpose of this paper is to present a selection of methods for ceramic analysis. The approach will be that of a petrographer and mineralogist who has examined thin sections of ceramics from the Near East and the Americas. The methods that are suggested do not exhaust the possibilities, but the combination has proved to be effective.

Several parts of a potsherd can be studied to determine what the ceramic is made of and how it was fired. The paste consists of a coarse mineral fraction, a fine-grained matrix, and perhaps some plant material or chaff. On the paste body there may be paint, slip, or glaze. This study will take a single sample (UR 29-46-64) that has some but not all of these features and will demonstrate the information that can be obtained using only a polished thin section and a small chip of the material. It must be remembered that such information, when multiplied by that from several hundred other samples from an area, allows for overlapping data assemblages. These data assemblages, coordinated with local geologic maps, may suggest ceramic provenances, routes of trade and communication, and other cultural patterns, as well as indicate the technology used in manufacture.

Much of the following discussion arises from a project done at the Massachusetts Institute of Technology and the Peabody Museum at Harvard. This project involved determining the provenance of over 500 Samarran-, 'Ubaid-, and Halaf-style sherds from southern, central, and northern Mesopotamia and the technology used in their manufacture. These materials range in age from about 5500 B.C. to 3500 B.C. Further information from this larger study may be obtained by referring to other papers (Oates et al., 1977; Kamilli and Lamberg-Karlovsky, 1979; Steinberg and Kamilli, in press; Kamilli, Steinberg, and Wright, in preparation) and abstracts (Kamilli, 1976, 1977, 1978; Kamilli and Oates, 1977, 1981; Kamilli and Steinberg, 1979). Several additional papers will be published in the future.

'Ubaid ware was originally restricted to parts of southern and central Mesopotamia (fig. 12.16), but later spread to all of central Mesopotamia and throughout the north. Pieces have been found in Saudi Arabia (Oates et al., 1977) and at Tepe Yahya, in the Iranian Zagros (Kamilli and Lambert-Karlovsky, 1979). The time range represented by changing 'Ubaid designs and pot shapes has been divided into the 'Ubaid-1, -2, -3, and -4 periods. Considerable confusion exists as these designations have been used in turn to classify the designs (e.g., an 'Ubaid-2 design) and the cultural events that are supposed to have happened within these time periods.

The painted designs on 'Ubaid sherds are usually geometric patterns, and these patterns appear throughout Mesopotamia. Figure 12.1 shows an example of a whole 'Ubaid-style bowl from the latest period ('Ubaid-4) at Ur. This black, geometric design on a buff, well-fired body is common. Some have suggested a wide-ranging ceramic trade (or at least transport) distributing goods from a few manufacturing sites; others have suggested local manufacture taking place at many sites with the help of itinerant potters. This is the kind of problem that petrographic analysis of a ceramic assemblage in conjunction with knowledge of local geology can help to solve.

The color of the paint on 'Ubaid wares may be red, brown, or black. The paste body may be red, buff, white, or green. A distinctive feature of southern Mesopotamian 'Ubaid pottery is that some of it was fired at unusually high temperatures (above 900°C) and under reducing conditions that frequently fused paste and paint and warped the paint body. This is surprising considering the early date and supposedly primitive furnaces used. The 'Ubaid period, during which urban centers were emerging and some trade routes were being estab-

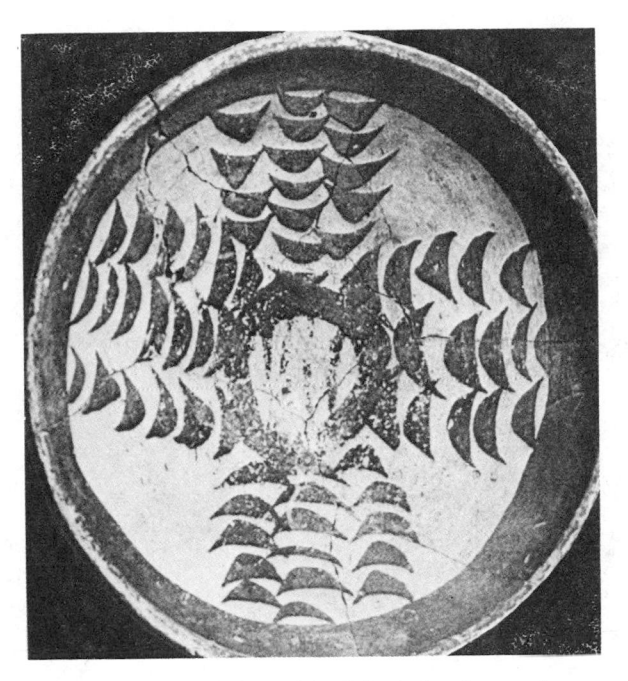

Fig. 12.1. A bowl from the last phase (4) of the 'Ubaid period, excavated from the site of Ur.

lished, just preceded the advent of high-temperature metallurgy in the Near East. The development of 'Ubaid technology and dispersal should be considered in this light.

The 'Ubaid-style sample analyzed in this study (fig. 12.2), number 29-46-64 from Ur, was borrowed from the teaching set stored at the museum of the University of Pennsylvania. It has bubbled, greenish-black paint resembling slag on a greenish-gray, very hard body and is clearly an example of one of the overfired wasters just mentioned.

Sample preparation consisted of having a single, polished thin section made from the sherd. This section was cut to standard (0.03 mm) thickness. Special care was used in slicing the sherd, then grinding down and polishing the slice, in order to preserve the extremely thin paint layer (fig. 12.3). Such a problem is less apt to occur with a well-fired sample because a recrystallized transition zone usually bonds the paint to the paste (fig. 12.10). Preparation of a polished, uncovered thin section allows one to use standard transmitted, unpolarized and polarized light microscopy, reflected light microscopy (on the opaque grains), and electron-microprobe analysis to determine the elemental composition of specific mineral grains of interest. It is especially helpful to be able to do both transmitted and reflected light optical tests on a single mineral grain in the paint layer and then to obtain a chemical analysis of the same grain using the microprobe.

Fig. 12.2.   Two overfired, painted potsherds from Ur. Sample 29-46-64 is on the right.

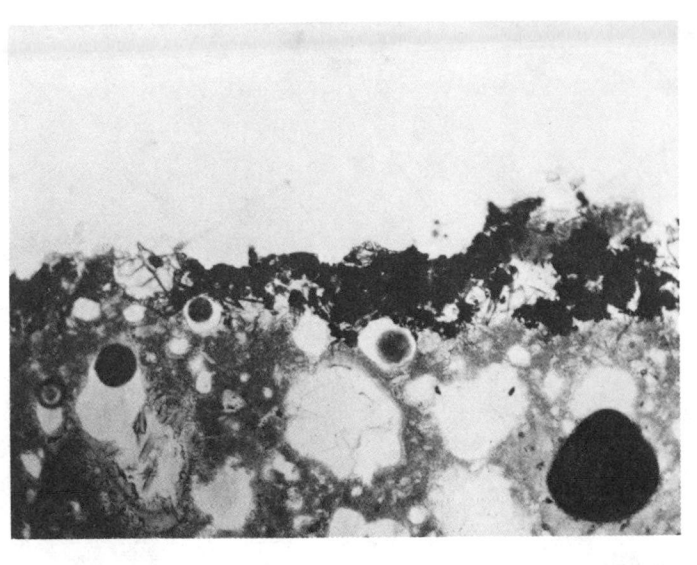

Fig. 12.3. Photomicrograph of a thin section showing the narrow paint layer on the paste body of sample 29-46-64 (25× magnification, uncrossed polars, substage condenser lens in).

PASTE ANALYSIS

The first step in the analysis was to identify the coarser mineral and rock fragments in the paste (fig. 12.4). Transmitted light indicated the presence of twinned plagioclase, clinopyroxene, spherulitic chert, quartz, and minor biotite. Use of the 5-axis universal stage suggested that the plagioclase was calcic and that the clinopyroxene was augite (Emmons, 1943). Finally, the electron microprobe was used to determine the exact composition of both the calcic plagioclase and the augite in order to compare their compositions among sherds from different sites. A modal analysis (Chayes, 1956) of 500 points showed the coarse fraction to be about 15 percent and the matrix to be about 85 percent of the paste. The angularity of the coarse grains noted in sample 29-46-64 is typical of the local Tigris and Euphrates sands and silts. No fragments of sherd grog or plant chaff exist in this sample.

The paste was next illuminated with incident light (by positioning the light source at an angle to the side of the microscope stage) to scan for metallic oxides. Only fine traces of magnetite and a few hematite grains were present along with rare grains that under reflected light were identified as ilmenite and chromite. These last minerals were verified by electron-microprobe analysis. The presence of ilmenite and chromite is unusual and served as additional evidence for comparison of samples among sites. Also, the oxidation state of the iron oxide helps to define the firing atmosphere and thoroughness of firing.

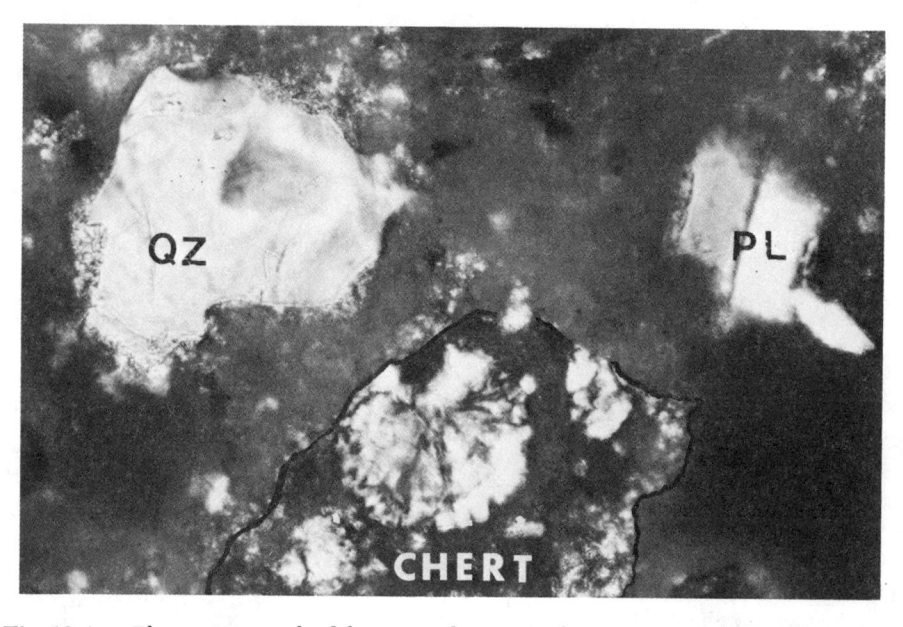

Fig. 12.4. Photomicrograph of the coarse fraction in the paste. Note the twinned calcic plagioclase, quartz, and spherulitic chert, and the angularity of the grains (250× magnification, crossed polars).

The matrix surrounding the coarser grains in this paste has an overall greenish color in unpolarized light and appears to have been mostly melted and recrystallized. The only grains identifiable under transmitted light are a few fine quartz and untwinned feldspar fragments and rare mica flakes. The rest is isotropic and featureless. Gas-bubble holes give it a scoriaceous appearance (fig. 12.5).

Under higher power (500×) and unpolarized light, one can see an intergrown network of fine, greenish crystals, clear crystals, and elongate clear needles, all too fine to be identifiable optically or by microprobe. In order to find out what these crystals were, it was necessary to take an additional small chip from the sherd and turn to scanning electron microscope and X-ray diffraction methods. No energy-dispersive scanning electron microscope (SEM) equipment was available. Figure 12.6 is a SEM photomicrograph of a bubble hole in the paste. It shows the presence of many long crystals (the elongate, clear crystals is transmitted light) that appear to be orthorhombic and might be aluminosilicates, such as sillimanite or mullite. The chip was ground up, X-rayed, and found to contain mullite. The pattern also showed the presence of hedenbergite (as iron-rich pyroxene) and fayalite (an iron-rich olivine), which explained the green crystals and greenish color of the paste. Melting of the original paste would have released elements such as iron, magnesium, calcium, aluminum, and silicon, which then recrystallized to form mullite, hedenbergite, fayalite, and what is probably a

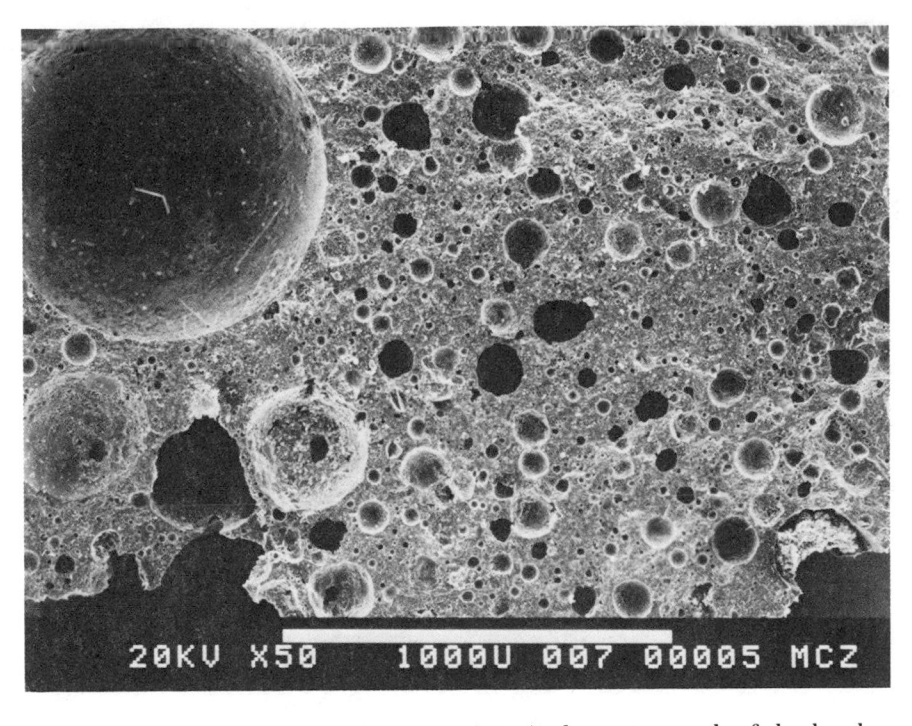

Fig. 12.5. Scanning electron microscope (SEM) photomicrograph of the breakage surface of sample 29-46-64. Note bubble holes (50× magnification).

secondary calcic plagioclase (the clear, short crystals). Mullite ($3Al_2O_3 \cdot 2SiO_2$) formation suggests firing temperatures in excess of 900°C; the hedenbergite ($CaFeSi_2O_6$) and fayalite (Fe_2SiO_4) are formed with the Fe^{2+} or reduced form of iron oxide, suggesting that high firing temperatures were combined with reducing conditions, at least in the high-temperature part of the firing cycle. This method would allow the reduced iron oxide and available calcium carbonate to act as fluxes, which would lower melting and recrystallization temperatures (Tite and Maniatis, 1975b). Broad-beam microprobe scanning did not identify sufficient amounts of alkali elements to have acted as fluxes as it did in certain paints. It is interesting that two generations of clinopyroxene exist in this sample: coarse-grained augite from the original paste source and fine-grained, interlocking hedenbergite recrystallized from the melted matrix.

It is also interesting that most of the coarse mineral and rock fragments in the paste are optically and crystallographically practically unchanged even though this ceramic may have been fired at temperatures above 900°C, perhaps as high as 1,150°C. The biotite has lost some of its pleochroism and some quartz has traces of concentric cracks, but the twinned plagioclase, pyroxene, and even the spherulitic chert appear unaffected.

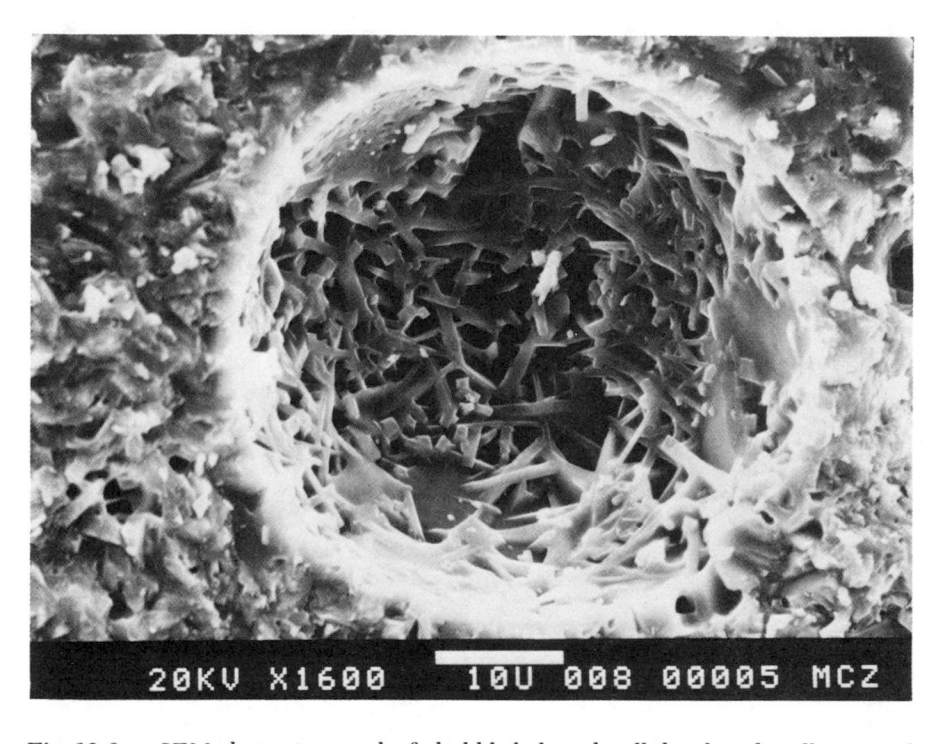

Fig. 12.6. SEM photomicrograph of a bubble hole and well-developed mullite crystals (1,600× magnification).

The finer grains of the matrix were, of course, more susceptible to alteration. Figure 12.7 is a SEM photomicrograph of the chip taken at 6,000× magnification. It shows the glassy breakage, a bubble hold indicating semi−melt conditions, and mullite crystals on the left. For comparison, figure 12.8 shows the flaky breakage surface of a chip from a much lower fired, unpainted sherd from Ur. The sheet-silicate flakes are still quite apparent.

Such SEM photomicrographs have been used (notably by Tite in England and Noll in Germany) to study paint-to-paste interfaces and their relation to composition and firing level. SEM photomicrographs may also be used to record details of plant fragments included in the original paste.

This sample and many others from the larger Mesopotamian study have been analyzed for trace-element content in the paste by McKerrell and Davidson in Edinburgh (see, e.g., Oates et al., 1977). Their analytical data and the writers' were combined to determine which samples had been made locally and which transported.

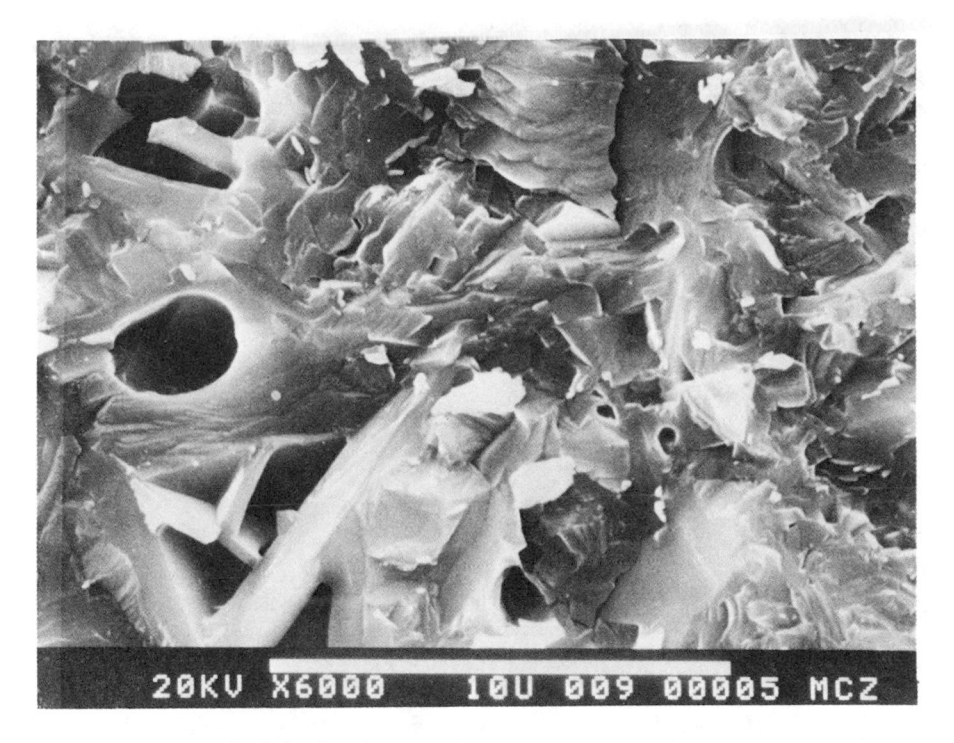

Fig. 12.7. Detail of the breakage surface (SEM photomicrograph taken at 6,000× magnification). Note glassy breakage and mullite crystals and bubble hole to the left.

PAINT ANALYSIS

The final part of the study was the analysis of the paint layer, an aspect badly neglected and practically never carried out in coordination with paste studies.

Paints used on ancient ceramics are either mineral mixtures or organic mixtures (such as those used by many Indian tribes of the southwestern United States). The most common mixture throughout history, however, seems to be mineral, especially a combination of iron oxide, clay minerals, and micas that can produce colors ranging from orange to red, brown, black, and green. This color variation can depend on the species of iron oxide and its amount relative to the silicates, but it also can depend on whether the iron is oxidized or reduced and whether it has melted and combined with the silicates to form green iron silicates. Potters can control the oxidizing and reducing phases in the firing cycle (Shepard, 1954), and mixtures tend to have lower melting temperatures (and recrystallization points) under reducing conditions than under oxidizing condi-

Fig. 12.8. SEM photomicrograph showing the breakage surface of a much lower fired sample from Ur. Note flakes of unmelted clay minerals and micas (5,100× magnification).

tions. In addition, mineral paint mixtures can be fluxed (caused to melt at lower temperatures) by the presence of small amounts of certain elements, such as potassium (as in some Halaf paints), sodium (common in Byzantine glazes), and lead (also popular in Byzantine glazes). FeO can act as a flux at higher temperatures, as can CaO (Shepard, 1954:22–23). Fluxing agents may be present in a paint mixture naturally, as is the potassium in some black, glassy paints from Seh Gabi (Kamilli, unpublished), or may have been added purposefully by the potters, as in certain Halaf paints (Steinberg and Kamilli, in press).

Occasionally, one finds a mineral paint dominated by some other color-producing element. For example, a manganese-rich mixture can produce an especially black paint (Kamilli and Lamberg-Karlovsky, 1979). Unusual mineral assemblages in paint are especially useful in establishing a ceramic sample's provenance. The paint on sample 29-46-64 has a mineral assemblage that includes hedenbergite, fayalite, pseudobrookite, and chromite. There are no common iron-oxide phases in this sample. This assemblage is similar to the paint assemblages on sherds from certain other sites (e.g., Al ʿUbaid, Telloh, Warka site 298), but it appears to be restricted to pots made in southern Mesopotamia. Possibly transported fragments have been found in Saudi Arabia (Oates et al.,

1977) and at Tepe Yahya, Iran (Kamilli and Lamberg-Karlovsky, 1979). The source of this exotic paint remains to be determined.

Figure 12.9 is a photomicrograph of the paint layer taken under transmitted light. Most of the layer apparently melted and recrystallized to an interlocking mass of greenish crystals and long, brown, possibly orthorhombic needles. There are also some round to subangular, nearly opaque grains (possibly relicts from the original paint mixture) that are translucent red under strong transmitted light. These minerals are all too small for reliable optical tests.

Figure 12.10, photographed under reflected light, demonstrates the striking texture of the interlocking crystals. The crystals that appeared green in transmitted light now appear as pale gray rosettes. The long brown needles are here white-reflecting, and the translucent red relicts appear here as subangular white grains. Under higher-power reflecting light, it becomes evident that these latter are zoned, with darker gray in the centers and white on the rims.

After the initial optical examination of this paint, the thin section was recoated with a layer of carbon just sufficient to allow electrical conductivity, enabling a specific grain to be located with a narrow microprobe beam. Probe analysis confirmed the green crystals as hedenbergite and fayalite, the long brown needles as an iron- and titanium-bearing mineral, and the rimmed grains as chromite. It also gave the bulk elemental composition of the chromite.

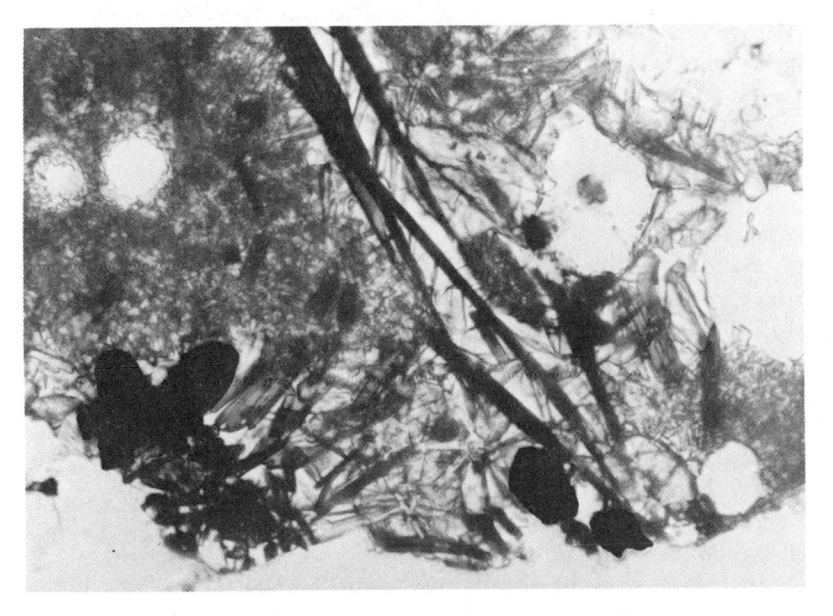

Fig. 12.9. Photomicrograph of the paint layer. Note the long, dark needles of pseudobrookite, the clear crystals of ferromagnesian silicates, and the opaque chromite grains (400× magnification, uncrossed polars, substage condenser lens in).

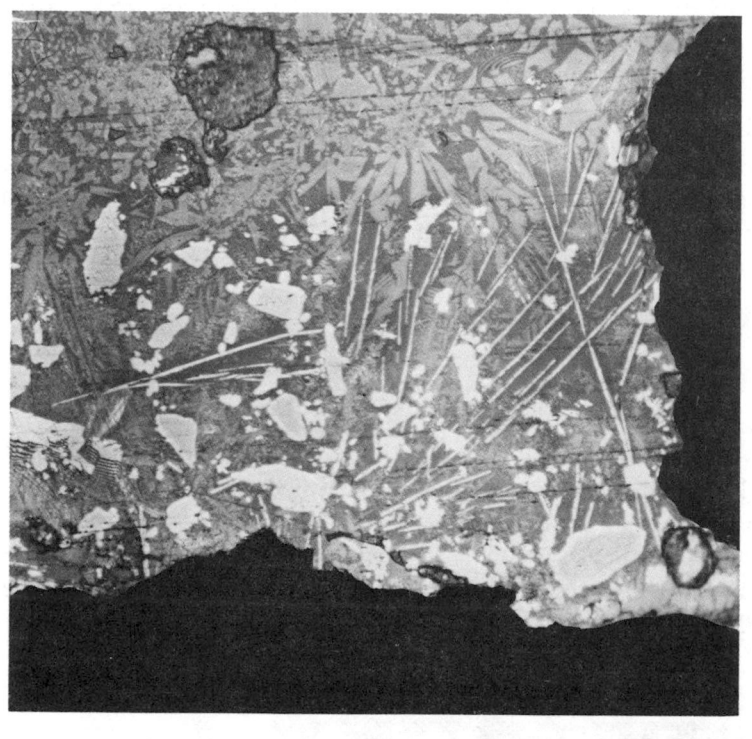

Fig. 12.10.   Photomicrograph of the paint layer. Note the long, white pseudobrookite needles, the gray, stubby ferromagnesian crystals, and the white-rimmed chromite grains; also the wide paint-to-paste transition zone (500× magnification, under reflected light).

The microprobe beam was not narrow enough to discern the compositional differences between the rim and center of the chromite; however, the microprobe element-density scanning method worked quite nicely, as shown in figures 12.11 and 12.12. The former shows the rim of one of the chromite grains to be enriched in iron, and the latter shows it to be depleted in aluminum. Such rims are not well developed in chromite grains in equivalent low-fired paints from Ur and are probably the result of a reaction occurring during firing of the ceramic. Element-density scanning was also used to pinpoint the elements present in an individual brown needle. Figures 12.13 and 12.14 show concentrations of iron and titanium; figure 12.15 indicates absence of silica. Note the skeletal crystal habit demonstrated by the element point distribution.

Note that these element-density scan photographs demonstrate the skeletal form of the crystal. Taken together, however, the results of all these tests point to pseudobrookite. The melted forms of the pseudobrookite and the green silicates, along with the rounding of the relict, refractory chromite, required extremely high firing temperatures—in excess of 900°C—and reducing conditions (Shep-

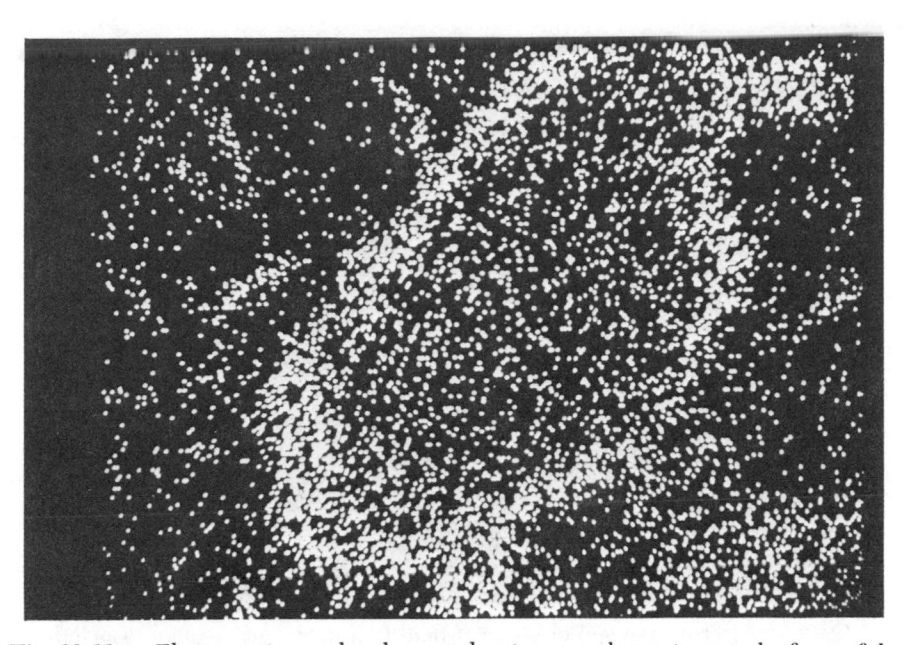

Fig. 12.11. Electron microprobe element-density scan photomicrograph of one of the rimmed chromite grains in the paint showing iron to be concentrated in the rim (or depleted in the center; magnification not recorded).

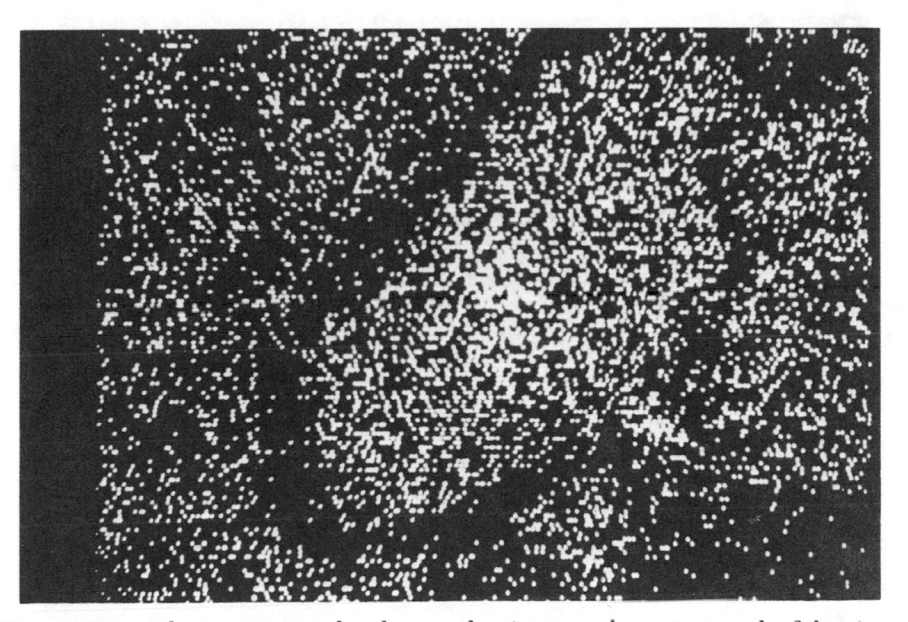

Fig. 12.12. Electron microprobe element-density scan photomicrograph of the same chromite grain showing aluminum to be depleted in the rim (or concentrated in the center; magnification not recorded).

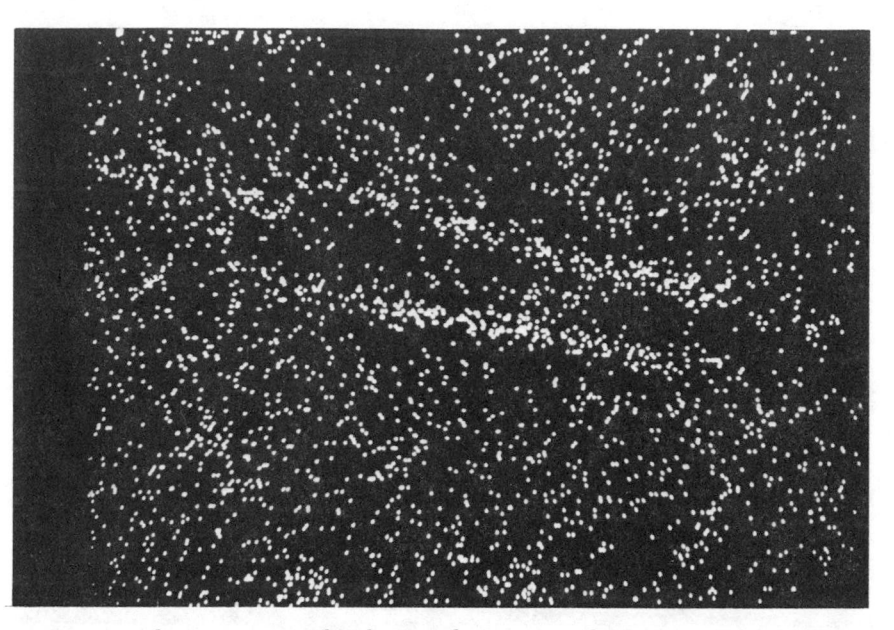

Fig. 12.13. Electron microprobe element-density scan photomicrograph of an individual pseudobrookite needle, showing concentration of iron. Note the skeletal crystal structure (magnification not recorded).

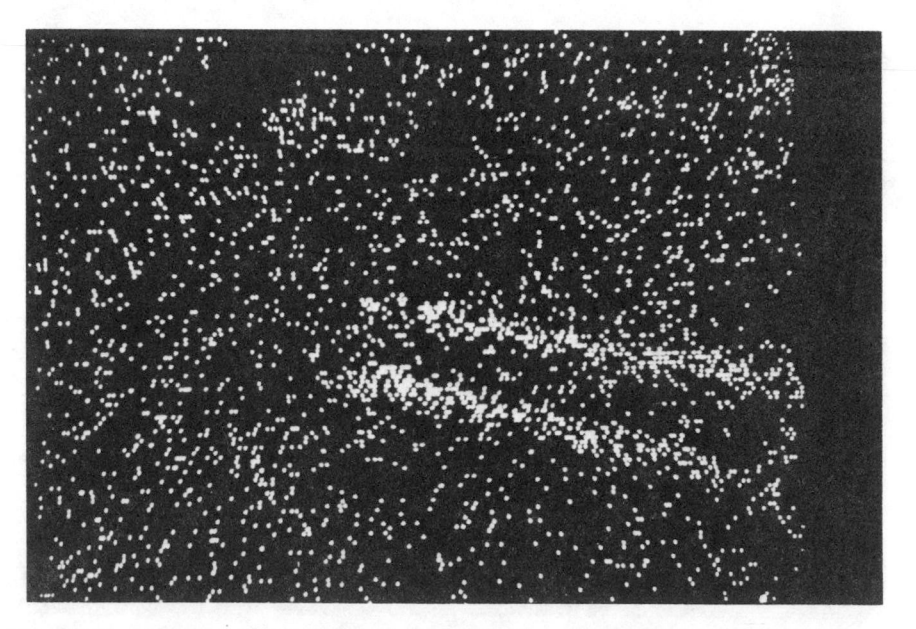

Fig. 12.14. Electron microprobe element-density scan photomicrograph of the same pseudobrookite needle, showing concentration of titanium (magnification not recorded).

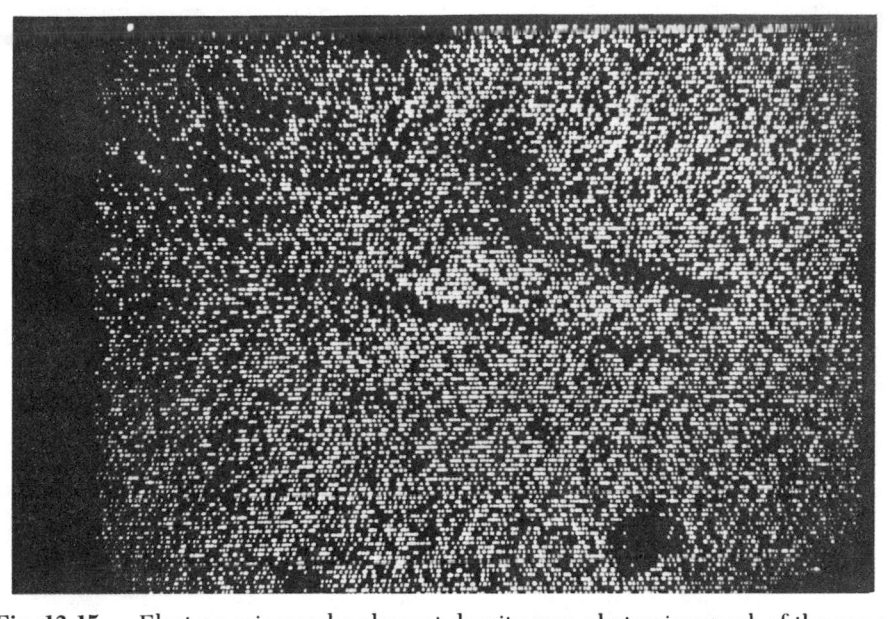

Fig. 12.15.   Electron microprobe element-density scan photomicrograph of the same pseudobrookite needle, showing lack of silicon (magnification not recorded).

ard, 1954:23). Equivalent samples from the same site fired to a much lower temperature have only slightly sintered paint. The low-fired paint is composed of ilmenite, chromite grains, clay minerals, micas, minor quartz, feldspar, and iron oxide. All these minerals (except chromite) would melt under higher temperatures and re-form as the green silicates and pseudobrookite.

The color that results from this unusual melted-paint mixture is a brownish to greenish black, almost identical to that produced by iron oxide–rich paints fired to high temperatures under reducing conditions. Optical or chemical examination is required to tell the difference. Both types of paint are present on sherds found in southern Mesopotamia (see fig. 12.16) and it is not known why one paint was preferred over the other at a given site—it certainly was not for the appearance. Availability is an obvious answer, but no local sources for the necessary minerals have been found.

The paint on sherd 29-46-64 was analyzed by a combination of methods that established mineral assemblage, exact composition of these minerals, textural relationships resulting from firing, and, indirectly, the temperature and atmosphere of the firing. The reflected-light and SEM study also determined the nature of the paint-to-paste interface (fig. 12.9). This zone is commonly one of chemical interaction during firing and can give information concerning firing conditions. How well the paint is attached to the paste affects the appearance of

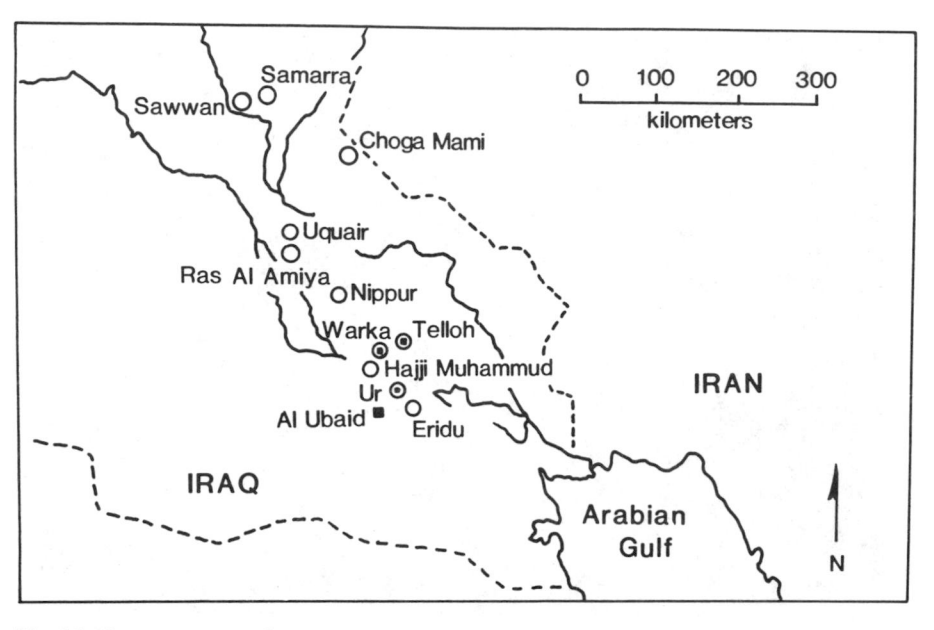

Fig. 12.16.   Paint type by site, southern and central Mesopotamia. O = iron oxide and silicate paint; ■ = chromite, titanium oxide, iron oxide and silicate paint; ◉ = both paint types present. Modern national boundaries are indicated by dashed line.

the ceramic; it is a feature that changes with the technical ability of the potters through time and may differ among ware types.

  None of this information could have been obtained by bulk chemical, X-ray diffraction, optical, or microprobe analysis alone. In addition, chemical and X-ray diffraction methods would have required the scraping of most of the paint from this sherd, resulting in the destruction of the sample.

CONCLUSIONS

Sherd 29-46-64 is admittedly a sample with unusual paint and paste characteristics as well as an unusual firing history; however, it was chosen to demonstrate the usefulness of a certain approach and set of analytical techniques.

  Two sets of data—one involving paste, the other paint—were produced from the polished thin section and chip from this single sample. Over 500 other samples from the Tigris-Euphrates region, and considerably more from the Arabian Gulf Coast and Iran, were analyzed. The results have been applied to several major archaeological questions. For example, the data suggest that most of the Samarran, ʿUbaid, and Halaf ceramics from these periods in southern and central Mesopotamia were made at or near the sites where they were found. This conclusion, combined with certain cultural information, suggests that there were strongly centralized organizations of pottery production in a number of places that apparently had good communication among themselves (probably based on their common religious and business interests). These centralized production organizations were responsible for seeing to it that a uniform style of pottery was manufactured at a number of localities. In addition, many of the 'Ubaid ceramics found at sites along the Arabian Gulf Coast are identical in paste and paint characteristics to those from the southern Mesopotamia area (and to sherd 29-46-64). They may have been brought south, perhaps by boat (Oates et al., 1977). Also, a preliminary study (Kamilli and Lamberg-Karlovsky, 1979) on the ceramics from Tepe Yahya in Iran, a site considerably to the east, has shown that the common red and buff wares were made locally but that certain 'Ubaid-style pieces were transported from southern Mesopotamia. Finally, these analytical data have helped to trace changes in ceramic technology in Mesopotamia from about 5500 to 3500 B.C.

Several independent methods of ceramic analysis should be used if cultural interpretations are to be made. Certainly bulk chemical, heavy-mineral, or binocular microscope analyses are not adequate in themselves. Atomic-absorption and neutron-activation analyses of the powdered pastes and paints give only elemental data without textural information. SEM photomicrographs alone show the grain form and surface textural relationships, but give little data on composition or mineral type. X-ray diffraction analysis can establish mineral species but not texture. Thin-section optical analysis is versatile and inexpensive and can supply textural information as well as mineral-species identification; however, it is most useful if combined with bulk chemical data such as that from spread-beam microprobe analysis.

If only a single method of analysis can be used on a suite of ceramic samples, it is important that the analyst realize its limitations, state them clearly, and suggest other methods that would complement the data. In addition, the use of photographs or photomicrographs in a report help to clarify complex descriptions, especially of petrographic relationships. Above all, the analyst should offer the repeatable data on which conclusions are based and describe the methods by which they were obtained. This would do much to help archaeologists and geologists evaluate published information, communicate with each other, and plan future research.

Acknowledgments

Much of the work on the larger project described in this chapter was done under NSF Grant #SOC75-17438. All electron microprobe analyses were obtained using an automated MAC-5 unit in the Department of Earth and Planetary Sciences at MIT. Special thanks to go Ed Seling at Harvard in the Museum of Comparative Zoology SEM photomicroscopy laboratory (established under NSF Grant #BMS-7412494) for running the equipment and offering advice; also to Harold Thompson in the Harvard thin-sectioning laboratory for taking the special care and time to prepare the polished thin sections for the paint analyses; finally, to Robert Kamilli for his counsel.

REFERENCES

Chayes, F. 1956. *Petrographic modal analysis.* New York: John Wiley and Sons.

Emmons, R. C. 1943. *The Universal stage.* Geological Society of America, Memoir No. 8.

Kamilli, D. 1976. Applications of petrographic analysis to the study of ancient Near Eastern ceramics. *Geological Society of America, Abstracts with Programs* 8 (6):945.

——. 1977. Ceramic technologies of 'Ubaid, Samarran and Halaf wares, Mesopotamia. *International Symposium in Archaeometry and Archaeological Prospection.* Univ. of Pennsylvania, Abstract Volume, 24.

——. 1978. Applications of new approaches to petrography of ancient ceramics. *Geological Society of America, Abstracts with Programs* 10 (7):431.

——. 1981. Implications of petrographic and electron microprobe analyses of samples from the Samarran and 'Ubaid ceramic sequence at the central Mesopotamian site of Choga Mami. *Geological Society of America, Abstracts with Programs* 13:482.

Kamilli, D., and J. Oates. 1977. Petrographic and electron microprobe analysis of 'Ubaid ceramics, and the nature of southern Mesopotamian contact with Saudi Arabia, 6th millenium, B.C. *Geological Society of America, Abstracts with Programs* 9 (7):1043.

Kamilli, D., and C. C. Lamberg-Karlovsky. 1979. Petrographic and electron microprobe analysis of ceramics from Tepe Yahya, Iran. *Archaeometry* 21:47–60.

——. 1981. Implications of petrographic and electron microprobe analysis of samples from the Samarran and 'Ubaid sequence at the central Mesopotamian site of Choga Mami: *Geological Society of America, Abstracts with Programs* 13(7):482.

Kamilli, D., and A. Steinberg. 1979. Petrographic and electron microprobe analyses of paints on Halaf ceramics from northern and central Mesopotamia. *Geological Society of America, Abstracts with Programs* 11:453–54.

Kamilli, D., A. Steinberg, and R. Wright. In preparation. An early pottery industry: Development and distribution of 'Ubaid wares.

Matson, F. 1971. A study of temperatures used in firing ancient Mesopotamian pottery. In *Science and archaeology,* ed. R. Brill, 65–80. Cambridge, MA: MIT Press.

Noll, W., R. Holm, and L. Born. 1975. Painting of ancient ceramics. *Angewandte Chemie, International Edition* 14:602–13.

Oates, J., T. Davidson, D. Kamilli, and H. McKerrell. 1977. Seafaring merchants of Ur? *Antiquity* 51:221–34.

Peacock, D. P. S. 1970. The scientific analysis of ancient ceramics: A review. *World Archaeology* 1:375–89.

Shepard, A. 1954. *Ceramics for the archaeologist.* Carnegie Institution of Washington Publication no. 609.

Steinberg, A., and D. Kamilli. In press. Paint and paste studies of selected Halaf sherds from Mesopotamia. In *Pots and potters: Current approaches in ceramic archaeology,* ed. P. Rice. UCLA Institute of Archaeology.

Tite, M. S. 1969. Determination of the firing temperature of ancient ceramics by measurement of thermal expansion: A reassessment. *Archaeometry* 11:131–43.

Tite, M. S., and Y. Maniatis. 1975a. Examination of ancient pottery using the scanning electron microscope. *Nature* 257:122–23.

——. 1975b. Scanning electron microscopy of fired calcareous clays. *Transactions of the British Ceramic Society* 74:19–22.

13

Isotopic Analysis of Marble

NORMAN HERZ

ABSTRACT

The first serious attempts to assign a provenance to Greek marble artifacts at the end of the last century relied exclusively on petrographic methods. Correct determinations of provenance are clearly important in detecting forgeries, but, since the approximate times of use of many classical quarries are documented, they can also assist in determining trade patterns and dating objects. Petrofabric analysis has been tried with some success, but it requires a large data base on original sources as well as destruction of enough material for a thin section. Trace-element compositions of marble are too highly variable for unique characterizations, but are sometimes useful when combined with other types of analysis. The electron-spin resonance signal intensity of Mn^{+2} recently has provided reliable signatures for some quarries, especially when combined with information on calcite vs. dolomite content.

The stable-isotopic signature technique, using $\delta^{13}C$ to $\delta^{18}O$, has proven the most promising method yet devised both for determining provenance and associating broken fragments of artifacts. Isotopic analysis alone has proven diagnostic in studies of provenance and association carried out in the field and with museum collections. A large data base of Greek, Aegean, and Turkish quarries has been built up, with the result that data comparisons are now possible. Variations from a mean up to several tenths per mil in isotopic values are caused to some extent by weathering and to a great extent by exchange with fluids and country rock during metamorphism. Large variations in isotopic signatures, up to five per mil, from Naxos are attributed to exchange during metamorphism.

Association of fragmentary marble artifacts by isotopic methods has been proved successful, especially for statuary and inscriptions. Stable isotopes have proven less successful in studies of malachite patinas formed on bronze.

Attempts to characterize classical marble according to its source date back to the earliest writings of ancient Greece and Rome. Theophrastus (1956:70–76) described the marbles of Paros, Mount Pendelikon, Chios, and Thebes and added that, except for its lower density, a variety of limestone called *poros* could be confused with Parian marble. Pliny (1962) commented at greater length on the characteristics of the more popular marbles, which by Roman times were extensively exploited throughout Greece and western Asia Minor (fig. 13.1). The Romans systematized the marble trade and developed many modern concepts, including standardization of qualities and dimensions as well as mass production and stockpiling, making it possible to accomplish such feats as building the Pantheon in less than ten years (A.D. 118 to ca. 125–128) and the baths of Diocletian in less than eight (A.D. 298 to ca. 305–306; Ward-Perkins, 1977). Such organization in exploitation, trade, and use of building stones was not seen again in Europe for well over a millennium.

During the nineteenth century, several accounts (including petrographic descriptions) of ancient marble quarries by geologists appeared. In 1837, L. Ross (Lepsius, 1890) described the quarries of Mount Pendelikon, near Athens. In the 1880s, G. R. Lepsius, working closely with archaeologists, collected and studied samples from the principal classical quarries. His petrographic descriptions were based on both hand-specimen and microscopic study. Lepsius (1890) described methods for characterizing marble artifacts and identified the sources of over 400 well-known artifacts in museums in Athens and elsewhere in Greece. Many coarse-grained specimens were assigned to a general "island marble" category, but most, he thought, could be given a definite provenance.

In 1898, the great American petrologist H. S. Washington warned that, contrary to Lepsius's pronouncements, the correct provenance and association of marble fragments was quite difficult. Within each quarry could be found a great range of textures, fabric, and mineralogy, making a blanket petrographic description of each marble type impossible. He cited disputed identifications of famous pieces in European museums, including the Satyr in the Louvre, which had been called Parian marble by one art historian and "possibly Thasian, but certainly not Parian" by another. A marble torso in the Medici collection had been identified as Carrara by one and "certainly not Carrara, clearly Pentelic" by another authority. Washington made an appeal for more detailed study of the original quarries, including extensive petrographic examination of all marble types found in each.

By the mid-twentieth century, these warnings had been completely forgotten, and descriptions of Greek inscriptions, for instance, generally included a statement as to provenance. These statements were based on each author's subjective application of Lepsius's criteria and never included petrographic descriptions, except for general observations on color and, occasionally, granularity. Herz and Pritchett (1953) showed that identical marble inscriptions were called Hymettian by one author and Pentelic by another and that the terminology, including descriptions of color, was not consistent among authors. They

Fig. 13.1.   Index map of Greece, the Aegean Sea, and western Turkey.

suggested that archaeologists should describe in detail the physical characteristics of the marble: its structures, textures, accessory minerals, weathering patina, etc. This, at least, would make it possible to compare separated pieces based on meaningful characteristics, and each epigrapher could attempt the relative association of fragments without worrying about an assignment of provenance.

Other studies, including one on Aegean marble by Renfrew and Peacey (1968), also pointed out the futility of assigning provenance without adequate data. All geological studies, apparently, were summarily rejected. Bernard Ashmole (1970), the distinguished Oxford art historian, concluded that "archaeologists would have to trust to common sense" until, at least, the scientists produced a convincing method to distinguish provenance. That method—stable-isotopic signatures of marble (Craig and Craig, 1972)—is now almost in hand.

MARBLE SOURCES

The importance of assigning a correct provenance to classical marble artifacts, in detecting forgeries, and in associating broken or separated fragments is obvious. However, since the approximate periods of operation and location of the major quarrying localities in Greece and Turkey are well known, the correct assignment of provenance can usually help both in dating the time of fabrication of an artifact and in determining the contemporary trade patterns. Starting in the seventh century B.C., the marble exploited for Greek statuary apparently came from the island of Naxos in the Cyclades (fig. 13.1). This very coarse grained marble fell into disfavor after the discovery of Parian marble, from Paros, just west of Naxos. Parian marble became the preferred statuary marble throughout Greek and Roman times and was also used during the Renaissance (Riederer and Hoefs, 1980). From the early fifth century B.C., the marble of Mount Pendelikon, near Athens, was extensively quarried and substituted for the more expensive, imported Parian marble, especially in the construction of the buildings of the Acropolis. Later in the fifth century the finer-grained, bluish marble of Mount Hymettus, also near Athens, was first exploited, and it remained very popular through Hellenistic and Roman times. Other important classical marble quarries include those at Doliana, used by the Spartans and their allies to construct the Tegea temple; at Aliki, on the island of Thasos in the northern Aegean, which produced a coarse-grained, white marble; at Proconnesus, in the Sea of Marmara; and at Aphrodisias, in southwest Asia Minor, all of which served as major Roman sources (Herz and Wenner, 1981).

Detailed descriptions of quarries and marble types have appeared for Greece (Dworakowska, 1975), the Aegean (Renfrew and Peacey, 1968), and Asia Minor (Monna and Pensabene, 1977). The normative physical characteristics of marbles used for statuary (table 13.1) have so many overlapping values that most cannot be distinguished by visual techniques alone. Compounding the problem is the variety of marbles found in each locality that were not considered of statuary quality yet were quarried for other purposes, such as construction.

ANALYTICAL TECHNIQUES ATTEMPTED

Before about 1954, only petrographic techniques were used to associate marble fragments or to assign a provenance; they involved either hand specimens or thin sections and were based in large part on the criteria listed by Lepsius (1890). Weiss (1954) and Herz (1955) showed that the petrofabrics of some statuary marbles were sufficiently different to distinguish Naxian, Parian, Pentelic, Hymettian, and Delian sources. This petrofabric study involved a plot of the poles to (0112) calcite deformation lamellae. Such a technique unfortunately requires a large, oriented piece of marble to make a thin section, generally greater than any museum curator would allow to be taken. Another problem in assigning provenance is that a detailed knowledge of the variation in fabric within each quarry is lacking. The technique could be applied for association of fragments of broken artifacts, however.

Trace-element analysis has also been undertaken in an attempt to characterize ancient marbles. Rybach and Nissen (1965) analyzed for sodium and manganese using neutron-activation techniques and were able to characterize Marmara and Parian marbles as containing little of the latter element. In this study, 230 samples from Greece, the Aegean, and western Turkey were analyzed and found to contain 0.5−200 parts per million (ppm) of manganese and 2−300 ppm of

Table 13.1. Normative Characteristics of the Principal Classical Marbles

Provenance	Color*	Accessory Minerals†	Average Grain Size (in mm)	Structures and Other Notes
Pendelikon	white, 9.5	qz, wm, gr, or	0.4–0.7	Strong foliation
Hymettus	medium–light gray, 5–6.5	qz, wm, gr, or	0.1–0.5	Well layered by color
Doliana	light gray–bluish white	qz, gr, or	< 0.5, 2–4	Layered, coarse crystals in finer matrix
Agrileza	light gray–white	qz, gr	0.1–0.5	Well layered by color
Apollona	very light gray–white, 8–9	qz, wm, ep, or, gr	1.1–2	Massive, embayed crystals
Paros	white, 9.5	qz, wm, gr	1.0	Massive, translucent
Aliki	white, 9	wm, gr	0.9	Foliated, some dolomite
Proconnesus	very light gray–white, 8.5–9.5	ep, gr	0.9–1.6	Strong mortar texture, recrystallized
Ephesos	white, 9	wm, or, gr	1.0	Layered, strong mortar texture
Aphrodisias	light gray–white, 7–9	qz, wm, or, gr	0.6–2.0	Layered, recrystallized
Dokimeion (Afyon)	medium light gray–white, 5.5–9.5	qz, wm, or, gr	0.6–1.3	Massive, some color banded
Carrara	very light gray–white, 9–9.5	qz, pl, wm, or, gr	0.6–1.1	Layered and massive, recrystallized

*Numbers refer to the Munsell Rock Color Chart.

†qz = quartz, wm = white micas, or = iron ores, gr = graphite, ep = epidote, pl = plagioclase

SOURCE: Data from Lepsius (1890), Renfrew and Peacey (1968), Lazzarini Moschini, and Stievano (1980), Monna and Pensabene (1977), and unpublished material.

sodium. However, the variation, even within the same hand specimen, was found to be very large. Pentelic marble, for example, showed 42–100 ppm of manganese and 21–66 ppm of sodium. A more detailed study, using both emission spectroscopy and X-ray fluorescence spectroscopy, also did not produce promising results (Conforto et al., 1975). Analysis for potassium, strontium, barium, aluminum, iron, manganese, silicon, and titanium in the ppm range led to the conclusion that trace elements alone could not distinguish the different marbles of Carrara, the Aegean, and western Turkey. It is apparent that trace-element analysis cannot characterize classical marbles due to the large variations in elemental concentration, commonly two orders of magnitude for each element within the same quarry. Trace-element studies on materials such as ceramics generally require many samples and a statistical handling of the data to overcome concentration variability in the material. Thus, the chances of characterizing the provenance of a single marble artifact or associating broken pieces by trace-element analysis alone seem remote.

Both natural and artifical thermoluminescence (TL) analyses have been done on marbles and limestones of classical Greece (Afordakos, Alexopoulos, and Miliotis, 1974). Unfortunately, the differences among TL curves of samples from the same quarry were about as great as the differences observed among quarries. Variations in any one marble block were small, suggesting that the method could be useful in restoration of fragments of statuary, as was done in this study. This type of analysis has a strong negative value: if two pieces show different TL glow curves, they could not be associated, but if the curves matched, association was permissible, but far from certain. The method is of no value for provenance studies.

Cordischi, Monna, and Segre (1981) attempted electron-spin resonance (ESR) analysis of marbles from Mediterranean quarries to test its applicability in provenance studies. They analyzed for the manganese ion Mn^{2+}, which is present in both calcite and dolomite phases. About 80 percent of the spectra showed only calcite, the rest, both calcite and dolomite. They were able to distinguish some of the marble types by a two-step analysis: first, the phases were identified as calcite alone, calcite plus traces of dolomite, or calcite plus dolomite; second, the ESR signal intensity was analyzed (fig. 13.2). The authors concluded that some quarries or groups of quarries were distinguishable from each other and that further ESR analysis was warranted. They suggest that ESR analysis together with mineralogical criteria and isotopic signatures might eventually yield definitive provenance assignments.

In 1972, Harmon and Valerie Craig suggested using stable isotopes of light elements as a signature for marble fragments. Of the ten archaeological samples they tested, unambiguous results were obtained for five; the other five samples could not be assigned a provenance. Despite the limited data base of isotopic compositions then available, the Craigs demonstrated the viability of the method both for studies of provenance and association of classical marble artifacts.

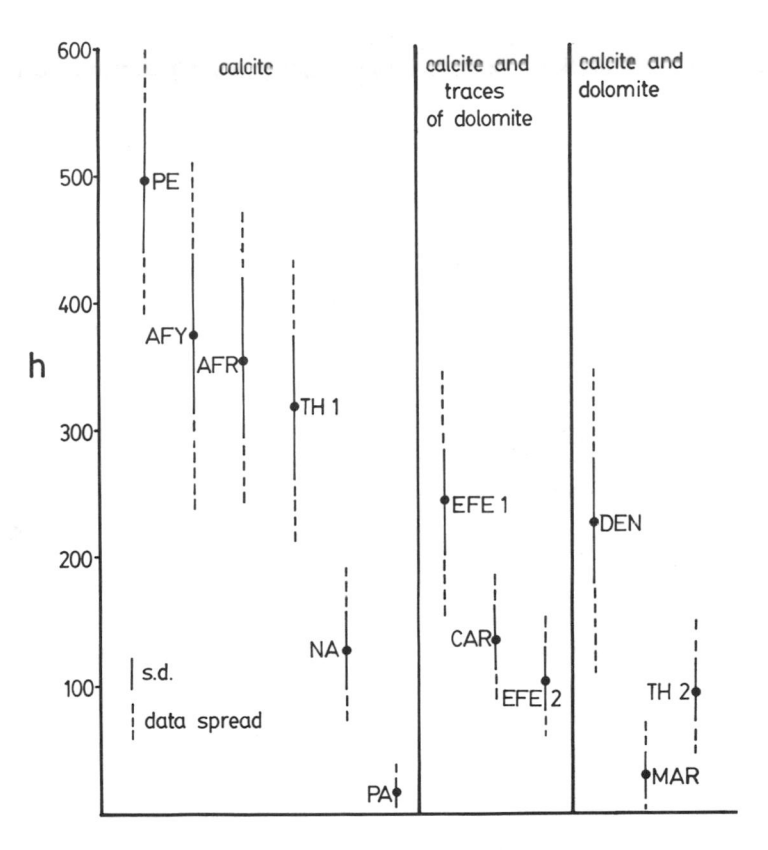

Fig. 13.2. Signal intensity (h) of electron-spin resonance spectra of Mn^{2+} in marble. Solid lines indicate the standard deviation, dashed lines the data spread, and the large dot is the mean. PE = Pendelikon, AFY = Afyon, AFR = Aphrodisias, TH = Thasos, NA = Naxos, PA = Paros, EFE = Ephesus, CAR = Carrara, DEN = Denizli, MAR = Marmara (from Cordischi, Monna, and Segre, 1981).

ISOTOPIC ANALYSIS: METHODOLOGY AND SYSTEMATICS

Isotopic analyses of carbon and oxygen are carried out with a mass spectrometer, an instrument that measures the proportions, in very small samples, of different isotopic masses of several elements. For marble analysis, this involves the precise measurement of the ratios of oxygen and carbon isotopes $^{18}O/^{16}O$ and $^{13}C/^{12}C$ after suitable chemical treatment has separated these elements from the calcium carbonate of the marble. The data are expressed in the form of δ-values, either $\delta^{13}C$ or $\delta^{18}O$. $\delta = (R \text{ sample}/R \text{ standard} - 1) 1000$, where $R = {}^{13}C/{}^{12}C$ or ${}^{18}O/{}^{16}O$. Thus, if a marble sample has a $\delta^{18}O$ of +10, we can say that the isotopic ratio of the oxygen is 10 parts ten thousand (or 10 per mil, expressed as ‰), enriched in the heavy isotope of oxygen ^{18}O as compared with a standard. The standard for this kind of analysis has been arbitrarily chosen as a calcareous marine fossil from South Carolina, the Pee Dee belemnite (PDB). The isotopic-variability data usually are presented as a scatter plot showing $\delta^{18}O$ versus $\delta^{13}C$. Figure 13.3 illustrates how groups of samples from different quarries occupy characteristic "fields" of the diagram.

For any technique to be successful in determining the provenance of marble, the characteristic "signature" must be homogeneous at least over the area of a quarry, and preferably within the limits of a geological district. At present, the known variations in $\delta^{13}C$ and $\delta^{18}O$ within a single marble slab or statue appear to be less than one-half part per mil; within an outcrop or most quarries it is generally $\pm$ 2 ‰. Detailed studies of marble units in Vermont and Ontario (Sheppard and Schwarcz, 1970) and Georgia (Herz and Wenner, 1979) suggest that carbon and oxygen isotopic variations are about 1.5‰ in $\delta^{18}O$ and 1‰ in $\delta^{13}C$ for samples from outcrops on the order of 30 m on a side (fig. 13.4). Variations are smallest along original bedding planes and larger and more irregular across original beds.

The isotopic composition of oxygen and carbon in the carbonate of marbles depends on (1) its mode of origin, as a chemical precipitate or as organic-shell

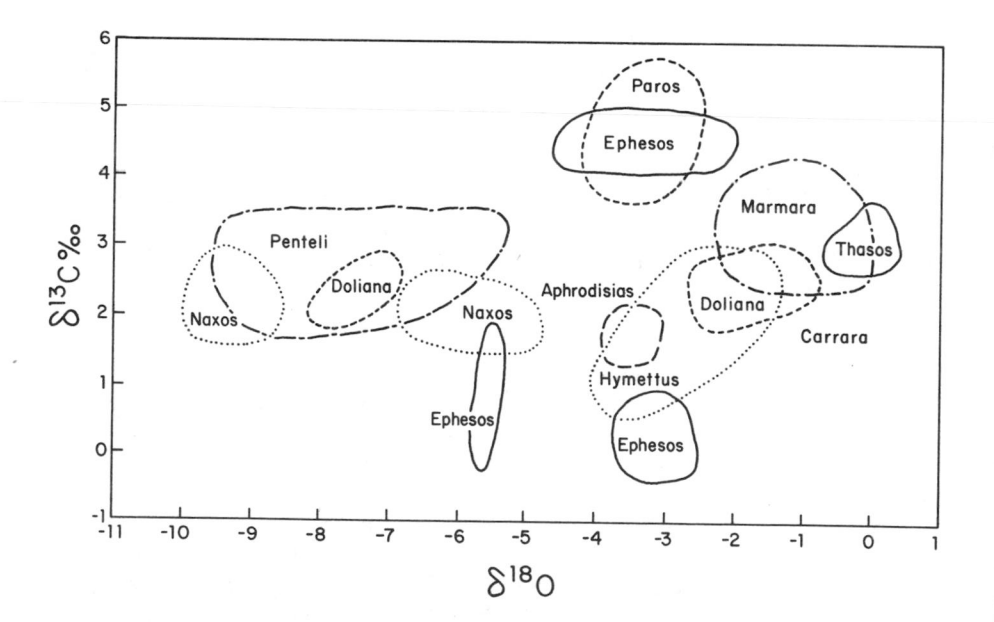

Fig. 13.3. $\delta^{13}C$ vs. $\delta^{18}O$ for the principal classical marble quarries, in per mil, relative to the PDB standard (data from Craig and Craig, 1972; Manfra, Masi, and Turi, 1975; Riederer and Hoefs, 1980; and unpublished material).

remains; (2) the composition of water associated with the limestone during its formation and later history; (3) the temperature of metamorphism of the limestone into marble; and (4) isotopic fractionation with pore waters and other mineral phases during metamorphism (Faure, 1977). Uniform isotopic compositions will be attained over a wide area if the following conditions are met: (1) isotopic equilibrium was attained during formation or metamorphism; (2) the marble unit is relatively pure and thick; and (3) the metamorphic gradient was not too steep. The principal sources of isotopic variations appear to be exchange between the carbonates and other mineral species during metamorphism or a steep metamorphic temperature gradient. Since only the purest white marble, free of accessory minerals, was quarried, exchange with other minerals is generally not a serious problem. In many marbles, even those that are relatively graphite-rich (table 13.1), the amount of graphite or organic matter is less than several tenths of a percent. Studies by Valley and O'Neil (1981) have shown that the effect of metamorphic fractionation of $\delta^{13}C$ in calcite-graphite pairs at

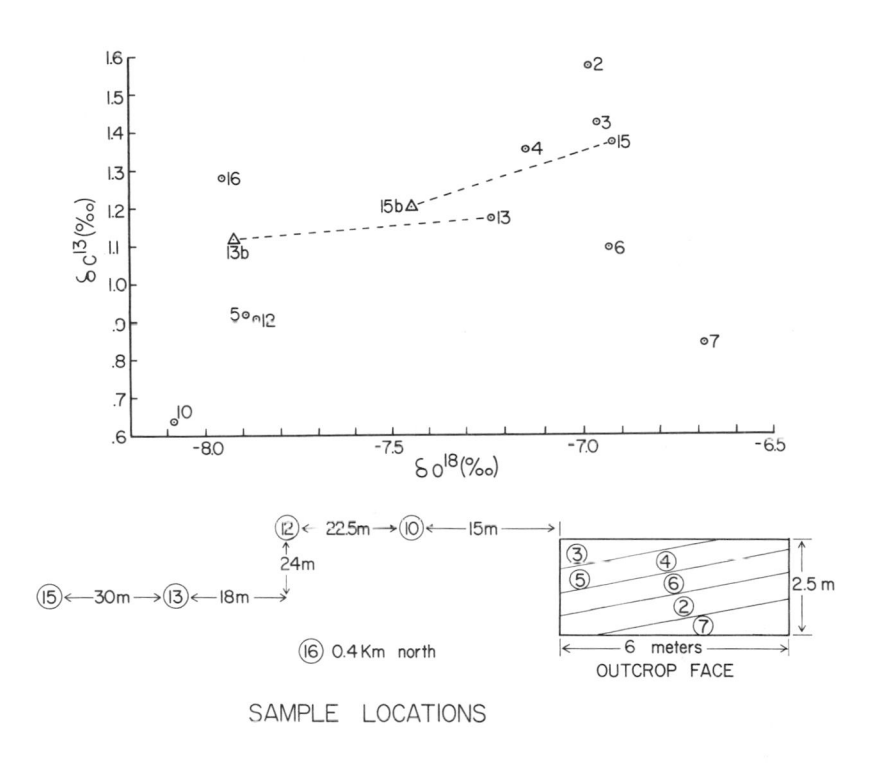

SAMPLE LOCATIONS

Fig. 13.4. $\delta^{13}C$ vs. $\delta^{18}O$ variation in the Tate Marble Quarry, Georgia (upper part); locations for samples taken from an outcrop face (lower right) and over a quarry face (lower left). Δ samples are weathered equivalents of fresh samples. δ-values in per mil, relative to the PDB standard.

temperatures above 500–600°C are consistently small (2.6–4.8‰). The effect on these marbles probably would be to raise the $\delta^{13}C$ of the small amount of graphite by several parts per mil and lower that of calcite by a negligible amount. Interlayered schistose beds or adjacent country rock would foster fractionation on a greater scale than is possible in pure marble, but this is rarely a problem in ancient quarries.

The combined effects of a steep metamorphic gradient, exchange with metamorphic pore fluids, and equilibration with other country rock can introduce very high variations in $\delta^{18}O$, but only moderate to low variations in $\delta^{13}C$. Marbles from the island of Naxos show a greater variation in $\delta^{18}O$ than any other important classical marble due to these factors (fig. 13.3). Estimated metamorphic temperatures experienced by Naxian marbles vary from 400°C in the southeast of the island to 700°C in the central part, around a migmatite complex. The horizontal geothermal gradient reached about 30°C per km in the western part, which is much steeper than normal for the Cycladic Islands massif: other islands, including Ios, Siphnos, Syros, and presumably also Paros, show less than 15°C per km (Jansen and Schuiling, 1976). Stable-isotope analysis showed clearly that $\delta^{18}O$ in marbles varied with metamorphic temperature and distance to the contact of the beds (Rye et al., 1976). For the marble around the village of Kinidaros, within the migmatite complex where temperatures reached 680–700°C, $\delta^{18}O$ values are as high as about −4.7 per mil. Metamorphic temperatures were 640°C around the important quarries at Apollona. Isotopic variations there within a 20-m thick marble bed were clearly influenced by distance to the contact with the neighboring schist (table 13.2) and explain why Naxian marble does not show a uniform $\delta^{18}O$ signature.

Another important mechanism for isotopic variation is weathering. If the broken fragments of an artifact have different weathering histories—for example,

Table 13.2. $\delta^{13}C$ and $\delta^{18}O$ of Marble from Apollona, Northern Naxos

Distance* (in cm)	$\delta^{18}O^{\dagger}$ (in ‰)	$\delta^{13}C^{\dagger}$ (in ‰)	Marble Type
T	−13.4	+1.7	Dolomite
3	−13.5	+1.9	Dolomite
30	−13.2	+2.0	Dolomite
300	−9.5	+2.0	Dolomite
1,000	−11.0	+2.6	Calcite
2,000	−7.2	+2.0	Calcite
1,000	−9.3	+2.7	Calcite
300	−8.3	+2.6	Calcite
30	−8.6	+4.1	Calcite
3	−13.5	+1.1	Calcite
B	−12.3	+1.8	Calcite

*Distances are from top (T) to bottom (B) contacts with schist.
†Metamorphic temperature of 640°C. Values relative to PDB standard.
SOURCE: Values extrapolated from locality 362, fig. 10 in Rye et al., 1976.

one piece buried in soil sediment, another submerged in a well, and a third used as a doorjamb—each might exchange carbon and oxygen isotopes with waters that are of far different composition from the pore waters of sedimentation or metamorphism. Since many museum curators are reluctant to allow sampling of material other than surface scrapings, it is important to assess possible alteration due to weathering. The results of analysis of fresh and weathered samples (table 13.3) show no significant change in $\delta^{13}C$, but a decrease in $\delta^{18}O$ to about 0.6 ‰ in the weathered material (see also fig. 13.4). This is the change to be expected if oxygen in the marble had undergone exchange with oxygen of meteoric waters (Faure, 1977).

The $\delta^{18}O$ and $\delta^{13}C$ signatures for some classical marble quarries overlap (fig. 13.4), suggesting the need for another discriminant. Analysis of strontium isotopic ratios, $^{87}Sr/^{86}Sr$, has recently been tried as an additional discriminant for determining the provenance of marble (Herz, Mose, and Wenner, 1982a). Initial results were promising, with three samples from the Thasos Aliki quarries showing 0.70769 to 0.70792 and single samples from Paros and Pendelikon showing 0.70757 and 0.70830, respectively. If one assumes (1) a uniform $^{87}Sr/^{86}Sr$ ratio for Tethyan seawater at any given time, (2) a changing ratio with time, and (3) similar metamorphic conditions over the area of a quarry site, then strontium-isotopic ratios may provide another reliable signature for testing marble provenance and association.

Isotopic analysis of archaeological marble has had two general goals: first, the accumulation of a primary data base that would include all the principal quarrying areas of classical Greece and Rome, and, second, to assist in the correct association of fragments of artifacts, especially statuary or inscriptions. The method was originally suggested as a way to develop definitive criteria for provenance; its success can be attested by the fact that , with the present primary data base, provenance can now be assigned to many marble artifacts. The following review highlights some of the principal work carried out on the use of stable isotopes to characterize marble provenance and to test the association of artifactual marble fragments.

Table 13.3. Stable-Isotopic Analysis of Fresh and Weathered Marble

Sample Number*	$\delta^{13}C$	$\delta^{18}O$
1 (fresh)	+1.17	−7.24
1 (weathered)	+1.12	−7.93
2 (fresh)	+1.37	−6.93
2 (weathered)	+1.20	−7.45
3 (fresh)	+2.57	−7.83
3 (weathered)	+2.64	−8.16

*Numbers 1 and 2 from the Tate Quarry, Georgia (see fig. 13.4), number 3 from a sarcophagus in the British Museum (Coleman and Walker, 1979).

MARBLE PROVENANCE

The first use of isotopes for marble-provenance studies was that of Craig and Craig (1972). The Craigs pointed out that marbles from various areas outside Greece differed by 10 and 20 per mil in $\delta^{13}C$ and $\delta^{18}O$, respectively, so it was reasonable to expect that significant differences could be obtained by analyzing archaeological marbles. They collected and analyzed a total of 170 samples from ancient quarries on Naxos and Paros and at Pendelikon and Hymettus and found that the marbles fell into well-defined isotopic clusters. The Naxian samples fell into two groups (fig. 13.3), now understandable as an effect of metamorphic equilibration. The high $\delta^{13}C$ in Parian marble compared with other marbles was attributed to its probable origin as a chemical precipitate. The lower $\delta^{13}C$ of other marbles suggests that they are organic (bioclastic) in origin. Lower $\delta^{18}O$ in Naxian and Pentelic marbles compared with Parian and Hymettian marbles may be due to interaction with meteoric waters at elevated metamorphic temperatures. Ten archaeological samples were analyzed and compared with this data base. A column and a basal slab from the Hephaesteion in the Athenian Agora are of Pentelic marble, a block in the Treasury of Siphnos at Delphi is Parian, and the Apollon Gate in Naxos is Naxian. Other samples from Caesarea (Israel), Delphi, and Epidaurus could not be assigned a provenance.

In 1975, Manfra and co-workers extended the method to western Anatolia. They sampled and analyzed marbles from Marmara, Ephesos, Aphrodisias, Denizli, and Afyon (fig. 13.1). Their study included both white or slightly colored marble as well as colored varieties; the classical Greeks used only the former, but the Romans used both. A plot of the entire data set showed a great scatter; however, when only the white or slightly colored marbles were plotted, the Marmara and Aphrodisias samples fell into well-defined groups (fig. 13.3). Samples from Ephesos were widely scattered, some with $\delta^{13}C$ values equal to that of Parian marble—presumably chemical precipitates—and others with values much lower than any seen in Greek marbles. The Denizli marbles had a uniform $\delta^{13}C$ but highly variable $\delta^{18}O$ values. The lack of other geological information made it difficult to interpret the significance of these isotopic data. Manfra and colleagues concluded that the problem of uniquely identifying the provenance of a given marble specimen could not be solved by isotopic methods alone. However, if information from petrographic and trace-element analyses were also available, then the method could be very useful, and eventually a geochemical "identification card" for each quarry of archaeological interest in the Mediterranean area could be developed.

Isotopic signatures were used to verify the provenance of 13 fragments in a collection of Hellenistic and Roman Isis and grave reliefs at the Agora Museum in Athens (Herz, Wenner, and Walters, 1977). Seven could be assigned sources that included Hymettus, Pendelikon, and Paros; the rest, whose signatures did not match any known source, were all very coarse-grained and represented the "island marble" type described by Lepsius. The study showed, first, that a variety of marble sources were used at this time and, second, that some important ancient quarries had not yet been characterized isotopically.

Coleman and Walker (1979) analyzed marble sarcophagi in the British Museum, who undertook this study primarily to determine the provenance of the sarcophagi and also to test the isotopic method. Comparing their analytical results to the data fields of Craig and Craig (1972) for the Greek quarries and Manfra, Masi, and Turi (1975) for the Turkish, they found that one sarcophagus from Crete was made of Pentelic marble, but another did not match any known isotopic data field and could have been made of a local marble. A third sarcophagus from Crete, as well as others from Sidamara, Turkey, were made of Proconnesan marble from Marmara. Restoration was tested on another sarcophagus, described below. Coleman and Walker concluded that since isotopic analysis provided provenance information, it would be invaluable for future studies of the ancient marble trade.

Germann, Holzmann, and Winkler (1980) found in a study of marble quarries and artifacts from Thessaly that isotopic analysis was not a panacea and that additional petrographic and trace-element information was needed. In eastern Thessaly in classical times four principal marble quarries existed: Atrax, Kastrion, Tempi, and Gonnos. These quarries were studied and characterized chemically, isotopically, and petrographically. Isotopically the entire Gonnos field fell within the larger Tempi field; Atrax had a distinctively higher $\delta^{13}C$ than the others; and Kastrion could be distinguished from Gonnos-Tempi by a lower $\delta^{18}O$ (fig. 13.5). However, comparing these data with information on other classical marbles revealed much overlap, as between the Gonnos and Hymettus sources. In one Atragian quarry, they found a variation of 2 ‰ in $\delta^{18}O$ and 1 ‰ in $\delta^{13}C$. Analysis of magnesium, manganese, iron, and strontium showed that Gonnos had less than 500 ppm magnesium, compared with greater than 1,000 ppm for the others, and about 350–550 ppm strontium, compared with 120–250 ppm for the others. Petrographic criteria were the least variable and the most significant for distinguishing the marbles. They concluded that provenance determinations should always include a knowledge of the range of possible sources and the variability within these sources. For all studies, geological fieldwork of the possible sources should be carried out first, followed by petrographic study of each marble type. Then, trace elements and isotopic ratios can be determined. Multivariate statistical analysis of all data could be helpful in bringing out source-typical geological patterns.

Riederer and Hoefs (1980) analyzed isotopically 51 Greek, Roman, and Renaissance copies of marble busts in the collections of the Antiquarium of the Residenz in Munich and could positively assign sources to 36 of them. Samples from Penteli and Naxos could be identified easily by the diagram in figure 13.3. The Paros and Ephesos isotopic fields overlapped, but the marbles themselves could be distinguished texturally, as could the samples from the overlapping Hymettus and Carrara fields. The authors concluded that microscopic study and isotopic analysis were the most promising techniques; trace-element study of the

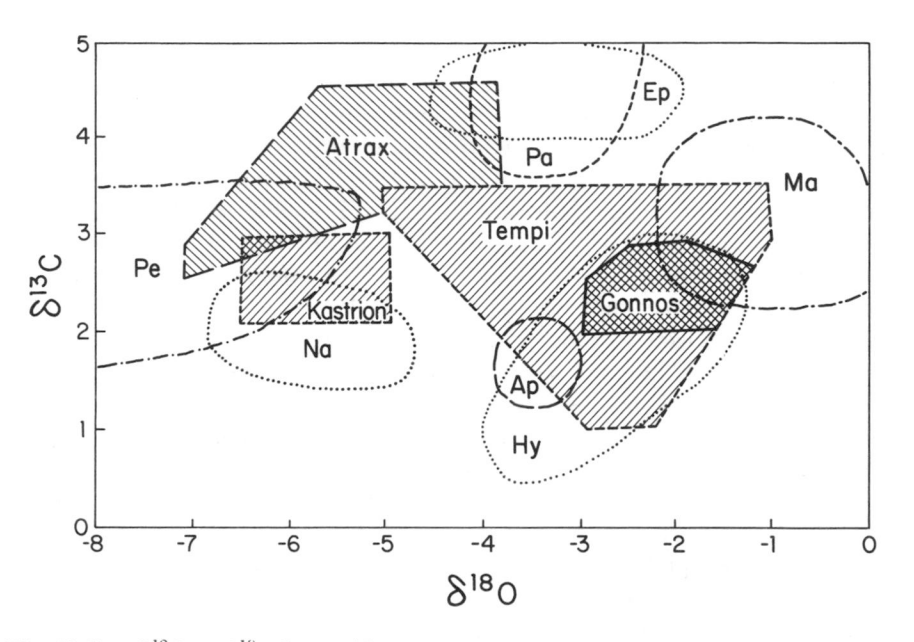

Fig. 13.5. δ^{13}C vs. δ^{18}O for marble quarries of ancient Thessaly, in per mil, relative to the PDB standard (from Germann, Holzmann, and Winkler, 1980).

carbonate and insoluble minerals was not helpful. This study demonstrated that different marble was traded throughout classical and Renaissance times and that during the Italian Renaissance Greek marbles continued to be as popular as in Roman times, but in Germany only Carrara marbles were imported.

The marble used to construct the temple of Apollo at Bassae in the Peloponnesus late in the fifth century B.C. has been attributed by various scholars to quarries at Paros, Pendelikon, and Doliana. However, Cooper (1981), who had been working on the temple site, found that the petrographic features of these alleged sources were quite different from the actual marble used on the site. He discovered a new group of ancient quarries near the tip of the lower Mani, the southernmost peninsula of the Peloponnesus (fig. 13.1). The Mani marbles appear similar to those used at Bassae, with an uneven grain size and thin streaks of dolomite. Their isotopic analysis (fig. 13.6) shows that the Mani quarries could have been the source for the marble used in Bassae; Doliana marble has similar isotopic values but is much coarser-grained than that at Bassae, and the divergent isotopic values of the marble of both Paros and Penteli ruled them out as possible sources (Herz et al., 1982b).

RECONSTRUCTION OF ARTIFACTS

For the reconstruction of marble artifacts and the testing of association of broken or separated fragments, the isotopic-signature technique is by far the best yet

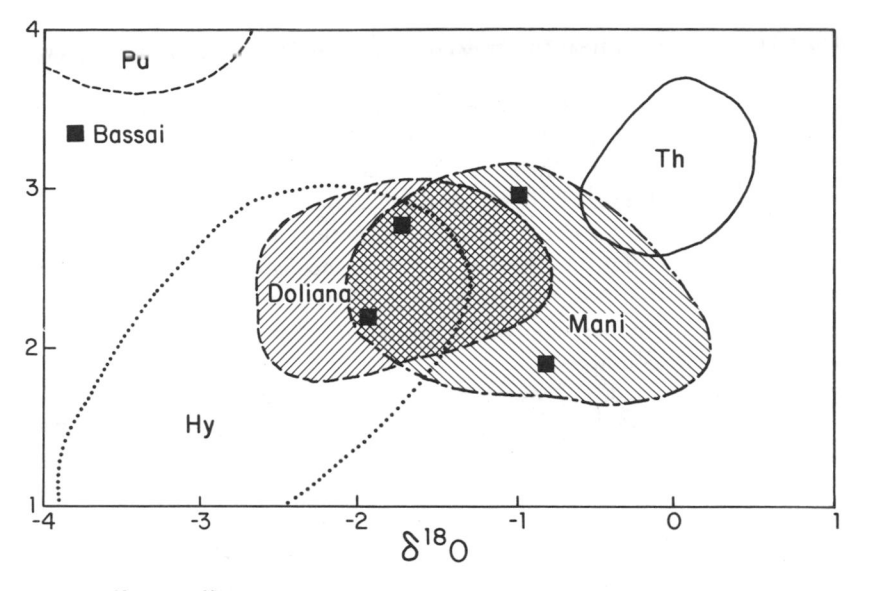

Fig. 13.6. $\delta^{13}C$ vs. $\delta^{18}O$ for marble quarries of the Mani and Doliana compared with roof tiles of the temple at Bassae, Peloponnesus (in per mil, relative to the PDB standard).

devised. In addition to yielding precise measurements that can be easily compared, the method has the important attribute of requiring only minute quantities of material for analysis—20 mg, or the volume of a pencil point. In reconstructions, all geological features of the sample should be noted first. In many published descriptions of marble inscriptions, for example, fragments with a strong lineation and good foliation are associated with pieces without any apparent structures (Herz and Pritchett, 1953). If the rock structures, granularity, and mineralogy in each appear to be similar, then the fragments should be tested further by isotopic analysis.

Samples taken from the same blocks in marble quarries have been found to differ up to 0.4 per mil in both $\delta^{18}O$ and $\delta^{13}C$ (Herz and Wenner, 1978). If weathered and fresh samples are compared, this difference can almost double for $\delta^{18}O$ but hardly changes for $\delta^{13}C$ (fig. 13.4 and table 13.3). If the measured differences are greater than these values, then the fragments could not have been part of the same original. In comparing geological features, it must be remembered that the ancient quarrymen took advantage of the fact that marble tends to split most easily along its foliation planes as well as along its more prominent lineations. A similarity of geological features would indicate only that an association was permissible on these grounds, whereas dissimilarity would be strong evidence against association.

Six stela inscriptions stored in the Epigraphical and Agora Museums, Athens, whose association had been debated by archaeologists, were analyzed isotopically after an examination of their geological features (Herz and Wenner,

1978). The results allowed the association of the fragments of three stelae, but showed that the others could never have been part of the same inscription. In the case of samples EM 8682, EM 8680, and I 5810, it was found that all were composed of a fine-grained, bluish-gray marble. A well-developed color layering dipped about 20° to the right in EM 8682 (Herz and Wenner, 1978, fig. 1) but about 10° to the left in the other two pieces. The δ-values of all three pieces are close—within 0.1 ‰ for both $\delta^{18}O$ and $\delta^{13}C$ in the two EM fragments and 0.3 ‰ between the EM and the I. This greater difference can be explained by different burial histories; the two EM fragments in the Epigraphical Museum were excavated from a different location from where the I fragment of the Agora Museum was excavated.

The difference in dip of color banding makes it highly unlikely that EM 8682 was ever part of the same stela as the other pieces, but the close similarity of the isotopic ratios allows the association of all three. Both the geological features and isotopic ratios indicate the marble is Hymettian. Two separate stelae were made about the same time from marble obtained from the same quarry; perhaps the same quarried slab was split to make the two stelae. One archaeologist had theorized that the pieces represented the same inscription, but cut on two separate blocks.

In their study of sarcophagi in the British Museum, Coleman and Walker (1979) tested fragments of a Sidamara sarcophagus whose association had been argued by archaeologists. The marble used was Turkish, but its source could not be identified by either geological or isotopic criteria. The marble was similar petrographically in all fragments, but arguments had centered on the identity of the style: technique, finish, and size of the carved figures and ornamental decoration. Replicate analysis gave highly discordant values for $\delta^{13}C$, in one case 0.4‰, or several orders of magnitude greater than the analytical precision. Oxygen analysis did yield results reproducible within 0.05‰. The great variation in carbon isotopes indicated that the samples were inhomogeneous, with an unidentified component having a carbon-isotopic composition very different from the average. Variable amounts of graphite or organic material in each sample aliquot could have been responsible for this variation (Valley and O'Neil, 1981). Oxygen alone was used to test association and was found to vary by less than 0.11 ‰ within the same fragment. It was concluded that fragments with a difference less than that value could be associated, but that those much greater belonged to separate sarcophagi. Values of about 0.15 ‰ were not diagnostic either way.

Craig and Craig (1979) noted that substitutions were commonly made in parts of ancient statutes because of the scarcity of Parian marble. Many statutes were composites, with the head made of Parian but the rest of the body of a common marble. Analysis of the Mausolus statue, recovered from a shipwreck where it had been badly corroded by seawater, showed a Parian marble head affixed to a body of a lesser-quality marble.

The authenticity of the five fragments that comprise the Antonia Minor portrait in the Fogg Museum at Harvard University (fig. 13.7) has been argued for some time. Erhart (1978) suggested that the portrait might have been

Fig. 13.7. Marble portrait of Antonia Minor (height, 58 cm; acquisition 1972.306; photo courtesy Fogg Art Museum, Harvard University).

assembled from completely different statues and that some of the parts even could have been very late fabrications. Antonia's pedigree, by the way, included her parents Marc Antony and Octavia, her uncle Augustus, and her sons Germanicus and Claudius Caesar.

The portrait was purchased by the Earl of Pembroke in 1678; a description in 1758 shows that its restoration was complete by that date. The debate over the authenticity and manner of restoration has raged ever since. The portrait consists of five parts: the head (I), the end of the ponytail (II), the right shoulder and breast (III), the lower left shoulder (IV), and the upper left shoulder and breast (V). Each piece had been tested under ultraviolet light, analyzed for major and trace elements, and studied with thin sections under the polarizing microscope. None of these tests could either prove or disprove any association or provenance of the individual pieces. Isotopic analysis of the fragments shows:

Piece Number	$\delta^{13}C$	$\delta^{18}O$	Provenance
I	+5.14	−3.27	Parian
II	+2.05	−2.55	Carrara
III	+4.71	−2.53	Parian
IV	+2.14	−1.53	Carrara
V	+4.58	−2.82	Parian

Tentative conclusions are: the head is authentic, but unrelated to the other pieces; III and V may have been taken from the same ancient portrait; II and IV are probably comparatively recent additions (Herz & Wenner, 1981).

SUMMARY AND CONCLUSIONS

The use of stable-isotopic signatures of $\delta^{13}C$ and $\delta^{18}O$ has now been thoroughly tested in provenance and association studies of classical Greek and Roman marbles. A large data base has been built up for the ancient quarries of Greece, the Aegean, and western Turkey, and the variation in stable-isotopic ratios for each area has been documented. Techniques such as trace-element analysis and thermoluminescence, which have been successfully used with other materials, do not work with marble because of its inherent variability. Electron-spin resonance (ESR) spectroscopy appears to hold some promise, based on preliminary tests of various classical marbles.

The inherent variation of marble isotopes in most quarries is less than 2‰; in single blocks much less than 0.4‰. Weathering apparently has a very slight effect on the $\delta^{13}C$ but can reach values of up to 0.6‰ in $\delta^{18}O$. Isotopic fractionation during metamorphism causes a much more serious variability, which is compounded by the presence of other mineral phases in or near the marble and the development of a steep thermal gradient.

Before isotopic analysis is done, samples should be studied petrographically. Structural and fabric features will immediately rule out some sources and may even suggest which are the most probable sources. Some studies have also used trace elements as a third determinant, and for some quarries this has proved useful.

Inscriptions, statues, and mausoleums have been reconstructed in the past solely on aesthetic criteria, resulting in scholarly debates on the validity of such reconstructions. The isotopic variation among the fragments of an artifact made from a single block should be very low, depending on the burial history of each fragment. One study used a variation among fragments of less than 0.15 ‰ as positive evidence for association. Other studies have shown that many ancient statues initially were constructed of more than one marble type or that several types were used in later reconstructions.

Attempts to extend this methodology to other materials, such as malachite weathering patinas on bronze objects (Smith, 1978), have not been successful. It is apparent from such studies that the geochemical controls on isotopic fractionation must be better known before provenance or authenticity can be demonstrated. The isotopic methods together with petrography and trace-element analysis have the potential for establishing diagnostic geochemical fingerprints for all classical marble used in the Mediterranean area.

Acknowledgments

I thank David B. Wenner for invaluable assistance in carrying out much of the project, for the support of his stable-isotope laboratory, and for discussions and critical reading of this manuscript. The work was supported by grants from the National Endowment for the Humanities, the National Geographic Society, and the Samuel H. Kress Foundation.

REFERENCES

Afordakos, G., K. Alexopoulos, and D. Miliotis. 1974. Using artificial thermoluminescence to reassemble statues from fragments. *Nature* 250:47–48.

Ashmole, B. 1970. Aegean marble: Science and common sense. *Annual of the British School at Athens* 65:1–2.

Coleman, M., and S. Walker. 1979. Stable isotope identification of Greek and Turkish marbles. *Archaeometry* 21:107–12.

Conforto, L., M. Felici, D. Monna, L. Serva, and A. Taddeucci. 1975. A preliminary evaluation of chemical data (trace element) from classical marble quarries in the Mediterranean. *Archaeometry* 17:201–13.

Cooper, F. A. 1981. A source of ancient marble in the southern Peloponnesos. *American Journal of Archaeology* 85:190–91.

Cordischi, D., D. Monna, and A. L. Segre. 1981. ESR analysis of marble samples from Mediterranean quarries of archaeological interest. *Archaeometry 21 Abstracts*, Brookhaven National Laboratory, New York, p. 17.

Craig, H., and V. Craig. 1972. Greek marbles: Determination of provenance by isotopic analysis. *Science* 176:401–03.

———. 1979. Isotopic studies of classical marble provenance. *Geological Society of America, Abstracts with Programs* 7:406.

Dworakowska, A. 1975. Quarries in Ancient Greece. *Bibliotheca Antiqua* (Polish Academy of Sciences), vol. 14.

Erhart, K. P. 1978. A portrait of Antonia Minor in the Fogg Art Museum and its iconographical tradition. *American Journal of Archaeology* 82:193–212.

Faure, G. 1977. *Principles of isotope geology.* New York: John Wiley.

Germann, K., G. Holzmann, and F. J. Winkler. 1980. Determination of marble provenance: Limits of isotopic analysis. *Archaeometry* 22:99–106.

Herz, N. 1955. Petrofabrics and classical archaeology. *American Journal of Science* 253:299–305.

Herz, N., and W. K. Pritchett. 1953. Marble in Attic epigraphy. *American Journal of Archaeology* 57:71–83.

Herz, N., D. B. Wenner, and E. J. Walters. 1977. Provenance of Greek statuary by isotopic methods. *Archaeological Institute of America Abstracts, 79th Meeting*, pp. 39–40.

Herz, N., and D. B. Wenner. 1978. Assembly of Greek marble inscriptions by isotopic methods. *Science* 199:1070–72.

———. 1979. The use of oxygen and carbon isotopic signatures on archaeological marble. *Geological Society of America, Abstracts with Programs* 7:443.

———. 1981. Tracing the origins of marble. *Archaeology* 34(5):14–21.

Herz, N., D. G. Mose, and D. B. Wenner. 1982a. $^{87}Sr/^{86}Sr$ ratios: A possible discriminant for classical marble provenance. *Geological Society of America, Abstracts with Programs* 14:514.

Herz, N., F. A. Cooper, and D. B. Wenner. 1982b. The Mani quarries: Marble source for the Bassai temple in the Peloponnesos. *American Journal of Archaeology* 86:270.

Jansen, J. B. H., and R. D. Schuiling. 1976. Metamorphism on Naxos: Petrology and geothermal gradients. *American Journal of Science* 276:1225–53.

Lazzarini, L., G. Moschini, and B. M. Stievano. 1980. A contribution to the identification of Italian, Greek and Anatolian marbles through a petrological study and the evaluation of Ca/Sr ratio. *Archaeometry* 22:173–83.

Lepsius, G. R. 1890. *Griechische Marmorstudien*. Berlin: Konigl. Akademie der Wissenschaften.

Manfra, L., U. Masi, and B. Turi. 1975. Carbon and oxygen isotope ratios of marbles from ancient quarries of western Anatolia and their archaeological significance. *Archaeometry* 17:215–21.

Monna, D., and P. Pensabene. 1977. *Marmi dell'Asia Minore*. Rome: Consiglio Nazionale delle Ricerche.

Pliny. 1962. *Natural History*, Book 36, trans. D. E. Eichholz. Vol. 10, Loeb Classical Library. Cambridge, MA: Harvard Univ. Press.

Renfrew, C., and J. S. Peacey. 1968. Aegean marble: A petrological study. *Annual of the British School at Athens* 63:45–66.

Riederer, J., and J. Hoefs. 1980. Die Bestimmung der Herkunft der Marmore von Büsten der Münchener Residenz. *Naturwissenschaften* 67:446–51.

Rybach, L., and H.-U. Nissen. 1965. Neutron activation of Mn and Na traces in marbles worked by the ancient Greeks. In *Radiochemical methods of analysis*, vol. 1, 105–17. Vienna: International Atomic Energy Agency.

Rye, R. O., R. D. Schuiling, D. M. Rye, and J. B. H. Jansen. 1976. Carbon, hydrogen, and oxygen isotope studies of the regional metamorphic complex at Naxos, Greece. *Geochimica et Cosmochimica Acta* 40:1031–49.

Sheppard, S. M. F., and H. P. Schwarcz. 1970. Fractionation of carbon and magnesium between coexisting metamorphic calcite and dolomite. *Contributions to Mineralogy and Petrology* 26:161–98.

Smith, A. W. 1978. Stable carbon and oxygen isotope ratios of malachite from the patinas of ancient bronze objects. *Archaeometry* 20:123–33.

Theophrastus. 1956. *On Stones*, trans. E. R. Caley and J. F. C. Richards. Columbus: Ohio State Univ. Press.

Valley, J. W., and J. R. O'Neil. 1981. $^{13}C/^{12}C$ exchange between calcite and graphite: A possible thermometer in Grenville marbles. *Geochimica et Cosmochimica Acta* 45:411–19.

Ward-Perkins, J. B. 1977. *Roman architecture*. New York: Abrams.

Washington, H. S. 1898. The identification of the marbles used in Greek sculpture. *American Journal of Archaeology* 2:1–18.

Weiss, L. E. 1954. Fabric analysis of some Greek marble and its applications to archaeology. *American Journal of Science* 252:641–62.

14

The Provenance of Artifactual Raw Materials

GEORGE RAPP, JR.

ABSTRACT

Numerous geochemical and geological techniques have been used to determine the geographic and geologic source of exotic materials uncovered in archaeological excavations. Trace-element and stable-isotope patterns in raw materials provide the best "fingerprints" of geologic deposits. Lithics, ceramics, and metals have been traced successfully to their source. Sensitive analytic techniques and discriminating statistical methods are necessary for such provenance studies. Neutron-activation analysis and mass spectrometry have been the most successful techniques for chemical patterning. Discriminant functions and the most advanced clustering techniques have provided the most powerful decision rules for distinguishing between disparate sources.

The archaeometric goal of provenance studies is to be able to specify with confidence the geographic source of the deposits that provide the raw materials for the manufacture of a specific artifact or set of artifacts. Such studies do not address the question of where the artifact was manufactured but only the source of the raw material.

Most of the first successes in determining the source of geologic materials by chemical analyses were through the use of trace-element concentrations. Trace elements are those elements found in concentrations below 100 ppm (i.e., below the 0.01 percent lower limit of classical rock and mineral analyses). Trace elements play no significant role in the physicochemical reactions that produce a geologic deposit. However, by virtue of their residence in the bulk chemical system they will be either concentrated or dispersed in certain phases of the geochemical processes that form a rock, mineral, or ore deposit. Unless two artifacts (which were not smelted or in any other way chemically altered) were formed from the same rock or ore body, it would be fortuitous for them to have

353

coincident trace-element concentrations of eight or more geochemically independent elements.

A wide variety of rock and mineral raw materials are amenable to such trace-element "fingerprinting." This chapter provides illustrations of trace-element fingerprinting of metal, lithic, ceramic, and other artifactual materials. In addition to trace-element techniques, analyses for diagnostic differences in stable-isotope composition—particularly of lead, carbon, and oxygen—have been successful in determining provenance. The preceding chapter in this book provides an excellent review of isotope studies of marble provenance.

METHODOLOGIES

Geochemical and geological methods of provenance determination can be considered under two topic headings: method of analysis and statistical considerations. Most analytical techniques in use today in provenance studies are chemical or physical. Neutron-activation analysis has been the preferred trace-element technique because it (1) provides data on concentrations of a large number of chemical elements at one time; (2) requires no complex sample-preparation techniques and only very small samples; (3) is sensitive to parts per billion concentrations of most metallic elements; and (4) can handle large numbers of samples on a routine basis. Numerous examples of the use of neutron-activation analyses are presented in the sections below on metals, lithics, and ceramics. Other somewhat comparable but less utilized analytical techniques are spark-source mass spectrometry (Friedman and Lerner, 1977), atomic-absorption analysis (Gritton and Magalousis, 1977; Hughes, Crowell, and Craddock, 1976) and X-ray flourescence analysis (Nelson, D'Auria, and Bennett, 1975). Determination of chemical patterning can also be accomplished by proton microprobe and milliprobe analyses, Mossbauer spectroscopy (Kostikas, Simopoulos, and Gangas, 1974, 1976; Longworth and Warren, 1979), X-ray photoelectron spectroscopy (Lambert and McLaughlin, 1976), infrared spectroscopy (Beck et al., 1971) and thermoluminescence (Huntley and Bailey, 1978), as well as other physical and chemical methods.

Each trace-element analytical technique has its own set of problems with adequate standards, differing sensitivities to different elements, lack of comparability between laboratories, and so forth. All techniques have the same problem of inhomogeneities in the samples and all require careful prevention of postsampling contamination at trace levels. Each technique has advantages for certain types of materials. Some techniques are largely nondestructive. It should be pointed out that emission spectroscopy, historically the first method utilized, proved not to have the sensitivity necessary for most trace-element fingerprinting of copper or ceramic materials.

In addition to the neutron-activation technique, stable-isotope analysis has proved to be very successful in provenance determination. Shackleton and Renfrew (1970) used oxygen-isotope analyses of *Spondylus* shells from Neolithic sites to indicate that the source for shells used as ornaments in the Balkans and central Europe during that period was the Aegean, rather than the Black, Sea. The pioneer work of Brill, Barnes, and Adams (1973) on the isotopic composition of lead in Egyptian artifacts paved the way for the present rapid advances in isotope investigations. Gale and Stos-Gale (1982) have successfully extended the use of isotope analyses to lead in copper ores and artifacts.

Statistical studies of analytical data on artifacts began with attempts to establish patterning that had significant meaning in terms of provenance. In the last decade or so, an important conceptual addition has been made to provenance studies, that of determining the fingerprints, signatures, or diagnostic chemical patterns of the source deposits themselves, thus allowing direct comparisons of artifacts with potential source materials. Attempts to assign artifactual materials to particular geographical source deposits have two inherent problems. First, it must be established that the artifact has not undergone any chemical or physical alteration that would invalidate direct comparison of the artifact with the same material from known deposits. Second, all potential source deposits must be adequately represented in the data base for a confident assignment of provenance based on chemical or physical patterning.

Statistical problems inherent in provenance studies fall into two categories: (1) the necessity of having complete and accurate analytical data on all the potential source deposits and (2) the applicability and power of known statistical methods. This chapter will not discuss many of the typical problems of analytical methodology, such as the effect of sampling size on analytical reliability, quality assurance, batch-dependent systematic variations, and the ever-present problem of major inhomogeneities in most geological and artifactual materials. An extensive literature is available on the problems of analytical reliability. In relation to archaeological provenance studies, one should consult particularly de Bruin et al. (1976), Charles (1973), Warren (1973), Bowman, Asaro, and Perlman (1973), Bromund, Bower, and Smith (1976), Wilson (1978), Michels (1982), Ives (1975), and Carriveau (1980).

Analysts using trace-element techniques have the special problem of fixing adequate and reliable standards. Three laboratories, those of Berkeley, under Frank Asaro, Brookhaven National Laboratory, under Garman Harbottle, and Hebrew University, under Isadore Perlman, have done extensive work on establishing reliable standards for the neutron-activation analysis of ceramic materials. Interlaboratory comparisons of trace-element standards among these three laboratories have provided a firm base for studies of ceramic provenance. However, for most materials the problem remains one of obtaining adequate standards. The National Bureau of Standards has a few standards applicable to copper-based alloys.

A wide variety of statistical techniques has been used to assign artifacts to sources or to similarity groups based on trace-chemical data. Discriminant analysis can be used when one has a priori knowledge of the specific groups (sources) an unknown must belong to—for example, when determining which regional obsidian deposit was the source of material for a settlement in a given archaeological period. The discriminant function is then a decision rule that assigns the unknown to one group on the basis of a set of measurements. One can test the coherence and trace-element uniqueness of each group (obsidian deposit) by removing members from groups to check if the decision rule returns them to their (known) group. Incorrect or overlapping groups can be uncovered by such tests.

Cluster analysis is a statistical technique for determining relationships in a large matrix of measurements. I have found that the common agglomerative-hierarchical dendrograms expressing relationships among trace-element patterns in copper deposits are not helpful in assigning unknowns (artifacts) to probable sources. However, if one uses cluster analysis as a form of correlation analysis, the results can be presented in a two-dimensional diagram in which group separations and overlaps are easily seen and unknowns can be compared to the groups (sources). Cluster analysis, therefore, can be an important tool in provenance studies. K-means cluster analysis, using standard Euclidean distance as a measure of similarity between reference groups or between unknowns and the reference groups, allows the operator to seek natural groups of any desired level of similarity.

In order to assign an artifact to the geologic deposit of most probable origin, it is necessary to use statistical techniques compatible with the data. In the Archaeometry Laboratory at the University of Minnesota we use a simple multivariate product function wherein the trace-element concentrations in an artifact are compared with all trace-element patterns from specified, coherent geographical sources, anywhere in size from a single deposit (mine) to a large region, each of which form a population defining a trace-element fingerprint. For each specified geographic source, the trace-element data are arranged in a two-dimensional matrix with the analytical data for each chemical element recorded in five concentration intervals, I_1 to I_5. The intervals are different for each chemical element and are chosen from an inspection of the data base to maximize the discriminating power of the function.

The product function d^* compares the unknown (artifact) with each possible source fingerprint by

$$d^* = \prod_{i=1}^{j} \left(\frac{N_{ij}}{n} \right),$$

where n is the number of analyses and N_{ij} is the number of analyses falling into the ith concentration interval of the jth element. The potential locality with the highest product value is the indicated source. The viability of this simple linear discriminant function *is totally dependent on a priori knowledge that any unknown is a member of one of the source sets.* In other words, if the correct source is not represented in the data base, the d^* function will assign as the source that locality where the product value is maximum, even if the similarity between the unknown and the source is slight. No part of the function is designed to reject a provenance determination because the unknown is insufficiently similar to all given sources in trace-element abundances.

From an investigation in the Archaeometry Laboratory of the sources of

Table 14.1. Six-Element Fingerprint of Nineteen Copper Nuggets from Snake River, Minnesota

	Co	Te	Fe	Hg	Sb	W
I_1	4	10	3	4	5	9
I_2	0	0	0	0	0	1
I_3	0	7	0	0	1	9
I_4	15	2	8	15	12	0
I_5	0	0	8	0	1	0
	19	19	19	19	19	19

prehistoric North American native copper, analyses of 19 copper nuggets from Snake River, Minnesota, gave the data shown in table 14.1. This table shows, for example, that of the 19 analyzed specimens from Snake River, four of the cobalt trace-element concentrations were in the range I_1 and 15 were in the range encompassed by I_4. No concentrations of cobalt fell within ranges I_2, I_3, and I_5. The reason for the bimodality in the Snake River element abundances shown above is not known. For the six-element Snake River fingerprint above, an unknown whose analysis placed cobalt in I_4, tellurium in I_3, iron in I_3, mercury in I_5, antimony in I_3, and tungsten in I_2 would have a d^*-value for Snake River of:

$$\frac{15}{19} \times \frac{7}{19} \times (0.005) \times (0.005) \times \frac{1}{19} \times \frac{1}{19} = .0000000201,$$

to be compared with d^*-values for all other potential sources. When any value in the product function is zero, an arbitrarily small number, such as 0.005, must be substituted to prevent the product from becoming zero. (It should be noted that the sample calculation given in Rapp et al. [1980] is incorrect.)

The test for the uniqueness of a fingerprint is accomplished by randomly removing specimen analyses one by one from a locality fingerprint, reconstituting the fingerprint without the test-specimen values in the fingerprint, then using d^*-values to assign the test specimen to a source. If all or nearly all are returned to the known source, then the fingerprint has a high level of uniqueness. Table 14.2 shows a typical example of the proper return by d^*-values of a Kingston Mine copper specimen to Kingston, Michigan, as the source.

Attempts to distinguish individual mines in northern Michigan provide the toughest test. The northern Michigan deposits are all of the same age and have similar geologic origins. Hence there is more overlap in fingerprint characteristics among these sources than between them and deposits in Alaska, Illinois, or Arizona. In a test of Isle Royale copper samples versus Kingston Mine samples, 17 of 20 randomly selected samples were returned to the known source.

We analyze by neutron activation analysis for 48 elements and usually use 16. For a 16-element fingerprint, 20 analyzed specimens from each source is a marginal number; 25 to 40 are recommended. Our work thus far indicates that the discriminant function seems to succeed if there are 20 or more analyzed samples in each source fingerprint.

Table 14.2. Summary of d^*-Values for Unknown vs. Fingerprints for Specimen Number 34-21-35DP (Kingston Mine Sample)

Area	d^*-Value
Kingston, MI	.000000625352*
Champion, MI	.000000000000
Centennial, MI	.000000000001
Isle Royale, MI	.000000004080
Alaska	.000000000741
Arizona	.000000000049
Snake River, MN	.000000000634
Wisconsin	.000000000000
Illinois	.000000000000
Lower Michigan	.000000000016

*Fingerprint of best fit.

The concentration intervals I_1 through I_5 for each element should be chosen to distribute the concentrations as evenly as possible throughout the five I cells. The ability of the d^* function to serve as a discriminator depends on the choice of trace elements and intervals used. For specific problems, one can maximize the effectiveness of the function by refining the intervals and the choice of elements, using only the data from relevant localities. An example of the use of this discriminant function is given in the section on metal provenance, below.

K-means cluster analysis is an iterative clustering technique with reallocation capability. (See Doran and Hodson, 1975:180−85 for a general discussion of the application of K-means procedures to archaeological data sets.) The most important advantage of K-means clustering over the hierarchical-aggregative techniques more often used in NAA data analysis (Bieber et al., 1976a) is its ability continuously to review cluster membership and to reallocate members by an optimizing criterion. In the PKM K-means program (BMDP, 1979), the reallocation algorithm minimizes the Euclidean distance between the members of a given cluster and the cluster centroid. Initially, all samples are members of a single cluster (i.e., K = 1); the cluster is subseqently subdivided until the final number of clusters specified by the user is attained. Samples are then iteratively reallocated to the cluster whose centroid is closest to them. The optimum K-number for any clustering operation is determined both quantitatively, by finding the K-number with the smallest mean of average intracluster Euclidean distances, and qualitatively, by examining the scatter plot of the orthogonal projection of samples onto the plane defined by the centroids of the three most populous clusters. K-means cluster analysis was found to be an effective means of discriminating among a limited number of possible provenances. An example of the use of K-means cluster analysis is given in the section on ceramic materials, below.

In addition to the multivariate techniques described above, provenance researchers have used principal-component analysis, factor analysis (de Bruin et al., 1972), many simple analyses of variance techniques, and a variety of binary and ternary plots to display the separation (or overlapping) of the fields of selected chemical or physical parameters (for an example, see fig. 14.1).

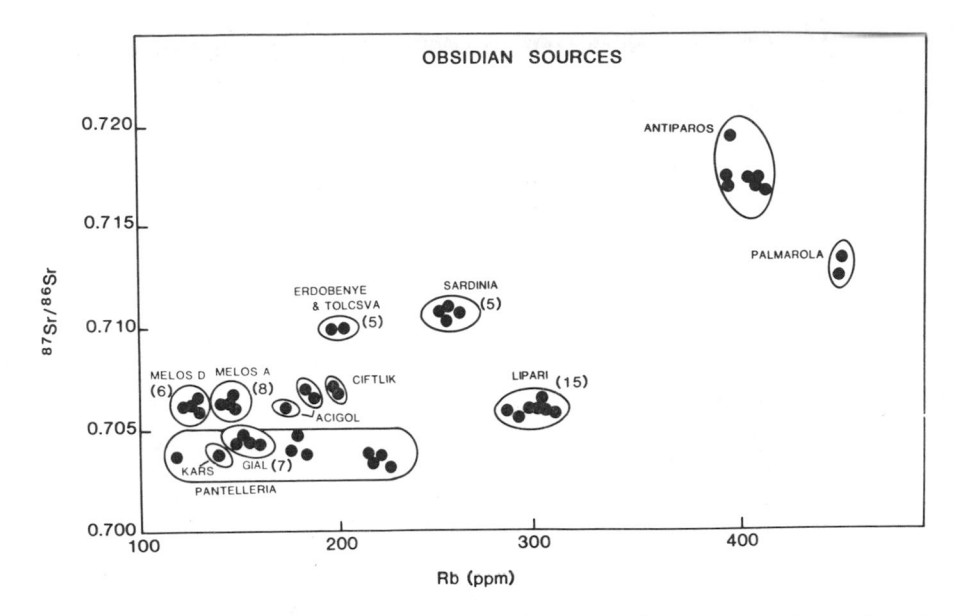

Fig. 14.1. Strontium-isotopic composition versus rubidium content for geological sources of obsidian important to Mediterranean cultures. To prevent confusion, closely coincident points have been plotted once only and the total number of points within an ellipse is given in parentheses (Gale, 1981).

LITHIC MATERIALS

The earliest and perhaps now the most successful trace-element sourcing has been accomplished for obsidian. Obsidian is not a common rock type and therefore the number of possible sources is normally limited. A good knowledge of the bedrock geology of a region will allow a quick determination of whether or not obsidian sources may occur there. In regions of silicic volcanic rocks, it could take a large amount of fieldwork to locate all possible sources of obsidian. In common with other stone materials, the manufacture of obsidian objects does not alter the chemical composition of the material.

Since the early (and only partially successful) work of Renfrew and his colleagues on Mediterranean obsidians (Cann and Renfrew, 1964; Renfrew, Cann, and Dixon, 1965; Dixon, Cann, and Renfew, 1968) and the work of Gordus and colleagues in the United States (Gordus, Wright, and Griffin, 1968), obsidian-provenance studies have increased in scope and success until successful applications are nearly worldwide. Regional applications have been done for Belize (Stross et al., 1978), Greece (Filippakis, Grimanis, and Perdikatsis, 1981), Guatemala (Asaro et al., 1978; Michels, 1982), middle western United States (Griffin, Gordus, and Wright, 1969), Mexico (Cobean et al., 1971; Ericson and

Kimberlin, 1977), Alaska (Patton and Miller, 1970), New Zealand (Leach and Anderson, 1978), and Ethiopia (Muir and Hivernel, 1976).

Gale (1981) has demonstrated that the most satisfactory discrimination of eastern Mediterranean obsidian deposits is accomplished by plotting ratios of isotope ^{87}Sr to ^{86}Sr against the trace-element concentration of rubidium (see fig. 14.1). Huntley and Bailey (1978) have used thermoluminescence to discriminate sources in the American Pacific Northwest that were insufficiently separated by X-ray fluorescence analysis (see Nelson, D'Auria, and Bennett, 1975).

Following the characterization of obsidians and their sources came major provenance studies of chert/flint and soapstone (soft stone) artifacts. Chert is microcrystalline quartz with very few chemical impurities substituting in the crystal structure but often with abundant impurities trapped as microinclusions. Cherts have been given many varietal names based on the colors imparted by the inclusions. Flint is gray to black because of included organic matter, jasper is red because of included hematite, and novaculite is white because of its high purity (except for relatively abundant extracrystalline water). The provenance of chert/ flint artifacts has been determined successfully by numerous researchers by means of trace-element fingerprinting. (See, particularly, Sieveking et al., 1970; de Bruin et al., 1972; and Luedtke, 1979.)

Many varieties of cryptocrystalline or microcrystalline quartz have been utilized by humans for thousands of years. Carnelian, with a translucent reddish color, is one of the materials used most abundantly by the ancients for gems and seals. Plasma (an opaque, deep green microcrystalline quartz), agate, and quartz crystals themselves (called rock crystal) are common from first and second millennia B.C. Mediterranean sites. All may be subject to trace-element provenance determination if all potential source deposits can be identified. Dennen (1967) used trace elements in sedimentary quartz to determine the source of the sedimentary rocks.

Soft-stone artifacts (soapstone, serpentinite, and related rocks) have proved to be amenable to provenance studies. Rocks that are refractory and cohesive yet easy to carve into desired shapes are abundant exotics in many archaeological sites. Allen, Luckenbach, and Holland (1975) have provided an introduction to the use of activation analysis for sourcing steatite artifacts. Kohl, Harbottle, and Sayre (1979) attempted to characterize the soft-stone vessels, carved in a highly diagnostic style, that have been found in sites from Mesopotamia to the Indus Valley. They were only partially successful in this characterization study. Allen and Pennell (1977) successfully used rare-earth elements to fingerprint soapstone deposits and determine the provenance of soapstone artifacts. Using X-ray fluorescence analysis of sanukite (a hypersthene andesite) samples of known source and prehistoric sanukite implements, Warashina, Kamaki, and Higashi-mua (1978) were able to assign more than 100 implements from 36 Japanese sites to four source districts.

Trace-element analysis has also been used successfully to determine where the turquoise was mined for artifacts from the Snaketown site in south-central

Arizona (Sigleo, 1975). Trace-element data do not provide discrimination among sources of Mediterranean marbles (Conforto et al., 1975; see also chap. 13 in this volume).

One of the earliest attempts to source archaeological remains by geological analysis was the effort by Stukeley (1740) to trace the origin of the megaliths at Stonehenge. Early observers realized that two types of stone were utilized in the construction of Stonehenge. The large "sarsens" were quartzose rocks of local origin, also used at the great circle of Avebury. The "blue stones" were doleritic masses. Maskelyne (1877) made the first careful petrographic descriptions of these rocks, and Thomas (1923) was able to trace the exotic blue stones by careful petrologic and petrographic analyses to Prescelly Mountains in Wales.

An excellent recent example of tracing the source of megaliths is the work of Heizer et al. (1973) on the origin of the rock used to construct the Colossi of Memnon on the plain near Thebes in Egypt. The statues were made of a very hard ferruginous quartzite. There were at least six sites from which the quartzite blocks could have been quarried. Using neutron-activation analysis for trace-element characterization, the team was able to show that the Gebel el Ahmar quarry, 676 km downstream on the Nile, was the likely source, not the nearer quarries at Aswan, 200 km upstream.

Volcanic tephra deposits provide marker beds that often allow the construction of a tephrochronology of regional significance (see chap. 10 in this volume). The provenance of artifactual volcanic pumice has been successfully determined by accurate determination of the index of refraction of the glass shards in the pumice. The most important example of this technique has been in the study of the effects of the catastrophic eruption of the volcanic island of Thera in the Aegean during the late Bronze Age (Rapp, Cooke, and Henrickson, 1973; Vitaliano and Vitaliano, 1974; Keller, 1980). Fornaseri, Malpieri, and Tolomco (1975) have determined the provenance of pumice recovered along the north coast of Cyprus.

An artifactual material of widespread significance in prehistoric Europe is amber, derived from the resin of the extinct pine species *Pinus succinifera*. This most common amber from European archaeological sites was formed during the early Tertiary in vast northern European conifer forests. Beck et al. (1971) have traced the distribution of Baltic amber in the first millennium B.C. from England to the Black Sea. In order to determine the origin of weathered amber from a Celtic site in Moravia, Beck et al. (1978) developed a gas chromatographic method to determine the succinic-acid content.

Numerous other lithic materials have been widely utilized from Neolithic times onward. The great deposits of lapis lazuli from Badakhshan and elsewhere in the Near East are abundant in archaeological sites from Mesopotamia to Egypt. Ocher, huntite (Barbieri et al., 1974), and many other minerals and rocks have been utilized extensively for pigments and cosmetics. Hassan and Hassan (1981), using lead-isotope ratios, have studied the provenance of galena in predynastic Egypt, where it served as an eye pigment. The use of jade for ornamental objects is at least 4,000 years old. All such lithic materials can be subject to provenance studies by the methods enumerated above.

Trace-element techniques can also be used to discriminate between small-object "look-alikes" found in excavations. Pollard, Bussell, and Baird (1981) used trace elements to distinguish jet from other black materials of early Bronze Age English origin.

CERAMIC MATERIALS

At approximately the same time as trace elements began to be used to discriminate among potential sources of obsidian, Perlman and Asaro (1969) began at Berkeley their pioneering efforts to determine the provenance of pottery by neutron-activation analyses. They were soon joined by Harbottle and Sayre at Brookhaven National Laboratory (Harbottle, 1970). By 1982 tens of thousands of sherds and pottery clays from around the world had been analyzed by them, primarily by the neutron-activation technique. These two laboratories have also engaged in the most comprehensive development of analytical standards yet available for trace-element provenance studies. Perlman has carried on this work in a new laboratory at Hebrew University. For a comparison of analyses between the Berkeley and Hebrew University laboratories, see Yellin et al. (1978).

While earlier attempts at characterization of pottery sources by emission spectrographic techniques (e.g., Catling, Blin-Stoyle, and Richards, 1961, 1963) proved ultimately to be unsuccessful, neutron-activation analysis has proved to be adequate under reasonably ideal conditions of availability of sherds and satisfactory sampling of potential pottery-clay sources. Emission spectrography, on the other hand, is insufficiently sensitive to the lower concentrations of many metallic elements in clays.

Since it is virtually impossible to analyze more than a small proportion of the pottery of any one type, a priori characterization by macroscopic features such as fabric is often utilized to narrow the problem. Thin-section petrographic techniques are in common use to study the composition and fabric of ceramics and can be used to determine the source of the temper (Dickinson and Shutler, 1974, 1979). A more detailed description of ceramic petrography is given in chapter 12 of this volume.

Examples of successful trace-element studies of ceramic materials include Farnsworth, Perlman, and Asaro (1977) on pottery from Corinth and Corfu in Greece. Corinthian fineware proved to have a nearly constant chemical composition over several hundred years. The clays closest in composition to that of the pottery clay were found near the Potters' Quarter and near the Tile Works, about 2.7 km distant from Corinth. Corinthian-style pottery had been made on Corfu, whereas the Corinthian-style pottery found on the island of Aegina had been made in nearby Corinth.

Neutron activation has been used to trace the source of coarse wares from western Cyrenaica and southern Crete that are similar in shape and workmanship but are not distinguishable petrologically (Krywonos et al., 1982). Harbottle, Sayre, and colleagues have also worked to characterize Mediterranean pottery. They have analyzed a large body of ceramic artifacts and clays from Greece, Cyprus, and the Levant. One study (Bieber et al., 1976b) presents general compositional groupings, while another (Brooks et al., 1975) attempts to distinguish between local and imported pottery at Tell el-Hesi, Israel. A few of the Tell el-Hesi sherds matched the trace-element fingerprint of a clay deposit near Jerusalem. Other groups from this site exhibited trace-element patterns indicating importation from the Aegean, Cyprus, and Mesopotamia. Bieber and colleagues (1976a) review some aspects of the application of multivariate statistical techniques to trace-element data on ceramics.

Other ceramic-provenance studies of note include a study by Hammond, Harbottle, and Gazard (1976) on Mayan ceramics and clays from Belize, in which their initial assumption—that the numerically dominant ceramic types were of local manufacture—was confirmed. Trace-element data suggested different sources of raw material for quotidian and elaborately decorated pottery. Poole and Finch (1972), using trace-chemical characterization, successfully correlated British postmedieval pottery with appropriate kiln-site material from Europe.

Tubb, Parker, and Nickless (1980) provide data on the problems and effects of weathering and firing conditions on observed elemental concentrations of major and minor (rather than trace) elements in a study of Romano-British pottery. Weathering and firing conditions affected barium, calcium, manganese, sodium, and titanium more than potassium, magnesium, iron, and aluminum. Iron, magnesium, calcium, and potassium proved to be the diagnostic elements for the region studied.

In a provenance study of Minoan pottery from Gournia, Crete, and related potential clay sources, Rapp and Gifford (1984) used K-means cluster analysis to interpret the trace-element data. Figure 14.2 illustrates the relationship between clay samples and sherds from the Isthmus of Irapetra versus sherds from outside this area. Table 14.3 summarizes the results of this clustering run. In no instance does a clay sample appear in a cluster with sherds *other than* from Early Minoan III Gournia (clusters 5 and 7) or from Middle Minoan Gournia (cluster 2). Sherds from this site are compositionally more similar to the clay samples than are those from other sites.

In figure 14.2, cluster 7 shows a weak grouping of all the Early Minoan III Gournia sherds with some clay samples from the upper member of the Makrilia formation, while cluster 2 shows a relationship among some Middle Minoan Gournia sherds, almost the remainder of the clay samples from the Makrilia formation, and clay samples from the Phothia formation. Cluster 5 shows a mixture of Early Minoan III Gournia sherds plus clays from the Ammoudares formation, the Pakheia Ammos formation, and one Makrilia formation clay sample. Clusters 3, 4, and 6 are heterogeneous mixtures of sherds, with a few Gournia sherds appearing in all three. Clusters 4 and 6 (containing mostly non-Irapetra-region sherds) are distinct from clusters 2, 5, and 7, which contain only sherds from Gournia plus clay samples.

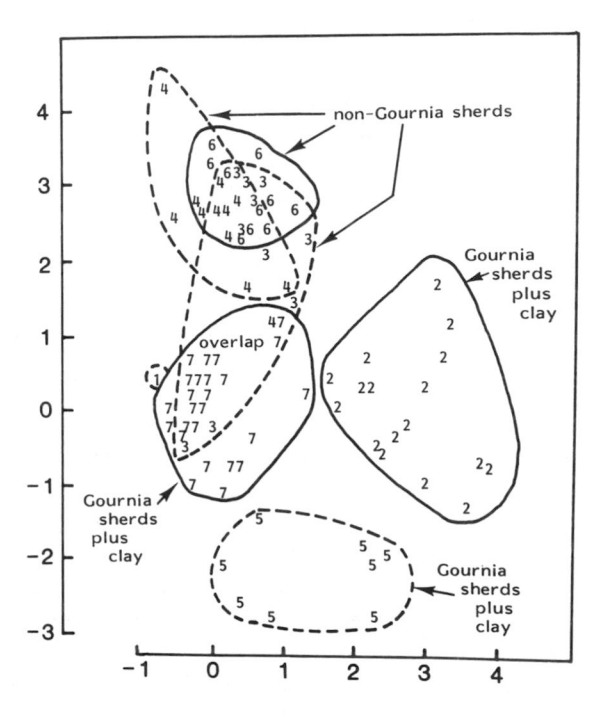

Fig. 14.2. Trace-element data from clay samples plus all sherds, K-means clustering (K = 7, 5 iterations) based on eight elements (Ba, Ce, Co, Cr, Eu, Fe, Sc, Th). Scatter plot of the seven clusters is projected from eight dimensions onto the plane defined by the centroids of the three most populous clusters (2, 6, and 7). See table 14.3 for cluster membership. Abscissa and ordinate are Mahalanobis distances from the central cluster.

METAL STUDIES

Fewer successful provenance determinations have been made for metals than for lithics and ceramics, largely because most metals recovered from archaeological sites outside North America had undergone smelting and/or alloying, resulting in significant changes in their trace-chemical compositions (McKerrell and Tylecote, 1972; Tylecote, Ghaznavi, and Boydell, 1977). Alloying elements may be purposefully added to metals in contentrations large enough, normally 1.5 to 2.0 percent, to impart definite changes to the physical properties of the metals. Such alterations in composition add sufficient trace-chemical complexity as to make the determination of raw material sources from trace-element patterns almost impossible. Although the chemical composition of smelted products may provide valuable clues to the technology used in the manufacture of artifacts, this aspect of composition studies lies outside the scope of this chapter.

Junghans, Sanmeister, and Schroder (1960, 1968, 1974) analyzed over 40,000 European prehistoric metal artifacts to develop a corpus of chemical character-

Table 14.3. Trace-Element Data from Clay Samples plus All Sherds

Sample	I.D.	D.C.	Sample	I.D.	D.C.	Sample	I.D.	D.C.
Cluster #1			Cluster #4			MS-4237	VAS	1.03
4615-33		0.00	MS-4894	PYRG	0.48	AE.752	KNOS	1.07
			AE.753.4	KNOS	0.50	MFA9.574	MOCH	1.14
Cluster #2			AMO9.334	MOCH	0.51			
JAG-5A	MAKL	0.75	MS-4234	VAS	0.51	Cluster #7		
JAG-16B	MAKU	0.90	MFA9.557	MOCH	0.55	4615-19	EMG	0.46
JAG-5C	MAKL	0.98	MFA9.575	MOCH	0.70	4615-41	EMG	0.67
JAG-16A	MAKU	1.03	AE.1672	KNOS	0.97	4615-9	EMG	0.68
JAG-8B	MAKL	1.31	BAI-14-A	VAS	1.33	4615-23	EMG	0.73
JAG-5B	MAKL	1.35	MS-4238	VAS	1.37	4615-44	EMG	0.80
JAG-5D	MAKL	1.36	4628-3	MMG	1.50	4615-36	EMG	0.85
JAG-8A	MAKL	1.37	MS-4236	VAS	1.63	4615-23	EMG	0.87
JAG-12	PHOT	1.75	MFA.575.2	MOCH	2.27	4615-32	EMG	0.96
JAG-7	MAKU	1.78				4615-13	EMG	1.01
4628-13	MMG	1.78	Cluster #5			4615-29	EMG	1.02
JAG-9	PHOT	1.78	JAG-14	AMMO	1.45	4615-10	EMG	1.12
JAG-1	MAKU	1.85	JAG-15	AMMO	1.50	4615-6	EMG	1.16
JAG-13	PHOT	2.08	JAG-3	MAKU	1.74	4615-2	EMG	1.18
4700-3	MMG	2.27	JAG-17	PAKH	2.04	4615-39	EMG	1.20
4628-19	MMG	2.40	4615-11	EMG	2.23	4615-35	EMG	1.32
JAG-6	PHOT	2.53	4615-22	EMG	2.35	4615-31	EMG	1.43
4628-14	MMG	2.68	4615-15	EMG	2.35	4615-18	EMG	1.48
4628-21	MMG	3.37	JAG-11	AMMO	2.47	4615-26	EMG	1.52
			4615-27	EMG	2.89	4615-7	EMG	1.59
Cluster #3						4615-47	EMG	1.60
4628-9	MMG	0.91	Cluster #6			4615-8	EMG	1.64
1938-21	KNOS	1.10	4628-15	MMG	0.40	4615-16	EMG	1.66
MFA575.1	MOCH	1.13	4700-1	MMG	0.40	4615-45	EMG	1.67
4700-10	MMG	1.13	60-19-14	ECRE	0.49	4615-40	EMG	1.67
4628-22	MMG	1.29	BAI-14-B	VAS	0.67	4615-21	EMG	1.68
4628-20	MMG	1.29	4700-9	MMG	0.68	4615-12	EMG	1.73
MFA575.3	MOCH	1.42	4628-7	MMG	0.72	JAG-2	MAKU	1.96
4700-5	MMG	1.84	1938.821	KNOS	0.88	JAG-4	MAKU	2.07
4615-17	EMG	2.45	MFA.9.573	MOCH	0.92	JAG-10	MAKU	2.14
4615-34	EMG	3.19	4628-6	MMG	0.99			

NOTE: I.D. = Locality (ident), D. C. = Euclidean distance from centroid. K-means clustering (K = 7, 5 iterations) based on eight elements (Ba, Ce, Co, Cr, Eu, Fe, Sc, Th). Cluster members are arranged in order of increasing Euclidean distance from the centroid.

izations. However, most of these analyses were done by emission spectroscopy, a technique much less sensitive to low trace-element concentrations than is neutron-activation analysis. The method used by Junghans's team, analyzing a very large number of artifacts and classifying them statistically, approaches only indirectly the question of the geographic source of the copper used in artifacts. Gilmore and Ottaway (1980) have described a combination of atomic-absorption spectroscopy and neutron-activation analyses for the chemical analysis of copper and copper-based alloys.

Friedman et al. (1966), Fields et al. (1971), and Bowman et al. (1975) used trace-element analyses to determine the type of geological ore from which copper was extracted. They divided ore types into native copper, oxide ores, and

Table 14.4. Summary of d^*-Values for Unknown vs. Fingerprints for Specimen Number 34-01-15D

Area	d^*-Value
Kingston, MI	.000000000034339479
Champion, MI	.00000000000000000
Centennial, MI	.00000000000000000
Isle Royale, MI	.00000000000060424
Alaska	.000000000002143598
Arizona	.000000000092267011
Snake River, MN	.000000000000056335
Wisconsin	.000000023223644233
Illinois	.00000000000000002
Lower Michigan	.004308931235278951*

*Fingerprint of best fit.

sulfide ores. Although their results have provided some potentially useful information, the experiments only indirectly addressed the question of the geographic origin of the copper.

Pre-Columbian inhabitants of North America did not smelt or alloy the relatively plentiful native copper used for the manufacture of weapons, implements, and jewelry. Goad and Noakes (1977) and our Archaeometry Laboratory (Rapp et al., 1980; Rapp, Allert, and Henrickson, 1983) have been successful in determining the native copper ore sources of North American prehistoric artifacts.

Another example of the use of d^* by our group is given in table 14.4. From a collection of what we have called float copper (surface finds) from Michigan, the trace-element fingerprint of the unknown is compared by the discriminant d^* with samples from: four Michigan mining localities (Kingston, Champion, Centennial, and Isle Royale); Alaska; Arizona; Snake River, Minnesota; Illinois; Wisconsin; and with Michigan float copper. As is evident from the values of the discriminant function, the unknown is assigned to Lake Michigan float copper (as it indeed should have been).

Table 14.5 illustrates the same system with an artifact of unknown provenance, a copper bar from the archaeological excavation at the McKinstry,

Table 14.5. Summary of d^*-Values for Unknown vs. Fingerprints for Specimen Number 34-1-82A (McKinstry MN, Bar Artifact)

Area	d^*-Value
Kingston, MI	.000006439059
Champion, MI	.000000000000
Centennial, MI	.000000000000
Isle Royale, MI	.000000028666
Snake River, MN	.000105815251*
Arizona	.000003576278
Alaska	.000000000111

*Fingerprint of best fit.

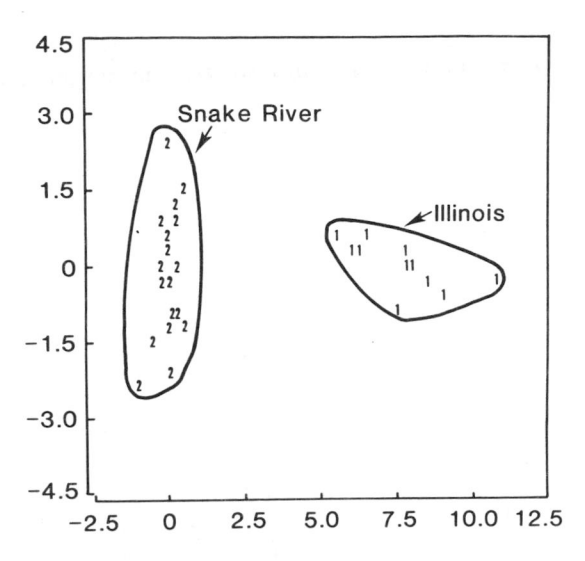

Fig. 14.3. Native copper samples from two localities, the Snake River area in Minnesota and a broad region in Illinois. Scatter plot of the trace-element composition clusters is projected onto the plane through the center of the clusters. Abscissa and ordinate are Mahalanobis distances from the central cluster.

Minnesota, site. As can be seen, the trace-element fingerprint would assign the source to the native copper deposits along Snake River, Minnesota.

Figure 14.3 illustrates the clarity with which cluster analysis can separate native copper samples from two regions on the basis of trace-element concentrations. Group 1 is composed of float-copper specimens from various locations within Illinois and group 2 is native copper specimens from the Snake River area in central Minnesota. Figure 14.3 is an orthogonal projection of the cluster data onto the plane passing through the center of the three most populous clusters. The ordinate and the abscissa represent positive and negative distances from the centroid of the initial cluster.

After copper and copper-based alloys, the most-studied metals are lead and silver. Lead ores, especially galena and cerussite, often contain a substantial amount of silver and probably were the chief source of silver in antiquity. Stos-Gale and Gale (1983) report lead-isotope and neutron-activation analyses for Bronze Age lead and silver artifacts from various sites in Greece. The analyses indicated that the mines at Lávrion were being exploited for silver and lead at least as early as the Middle Bronze Age (early second millennium B.C.). Gale and Stos-Gale (1981) show that nearly all lead and silver in the ancient Aegean came from either Lávrion, in Attica, or the Cycladic island of Siphnos.

In a slightly earlier study, Gentner et al. (1978) showed that argentiferous lead ores in the Aegean region can be distinguished from one another by lead-isotope and chemical composition. They identified mines in Lávrion, Macedonia, Thrace, Thasos, Siphnos, and Asia Minor. The source of the silver for all Athenian coins of the Archaic period analyzed was Lávrion, but coins from the nearby island of Aegina showed a variety of sources. Corinthian coins showed a pattern similar to those from Aegina.

Wyttenbach and Schubiger (1973) used neutron-activation analyses to study lead used in Roman times. Analysis of thirteen lead casting pigs and twenty water pipes showed no correlation in trace-element content between the pigs and mine samples, or between the water pipes and samples from the place where they were excavated. Remelting, alloying, or mixing may have destroyed any chemical patterns derived from the ores. With the success of lead-isotope analyses, the use of trace-element characterization cannot be recommended for lead.

Inclusions of gold alloys with other noble metals have been noted in ancient gold artifacts for the last century. In an intensive study, using electron-microprobe and related techniques, of platinum-group metal inclusions in ancient gold artifacts, Ogden (1977) concluded that the high frequency of occurrence of platinoid inclusions may preclude provenance determination based on the chemical characterization of these inclusions. Of the important placer-gold sources in antiquity, he found that only those in Wales and the Eastern Desert of Egypt were without recorded platinum-group associations. Using an energy-dispersive X-ray fluorescence spectrometer, Meeks and Tite (1980) examined platinum-group inclusions in gold jewelry and coins from the ancient Near East and eastern Mediterranean. They found that all the inclusions analyzed were of the iridium-osmium-ruthenium alloy type; no platinum-iridium inclusions were detected. They also conclude that these inclusions do not provide a basis for characterizing the source of the gold. On the other hand, Warren and Thompson (1944) determined the concentrations of 17 minor elements in gold from 66 source deposits throughout the world. The results showed that concentrations vary significantly from region to region. The differences can more readily be related to metallogenetic provinces than to the type of deposit. Since no large-scale attempt has been made to determine trace-element patterns using 15 to 20 elements in gold from placers important in antiquity, it may yet be possible to determine the source of artifact gold.

For the Bronze Age of the Old World, tin was the important alloy metal. Although some work has been done (Rapp, 1979), it is presently not possible to identify tin sources from trace-element analyses of either bronze or metallic tin. However, with the data base available at the Archaeometry Laboratory I can determine the source of most cassiterite (SnO_2), if this tin ore mineral is found at an ancient metallurgical site.

OTHER MATERIALS

The analytical and statistical methods used in provenance studies of archaeological materials are essentially the same as those used in authenticity studies in the art world of painting and sculpture. Indeed, the two fields grade into one another. Faience, a pre-glass or sintered glass made by coating a core material of powdered quartz with a vitreous alkaline glaze, has its origins in predynastic Egypt. Aspinall et al. (1972), McKerrell (1972), and others have used chemical characterization in provenance studies of faience beads.

A dramatic example of the use of chemical characterization in authenticity studies is the case of the Vinland Map. This map, the object of a publication by Skelton, Marston, and Painter (1965), was purported to have been made some fifty years before Columbus's first voyage to the New World. Using recently developed microanalytical sampling tools, McCrone and McCrone (1974) applied microprobe analysis to determine that the ink from the related source document, the Tartar Relation, was gallo-tannate based, but the ink from the Vinland Map was different from any early ink tested: its pigment was principally a refined product consisting of the mineral anatase (TiO_2). Anatase in this form has been available only since about 1920.

CONCLUSIONS AND CONSIDERATIONS

Provenance studies supply archaeologists, anthropologists, and historians with whole new data sets not available from other investigations. These data give evidence not only about trade and supply routes, but also somewhat less direct evidence about the technology of mining and transportation. Provenance studies grade imperceptibly into authenticity studies and also into compositional studies designed to investigate the technology of manufacturing and the utilization of raw materials.

These research problems require both artifactual and source materials, which are often difficult to procure. Critical artifacts are housed in museums, and permission for even small samples of these objects for destructive testing is invariably difficult to get. All possible source deposits must be carefully sought out and adequately sampled.

Although they are not new, provenance studies have in the last two decades reached an important level of sophistication with new and improved analytical and statistical techniques. Additional analytical methods will become available in the coming decades, but the development of more powerful statistical techniques probably will lag behind. It will be necessary to increase by a great deal our sophistication in establishing confidence and ambiguity levels in the assignment of sources from trace-element and isotope patterns.

Even with the recent proliferation of provenance studies, research opportunities for archaeological geologists and geochemists are essentially unlimited. Detailed research must be done to establish the best analytical and statistical methods for each exotic artifactual material regularly recovered in archaeological excavation. Frequently, each region presents a different array of patterns and therefore needs unique solutions.

Geologists and geochemists can work with chemists and others in archaeometry to increase the reliability of the methods and to increase the size and scope of the worldwide data base of quality-assured data. The usefulness and applicability of provenance investigations by chemical characterization will grow in direct proportion to the size and availability of the data base. This may require the establishment of additional laboratories in research universities and research museums. Although the results of provenance studies are published in a wide variety of journals and books, to date the main outlets have been two British journals: *Archaeometry* and *Journal of Archaeological Science*.

REFERENCES

Allen, R. O., A. H. Luckenbach, and C. G. Holland. 1975. The application of instrumental neutron activation analysis to a study of prehistoric steatite artifacts and source materials. *Archaeometry* 17:69–83.

Allen, R. O., and S. E. Pennell. 1977. Rare earth element distribution patterns to characterize soapstone artifacts. In *Archaeological chemistry II*, ed. G. Carter 230–57. Advances in Chemistry Series, no. 171. New York: American Chemical Society.

Asaro, F., H. V. Michel, R. Sidrys, and F. Stross. 1978. High-precision chemical characterization of major obsidian sources in Guatemala. *American Antiquity* 43:436–43.

Aspinall, A., S. E. Warren, J. G. Crummett, and R. G. Newton. 1972. Neutron activation analysis of faience beads. *Archaeometry* 14:27–40.

Barbieri, M., G. Calderoni, C. Cortesi, and M. Fornaseri. 1974. Huntite, a mineral used in antiquity. *Archaeometry* 16:211–20.

Beck, C. W., A. B. Adams, G. C. Southard, and C. Fellows. 1971. Determination of the origin of Greek amber artifacts by computer classification of infrared spectra. In *Science and archaeology*, ed. R.H. Brill, 235–40. Cambridge, MA: MIT Press.

Beck, C. W., J. Greenlie, M. P. Diamond, A. M. Macchiarulo, A. A. Hannenberg, and M. S. Hauck. 1978. The chemical identification of Baltic amber at the Celtic oppidum Stare Hradisko in Moravia. *Journal of Archaeological Science* 5:343–54.

Bieber, A. M., Jr., D. W. Brooks, G. Harbottle, and E. V. Sayre, 1976a. Application of multivariate techniques to analytical data on Aegean ceramics. *Archaeometry* 18:59–74.

———. 1976b. Compositional groupings of some ancient Aegean and eastern Mediterranean pottery. *Atti Dei Convegni Lincei* (Rome) 11:111–43.

BMDP. 1979. *Biomedical computer programs*, P-Series. Berkeley: Univ. of California.

Bowman, H. R., F. Asaro, and I. Perlman. 1973. Composition variations in obsidian sources and the archaeological implications. *Archaeometry* 15:123–27.

Bowman, R., A. M. Friedman, J. Lerner, and J. Milsted. 1975. A statistical study of the impurity occurrences in copper ores and their relationship to ore types. *Archaeometry* 17:157–63.

Brill, R. H., I. L. Barnes, and B. Adams. 1973. Lead isotopes in some ancient Egyptian objects. In *Recent advances in science and technology of materials*, vol 3, ed. A. Beshay, 9–27. New York: Plenum.

Bromund, R. H., N. W. Bower, and R. H. Smith. 1976. Inclusions in ancient ceramics: An approach to the problem of sampling for chemical analysis. *Archaeometry* 18:218–21.

Brooks, D., A. M. Bieber, Jr., G. Harbottle, and E. V. Sayre. 1975. Biblical studies through activation analysis of ancient pottery. In *Archaeological chemistry*, ed. C. Beck. Advances in Chemistry Series, no. 138. New York: American Chemical Society.

Cann, J. R., and C. Renfrew. 1964. The characterization of obsidian and its application to the Mediterranean region. *Proceedings of the Prehistoric Society* 30:111–23.

Carriveau, G. W. 1980. Contamination of hard-fired ceramics during sampling with diamond burrs. *Archaeometry* 22:209–10.

Catling, H. W., A. E. Blin-Stoyle, and E. E. Richards. 1961. Spectrographic analyses of Mycenaean and Minoan pottery. *Archaeometry* 4:31–38.

Catling, H. W., E. E. Richards, and A. E. Blin-Stoyle. 1963. Correlations between composition and provenance of Mycenaean and Minoan pottery. *Annual of the British School of Archaeology at Athens* 58:94–115.

Charles, J. A. 1973. Heterogeneity in metals. *Archaeometry* 15:105–14.

Cobean, R. H., M. D. Coe, E. A. Perry, Jr., K. K. Turekian, and D. P. Kharkar. 1971. Obsidian trade at San Lorenzo Tenochtitlán, Mexico. *Science* 174:666–71.

Conforto, L., M. Felici, D. Monna, L. Serva, and A. Taddeucci. 1975. A preliminary evaluation of chemical data (trace element) from Classical marble quarries in the Mediterranean. *Archaeometry* 17:201–13.

de Bruin, M., P. J. M. Korthoven, C. C. Bakels, and F. C. A. Groen. 1972. The use of non-destructive activation analysis and pattern recognition in the study of flint artefacts. *Archaeometry* 14:55–63.

de Bruin, M., P. J. M. Korthoven, A. J. v.d. Steen, J. P. W. Houtman, and R. P. W. Duin. 1976. The use of trace element concentrations in the identification of objects. *Archaeometry* 18:75–83.

Dennen, W. H. 1967. Trace elements in quartz as indicators of provenance. *Bulletin of the Geological Society of America* 78:125–30.

Dickinson, W. R., and R. Shutler, Jr. 1974. Probable Fijian origin of quartzose temper sands in prehistoric pottery from Tong and the Marquesas. *Science* 185:454–57.

———. 1979. Petrography of sand tempers in Pacific island potsherds: Summary: *Bulletin of the Geological Society of America* 90:993–95.

Dixon, J. E., J. R. Cann, and C. Renfrew. 1968. Obsidian and the origins of trade. *Scientific American* 218:38–46.

Doran, J. E., and F. R. Hodson. 1975. *Mathematics and computers in archaeology*. Cambridge, MA.: Harvard Univ. Press.

Ericson, J. E., and J. Kimberlin. 1977. Obsidian sources, their chemical characterization, and hydration rates in west Mexico. *Archaeometry* 19:157–66.

Farnsworth, M., I. Perlman, and F. Asaro. 1977. Corinth and Corfu: A neutron activation study of their pottery. *American Journal of Archaeology* 81:455–68.

Fields, P. R., J. Milsted, E. Henrickson, and R. Ramette. 1971. Trace impurity patterns in copper ores and artifacts. In *Science and archaeology*, ed. R. Brill, 131–43. Cambridge, MA: MIT Press.

Filippakis, S. E., A. P. Grimanis, and B. Perdikatsis. 1981. X-ray and neutron activation analysis of obsidians from Kitsos Cave. *Science and Archaeology* 23:21–26.

Fornaseri, M., L. Malpieri, and L. Tolomeo. 1975. Provenance of pumices in the north coast of Cyprus. *Archaeometry* 17:112–16.

Friedman, A. M., M. Conway, M. Kastner, J. Milsted, D. Metta, P. R. Fields, and E. Olson. 1966. Copper artifacts: Correlation with source types of copper ores. *Science* 152:1504–06.

Friedman, A. M., and J. Lerner. 1977. Spark source mass spectrometry in archaeological chemistry. In *Archaeological chemistry II*, ed. G. Carter, 70–78. Advances in Chemistry Series, no. 171. New York: American Chemical Society.

Gale, N. H. 1981. Mediterranean obsidian source characterization by strontium isotope analysis. *Archaeometry* 23:41–51.

Gale, N. H., and Z. A. Stos-Gale. 1981. Lead and silver in the ancient Aegean. *Scientific American* 245:176–91.

———. 1982. Bronze Age copper sources in the Mediterranean: A new approach. *Science* 216:11–19.

Gentner, W., O. Müller, G. A. Wagner, and N. H. Gale. 1978. Silver sources of Archaic Greek coinage. *Naturwissenschaften* 65:273–84.

Gilmore, G. R., and B. S. Ottaway. 1980. Micromethods for the determination of trace-elements in copper-based metal artifacts. *Journal of Archaeological Science* 7:241–54.

Goad, S., and J. Noakes. 1977. Prehistoric copper artifacts in the eastern United States. In *Archaeological chemistry II*, ed. G. Carter, 335–46. Advances in Chemistry Series no. 171. New York: American Chemical Society.

Gordus, A. A., G. A. Wright, and J. B. Griffin. 1968. Obsidian sources characterized by neutron activation analysis. *Science* 161:382–84.

Griffin, J. B., A. A. Gordus, and G. A. Wright. 1969. Identification of the sources of Hopewellian obsidian in the Middle West. *American Antiquity* 34:1–14.

Gritton, V., and N. M. Magalousis. 1977. Atomic absorption spectroscopy of archaeological ceramic materials. In *Archaeological chemistry II*, ed. G. Carter 258–70. Advances in Chemistry Series, no. 171. New York: American Chemical Society.

Hammond, N., G. Harbottle, and T. Gazard. 1976. Neutron activation and statistical analysis of Maya ceramics and clays from Lubaantun, Belize. *Archaeometry* 18: 147–68.

Harbottle, G. 1970. Neutron activation analysis of potsherds from Knossos and Mycenae. *Archaeometry* 12:23–34.

Hassan, A. A., and F. A. Hassan. 1981. Source of galena in predynastic Egypt at Nagada. *Archaeometry* 23:77–82.

Heizer, R. F., F. Stross, T. R. Hester, A. Albee, I. Perlman, F. Asaro, and H. Bowman. 1973. The Colossi of Memnon revisited. *Science* 182:1219–25.

Hughes, M. J., M. R. Crowell, and P. T. Craddock. 1976. Atomic absorption techniques in archaeology. *Archaeometry* 18:19–37.

Huntley, D. J., and D. C. Bailey. 1978. Obsidian source identification by thermoluminescence. *Archaeometry* 20:159–70.

Ives, D. J. 1975. Trace element analyses of archaeological materials. *American Antiquity* 40:235–36.

Junghans, S., E. Sanmeister, and M. Schroder. 1960. *Metallanalysen Kupferzeitlicher und Frühbronzezeitlicher Bodenfunde aus Europa*. Berlin: Mann.

———. 1968. *Kupfer und Bronze in der Frühen Metallzeit Europas*. Berlin: Mann.

———. 1974. *Kupfer und Bronze in der Frühen Metallzeit Europas*. Berlin: Mann.

Keller, J. 1980. Prehistoric pumice tephra on Aegean islands. In *Thera and the Aegean world, II*, ed. C. Doumas. London: Thera and the Aegean World.

Kohl, P. L., G. Harbottle, and E. V. Sayre. 1979. Physical and chemical analyses of soft stone vessels from Southwest Asia. *Archaeometry* 21:131–59.

Kostikas, A., A. Simopoulos, and N. H. Gangas. 1974. Mossbauer studies of ancient pottery. *Journal de Physique Colloque C1* 35:107–15.

———. 1976. Analyses of archaeological artifacts. In *Applications of Mossbauer spectroscopy*, vol. 1, 241–61.

Krywonos, W., G. W. A. Newton, V. J. Robinson, and J. A. Riley. 1982. Neutron activation analyses of some Roman and Islamic coarse wares of western Cyrenaica and Crete. *Journal of Archaeological Science* 9:63–78.

Lambert, J. B., and C. D. McLaughlin. 1976. X-ray photoelectron spectroscopy: A new analytical method for the examination of archaeololgical artifacts. *Archaeometry* 18:169–80.

Leach, B. F., and A. J. Anderson. 1978. The prehistoric soures of Palliser Bay obsidian. *Journal of Archaeological Science* 5:301–07.

Longworth, G., and S. E. Warren. 1979. The application of Mossbauer spectroscopy to the characterization of western Mediterranean obsidian. *Journal of Archaeological Science* 6:179–93.

Luedtke, B. E. 1979. The identification of sources of chert artifacts. *American Antiquity* 44:744–57.

McCrone, W. C., and L. B. McCrone. 1974. The strange case of the Vinland Map, VIII: The Vinland Map ink. *Geographical Journal* 140:212–14.

McKerrell, H. 1972. On the origins of British faience beads and some aspects of the Wessex-Mycenae relationship. *Proceedings of the Prehistoric Society* 38:286–301.

McKerrell, H., and R. F. Tylecote. 1972. The working of copper-arsenic alloys in the early Bronze Age and the effect on the determination of provenance. *Proceedings of the Prehistoric Society* 38:209–18.

Maskelyne, N. S. 1877. Stonehenge: The petrology of its stones. *Wiltshire Archaeological and Natural History Magazine* 17:147–60.

Meeks, N. D., and M. S. Tite. 1980. The analyses of platinum-group element inclusions in gold antiquities. *Journal of Archaeological Science* 7:267–75.

Michels, J. W. 1982. Bulk element composition versus trace element composition in the reconstruction of an obsidian source system. *Journal of Archaeological Science* 9:113–23.

Muir, I. D., and F. Hivernel. 1976. Obsidians from the Melka-Konture prehistoric site, Ethiopia. *Journal of Archaeological Science* 3:211–17.

Nelson, D. E., J. M. D'Auria, and R. B. Bennett. 1975. Characterization of Pacific Northwest Coast obsidian by X-ray fluorescence analysis. *Archaeometry* 17:85–97.

Ogden, J. M. 1977. Platinum group metal inclusions in ancient gold artifacts. *Journal of the History of Metallurgy Society* 11:53–72.

Patton, W. W., and T. P. Miller. 1970. A possible bedrock source for obsidian found in archaeological sites in northwestern Alaska. *Science* 169:760–61.

Perlman, I., and F. Asaro. 1969. Pottery analysis by neutron activation. *Archaeometry* 11:21–52.

Pollard, A. M., G. D. Bussell, and D. C. Baird. 1981. The analytical investigation of early Bronze Age jet-like material from the Devizes Museum. *Archaeometry* 23:139–67.

Poole, A. B., and L. R. Finch. 1972. The utilization of trace chemical composition to correlate British post-Medieval pottery with European kiln site materials. *Archaeometry* 14:79−91.

Rapp, G., Jr. 1979. Trace elements as a guide to the geographical source of tin ore: Smelting experiments. In *The search for ancient tin*, ed. A. D. Franklin, J. S. Olin, and T. A. Wertime, 59−63. Washington, DC: Smithsonian Institution.

Rapp, G., Jr., S. R. B. Cooke, and E. Henrickson. 1973. Pumice from Thera (Santorini) identified from a Greek mainland archaeological excavation. *Science* 179:471−73.

Rapp, G., Jr., E. Henrickson, M. Miller, and S. Aschenbrenner. 1980. Trace-element fingerprinting as a guide to the geographic sources of native copper. *Journal of Metals* 32:35−45.

Rapp, G., Jr., J. Allert, and E. Henrickson. 1984. Trace-element discrimination of discrete sources of native copper. In *Archaeological chemistry III*, ed. J. Lambert. Advances in Chemistry Series no. 205. New York: American Chemical Society.

Rapp, G., Jr., and J. A. Gifford. 1984. Neutron activation and cluster analysis. In *East Cretan white-on-dark ware*, ed. P. Betancourt. Monograph of the University Museum, University of Pennsylvannia.

Renfrew, C., J. R. Cann, and J. E. Dixon. 1965. Obsidian in the Aegean. *Annual of the British School of Archaeology in Athens* 60:225−47.

Shackleton, N., and C. Renfrew. 1970. Neolithic trade routes re-aligned by oxygen isotope analyses. *Nature* 228:1062−65.

Sieveking, G. de G., P. T. Craddock, M. J. Hughes, P. Bush, and J. Ferguson. 1970. Characterization of prehistoric flint mine products. *Nature* 228:251−54.

Sigleo, A. C. 1975. Turquoise mine and artifact correlation for Snaketown, Arizona. *Science* 189:459−60.

Skelton, R. A., T. E. Marston, and G. D. Painter. 1965. *The Vinland Map and the Tartar Relation*. New Haven: Yale Univ. Press.

Stos-Gale, Z. A., and N. H. Gale. 1982. Analyses of Mycenaean lead and silver artefacts from Mycenae, Perati, Vapheio and the Athenian Acropolis and Agora: Evidence for Bronze Age working of Lávrion. *Journal of Field Archaeology* 9:467−85.

Stross, F. H., H. R. Bowman, H. V. Michek, F. Asaro, and N. Hammond. 1978. Mayan obsidian: Source correlation for southern Belize artifacts. *Archaeometry* 20:89−93.

Stukeley, W. 1740. *Stonehenge: A temple restored to the British Druids*. London.

Thomas, H. H. 1923. The source of the stones of Stonehenge. *Antiquaries Journal* 3:239−60.

Tubb, A., A. J. Parker, and G. Nickless. 1980. The analysis of Romano-British pottery by atomic absorption spectrophotometry. *Archaeometry* 22:153−171.

Tylecote, R. F., H. A. Ghaznavi, and P. J. Boydell. 1977. Partitioning of trace-elements between the ores, fluxes, slags and metal during the smelting of copper. *Journal of Archaeological Science* 4:305−33.

Vitaliano, C. J., and D. B. Vitaliano. 1974. Volcanic tephra on Crete. *American Journal of Archaeology* 78:19−24.

Warashina, T., Y. Kamaki, and T. Higashimua. 1978. Sourcing of sanukite implements by X-ray fluorescence analysis, II. *Journal of Archaeological Science* 5:283−91.

Warren, H. V., and R. M. Thompson. 1944. Minor elements in gold. *Economic Geology* 39:457−71.

Warren, S. E. 1973. Geometrical factors in the neutron activation analysis of archaeological specimens. *Archaeometry* 15:115−22.

Wilson, A. L. 1978. Elemental analysis of pottery in the study of its provenance: A review. *Journal of Archaeological Science* 5:219–36.

Wyttenbach, A., and P. A. Schubiger. 1973. Trace element content of Roman lead by neutron activation analysis. *Archaeometry* 15:199–207.

Yellin, J., I. Perlman, F. Asaro, H. V. Michel, and D. F. Mosier. 1978. Comparison of neutron activation analyses from the Lawrence Berkeley Laboratory and the Hebrew University. *Archaeometry* 20:95–100.

Appendix:
A Selective Bibliography of
Archaeological Geology

GEORGE RAPP, JR., and JOHN A. GIFFORD

The literature of this subfield is scattered among a wide variety of journals, excavation reports, and books whose focus is other than archaeological geology. The quantity of publications in English alone would put a comprehensive bibliography well beyond the scope of an appendix. In this selective bibliography we have excluded (1) all references included in the chapters in this book and (2) the references included in E. H. Sellards's two readily accessible bibliographies on the antiquity of man in America: (1) "Early Man in America, Index to Localities and Selected Bibliography," *Bulletin of the Geological Society of America* 51(1940):373–432, and (2) "Early Man in America, Index to Localities and Selected Bibliography, 1940–1945," *Bulletin of the Geological Society of America* 58(1947):955–78.

We have concentrated on those publications of historical, theoretical, or practical importance. Clearly there will be important omissions in any bibliography of this nature; also, regrettably, time and resources did not allow serious consideration of most foreign publications. We have tried to provide at least a representative selection of publications in each topic area. We have not used a rigorous definition of archaeological geology: some peripheral items have been included chiefly because the investigator/author was a geologist. Much archaeological geology of regional importance is buried in site reports of limited distribution; the archaeological literature should be consulted for such items. Many individuals have contributed entries for this selective bibliography and we appreciate their assistance. If in some cases this has led to errors in the citations, we apologize in advance.

The bibliography is arranged first by topic heading and then alphabetically by author. There is a special topical section for those items published before 1900 and there is also a Miscellaneous section. Many of the topical sections need some explanation. Paleontology, for the purpose of this bibliography, is a broad category designed to include both vertebrate and invertebrate paleontology and a brief selection of items in palynology, geobotany, and phytolith studies. This is a nonstandard definition of paleontology, but it serves to indicate the role of geoscientists trained in the analyses of faunal and floral remains. Archaeologists use the term *zooarchaeology* to include much of what is included here. As zooarchaeological references number in the many hundreds, this selection is somewhere between a limited introduction to the field and a representative cross section. Very little is included about the direct study of human physical remains. This is a distinct and well-developed field usually called *physical anthropology* or *paleoanthropology*. For an introduction to one aspect of this field, see D. Morse, *Ancient Disease in the Midwest* (Reports of Investigations, no. 15, Illinois State Museum, 1978).

For provenance studies we have a reference library of over five times as many items as have been included here and in chapter 14. Some were not included because they are repetitious, many others because the information or methodologies have been superseded as better techniques were developed.

Although none of the topical sections is comprehensive, some, such as Paleoclimatology and Geochronology, are extremely limited introductions to the scope of the interactions between geology and archaeology. The field of radiocarbon analysis and chronology is extensive and no effort has been made here to present even a cross section of the important references.

The topic headings in this selective bibliography are, in order, Nineteenth-Century Sources, Ceramic Petrography, Environmental Geoarchaeology, Geochronology and Dating, Geomorphic Studies, Geophysical Prospecting, Hominid Studies, Lithic Materials, Metals and Mining, Miscellaneous, Paleoclimatology, Paleontology, Provenance Studies, Archaeological Sediments and Site Geology.

NINETEENTH-CENTURY SOURCES

Appy, E. P. 1889. Ancient mining in America. *American Antiquarian* 2:92–99.

Babbitt, F. E. 1880. Ancient quartz workers and their quarries in Minnesota: *American Antiquarian* 3(1):18–23.

Becker, G. F. 1891. Antiquities from under Table Mountain, Tuolumne County, in California. *Bulletin of the Geological Society of America* 2:189–200.

Bent, J. T. 1885. On the gold and silver mines of Siphnos. *Journal of Helladic Studies*, pp. 195–98.

Blake, W. P. 1899a. Aboriginal turquoise mining in Arizona and New Mexico. *American Antiquarian* 21 (5):278–84.

———. 1899b. The Pliocene skull of California and the flint implements of Table Mountain. *Journal of Geology* 7:631–37.

Borlase, W. 1753. Of the great alterations which the islands of Scilly have undergone since the time of the ancients. *Royal Society Philosophical Transactions* (London) 48: 55–69.

Desnoyers, J. 1863. Notes sur des indices matériels de la coexistence de l'homme avec *l'Elephas meridionalis* dans un terrain des environs de Chartres, plus ancien que les terrains transport quaternaires des vallées de la Somme et de la Seine. *Comptes rendus d'Académie de Sciences* (Paris) 56:1073–83.

Foster, J. W. 1870. On recent discoveries in ethnology as connected with geology. *Transactions of the American Association for the Advancement of Science, 1870* (Presidential Address).

———. 1874. *Pre-historic races of the United States of America.* 3d ed. Chicago: S. C. Griggs.

Fowke, G. 1894. Material for aboriginal stone implements. *Archaeologist* 2:328–35.

Geikie, J. 1894. *The Great Ice Age and its relation to the antiquity of man.* 3d ed. London: John Murray.

Holmes, W. H. 1879. Notes on an extensive deposit of obsidian in Yellowstone National Park. *American Naturalist* 13 (4):247–50.

———. 1891. Aboriginal novaculite quarries in Garland County, Arkansas. *American Anthropologist* o.s. 4:313–15.

———. 1892. Modern quarry refuse and the Paleolithic theory. *Science* 20:295–97.

———. 1893. Distribution of stone implements in the Tidewater country. *American Anthropologist* 6:1–14.

Jackson, W. H. 1876. Ancient ruins in southwest Colorado. *Annual Report of the U.S. Geological and Geographical Survey of the Territories, 1874*, pp. 367–81. Washington, DC.

Judd, J. W. 1887. On the unmaking of flints. *Proceedings of the Geologists Association* 10:217–26.

Koch, H. C. 1860. Mastodon remains, in the State of Missouri, together with evidence of the existence of man contemporaneously with the mastodon. *Transactions of the Academy of Science of St. Louis*, vol. 1, 1856–1860, pp. 61–64.

Lubbock, J. 1862. On the evidences of the antiquity of man afforded by the physical structure of the Somme Valley. *Natural History Review*, pp. 244–69.

McGee, W. J. 1889. An obsidian implement from a Pleistocene deposit in Nevada. *American Anthropologist* 2:301–12.

Maskelyne, N. S. 1877. Petrology of Stonehenge. *Wiltshire Archaeological and Natural History Magazine* 17:147–60.

Mercer, H. C. 1893. Trenton and Somme gravel specimens compared with ancient quarry refuse in America and Europe. *American Naturalist* 27:962–78.

———. 1894. Indian jasper mines in the Lehigh Hills. *American Anthropologist* 7:80–92.

Rau, C. 1881. Aboriginal stone drilling. *American Naturalist* 15:536–42.

Sellers, G. E. 1886. Observations on stone-chipping. *Annual Report, Smithsonian Institution, for 1885*, pp. 871–91.

Spurrell, F. C. J. 1885. Early embankments on the margins of the Thames Estuary. *Archaeological Journal* 42:269–302.

Stevens, E. T. 1870. Flint chips. In *A guide to prehistoric archaeology*. London: Bell and Daly.

Taylor, W. 1898. The pueblos and ancient mines near Allison, New Mexico. *American Antiquarian* 22:258–61.

de Vibraye, M. 1864. Note sur des nouvelles prévues de l'existence de l'homme dans le centre de la France à une époque ou s'y trouvaient aussi divers animaux qui de nos jours n'habitent pas cette contrée. *Comptes rendus d'Académie de Sciences* (Paris) 58:409–16.

Vilanova y Piera, J., and J. de la Rada y Delgado. 1894. *Geología y protohistoria ibéricas*. Madrid.

CERAMIC PETROGRAPHY

Arnold, D. E. 1972. Mineralogical analysis of ceramic materials from Quinua, Department of Ayacucho, Peru. *Archaeometry* 14:93–102.

Boardman, J., and F. Schweizer. 1973. Clay analyses of archaic Greek pottery. *Annual of the British School at Athens* 68:267–83.

Cornwall, I.W., and H. W. M. Hodges. 1964. Thin sections of British Neolithic pottery: Windmill Hill—a test site. *Bulletin of the University of London, Institute of Archaeology* 4:29–33.

Courtois, L. 1976. Examen au microscope petrographique des céramiques archéologiques. *Notes et monographies archéologiques*, no. 8 (Centre de Recherches Archéologiques, Paris).

Cowgill, U. M., and G. E. Hutchinson. 1969. A chemical and mineralogical examination of the ceramic sequence from Tikal, El Petén, Guatemala. *American Journal of Science* 267:465–77.

Dickinson, W. R. and R. Shutler, Jr. 1971. Temper sands in prehistoric pottery of the Pacific Islands. *Archaeology and Physical Anthropology in Oceania* 6:191–203.

———. 1979. Petrography of sand tempers in Pacific Islands potsherds. *Bulletin of the Geological Society of America* 90:1644–1701.

Einfalt, H. C. 1978. Chemical and mineralogical investigations of sherds from the Akrotiri excavations. In *Thera and the Aegean World*, vol. 1, ed. C. Doumas, 459–69. London: Thera and the Aegean World.

Farnsworth, M. 1964. Greek pottery: A mineralogical study. *American Journal of Archaeology* 68:221–28.

———. 1970. Corinthian pottery: Technical studies. *American Journal of Archaeology* 74:9–20.

Felts, W. M. 1942. A petrographic examination of potsherds from ancient Troy. *American Journal of Archaeology* 46:237–44.

Heimann, R. 1977. A simple method for estimation of the macroporosity of ceramic sherds by a replica technique. *Archaeometry* 19:55.

Hodges, H. W. M. 1962. Thin sections of prehistoric pottery: An empirical study. *Bulletin of the University of London, Institute of Archaeology* 3:58–68.

———. 1963. The examination of ceramic materials in thin section. In *The scientist and archaeology*, ed. E. W. Pyddoke, 101–10. London: Phoenix House.

Lazzarini, L., S. Calogero, N. Burriesci, and M. Petrera. 1980. Chemical, mineralogical and Mossbauer studies of Venetian and Paduan Renaissance sgraffito ceramics. *Archaeometry* 22:57–68.

Maggetti, M., and G. Galetti. 1980. Compostion of Iron Age fine ceramics from Châtillon-s-Glâne (Kt. Fribourg, Switzerland) and the Heuneburg (Kr. Sigmaringen, West Germany). *Journal of Archaeological Science* 7:87–91.

Maniatis, Y., and M. S. Tite. 1978. Ceramic technology in the Aegean World during the Bronze Age. In *Thera and the Aegean World*, vol. 1, ed. C. Doumas, 483–92. London: Thera and the Aegean World.

———. 1981. Technological examination of Neolithic–Bronze Age pottery from central and southeast Europe and from the Near East. *Journal of Archaeological Science* 8:59–76.

Matson, F. R. 1943. Technologic notes on pottery. In *The excavations at Dura Europas*, ed. N. Toll. New Haven: Yale Univ. Press.

———. 1955. Ceramic archaeology. *American Ceramic Society Bulletin* 34:33–44.

———. 1960. The quantitative study of ceramic materials. *Viking Fund Publications in Anthropology* 28:34–59.

———. ed. 1965. *Ceramics and man*. Chicago: Aldine.

———. 1972. Ceramic studies. In *The Minnesota Messenia expedition: Reconstructing a*

Bronze Age regional environment, ed. W. A. McDonald and G. Rapp, Jr., 200–24. Minneapolis. Univ. of Minnesota Press.

Noll, W. 1978. Material and techniques of the Minoan ceramics of Thera and Crete. In *Thera and the Aegean World*, vol. 1, ed. C. Doumas, 493–505. London: Thera and the Aegean World.

Peacock, D. P. S. 1968. A petrological study of certain Iron Age pottery from western England. *Proceedings of the Prehistoric Society* 34:414–27.

Rice, P. M., ed. 1982. *Pots and potters: Current approaches to ceramic archaeology*. University Park: Pennsylvania State Univ. Press.

Van der Leuw, S. E. 1976. *Studies in the technology of ancient pottery*. Amsterdam: Univ. of Amsterdam Press.

Williams, D. F. 1978. A petrological examination of pottery from Thera. In *Thera and the Aegean World*, vol. 1, ed. C. Doumas, 507–14. London: Thera and the Aegean World.

Williams, H. 1956. Petrographic notes on tempers of pottery from Chupicuaro, Cerro del Telpalcate and Tecomán, Mexico. *American Philosophical Society Transactions* 46: 576–82.

Williams, J. L., and D. A. Jenkins. 1976. The use of petrographic, heavy mineral and arc spectrographic techniques in assessing the provenance of sediments used in ceramics. In *Geoarchaeology*. ed. D. A. Davidson and M. L. Shackley, 115–35. London: Duckworth.

ENVIRONMENTAL GEOARCHAEOLOGY

Adams, R. M., and H. J. Nissen. 1972. *The Uruk countryside: The natural setting of urban societies*. Chicago: Univ. of Chicago Press.

Agrawal, D. P., and B. M. Pande, eds. 1977. *Ecology and archaeology of western India*. Delhi: Concept Publishing Co.

Allchin, B, A. Goudie, and K. T. M. Hegde. 1972. Prehistory and environmental change in western India: A note on the Budha Pushkar Basin, Rajasthan. *Man* 7 (4):541–64.

Bar-Yosef, O., and E. Tchernov. 1972. *On the palaeoecological history of the site of 'Ubeidiya*. Jerusalem: Israel Academy of Sciences and Humanities.

Baumhoff, M. A. 1963. Ecological determinants of aboriginal California populations. *Univ. of California Publications in American Archaeology and Ethnology* 49:155–236.

Bennett, J. W. 1976. *The ecological transition: Cultural anthropology and human adaptation*. New York: Pergamon.

Blanc, A. C. 1957. On the Pleistocene sequence of Rome: Paleoecologic and archeologic correlations. *Quaternaria* 4:108–09.

Bryan, A. L. 1975. Paleoenvironments and cultural diversity in late Pleistocene South America: A rejoinder to Vance Haynes and a reply to Thomas Lynch. *Quaternary Research* 5:151–59.

Buckland, P. C. 1974. Archaeology and environment in York. *Journal of Archaeological Science* 1:303–16.

Butzer, K. W. 1974. Geological and ecological perspectives on the middle Pleistocene. *Quaternary Research* 4:136–48.

————. 1975a. Paleo-ecology of South African Australopithecines: Taung revisited. *Current Anthropology* 15:367–82, 420–26.

————. 1975b. Patterns of environmental change in the Near East during late Pleistocene and early Holocene times. In *Problems in prehistory: North Africa and the Levant*, ed. F. Wendorf and A. E. Marks, 389–410. Dallas: SMU Press.

————. 1977. Environment, culture and human evolution. *American Scientist* 65: 572–84.

————. 1978a. The late prehistoric environmental history of the Near East. In *The environmental history of the Near and Middle East*, ed. W. C. Brice, 5–12. New York: Academic Press.

———— 1978b. Toward an integrated, contextual approach in archaeology. *Journal of Archaeological Science* 5:191–93.

————. 1981a. Long-term Nile flood variation and political discontinuities in pharaonic Egypt. In *The causes and consequences of food production in Africa*, ed. J. D. Clark, and S. Brandt. Berkeley: Univ. of California Press.

————. 1981b. Rise and fall of Axum, Ethiopia: A geo-archaeological interpretation. *American Antiquity* 46:471–95.

————. 1982. Archaeology as human ecology. Cambridge: Cambridge Univ. Press.

Butzer, K. W., and H. B. S. Cooke. 1981. Palaeoecology of the African continent. In *Cambridge history of Africa*, vol. 1, ed. J. D. Clark. Cambridge: Cambridge Univ. Press.

Clark, J. D. 1964. The influence of environment in inducing culture change at the Kalambo Falls prehistoric site. *South African Archaeological Bulletin* 19:93–101.

Costantini, L., and M. Tosi. 1978. The environment of southern Sistan in the third millennium B.C., and its exploitation by the protourban Hilmand civilization. In *The environmental history of the Near and Middle East*, ed. W. C. Brice, 165–83. New York: Academic Press.

Dakaris, S. I., E. S. Higgs, and R. W. Hey, 1964. The climate, environment and industries of Stone Age Greece: Part I. *Proceedings of the Prehistoric Society* 30:187–99.

Dimbleby, G. W. 1977. *Ecology and archaeology*. London: Edward Arnold.

Euler, R., G. Gumerman, T. Karlstrom, J. Dean and R. Hevly. 1979. The Colorado plateaus: Cultural dynamics and paleoenvironments. *Science* 205:1089–1101.

Evans, J. G. 1975. *The environment of early man in the British Isles*. London: Paul Elek.

————. 1978. *An introduction to environmental archaeology*. Ithaca, NY: Cornell Univ. Press.

Evans, J. G., and S. Limbrey, eds. 1975. *The effect of man on the landscape: The Highland Zone*. London: Council for British Archaeology.

Gill, E. D. 1968. Palaeoecology of fossil human skeletons. *Palaeogeography, Palaeoclimatology, Palaeoecology* 4:211–17.

Haynes, C. V. 1974. Paleoenvironments and culture diversity in late Pleistocene South America: A reply to A. L. Bryan. *Quaternary Research* 4:378–82.

Haynes, C. V., Jr. 1976. Ecology of early man in the New World. In *Ecology of the Pleistocene*, ed. B. F. Perkins, 71–76. *Geoscience and man*, vol. 13. Baton Rouge: Louisiana State Univ. School of Geoscience.

Higgs, E. 1978. Environmental changes in northern Greece. In *The environmental history of the Near and Middle East*, ed. W. C. Brice. New York: Academic Press.

Higgs, E. S., and C. Vita-Finzi. 1966. The climate, environment and industries of Stone Age Greece: Part II. *Proceedings of the Prehistoric Society* 32:1–29.

Higgs, E. S., C. Vita-Finzi, D. R. Harris, and A. Fagg. 1967. The climate, environment and industries of Stone Age Greece: Part III. *Proceedings of the Prehistoric Society* 33:1–29.

Higgs, E. S., and D. Webley. 1971. Further information concerning the environment of prehistoric man in Epirus. *Proceedings of the Prehistoric Society* 37:367–80.

Hobler, P. M., and J. J. Hester. 1969. Prehistory and environment in the Libyan desert. *South African Archaeological Bulletin* 23:120–30.

Holbrook, S. J., and J. C. Mackey. 1976. Prehistoric environmental change in northern New Mexico: Evidence from A. Gallina phase archaeological site. *Kiva* 41.

Hole, F., K. V. Flannery, and J. A. Neely. 1969. *Prehistory and human ecology of the Deh Luran Plain*. Memoirs of the Museum of Anthropology, no. 1. Ann Arbor, MI: Univ. of Michigan.

Horowitz, A. 1975. The Pleistocene paleoenvironments of Israel. In *Problems in prehistory: North Africa and the Levant*, F. Wendorf and A. E. Marks, 207–27. Dallas: SMU Press.

Hours, F. 1975. The lower Paleolithic of Lebanon and Syria. In *Problems in prehistory: North Africa and the Levant*, F. Wendorf and A. E. Marks, 249–71. Dallas: SMU Press

Hughes, J. D. 1975. *Ecology of ancient civilizations*. Albuquerque: Univ. of New Mexico Press.

Kirkby, A. V. 1973. *The use of land and water resources in the past and present valley of Oaxaca*. Memoirs of the Museum of Anthropology 5:1–174. Ann Arbor, MI: Univ. of Michigan.

Lubell, D., F. A. Hassan, A. Gautier, and J. L. Ballais. 1976. The Capsian escargotieres: An interdisciplinary study elucidates Holocene ecology and subsistence in North Africa. *Science* 191:910–20.

de Lumley, H. 1975. Cultural evolution in France in its paleoecological setting during the middle Pleistocene. In *After the Australopithecines*, ed. K. W. Butzer and G. L. Isaac, 745–808. Chicago: Aldine.

Mann, A. 1976. Ecology of early man in the Old World. In *Ecology of the Pleistocene*, ed. B. F. Perkins, 61–70. Baton Rouge: Louisiana State Univ. School of Geoscience.

Mehringer, P. J., Jr., and C. V. Haynes, Jr. 1965. The pollen evidence for the environment of early man and extinct mammals at the Lehner mammoth site, southeastern Arizona. *American Antiquity* 31 (1):17–23.

Pearson, R. 1977. Paleoenvironments and human settlements in Japan and Korea. *Science* 197:1239–46.

Rapp, G., Jr., S. E. Aschenbrenner, and J. C. Kraft. 1978. The Holocene environmental history of the Nichoria region. In *Excavations at Nichoria in southwest Greece*. vol. 1, *Site, environs, and techniques*, ed. G. Rapp, Jr., and S. E. Aschenbrenner, 13–25. Minneapolis: Univ. of Minnesota Press.

Sahlins, M. D. 1977. Culture and environment. In *Horizons of anthropology*, ed. S. Tax and L. G. Freeman, 215–31. Chicago: Aldine.

Sahni, M. R. 1955. Biogeological evidence bearing on the decline of the Indus valley civilization. *Journal of the Palaeontological Society of India* 1.

Said, R., and B. Issawi. 1965. Preliminary results of a geological expedition to lower

Nubia and to Kurkur and Dungul Oases. In *Contributions to the prehistory of Nubia*, ed. F. Wendorf, 1–28. Dallas: SMU Press.

Schoenwetter, J. A. 1981. Prologue to a contextual archaeology. *Journal of Archaeological Science* 8:367–79.

Shackleton, N. J. 1970. Stable isotopic study of the palaeoenvironments of the Neolithic site of Nea Nikomedeia, Greece. *Nature* 227:943–44.

———. 1973. Oxygen isotope analysis as a means of determining season of occupation of prehistoric midden sites. *Archaeometry* 15:133–43.

Simmons, I. G. 1969. Evidence for vegetation changes associated with Mesolithic man in Britain. In *The domestication and exploitation of plants and animals*, ed. P. J. Ucko, and G. W. Dimbleby, 110–19. London: Duckworth.

Simmons, I. G., and V. B. Proudfoot. 1969. Environment and early man on Dartmoor, Devon, England. *Proceedings of the Prehistoric Society* 35:203–19.

Simmons, I. G., and M. J. Tooley, eds. 1981. *The environment in British prehistory*. Ithaca, NY: Cornell Univ. Press.

Singh, G. 1971. The Indus valley culture seen in the context of postglacial climatic and ecological studies in northwest India. *Archaeology and Physical Anthropology in Oceania* 6 (2):177–89.

van Zeist, W. 1969. Reflections on prehistoric environments in the Near East. In *The domestication and exploitation of plants and animals*, ed. P. J. Ucko and G. W. Dimbleby. Chicago: Aldine.

Wendorf, F., and J. J. Hester. 1962. Early man's utilization of the Great Plains environment. *American Antiquity* 28:159–71.

West, R. C., ed. 1964. *Natural environment and early cultures. Handbook of middle American Indians*, vol. 1. Austin: Univ. of Texas Press.

Wright, H. E., Jr. 1968. Natural environment of early food production north of Mesopotamia. *Science* 161:334–39.

———. 1970. Environmental changes and the origin of agriculture in the Near East. *BioScience* 20:210–13.

GEOCHRONOLOGY AND DATING

Antevs, E. 1959. Geological age of the Lehrner Mammoth site. *American Antiquity* 25:31.

Aronson, J. L., T. J. Schmitt, R. C. Walter, M. Taieb, J. J. Tiercelin, D. C. Johanson, C. W. Naeser, and A. E. M. Nairn. 1977. New geochronologic and paleomagnetic data for the hominid-bearing Hadar Formation, Ethiopia. *Nature* 267:323–27.

Bada, J. L., and R. Protsch. 1973. Racemization reaction of aspartic acid and its use in dating fossil bones. *Proceedings of the National Academy of Sciences* 70:1331–34.

Bell, W. T. 1977. Thermoluminescence dating: Revised dose-rate data. *Archaeometry* 19:99–100.

Berger, R. 1979. Radiocarbon dating with accelerators. *Journal of Archaeological Science* 6:101–04.

Berggren, W. A. 1980. Towards a Quaternary time scale. *Quaternary Research* 13:277–302.

Bischoff, J., and R. Rosenbauer. 1981. Uranium series dating of human skeletal remains from the Del Mar and Sunnyvale sites, California. *Science* 213:1003–05.

Borden, C. E. 1966. Radiocarbon and geological dating of the lower Fraser Canyon archaeological sequence. *Proceedings of the 6th International Conference on Radio-carbon and Tritium Dating (Pullman, 1965).*

Brown, F. H., and K. R. Lajoie. 1971. Radiometric age determinations on Pliocene/Pleistocene formations in the lower Omo Basin, southern Ethiopia. *Nature* 229: 483–85.

Brown, F. H., and W. P. Nash. 1976. Radiometric dating and tuff mineralogy of Omo Group deposits. In *Earliest man and environments in the Lake Rudolf Basin*, ed. Y. Coppens, F. C. Howell, G. L. Isaac, and R. E. F. Leakey, 50–63. Chicago: Univ. of Chicago Press.

Brown, F. H., R. T. Shuey, and M. K. Croes. 1978. Magnetostratigraphy of the Shungura and Usno formations, southwestern Ethiopia: New data and comprehensive re-analysis. *Geophysical Journal of the Royal Astronomical Society* 54:519–38.

Bryan, K. 1941. *Correlation of the deposits of Sandia Cave, New Mexico, with the glacial chronology.* Smithsonian Miscellaneous Collections, vol. 99, no. 23.

Burchell, J. P. T. 1957. Land shells as a critical factor in the dating of post-Pleistocene deposits. *Proceedings of the Prehistoric Society* 23:236–38.

Burleigh, R. 1974. Radiocarbon dating: Some practical considerations for the archaeologist. *Journal of Archaeological Science* 1:69–88.

Camps, G., G. Delibrias, and J. Thommeret. 1968. Chronologie absolue et succession des civilisations préhistoriques dans le nord de l'Afrique. *Libyca* 16:9–28.

Curtis, G. H., R. E. Drake, T. E. Cerling, B. W. Cerling, and J. Hampel. 1975. Age of the KBS tuff in Koobi Fora formation, northern Kenya. *Nature* 258:395–98.

Douglass, A. E. 1921. Dating our prehistoric ruins. *Natural History* 21 (1):27–30.

Eighmy, J. L., R. S. Sternberg, and R. F. Butler. 1980. Archaeomagnetic dating in the American Southwest. *American Antiquity* 45:507–17.

Evernden, J. F., and G. H. Curtis. 1965. The potassium-argon dating of late Cenozoic rocks in east Africa and Italy. *Current Anthropology* 6:343–85.

Fairbridge, R. W. 1962. New radiocarbon dates of Nile sediments. *Nature* 196:108–10.

Ferguson, C. W. 1970. Concepts and techniques of dendrochronology. In *Scientific methods in medieval archaeology*, ed. R. Berger. Berkeley: Univ. of California Press.

Findlater, I. C., F. J. Fitch, J. A. Miller, and R. T. Watkins. 1974. Dating of the rock succession containing fossil hominids at East Rudolf, Kenya. *Nature* 251:213–15.

Fitch, F. J., and J. A. Miller. 1970. Radioisotopic age determinations of Lake Rudolf artifact site. *Nature* 226:226–28.

Fitch, F. J., P. J. Hooker, and J. A. Miller. 1976a. $^{40}Ar/^{39}Ar$ dating of the KBS tuff in the Koobi Fora formation, East Rudolf, Kenya. *Nature* 263:740–44.

Fitch, F. J., and J. A. Miller. 1976b. Conventional potassium-argon and argon-40/argon-39 dating of volcanic rocks from East Rudolf. In *Earliest man and environments in the Lake Rudolf Basin*, ed. Y. Coppens, F. C. Howell, G. L. Isaac, and R. E. F. Leakey, 123–47. Chicago: Univ. of Chicago Press.

Fleischler, R. L., P. B. Price, R. M. Walker, and L. S. B. Leakey. 1965. Fission-track dating of Bed I, Olduvai Gorge. *Science* 148:72–74.

Fleming, S. 1976. *Dating in archaeology: A guide to scientific techniques.* New York: St. Martin's Press.

———. 1980. *Thermoluminescence: Techniques in archaeology.* London: Oxford Univ. Press.

Flint, R. F. 1959. On the basis of Pleistocene correlation in East Africa. *Geological Magazine* 96:265–84.

Friedman, I., and F. Trembour. 1978. Obsidian, the dating stone. *American Scientist* 66:44–51.

Friedrich, W. L., and H. Pichler. 1976. Radiocarbon dates of Santorini volcanics. *Nature* 262:373–74.

Fritts, H. C. 1965. Dendrochronology. In *The Quaternary of the United States*, ed. H. E. Wright and D. G. Frey. Princeton, NJ: Princeton Univ. Press.

Frye, J. C. 1973. Pleistocene succession of central interior United States. *Quaternary Research* 3:275–83.

de Geer, A. 1910. A geochronology of the last 12,000 years. *Comptes rendus du XIme Congrès Géologique International*. Stockholm.

de Geer, G. 1934. Equatorial Paleolithic varves in East Africa. *Geografiska Annaler* 16:75–96.

Göksu, H. Y. 1979. The TL age determination of fossil human footprints. *Archaeo-Physica* 10:455–62. Proceedings of the 18th International Symposium on Archaeometry and Archaeological Prospection, Bonn.

Gromme, C. S., T. A. Reilly, A. E. Mussett, and R. L. Hay. 1970. Paleomagnetism and potassium-argon ages of volcanic rocks of Ngorongoro Caldera, Tanzania. *Geophysical Journal of the Royal Astronomical Society* 22:101–15.

Grootes, P. 1978. Carbon-14 time scale extended: Comparison of chronologies. *Science* 200:11–15.

Harbottle, G., E. Sayre, and R. Stoenner. 1979. Carbon-14 dating of small samples by proportional counting. *Science* 206:683–85.

Hare, P. E. 1974. Amino acid dating—a history and an evaluation. *MASCA Newsletter* 10 (1):4–7.

Hassan, A., J. Termine, and C. V. Haynes. 1977. Mineralogical studies on bone apatite and their implications for radiocarbon dating. *Radiocarbon* 19 (3):364–74.

Haynes, C. V., 1967. Carbon-14 dates and early man in the New World. In *Pleistocene extinctions: The search for a cause*, ed. P. S. Martin, and H. E. Wright, 267–86. New Haven: Yale Univ. Press.

———. 1968. Geochronology of late Quaternary alluvium. In *Means of correlation of Quaternary successions*, ed. R. B. Morrison and H. E. Wright, 591–631. Salt Lake City: Univ. of Utah Press.

———. 1980a. Geochronology of Wadi Tushka: Lost tributary of the Nile. *Science* 210:68–71.

———. 1980b. Radiocarbon evidence for Holocene recharge of groundwater, Western Desert, Egypt. *Radiocarbon* 22.

Hennig, G. J., and U. Bangert. 1979. Dating of Pleistocene calcite formations by disequilibria in the uranium decay series. *Archaeo-physica* 10:464–76. Proceedings of the 18th International Symposium on Archaeometry and Archaeological Prospection, Bonn.

Horowitz, A., G. Siedner, and O. Bar-Yosef. 1973. Radiometric dating of the Ubeidiya formation, Jordan Valley, Israel. *Nature* 242:186–87.

Howell, F. C., G. H. Cole, M. R. Kleindienst, B. J. Szabo, and K. P. Oakley. 1972. Uranium series dating of bone from the Isimila prehistoric site, Tanzania. *Nature* 237:51–52.

Hurford, A. J., A. J. W. Gleadow, and C. W. Naeser. 1976. Fission-track dating of pumice from the KBS tuff, East Rudolf, Kenya. *Nature* 263:738–40.

Libby, W. F. 1963. Accuracy of radiocarbon dates. *Science* 140:278–80.

Maugh, T. H. 1978. Radiodating: Direct detection extends range of the technique. *Science* 200:635–38.

Michels, J. W. 1973. *Dating methods in archaeology*. New York: Academic Press.

Morrison, I. A. 1976. Comparative stratigraphy and radiocarbon chronology of Holocene marine changes on the western seaboard of Europe. In *Geoarchaeology*, ed. D. A. Davidson, and M. L. Shackley, 159–73. London: Duckworth.

Movius, H. L. 1960. Radiocarbon dates and upper Paleolithic archaeology in central and western Europe. *Current Anthropology* 1:355–91.

Opdyke, N. D., et al. 1979. Magnetic polarity stratigraphy and vertebrate paleontology of the Upper Siwalik subgroup of northern Pakistan. *Palaeogeography, Palaeoclimatology, Palaeoecology* 27:1–34.

Pewe, T. L. 1954. The geological approach to dating archaeological sites. *American Antiquity* 20:51–61.

Preece, R. C. 1980. The biostratigraphy and dating of the tufa deposit at the Mesolithic site at Blashenwell, Dorset, England. *Journal of Archaeological Science* 7:345–62.

Purdy, B. A., and D. E. Clark. 1979. Weathering studies of chert: A potential solution to the chronology problem in Florida. *Archaeo-physica* 10:440–50. Proceedings of the 18th International Symposium on Archaeometry and Archaeological Prospection, Bonn.

Rubin, M., R. C. Likens, and E. G. Berry. 1963. On the validity of radiocarbon dates from snail shells. *Journal of Geology* 71:84–89.

Schwarcz, H. P. 1980. Absolute age determination of archaeological sites by uranium series dating of travertines. *Archaeometry* 22:3–24.

Schwarcz, H. P., P. D. Goldberg, and B. Blackwell. 1980. Uranium series dating of archaeological sites in Israel. *Israel Journal of Earth-Sciences* 29:157–65.

Seward, D., G. A. Wagner, and H. Pichler. 1980. Fission track ages of Santorini volcanics (Greece). In *Thera and the Aegean World II*, ed. C. Doumas, 101–08. London: Thera and the Aegean World.

Singer, R., J. Wymer, and B. Gladfelter. 1973. Radiocarbon dates from Hoxne, Suffolk. *Journal of Geology* 81:508–09.

Suess, H. E. 1967. Bristlecone pine calibration of the radiocarbon time scale from 4100 B.C. to 1500 B.C. *Proceedings of the Monaco Symposium on Radiocarbon Dating and Methods of Low-level Counting*, pp. 143–51.

Vita-Finzi, C. 1973. *Recent earth history*. London: Macmillan.

Wendorf, F., R. Schild, and R. Said. 1970. Problems of dating the late Paleolithic age in Egypt. In *Radiocarbon variations and absolute chronology*, ed. I. U. Olsson, 57–79. New York: Wiley Interscience.

Wintle, A. G. 1980. Thermoluminescence dating: A review of recent applications to non-pottery materials. *Archaeometry* 22:113–22.

Wright, H. E., Jr. 1958. Geologic dating in prehistory. *Archaeology* 11:19–25.

Zimmerman, D. W. 1971a. Thermoluminescence dating using fine grains from pottery. *Archaeometry* 13:29–52.

———. 1971b. Uranium distribution in archaeologic ceramics: Dating of radioactive inclusions. *Science* 174:818–19.

GEOMORPHIC STUDIES

Akeroyd, A. V. 1972. Archaeological and historical evidence for subsidence in southern Britain. *Philosophical Transactions of the Royal Society* (London) A272:151–71.

Albanese, J. 1977. Paleotopography and Paleoindian sites in Wyoming and Colorado. In *Paleoindian lifeways*, ed. E. Johnson, 28–47. *Texas Tech University Museum Journal*, no. 17.

van Andel, T., T. Jacobson, J. Jolly, and N. Lianos. 1980. Late Quaternary history of the coastal zone near Franchthi Cave, southern Argolid, Greece. *Journal of Field Archaeology* 7:389–402.

Arnold, B. A. 1957. Late Pleistocene and recent changes in landforms, climate, and archaeology in central Baja California. *Univ. of California Publications in Geography* 10:201–318.

Barker, G. 1981. *Landscape and society: Prehistoric central Italy*. New York: Academic Press.

Bayer, J. 1912. Das geologisch-archäologische Verhaltnis im Eiszeitalter. *Zeitschrift für Ethnologie* 44(1):1–22.

Benedict, J. 1981. *The Fourth of July Valley—glacial geology and archaeology of the timberline ecotone*. Center for Mountain Archaeology Research Report no. 2.

Besancon, J., and F. Hours. 1971. Préhistoire et géomorphologie: les formes du relief et les dépôts quaternaires dans la région de Joub Jannine (Beqaa meridionale, Liban). *Hannon* 6:29–135.

Boon, G. C. 1980. Caerleon and the Gwent levels in early historic times. In *Archaeology and coastal change*, ed. F. H. Thompson, 24–36. London: Society of Antiquaries.

Brooks, M., and D. Colquhoun. 1979. Preliminary archaeological and geological evidence for Holocene sea level fluctuations . . . . *Florida Anthropologist* 32(3):85–103.

Butzer, K. W. 1959. Contributions to the Pleistocene geology of the Nile Valley. *Erdkunde* 13:46–67.

———. 1960a. On the Pleistocene shorelines of Arabs' Gulf, Egypt. *Journal of Geology* 68:622–37.

———. 1960b. Remarks on the geography of settlement in the Nile Valley during Hellenistic times. *Bulletin, Société de Géographie d'Eqypte* 33:5–36.

———. 1963. Climatic-geomorphologic interpretation of Pleistocene sediments in the Eurafrican subtropics. *Viking Fund Publications in Anthropology* 36:1–27.

———. 1964. Pleistocene geomorphology and stratigraphy of the Costa Brava region, Catalonia. *Abhandlungen der Akademie der Wissenschaften Lit. (Mainz), Mathematisch-naturwissenshaftliche Klasse* 1:1–51.

———. 1965a. Desert landforms at the Kurkur Oasis, Egypt. *Annals of the Association of American Geographers* 55:578–91.

———. 1965b. Physical conditions in eastern Europe, western Asia, and Egypt before the period of agricultural and urban settlement. *Cambridge Ancient History*, rev. ed. fasc. 33, vol. 1, chap. 2.

———. 1966. Geologie und Paläogeographie archäologischer Fundstellen bei Sayala (Unternubien). *Denkschriften, Österreichische Akademie der Wissenschaften (Wien), Philosophische-historische Klasse* 92:89–98.

———. 1967. Late Pleistocene deposits of the Kom Ombo Plain, Upper Egypt. *Alfred Rust Festschrift, Fundamenta B-2*, pp. 213–27.

———. 1969. Geomorphological observations in the lower Omo Basin, southwestern Ethiopia. *Colloquium Geographicum, Troll-Festschrift* 12:177–92.

———. 1971. *Recent history of an Ethiopian delta*. Univ. of Chicago, Department of Geography, Research Paper 136.

————. 1979. Pleistocene history of the Nile Valley in Egypt and Lower Nubia. In *The Sahara and the Nile*, ed. Williams M. A. J. and H. Faure, 248–76. Rotterdam: A. Balkema.

Butzer, K. W., and D. M. Helgren. 1972. Late Cenozoic evolution of the Cape Coast between Knysna and Cape St. Francis, South Africa. *Quaternary Research* 2:143–69.

Caton-Thompson, G., and E. W. Gardner. 1932. The prehistoric geography of the Kharga Oasis. *Geographical Journal* 80:369–409.

Crabtree, K. 1971. The Overton Down experimental earthwork, Wiltshire, 1968: Geomorphology of the area. *Proceedings of the University of Bristol Speleological Society* 12:237–44.

Cunliffe, B. W. 1980. The evolution of Romney Marsh: A preliminary statement. In *Archaeology and coastal change*, ed. F. H. Thompson, 37–55. London: Society of Antiquaries.

Davidson, D. 1972. Terrain adjustment and prehistoric communities. In *Man, settlement and urbanism*, ed. P. J. Ucko, R. Tringham, and G. W. Dimbleby, 17–22. London: Duckworth.

Davidson, D. A., C. Renfrew, and C. Tasker. 1976. Erosion and prehistory in Melos. *Journal of Archaeological Science* 3:219–27.

Delano-Smith, C. 1976. The Tavoliere of Foggia (Italy): An aggrading coastland and its early settlement patterns. In *Geoarchaeology*, ed. D. A. Davidson and M. L. Shackley, 197–211. London: Duckworth.

————. 1978. Coastal sedimentation, lagoons and ports in Italy. In *Papers in Italian archaeology I*, ed. H. M. Blake, T. W. Potter, and D. B. Whitehouse, 25–33. British Archaeological Reports, Supplementary Series 4. Oxford.

Devoy, R. J. 1980. Post-galcial environmental change and man in the Thames Estuary: A synopsis. In *Archaeology and coastal change*, ed. F. H. Thompson, 134–48. London: Society of Antiquaries.

Dorrell, P. 1972. A note on the geomorphology of the country near Umm Dabaghiyah. *Iraq* 34:69–72.

Drew, D. L. 1979. Early man in North America and where to look for him: Geomorphic contexts. *Plains Anthropologist* 24:269–81.

Eisma, D. 1978. Stream deposition and erosion by the eastern shore of the Aegean. In *The environmental history of the Near and Middle East*, ed. W. C. Brice, 67–81. New York: Academic Press.

El-Baz, F., et al. 1980. Journey to the Gilf Debir and Uweinat, southwest Egypt, 1978. *Geographic Journal* 146:51–93.

Erinc, S. 1978. Changes in the physical environment in Turkey since the end of the last glacial. In *The environmental history of the Near and Middle East*, ed. W. C. Brice, 87–109. New York: Academic Press.

Erol, O. 1978. The Quaternary history of the lake basins of central and southern Anatolia. In *The environmental history of the Near and Middle East*, ed. W. C. Brice, 111–39. New York: Academic Press.

Farrand, W. R. 1965. Geology and physiography of the Beysehir-Sugla depression, western Taurus Lake district, Turkey. *Turk Arkeoloji Dergisi* 13:149–54.

Flemming, N. C. 1968. Holocene earth movements and eustatic sea level change in the Peloponnese. *Nature* 217:1031–32.

————. 1969. *Archaeological evidence for eustatic change of sea level and earth move-*

ments in the western Mediterranean during the last 2,000 years. Special Paper 109, The Geological Society of America.

―――. 1978. Thera as the tectonic focus of the South Aegean: Archaeological evidence from the Aegean margin. In Thera and the Aegean World, vol. 1, ed. C. Doumas, 81−84. London: Thera and the Aegean World.

Grove, A. T. 1958. The ancient erg of Hausaland and similar formations on the south side of the Sahara. Geographical Journal 124:528−33.

Hansen, C. L., and K. W. Butzer. 1966. Early Pleistocene deposits of the Nile Valley in Egyptian Nubia. Quaternaria 8:177−85.

Harris, S. A., and R. M. Adams. 1957. A note on canal and marsh stratigraphy near Zubediyah. Sumer 13:157−63.

Hassan, F. 1981. Historical Nile floods and their implications for climatic change. Science 212:1142−45.

Haynes, C. V., P. Mehringer, Jr., and S. Zaghloul. 1979. Pluvial lakes of northwestern Sudan. Geographical Journal 145:437−45.

Hobler, P. M., and J. J. Hester. 1969. Prehistory and environment in the Libyan Desert. South African Archaeologial Bulletin 33:120−30.

Jardine, W. G., and A. Morrison. 1976. The archaeological significance of Holocene coastal deposits in south-western Scotland. In Geoarchaeology, ed. D. A. Davidson and M. L. Shackley, 175−95. London: Duckworth.

Jelgersma, S., et al. 1970. The coastal dunes of the western Netherlands: Geology, vegetational history and archaeology. Mededelingen Rijks Geologische Dienst 21: 93−167.

Johnson, D. (with A. Junger). 1980. Was there a Quaternary land bridge to the northern Channel Islands? In The California islands: Proceedings of a multidisciplinary symposium, ed. D. M. Powers, 33−39. Santa Barbara Museum of Natural History, California.

Jones, G. D. B. 1980. Archaeology and coastal change in the Northwest. In Archaeology and coastal change, ed. F. H. Thompson, 87−102. London: Society of Antiquaries.

Judson, S. 1963. Erosion and deposition in Italian stream valleys during historic time. Science 140:898−99.

Larsen, C. E. 1975. The Mesopotamian Delta region: A reconsideration of Lees and Falcon. Journal of the American Oriental Society 95:43−57.

―――. 1983. Life and land use on the Bahran Islands: The geoarchaeology of an ancient society. Chicago: Univ. of Chicago Press.

Larsen, C. E., and G. Evans. 1978. The Holocene geological history of the Tigris-Euphrates-Karun Delta. In The environmental history of the Near and Middle East, ed. W. C. Brice, 227−44. New York: Academic Press.

Lees, G. M., and N. L. Falcon. 1952. The geographical history of the Mesopotamian plains. Geographical Journal 118:24−39.

Louwe Kooijmans, L. P. 1980. Archaeology and coastal change in the Netherlands. In Archaeology and coastal change, ed. F. H. Thompson, 106−33. London: Society of Antiquaries.

Loy, W. G. 1967. The land of Nestor. Office of Naval Research Report 34, National Research Council (U.S.A.).

―――. 1970. The land of Nestor: A physical geography of the southwest Peloponnese. Washington, DC: National Academy of Sciences Report no. 34.

Loy, W. G., and H. E. Wright, Jr. 1972. The physical setting. In The Minnesota Messenia expedition: Reconstructing a Bronze Age regional environment, ed. W. A. McDonald and G. Rapp, Jr., 36−46. Minneapolis: Univ. of Minnesota Press.

McIntire, W. C. 1958. *Prehistoric Indian settlements of the changing Mississippi River delta*. Baton Rouge: Louisiana State Univ. Coastal Studies Series, no. 1.

Michael, H. N., ed. 1964. *The archaeology and geomorphology of northern Asia*. Toronto: Univ. of Toronto Press.

Morrison, I. A. 1973. Geomorphological investigation of marine and lacustrine environments of archaeological sites, using diving techniques. In Science Diving, ed. N. C. Flemming. Proceedings of the 3rd Symposium of the Scientific Committee of the Confédération Mondiale des Activitiés Subaquatiques. London: BSAC.

Moss, J. H. 1951. Glaciation in the Wind River Mountains and its relation to early man in the Eden Valley, Wyoming. *University of Pennsylvania, Museum Monographs*, pp. 9−92.

Paepe, R. 1969. Geomorphic surfaces and Quaternary deposits of the Adami area (southeast Attica). *Thorikos* 4:7−52.

Pirazzoli, P. A. 1976. Sea level variations in the northwest Mediterranean during Roman times. *Science* 194:519−21.

Raikes, R. L. 1966. The physical evidence for Noah's flood. *Iraq* 28:52−63.

Raphael, C. N. 1973. Late Quaternary changes in coastal Elis, Greece. *Geographical Review* 63:73−89.

─────. 1978. The erosional history of the Plain of Elis in the Peloponnese. In *The environmental history of the Near and Middle East*, ed. W. C. Brice, 51−66. New York: Academic Press.

Rapp, G., Jr. 1978. The physiographic setting. In *Site, environs, and techniques*, ed. G. Rapp, Jr., and S. E. Aschenbrenner, 26−30. *Excavations At Nichoria in southwest Greece*, vol. 1. Minneapolis: Univ. of Minnesota Press.

Rapp, G., Jr., and J. C. Kraft. 1978. Aegean sea level changes in the Bronze Age. In *Thera and the Aegean World*, vol. 1, ed. C. Doumas, 183−94. London: Thera and the Aegean World.

Segovia, A. 1977. Archaeologic implications of the geomorphic evolution of the lower Shenandoah Valley, Virginia. *Annals of the New York Academy of Science* 288:189−93.

Seth, S. K. 1978. The desiccation of the Thar desert and its environs during the protohistorical and historical periods. In *The environmental history of the Near and Middle East*, ed. W. C. Brice, 279−305. New York: Academic Press.

Shepard, F. P. 1964. Sea level changes in the past 6000 years: Possible archaeological significance. *Science* 143:574−76.

Simmons, B. B. 1980. Iron age and Roman coasts around the Wash. In *Archaeology and coastal change*, ed. F. H. Thompson, 56−73. London: Society of Antiquaries.

Sneh, A., and T. Weissbrod. 1973. Nile delta: The defunct Pelusiac branch identified. *Science* 180:59−61.

Stafford, T., Jr. 1981. Alluvial geology and archaeological potential of the Texas southern High Plains. *American Antiquity* 46:548−66.

Vita-Finzi, C. 1966. Historical marine levels in Italy. *Nature* 209:906.

─────. 1978. Recent alluvial history in the catchment of the Arabo-Persian Gulf. In *The environmental history of the Near and Middle East*, ed. W. C. Brice, 255−61. New York: Academic Press.

Weide, D. L., and M. L. Weide. 1973. Application of geomorphic data to archaeology: A comment. *American Antiquity* 38:428−31.

Wilkinson, T. J. 1978. Erosion and sedimentation along the Euphrates Valley in Northern

Syria. In *The environmental history of the Near and Middle East*, ed. W. C. Brice, 215–26. New York: Academic Press.

Wright, H. E., Jr. 1952. The geological setting of four prehistoric sites in northeastern Iraq. *Bulletin of the American Schools of Oriental Research*, no. 128:11–24.

———. 1962. Late Pleistocene geology of coastal Lebanon. *Quaternaria* 6:525–39.

———. 1978. Glacial fluctuations, sea-level changes, and catastrophic floods. In *Atlantis: Fact or fiction*, ed. E. S. Ramage, 161–74. Bloomington: Indiana Univ. Press.

Zeuner, F., I. Cornwall, and D. Kirkbride. 1961. The shoreline chronology of the Palaeolithic of the Abri Zumoffen rockshelter near Adlun, South Lebanon. *Bulletin du Musée de Beyrouth* 16:49–60.

GEOPHYSICAL PROSPECTING

Aitken, M. J. 1961. *Physics and archaeology*. New York: Wiley Interscience.

———. 1969. Magnetic location. In *Science in archaeology*, ed. D. Brothwell, and E. Higgs. London: Thames and Hudson.

———. 1970. Magnetic prospecting. In *Scientific methods in medieval archaeology*, ed. R. Berger. Berkeley: Univ. of California Press.

Breiner, S., and M. D. Coe. 1972. Magnetic exploration of the Olmec civilization. *American Scientist* 60:566–75.

Carson, H. H. 1962. A seismic survey at Harper's Ferry. *Archaeometry* 5:119–23.

Hesse, A., and S. Renimel. 1979. Can we survey the limits of an archaeological site? The example of St. Romain en Gal (France). *Archaeo-Physica* 10:638–46. Proceedings of the 18th International Symposium on Archaeometry and Archaeological Prospection, Bonn.

Howell, M. I. 1968. The soil conductivity anomaly detector (SCM) in archaeological prospection. *Prospezione Archeologiche* 3:101–04.

Klasner, J. S., and P. Calengas. 1981. Electrical resistivity and soil studies at Orendorf archaeological site, Illinois: A case of study. *Journal of Field Archaeology* 8:167–74.

Morrison, H. F. 1971. High-sensitivity magnetometers in archaeological exploration. *University of California, Archaeological Research Facility, Contributions* 12:6–20.

Mullins, C. 1974. The magnetic properties of the soil and their application to archaeological prospecting. *Archaeo-Physica* 5:143–47.

Parchas, C., and A. Tabbagh. 1979. Simultaneous measurements of electrical conductivity and magnetic susceptibility of the ground in electromagnetic prospecting. *Archaeo-Physica* 10:682–91. Proceedings of the 18th International Symposium on Archaeometry and Archaeological Prospection, Bonn.

Ralph, E. K. 1969. Archaeological prospecting. *Expedition* 11(2):14–21.

Rapp, G., Jr., and E. Henrickson. 1972. Geophysical exploration. In *The Minnesota Messenia expedition: Reconstructing a Bronze Age regional environment*, ed. W. A. McDonald and G. Rapp, Jr., 234–39. Minneapolis: Univ. of Minnesota Press.

Scollar, I., and I. Graham. 1974. A method of determination of the total magnetic moment of soil samples in a constant field. *Prospezioni Archaeologiche* 7:85–92.

Stanley, J. M. 1979. The application of geophysical methods to hunter/gatherer prehistory in Australia. *Archaeo-Physica* 10:692–99. Proceedings of the 18th International Symposium on Archaeometry and Archaeological Prospection, Bonn.

Tite, M. S. 1961. Alternative instruments for magnetic surveying: Comparative tests at the Iron Age Hill-Fort at Rainsborough. *Archaeometry* 4:85–90.

———. 1972. The influence of geology on the magnetic susceptibility of soils on archaeological sites. *Archaeometry* 14:229–36.

Tite, M. S., and C. Mullins. 1971. Enhancement of the magnetic susceptibility of soils on archaeological sites. *Archaeometry* 13:209–19.

HOMINID STUDIES

Aronson, J. L., and M. Taieb. 1981. Geology and paleogeography of the Hadar hominid site, Ethiopia. In *Hominid sites*, 165–96. See Rapp and Vondra, 1981.

Bartholomew, G. A., and J. B. Birdsell. 1953. Ecology and the protohominids. *American Anthropologist* 55:481–98.

Bar-Yosef, O. 1975. Archeological occurrences in the middle Pleistocene of Israel. In *After the Australopithecines*, ed. K. W. Butzer and G. L. Isaac, 571–604. Chicago: Aldine.

Behrensmeyer, A. K. 1970. Preliminary geological interpretation of a new hominid site in the Lake Rudolf Basin. *Nature* 226:225–26.

———. 1974. Late Cenozoic sedimentation in the Lake Rudolf Basin, Kenya. *Annals of the Geological Survey of Egypt* 4:287–306.

———. 1976. Lothogam Hill, Kanapoi, and Ekora: A general summary of stratigraphy and faunas. In *Earliest man and environments in the Lake Rudolf Basin*, ed. Y. Coppens, F. C. Howell, G. L. Isaac, and R. E. F. Leakey, 163–70. Chicago: Univ. of Chicago Press.

———. 1978. Correlation of Plio-Pleistocene sequences in the northern Lake Turkana Basin—a summary of evidence and issues. In *Background to early man in Africa*, ed. W. W. Bishop, 421–40. Geological Society of London, Special Publication no. 5.

Bishop, W. W. 1963. The later Tertiary and Pleistocene in eastern equatorial Africa, with implications for primate and human distributions. In *African ecology and human evolution*, ed. F. C. Howell, and F. Bourleire, 246–75. Viking Fund Publications in Anthropology, vol. 36.

———. 1965. Quaternary geology and geomorphology in the Albertine Rift Valley, Uganda. In *International studies on the Quaternary*, ed. H. E. Wright, Jr., and D. G. Frey, 295–321. Geological Society of America Special Paper no. 84.

———. 1971. The Late Cenozoic history of East Africa in relation to hominoid evolution. In *The late Cenozoic glacial ages*, ed. K. Turekian, 493–527. New Haven: Yale Univ. Press.

Bishop, W. W., and M. Posnansky. 1960. Pleistocene environments and early man in Uganda. *Uganda Journal* 24:44–61.

Black, D., ed. 1933. Fossil man in China: The Choukoutien cave deposits with a synopsis of our present knowledge of the late Cenozoic in China. *Memoirs of the Geological Survey of China, Series A.*

Bonnefille, R. 1976. Palynological evidence for an important change in the vegetation of the Omo Basin between 2.5 and 2 million years. In *Earliest man and environments in the Lake Rudolf Basin*, ed. Y. Coppens, F. C. Howell, G. L. Isaac, and R. E. F. Leakey, 421–31. Chicago: Univ. of Chicago Press.

Bonnefille, R., and M. Taieb. 1971. Quaternaire de la région de Melka Kontoure: Géologie et palynologie. *VII Pan African Congress of Prehistory and Quaternary Studies, Addis Ababa* (mimeographed pamphlet).

Bonnefille, R., F. H. Brown, J. Chavaillon, Y. Coppens, P. Haesaerts, J. de Heinzelin, and F. C. Howell. 1973. Situation stratigraphique des localités a hominides des gisements Plio-Pleistocenes de l'Omo en Ethiopie. *Comptes rendus d'Académie des Science* 276:2781–84.

Bonnichsen, R. 1975. On faunal analysis and the Australopithecines. *Current Anthropology* 16(4):635–36.

Bordes, F. 1971. Physical evolution and technological evolution in man: A parallelism. *World Archaeology* 3:1–5.

Bowen, B. E., and C. F. Vondra. 1973. Stratigraphical relationships of Plio-Pleistocene deposits, East Rudolf, Kenya. *Nature* 242:391–93.

Brain, C. K. 1958. *The Transvaal ape-man-bearing cave deposits*. Transvaal Museum Memoir 11.

Brock, A., and G. L. Isaac. 1974. Paleomagnetic stratigraphy and chronology of the hominid-bearing sediments east of Lake Rudolf, Kenya. *Nature* 247:344–48.

Brown, F. H. 1981. Environments in the lower Omo Basin from one to four million years ago. In *Hominid sites*, 149–64. *See* Rapp and Vondra, 1981.

Brown, F. H., F. C. Howell, and G. G. Eck. 1978. Observations on problems of correlation of late Cenozoic hominid-bearing formations in the North Lake Turkana Basin. In *Geological background to fossil man*, ed. W. W. Bishop, 473–98. Edinburgh: Scottish Academic Press.

Burggraf, D. R., Jr., H. J. White, H. J. Frank, and C. F. Vondra. 1981. Hominid habitats in the Rift Valley: Part 2. In *Hominid sites*, 115–48. *See* Rapp and Vondra, 1981.

Butzer, K. W. 1971. The lower Omo Basin: Geology, fauna and hominids of Plio-Pleistocene age. *Naturwissenschaften* 55:7–16.

———. 1978. Geoecological perspectives on early hominid evolution. In *Early hominids of Africa*, ed. C. Johhy, 191–217. London: Duckworth.

Butzer, K. W., and C. L. Hansen. 1967. Upper Pleistocene stratigraphy in southern Egypt. In *Background to evolution in Africa*, ed. W. W. Bishop, and J. D. Clark, 329–56. Chicago: Univ. of Chicago Press.

Butzer, K. W., M. H. Day, and R. E. Leakey. 1969a. Early *Homo sapiens* remains from the Omo River region of southwest Ethiopia. *Nature* 222:1132–38.

Butzer, K. W., and D. L. Thurber. 1969b. Some late Cenozoic sedimentary formations of the lower Omo Basin. *Nature* 222:1138–43.

Cerling, T. E. 1979. Paleochemistry of the Plio-Pleistocene Lake Turkana, Kenya. *Paleogeography, Paleoclimatology, Paleoecology* 27:247–85.

Cerling, T., F. Brown, B. Cerling, G. Curtis, and R. Drake. 1979. Preliminary correlations between the Koobi Fora and Shungura formations, East Africa. *Nature* 279: 118–21.

Clark, J. D. 1967. *Atlas of African prehistory*. Chicago: Univ. of Chicago Press.

Coles, J. M., and E. S. Higgs. 1969. *The archaeology of early man*. New York: Praeger.

Conroy, G., C. Jolly, D. Cramer, and J. Kalb. 1978. Newly discovered hominoid skull from the Afar Depression, Ethiopia. *Nature* 275:67–70.

Cook, H. J. 1927. New geological and paleontological evidence bearing on the antiquity of man in America. *Natural History* 27:240–47.

Cooke, H. B. S., and V. J. Maglio. 1972. Plio-Pleistocene stratigraphy in East Africa in relation to proboscidean and suid evolution. In *Calibration of hominid evolution*, ed. W. W. Bishop and J. A. Miller, 303–29. Edinburgh: Scottish Academic Press.

Cutris, G. H., and R. L. Hay. 1972. Further geologic studies and K-Ar dating of Olduvai

Gorge and Ngorongoro Crater. In *Calibration of hominid Evolution*, ed. W. W. Bishop and J.A. Miller, 289–301. Edinburgh: Scottish Academic Press.

Curtis, G. H., T. Drake, T. E. Cerling, and J. H. Hampel. 1975. Age of KBS tuff in Koobi Fora formation, East Rudolf, Kenya. *Nature* 258:395–98.

DeHeinzelin, J., F. H. Brown, and F. C. Howell. 1971. Pliocene/Pleistocene formations in the lower Omo Basin, southern Ethiopia. *Quaternaria* 13:247–68.

Farrand, W. R. 1972. Geological correlation of prehistoric sites in the Levant. In *The origin of* Homo sapiens, ed. F. Bordes, 227–35. Proceedings of the Paris Symposium, 2–5 September 1969. Paris: UNESCO.

Findlater, I. C. 1978. Stratigraphy. In *Koobi Fora research project*, vol. 1, ed. M. G. Leakey and R. E. F. Leakey, 14–31. Oxford: Clarendon Press.

Flint, R. F. 1959. On the basis of Pleistocene correlation in East Africa. *Geological Magazine* 96:265–84.

Gladfelter, B. G. 1975. Middle Pleistocene sedimentary sequences in East Anglia (United Kingdom). In *After the Australopithecines*, ed. K. W. Butzer and G. L. Isaac, 225–58. Chicago: Aldine.

Gromme, C. S., and R. L. Hay. 1971. Geomagnetic polarity epochs: Age and duration of the Olduvai normal polarity epoch. *Earth and Planetary Science Letters* 10:179–85.

Hay, R. L. 1971. Geologic background of Beds I and II. In *Olduvai Gorge*, vol. 3, ed. M. D. Leakey, 9–18. Cambridge: Cambridge Univ. Press.

———. 1981a. Lithofacies and environments of Bed I, Olduvai Gorge, Tanzania. In *Hominid sites*, 25–56. *See* Rapp and Vondra, 1981.

———. 1981b. Paleoenvironment of the Laetolil Beds, northern Tanzania. In *Hominid sites*, 7–24. *See* Rapp and Vondra, 1981.

Hay, R. L., and R. J. Reeder. 1978. Calcretes of Olduvai Gorge and the Ndolanya Beds of northern Tanzania. *Sedimentology* 25:649–73.

Hillhouse, J. W., J. W. M. Ndombi, A. Cox, and A. Brock. 1977. Additional results on paleomagnetic stratigraphy of the Koobi Fora formation, east of Lake Turkana (Lake Rudolf), Kenya. *Nature* 265:411–15.

Howell, F. C. 1959. Upper Pleistocene stratigraphy and early man in the Levant. *Proceedings of the American Philosophical Society* 103:1–65.

Isaac, G. L. 1965. The stratigraphy of the Peninj Beds and the provenance of the Natron Australopithecine mandible. *Quaternaria* 7:101–30.

———. 1966. The geological history of the Olorgesailie area. *Publicaciones del Museo Arqueologico Santa Cruz de Tenerife* 6(2):125–33.

———. 1967. The stratigraphy of the Peninj Group—early middle Pleistocene formations west of Lake Natron, Tanzania. In *Background to evolution in Africa*, ed. W. W. Bishop and J. D. Clark, 229–57. Chicago: Univ. of Chicago Press.

———. 1975. Stratigraphy and cultural patterns in East Africa during the middle ranges of Pleistocene time. In *After the Australopithecines*, ed. K. W. Butzer and G. L. Isaac, 495–542. Chicago: Aldine.

———. 1977. *Olorgesailie*. Chicago: Univ. of Chicago Press.

Jaeger, J.-J. 1975. The Mammalian faunas and hominid fossils of the Middle Pleistocene of the Maghreb. In *After the Australopithecines*, ed. K. W. Butzer and G. L. Isaac, 399–414. Chicago: Aldine.

Johanson, D. C., M. Taieb, B. T. Gray, and Y. Coppens. 1978. Geological framework of the Pliocene Hadar formation (Afar, Ethiopia). In *Geological background to fossil man*, ed. W. W. Bishop, 549–64. Edinburgh: Scottish Academic Press.

Johnson, G. D. 1974. Cenozoic lacustrine stromatolites from hominid-bearing sediments east of Lake Rudolf, Kenya. *Nature* 247:520–22.

———. 1977. Paleopedology of *Ramapithecus*-bearing sediments, north India. *Geologische Rundschau* 66:192–216.

Johnson, G. D., and C. F. Vondra. 1972. Siwalik sediments in a portion of the Punjab re-entrant: The sequence at Haritalyangar, District Bilaspur, H. P. *Himalayan Geology* 2:118–44.

Johnson, G. D., and R. G. H. Raynolds. 1976. Late Cenozoic environments of the Koobi Fora formation: The upper member along the western Koobi Fora Ridge. In *Earliest man and environments in the Lake Rudolf Basin*, ed. Y. Coppens, F. C. Howell, G. L. Isaac, and R. E. F. Leakey, 115–22. Chicago: Univ. of Chicago Press.

Johnson, G. D., et al. 1979. Magnetic reversal stratigraphy and sedimentary tectonic history of the Upper Siwalik Group, eastern Salt Range and southwestern Kashmir. In *Geodynamics of Pakistan*, ed. A. Farah and K. A. DeJong, 149–65. Quetta: Geological Survey of Pakistan.

Johnson, G. D., P. H. Rey, R. H. Ardrey, C. F. Visser, N. D. Opdyke, and R. A. K. Tahirkheli. 1981. Paleoenvironments of the Siwalik Group, Pakistan and India. In *Hominid sites*, 197–254. *See* Rapp and Vondra, 1981.

Kalb, J. 1979. Mio-Pleistocene deposits in the Afar Depression in Ethiopia. *Sinet: Ethiopian Journal of Science* 1:87–98.

Kent, P. E. 1941. The recent history and Pleistocene deposits of the plateau north of Lake Eyasi, Tanganyika. *Geological Magazine* 78:173–84.

Lan-Po, C. 1975. *The cave home of Peking man*. Peking: Foreign Languages Press.

Leakey, L. S. B. 1965. *Olduvai Gorge, 1951–61*. Cambridge: Cambridge Univ. Press.

Leakey, M. D. 1971. *Olduvai Gorge*, vol. 3. Cambridge: Cambridge Univ. Press.

Leakey, M. D., R. L. Hay, R. Protsch, et al. 1972. Stratigraphy, archaeology, and age of the Ndutu and Naisiusiu Beds, Olduvai Gorge, Tanzania. *World Archaeology* 3: 328–41.

Leakey, M. D., and R. L. Hay. 1979. Pliocene footprints in the Laetolil Beds at Laetoli, northern Tanzania. *Nature* 278:317–23.

Leakey, R., and R. Lewin. 1978. *People of the lake: Mankind and its beginning*. New York: Anchor Press.

Movius, H. L., Jr. 1944. *Early man and Pleistocene stratigraphy in southern and eastern Asia*. Papers of the Peabody Museum, vol. 19.

Ozansoy, F. 1969. *Pleistocene fossil human footprints in Turkey*. Bulletin of Mineral Research and Exploration Institute of Turkey (foreign ed.), no. 72.

Partridge, T. C. 1978. Re-appraisal of lithostratigraphy of the Sterkfontein hominid site. *Nature* 275:282–87.

Patterson, B., A. K. Behrensmeyer, and W. D. Sill. 1970. Geology and fauna of a new Pliocene locality in northwest Kenya. *Nature* 226:918–21.

Penck, A. 1908. Das Alter des Menschengeschlechts: *Zeitschrift für Ethnologie* 40: 390–407.

Pilbeam, D. R., et al. 1977a. Geology and palaeontology of Neogene strata of Pakistan. *Nature* 270:684–89.

Pilbeam, D. R., et al. 1977b. New hominoid primates from the Siwaliks of Pakistan and their bearing on hominoid evolution. *Nature* 270:689–95.

Rapp, G., Jr., and C. F. Vondra, eds. 1981. *Hominid sites: Their geologic settings*. Boulder, CO: Westview Press.

Reck, H. 1951. A preliminary survey of the tectonics and stratigraphy of Olduvai. In

Olduvai Gorge, ed. L. S. B. Leakey, 5–19. Cambridge: Cambridge Univ. Press.

Sellards, E. H. 1952. *Early man in America*. Austin: Univ. of Texas Press.

Taieb, M., Y. Coppens, D. C. Johanson, and J. Kalb. 1972. Dépôts sédimentaires et faunes du plio-pleistocene de la basse vallée de l'Awash. *Comptes rendus d'Académie des Science* (Paris), series D, 275:819–22.

Taieb, M., and J. J. Tiercelin. 1979. Sédimentation pliocene et paleoenvironnements de Rift, exemple de la formation a hominides d'Hadar (Afar, Ethiopie). *Bulletin de la Société Geologique de France* 1:243–53.

de Terra, H. 1943. Pleistocene geology and early man in Java. *Transactions of the American Philosophical Society* 32:437–64.

Vondra, C. F., G. D. Johnson, A. K. Behrensmeyer, and B. E. Bowen. 1971. Preliminary stratigraphical studies of the East Rudolf Basin, Kenya. *Nature* 231:245–48.

Vondra, C. F., and B. E. Bowen. 1976. Plio-Pleistocene deposits and environments, East Rudolf, Kenya. In *Earliest man and environments in the Lake Rudolf Basin*, ed. Y. Coppens, F.C. Howell, G. L. Isaac, and R. E. F. Leakey, 79–93. Chicago: Univ. of Chicago Press.

Vondra, C. F., and B. E. Bowen. 1978. Stratigraphy, sedimentary facies, and paleo-environments, East Rudolf, Kenya. In *Background to early man in Africa*, ed. W. W. Bishop, 395–414. Geological Society of London, Special Publication no. 5.

Vondra, C. F., M. E. Mathisen, D. R. Burggraf, Jr., and E. P. Kvale. 1981. Plio-Pleistocene geology of northern Luzon, Philippines. In *Hominid sites*, 255–310. *See* Rapp and Vondra, 1981.

Walker, A., and R. Leakey. 1978. The hominids of East Turkana. *Scientific American* 239(2):54–66.

Washburn, S. L. 1957. Australopithecines: The hunters or the hunted? *American Anthropologist* 59:612–14.

White, H. J., D. R. Burggraf, Jr., R. B. Bainbridge, Jr., and C. F. Vondra. 1981. Hominid habitats in the Rift Valley: Part 1. In *Hominid sites*, 57–114. *See* Rapp and Vondra, 1981.

LITHIC MATERIALS

Abbott, W. J. L. 1915. Flint fracture. *Nature* 94:198.

Ackerman, R. E. 1964. Lichens and the patination of chert in Alaska. *American Antiquity* 29:386–87.

Barnes, A. S. 1939. The differences between natural and human flaking on prehistoric flint implements. *American Anthropologist* 41:99–112.

Barrow, T. 1962. An experiment in working nephrite. *Journal of Polynesian Society* 71.

Beck, L. 1966. Jade. *Anthropological Journal of Canada* 4 (1):12–22.

Bell, R. E. 1955. Lithic analysis and archaeological method. *American Anthropologist* 55:299–301.

Bixby, L. B. 1945. Flint chipping. *American Antiquity* 10:353–61.

Bjorkman, J. K. 1973. Meteors and meteorites in the ancient Near East. *Meteoritics* 8:91–132.

Bolton, R. P. 1930. An aboriginal chert-quarry in northern Vermont. *Indian Notes* 7:457–65. Museum of American Indian, Heye Foundation.

Borden, F. E. 1971. *The use of surface erosion observations to determine chronological sequence in artifacts from a Mojave Desert site*. Archaeological Survey Association of Southern California Paper no. 7.

Bordes, F. 1969. Reflections on typology and techniques in the Paleolithic. *Arctic Anthropology* 6:1–29.

Bordes, F., and D. Crabtree. 1969. The Corbiac blade technique and other experiments. *Tebiwa* 12:1–21.

Breuil, H. 1943. On the presence of quartzites mechanically broken in the Dwyka tillites and their derivation in the older gravels of the Vaal. *South African Journal of Science* 40:285–86.

Bryan, A. L., and D. R. Tuohy. 1960. A basalt quarry in northeastern Oregon. *Proceedings of the American Philosophical Society* 104:485–510.

Bryan, K. 1938. Prehistoric quarries and implements of pre-Amerindian aspect in New Mexico. *Science* 87:343–46.

———. 1939. Stone cultures near Cerro Pedernal and their geological antiquity. *Bulletin of the Texas Archaeological and Paleontological Society* 11:9–46.

———. 1950. *Flint quarries: The sources of tools, and at the same time, the factories of the American Indian.* Papers, Peabody Museum of American Archaeology and Ethnology, vol. 17.

Carter, G. F. 1967. Artifacts and naturifacts: Introduction. *Anthropological Journal of Canada* 5:2–5.

Clark, J. D. 1958. The natural fracture of pebbles from the Batoka Gorge, northern Rhodesia, and its bearing on the Kafuan industries of Africa. *Proceedings of the Prehistoric Society* 34:64–77.

Clarke, W. G. 1914. Some aspects of striation. *Proceedings of the Prehistoric Society East Anglia* 1:434–38.

Crabtree, D. E. 1966. A stoneworker's approach to analyzing and replicating the Lindenmeier Folsom. *Tebiwa* 9:3–39.

———. 1972. *An introduction to flintworking, part 1: An introduction to the technology of stone tools.* Occasional Papers of the Idaho State University, no. 28.

Crabtree, D. E., and B. R. Butler. 1964. Notes on experiments in flint-knapping: 1. Heat treatment of silica minerals. *Tebiwa* 7:1–6.

Crabtree, D. E., and E. H. Swanson, Jr. 1968. Edge-ground cobbles and blade-making in the Northwest. *Tebiwa* 11:50–58.

Curtis, G. H. 1959. The petrology of artifacts and architectural stone at La Venta. *Bureau of American Ethnology, Bulletin* 170:284–89.

Curwen, E. C. 1940. The white patination of black flint. *Antiquity* 14:435–37.

Einfalt, H. C. 1978. Stone materials in ancient Akrotiri—a short compilation. In *Thera and the Aegean world*, vol. 1, ed. C. Doumas, 523–27. London: Thera and the Aegean World.

Ellis, H. H. 1938. Lithic problems. *American Antiquity* 4:63–64.

Ellis, S. E. 1969. The petrography and provenance of Anglo-Saxon and medieval honestones, with notes on other hones. *Bulletin, British Museum (Natural History), Mineralogy* 2:133–87.

Feldman, L. H. 1971. Of the stone called Iztli. *American Antiquity* 36:213–14.

———. 1973. Stones for the archaeologist. *University of California, Archaeological Research Facility,Contributions* 18:87–104.

Foshag, W. F. 1955. Chalchihuitl—a study in jade. *American Mineralogist* 40 (11, 12):1062–70.

———. 1957. *Mineralogical studies on Guatemalan jade.* Smithsonian Miscellaneous Collections, vol. 135, no. 5.

Foshag, W. F., and R. Leslie. 1955. Jadeite from Manzanal, Guatemala. *American Antiquity* 21:81–83.

Giot, P.-R. 1951. A petrological investigation of Breton stone axes. *Proceedings of the Prehistoric Society* 17:228 ff.

Goodman, M. E. 1944. The physical properties of stone tool materials. *American Antiquity* 9:415–33.

Goodwin, A. J. H. 1960. Chemical alteration (patination) of stone. In *The application of quantitative methods in archeology*, ed. R. F. Heizer and S. Cook, 300–12. Viking Fund Publications in Anthropology, vol. 28.

Harner, J. J. 1956. Thermo-facts vs. artifacts: An experimental study of the Malpais industry. *University of California Archaeological Survey Reports* 33:39–43.

Hassan, F. 1972. Toward a definition of 'lames,' 'lamelles,' and 'microlamelles.' *Pan-African Congress on Prehistory and the Study of the Quaternary, Bulletin* 5: 51–59.

Haynes, C. V. 1964. Fluted projectile points: Their age and dispersal. *Science* 145: 1408–13.

———. 1973. The Calico site: Artifacts or geo-facts? *Science* 181:305–10.

Healy, J. A. 1966. Applying the ancient craft of knapping through controlled fracturing. *Archaeology in Montana* 6:5–21.

Hester, T. R., R. F. Heizer, and R. N. Jack. 1971. Technology and geologic sources of obsidian artifacts from Cerro de las Mesas, Veracruz, with observations on Olmec trade. *University of California, Archaeological Research Facility, Contributions* 13:133–41.

Holmes, W. H. 1919. *Handbook of aboriginal American antiquities: Part 1. The lithic industries*. Bureau of Amerian Ethnology Bulletin, no. 60.

Hurst, V. J., and A. R. Kelly. 1961. Patination of cultural flints. *Science* 134:251–56.

Johnston, C. S. 1939. Flint flaking. *Scentific American* 161:82–83.

Jope, E. M. 1953. History, archaeology and petrography. *Advancement of Science* 9:432–35.

Keely, L. H. Technique and methodology in microwear studies: A critical review. *World Archaeology* 5:323–36.

Knowles, F. H. S. 1944. *The manufacture of a flint arrowhead by quartzite hammerstone*. Occasional Papers on Technology, no. 1. Oxford: Pitt Rivers Museum.

Lal Guari, K. 1978. The preservation of stone. *Scientific American* (June):126–36.

Lemley, H. J. 1942. Prehistoric novaculite quarries of Arkansas. *Bulletin of the Texas Archaeological and Paleontological Society* 14:32–37.

Losey, T. C. 1971. The Stony Plain quarry site. *Plains Anthropologist* 16 (52):138–54.

McCrone, A. W., D. South, R. Ascher, and M. Ascher. 1965. Stone artifacts: Identification problems. *Science* 148:167–68.

MacDonald, G. F., and D. Sanger. 1968. Some aspects of microscope analysis and photomicrography of lithic artifacts. *American Antiquity* 33:237–40.

Melcher, C., and D. Zimmerman. 1977. Thermoluminescent determination of prehistoric heat treatment of chert artifacts. *Science* 197:1359–62.

Merritt, P. L. 1932. The identification of jade by means of X-ray diffraction patterns. *American Mineralogist* 17:497–508.

Michels, J. W. 1971. The colonial obsidian industry of the Valley of Mexico. In *Science and archaeology*, ed. R. H. Brill, 251–71. Cambridge, MA: MIT Press.

Moir, J. R. 1912a. The natural fracture of flint. *Nature* 90:461–63.

————. 1912b. The natural fracture of flint and its bearing upon rudimentary flint implements. *Proceedings of the Prehistoric Society, East Anglia* 1:171–84.

————. 1914. The striation of flint surfaces. *Man* 14:177–81.

————. 1915. The large non-conchoidal fracture surfaces of early flint implements. *Nature* 94:89, 227, 288.

————. 1928. Where nature imitates man. *Scientific American* 138:426–27.

Moore, D. T. 1978. The petrography and archaeology of English honestones. *Journal of Archaeological Science* 5:61–73.

Nelson, N. C. 1928. Pseudo-artifacts from the Pliocene of Nebraska. *Science* 67:316–17.

Norman, D., and W. W. A. Johnston. 1941. Note on a spectroscopic study of Central American and Asiatic jades. *Journal of the Optical Society of America* 31:85–86.

Oakley, K. P. 1939. The nature and origin of flint. *Science Progress* 34:277–86.

Olausson, D. S., and L. Larsson. 1982. Testing for the presence of thermal pretreatment of flint in the Mesolithic and Neolithic of Sweden. *Journal of Archaeological Science* 9:275–85.

Purdy, B. A., and H. K. Brooks. 1971. Thermal alteration of silica minerals: An archaeological approach. *Science* 173:322–25.

Ray, C. N. 1947. Chemical alteration of silicate artifacts. *Bulletin of the Texas Archaeological and Paleontological Society* 18:28–39.

Saraydar, S., and I. Shimade. 1971. A quantiative comparison of efficiency between a stone axe and a steel axe. *American Antiquity* 36:216–17.

Schmalz, R. F. 1960. Flint and the patination of flint artifacts. *Proceedings of the Prehistoric Society* 26:44–49.

Sedgley, J. P. 1970. Petrographic examination of stone artifacts. *Science and Archaeology* 2–3:10–12.

Semenov, S. A. 1964. *Prehistoric technology*. London: Cory, Adams and MacKay.

Shotton, F. W. 1959. New petrological groups based on axes from the West Midlands. *Proceedings of the Prehistoric Society* 25:135–43.

————. 1968. Prehistoric man's use of stone in Britain. *Proceedings of the Geological Association* 79:477–91.

————. 1969. Petrological examination. In *Science in archaeology*, 2d ed., ed. D. Brothwell and E. Higgs, 571–77. London: Thames and Hudson.

Shotton, F. W., and G. L. Hendry. 1979. The developing field of petrology in archaeology. *Journal of Archaeological Science* 6:75–84.

Speth, J. D. 1972. Mechanical basis of percussion flaking. *American Antiquity* 37:34–60.

Stirling, M. W. 1968. Aboriginal jade use in the New World. *Congreso Internacional de Americanistas, Actas y Memorias* 37 (4):19–28.

Tringham, R., et al. 1974. Experimentation in the formation of edge damage: A new approach to lithic analysis. *Journal of Field Archaeology* 1:171–76.

Wallis, F. S. 1955. Petrology as an aid to prehistoric and medieval archaeology. *Endeavour* 14:146–51.

————. 1963. Petrological examination. In *The scientist and archaeology*, ed. E. Pyddoke, 80–100. London: Phoenix House.

Warren, S. H. 1921. A natural 'eolith' factory beneath the Thanet Sand. *Quarterly Journal of the Geological Society* 76:238–53.

Watanabe, H. 1968. Flake production in a transitional industry from Amud Cave, Israel: A statistical approach to paleolithic technotypology. In *La Préhistoire: problèmes et tendances*, ed. F. Bordes, Paris: CNRS.

Weide, D. L. 1969. A petrographic examination of basalt artifacts from the Panamint Valley, California. In *The western lithic co-tradition*, ed. E. L. Davis, C. W. Brott, and D. Weide. San Diego Museum Papers, no. 6.

Weymouth, J. H., and W. O. Williamson. 1951. Some physical properties of raw and calcined flint. *Mineralogical Magazine* 29:573−93.

Wilmsen, E. N. 1968. Lithic analysis in paleo-anthropology. *Science* 161:982−87.

Winchell, N. H. 1913. *The weathering of aboriginal stone artifacts: A consideration of the paleoliths of Kansas*. Minnesota Historical Society Collections, vol. 16, pt. 1, no. 1. St. Paul.

Witthoft, J. 1969. Lithic materials and technology. *Proceedings, 25th Southeastern Archaeological Conference Bulletin* 9:3−15.

Witthoft, J., and E. S. Wilkins. 1967. Petrographic studies: Lateritic flints. *MASCA Newsletter* 3:3−4.

METALS AND MINING

Allan, J. C. 1970. *Considerations on the antiquity of mining in the Iberian Peninsula*. Royal Anthropological Institute Occasional Paper no. 27.

Ball, S. H. 1941. The mining of gems and ornamental stones by American Indians. *Bureau of American Ethnology Bulletin* 128:1−77.

Barnard, N., and S. Tamotsu. 1975. *Metallurgical remains of ancient China*. Tokyo: Nichiosha.

Becker, C. J. 1959. Flint-mining in Neolithic Denmark. *Antiquity* 33:87−92.

Benson, E., ed. 1980. *Pre-Columbian metallurgy of South America* (Papers from a conference, Washington, DC, Oct. 1975). Washington, DC: Dumbarton Oaks Research Library and Collections.

Boulakia, J. 1972. Lead in the Roman world. *American Journal of Archaeology* 76:139−44.

Coghlan, H. D. 1972. Some reflections on the prehistoric working of copper and bronze. *Archaeologica Austriaca* 52:93−104.

Coghlan, H. H. 1956. *Notes on prehistoric and early iron in the Old World: Pitt Rivers Museum, Oxford*. Occasional Paper on Technology no. 8.

Conophagos, C. E. 1980. *Le Laurium antique et la technique grecque de la production de l'argent*. Athens: Ekdotike Hellados.

Cooke, S. R. B., F. Henrickson, and G. Rapp, Jr. 1972. Metallurgical and geochemical studies. In *The Minnesota Messenia expedition: Reconstructing a Bronze Age regional environment*, ed. W. A. McDonald, and G. Rapp, Jr., 225−33. Minneapolis: Univ. of Minnesota Press.

Cooke, S. R. B., and B. V. Nielsen. 1978. Slags and other metallurgical products. In *Excavations at Nichoria in southwest Greece*. vol. 1, *Site, environs, and techniques*, ed. G. Rapp, Jr., and S. E. Aschenbrenner, 182−224. Minneapolis: Univ. of Minnesota Press.

Craddock. P. T., ed. 1980. *Scientific studies in early mining and extractive metallurgy*. British Museum Occasional Paper no. 20.

Davies, O. 1932a. Ancient mines in southern Macedonia. *Journal of the Royal Anthropological Institute* 62:15−162.

———. 1932b. Bronze Age mining around the Aegean. *Nature* 130:985−87.

———. 1932c. The copper mines of Cyprus. *Annual of the British School at Athens*, no. 30, sessions 1928–29, 1929–30, London.

———. 1935. *Roman mines in Europe.* Oxford: Clarendon Press.

De Jesus, P. S. 1978. Metal resources in ancient Anatolia. *Anatolian Studies* 28:97–102.

———. 1981. A survey of some ancient mines and smelting sites in Turkey. *Sonderdruck aus Archäologie und Naturwissenschaften* 2:95–105.

Dominian, L. 1911. History and geology of ancient goldfields in Turkey. *Transactions of the American Institute of Mining Engineers* 42:569–89.

Easby, D. T. 1966. Early metallurgy in the New World. *Scientific American* 214:73–81.

Forbes, R. J. 1964. *Studies in ancient technology.* Vol. 8, *Metallurgy.* Leiden: E. J. Brill.

———. 1972. *Studies in ancient technology.* Vol. 9, *Mining and metallurgy.* Leiden: E. J. Brill.

Franklin, U. M., J.-C. Grosjean, and M. J. Tinkler. 1976. A study of ancient slags from Oman. *Canadian Metallurgical Quarterly* 15:1–7.

Gale, N. H. 1980. Some aspects of lead and silver mining in the Aegean. In *Thera and the Aegean world II,* ed. C. Doumas, 161–95. London: Thera and the Aegean World.

Gettens, R. J. 1968. Mineral alteration products on ancient metal objects. In *Recent advances in conservation,* ed. G. Thomson. London: Butterworth.

Giles, D. L., and E. P. Kuijpers. 1974. Stratiform copper deposit, northern Anotolia, Turkey: evidence for early Bronze I (2800 B.C.) mining activity. *Science* 186:823–25.

Griffen, J. B., ed. 1961. *Lake Superior copper and the Indians: Miscellaneous studies of Great Lakes prehistory.* Anthropological Papers of the Museum of Anthropology no. 17, Univ. of Michigan.

Hall, E. T. 1961. Surface-enrichment of buried metals. *Archaeometry* 4:62–66.

Hall, E. T., and D. M. Metcalf, eds. 1972. *Methods of chemical and metallurgical investigation of ancient coinage.* London: Royal Numismatic Society.

Harrington, M. R. 1951. A colossal quarry. *Southwest Museum Masterkey* 25:14–18.

Heizer, R. F., and A. E. Treganza. 1944. Mines and quarries of the Indians of California. *California Journal of Mines and Geology* 40:291–359.

Khalil, L. A., and H.-G. Bachmann. 1981. Evidence of copper smelting in Bronze Age Jericho. *Journal of the History of Metallurgy Society* 15:103–06.

Klein, E. 1979. Chemical and mineralogical studies in Siphnos ores and slags. *Archaeo-Physica* 10:223–29. Proceedings of the 18th International Symposium on Archaeometry and Archaeological Prospection, Bonn.

Maddin, R., T. Stech Wheeler, and J. D. Muhly. 1980. Distinguishing artifacts made of native copper. *Journal of Archaeological Science* 7:211–25.

Milton, C., E. Dwornik, R. Finkelman, and P. Toulmin, III. 1976. Slag from an ancient copper smelter at Timna, Israel. *Journal of the History of Metallurgy Society* 10:24–33.

Muhly, J. 1973. Copper and tin. *Transactions of the Connecticut Academy of Arts and Sciences* 43:155–535.

Muhly, J., and T. Wertime. 1973. Evidence for the sources and uses of tin during the Bronze Age of the Near East. *World Archaeology* 5:111–22.

Oddy, W. A., ed. 1980. *Aspects of early metallurgy.* British Museum Occasional Paper no. 17.

Ottaway, B. S. 1979. Interpretation of prehistoric metal artifacts with the aid of cluster analysis. *Archaeo-Physica* 10:597–606. Proceedings of the 18th International Symposium on Archaeometry and Archaeological Prospection, Bonn.

Patterson, C. C. 1971. Native copper, silver and gold accessible to early metallurgists. *American Antiquity* 36.286–321.

Phillips, W. A. 1900. Aboriginal quarries and shops at Mill Creek, Illinois. *American Anthropologist* 2:37.

Rapp, G. Jr., R. E. Jones, S. R. B. Cooke, and E. L. Henrickson. 1978. Analysis of the metal artifacts. In *Excavations at Nichoria in southwest Greece*, vol. 1, *Site, environs, and techniques*, ed. G. Rapp, Jr., and S. E. Aschenbrenner, 166–81. Minneapolis: Univ. of Minnesota Press.

Rothenberg, B. 1972. *Timna, valley of the Biblical copper mines*. London: Thames and Hudson.

Rothenberg, B., R. F. Tylecote, and P. J. Boydell. 1978. *Chalcolith Copper smelting: Excavations and experiments*. Monograph no. 1, Institute for Archaeo-Metallurgical Studies, London.

Schubiger, P. A., O. Müller, and W. Gentner. 1977. Neutron activation analysis on ancient Greek silver coins and related materials. *Journal of Radioanalytical Chemistry* 39:99–112.

Shepherd, R. 1980. *Prehistoric mining and allied industries*. London: Academic Press.

Sherratt, A. G. 1976. Resources, technology and trade: An essay in early European metallurgy. In *Problems in economic and social archaeology*, ed. G. Sieveking, I. Longworth, and K. Wilson, 557–82. London: Duckworth.

Shimada, I., S. Epstein, and A. K. Craig. 1982. Batan Grande: A prehistoric metallurgical center in Peru. *Science* 216:952–59.

Sieveking, G. de G., et al. 1971. Characterization of prehistoric flint mine products. *Nature* 288:251–54.

———. 1972. Prehistoric flint mines and their identification as sources of raw materials. *Archaeometry* 14:151–76.

Slater, E. A., and A. McKenzie. 1980. Late Bronze Age Aegean metallurgy in the light of the Theran analyses. In *Thera and the Aegean world II*, ed. C. Doumas, 197–215. London: Thera and the Aegean World.

Steinberg, A., and F. L. Koucky. 1974. Preliminary metallurgical research in the ancient Cypriot copper industry. In *American expedition to Idalion, Cyprus*, ed. E. Stager, A. Walker, and E. Wright, 149–78. Cambridge: ASOR.

Tylecote, R. F. 1976. *A history of metallurgy*. London: The Metals Society.

———. 1979. The effect of soil conditions on the long-term corrosion of buried tin-bronzes and copper. *Journal of Archaeological Science* 6:345–68.

Tylecote, R. F., B. Rothenberg, and A. Lupu. 1967. A study of early copper smelting and working sites in Israel. *Journal of the Institute of Metals* 95:235–43.

Wagner, G. A., and G. Weisgerber. 1979. The ancient silver mine at Ayos Sostis on Siphnos, Greece. *Archaeo-Physica* 10:209–22. Proceedings of the 18th International Symposium on Archaeometry and Archaeological Prospection, Bonn.

Wallace, W. J. 1962. Two basalt quarries from Death Valley. *Archaeology* 15:46–49.

Wertime, T. A. 1973. The beginnings of metallurgy: A new look. *Science* 182:875–87.

Wertime, T. A., and J. D. Muhly, eds. 1980. *The coming of the Age of Iron*. New Haven and London: Yale Univ. Press.

White, J. R. 1980. Historic blast furnace slags: Archaeological and metallurgical analysis. *Journal of the History of Metallurgy Society* 14:55–64.

Worssam, B. C., and J. Gibson-Hill. 1976. Analyses of Wealden iron ores. *Journal of the History of Metallurgy Society* 10:77–82.

MISCELLANEOUS

Adams, R. E. W. 1975. Stratigraphy. In *Field methods in archaeology*, ed. T. R. Hester, R. F. Heizer, and J. A. Graham, 147–62. Palo Alto: Mayfield.

Agraival, D. P. 1971. *The Copper Bronze Age of India*. New Delhi: Munshiram Manoharlal.

Aitken, M. J. 1961. *Physics and archaeology*. London: Interscience.

Aitken, M. J., D. W. Zimmerman, and S. J. Fleming. 1968. Thermoluminescence dating of ancient pottery. *Nature* 219:442.

Aitken, M. J., P. Alcock, G. Bussell, and C. Shaw. 1981. Archaeomagnetic determination of the past geomagnetic intensity using ancient ceramics: Allowance for anisotropy. *Archaeometry* 23:53–64.

Allibone, T. E., et al. 1971. *Impact of the natural sciences on archaeology: A joint symposium of the Royal Society and the British Academy*. London: Oxford Univ. Press.

Ambraseys, N. N. 1971. Value of historial records of earthquakes. *Nature* 232:375.

———. 1973. Earth sciences in archaeology and history. *Antiquity* 47:229.

———. 1978. Studies in historical seismicity and tectonics. In *The environmental history of the Near and Middle East*, ed. W. C. Brice, 185–212. London: Academic Press.

Aschenbrenner, S. E., and S. R. B. Cooke. 1978. Screening and gravity concentration: Recovery of small-scale remains. In *Excavations at Nichoria in southwest Greece*. Vol. 1, *Site, environs, and techniques*, ed. G. Rapp, Jr., and S. E. Aschenbrenner, 156–65. Minneapolis: Univ. of Minnesota Press.

Atkinson, R. J. C. 1957. Worms and weathering. *Antiquity* 31:219–33.

Back, W. 1981. Hydromythology and ethnohydrology in the New World. *Water Resources Research* 17:257–87.

Ballantyne, B. B., S. C. Zoltai, and M. J. Tamplin. 1970. *Annotated bibliography of the Quaternary in Manitoba and the adjacent Lake Agassiz Region (including archaeology of Manitoba)*. Geological Paper 2/70, Geological Survey of Manitoba, Winnipeg.

Beck, C. W. 1973. *Archaeological chemistry*. Advances in Chemistry Series, no. 138. New York: American Chemical Society.

Bennett, J. G. 1963. Geophysics and human history. *Systematics* 1:127–56.

Berger, R., ed. 1971. *Scientific methods in medieval archaeology*. Berkeley: Univ. of California Press.

Biek, L. 1963. *Archaeology and the microscope*. London: Butterworth Press.

Bliss, W. L. 1938. An archaeological and geological reconnaissance of Alberta, MacKenzie Valley, and upper Yukon. *American Philosophical Society Yearbook*. pp. 136–39.

Boardman, J., and F. Schweizer. 1973. Clay analyses of archaic Greek pottery. *Annual of the British School at Athens* 68:267–83.

Boaz, N. T., and A. K. Behrensmeyer. 1976. Hominid taphonomy: Transport of human skeletal parts in an artificial fluviatile environment. *American Journal of Physical Anthropology* 45:53–60.

Bond, A., and Sparks, R. S. J. 1976. The Minoan eruption of Santorini, Greece. *Journal of the Geological Society, London* 131:1–16.

Bond, G. 1963. The Pleistocene in southern Africa with implications for primate and human distribution. *Viking Fund Publications in Anthropology* 36:308–34.

Bradford, J. 1957. *Ancient landscapes: Studies in field archaeology*. London: G. Bell.

Braidwood, R. J. 1958. Near Eastern prehistory. *Science* 127:1419−30.

———. 1960. Levels in prehistory: A model for the consideration of the evidence. In *Evolution after Darwin*, ed. S. Tax, 143−51. Chicago: Univ. of Chicago Press.

Braidwood, R. J., and C. A. Reed. 1957. The achievement and early consequences of food-production: A consideration of the archaeological and natural-historical evidence. *Cold Springs Harbour Symposia on Quantitative Biology* 22:19−31.

Braidwood, R. J., B. Howe, and C. A. Reed. 1961. The Iranian prehistoric project. *Science* 133:2008−10.

Breuil, H., and G. Zbyszewski. 1942 and 1945. Contribution à l'étude des industries paleolithiques du Portugal et de leurs rapports avec la géologie du Quaternaire. *Communicacoes dos Servicos Geologicos de Portugal* 23:3−369 and 26:3−662.

Brill, R. H., ed. 1971. *Science and archaeology*. 2d ed. Cambridge, MA: MIT Press.

Brothwell, D., and E. S. Higgs, eds. 1969. *Science in archaeology*. London: Thames and Hudson.

Bryan, K. 1937. Geology of the Folsom deposits in New Mexico and Colorado. In *Early man, as depicted by leading authorities at the International Symposium*, ed. G. G. MacCurdy, 139−52. Philadelphia: J. B. Lippincott and Co.

———. 1939. Stone cultures near Cerro Pedernal and their geological antiquity. *Texas Archaeological and Paleontological Society Bulletin* 11:9−42.

———. 1941. Geologic antiquity of man in America. *Science* 93:505−14.

———. 1950. The geology and archeology. In *Ventana Cave, Arizona*, ed. E. W. Haury, 75−126. Albuquerque: Univ. of New Mexico.

Burgess, R. L. 1978. Some results of a geo-archaeological study. *Newsletter, Society for Archaeological Sciences* 1 (3):1−2.

Butzer, K. W. 1961. Archäologische Fundstellen Ober- und Mittelägyptens in ihrer geologischen Landschaft. *Mitteilungen Deutsches Archaeologische Institut Abteilung Kairo* 17:54−68.

———. 1966. Geologie und Paläogeographie archäologischer Fundstellen bei Sayala (Unternubien). *Denkschriften, Österreichische Akademie der Wissenschaften (Wien), Philosophische-historische Klasse* 92:89−98.

———. 1974. Modern Egyptian pottery clays and predynastic buff ware. *Journal of Near Eastern Studies* 33:377−82.

Butzer, K. W., and G. L. Isaac, eds. 1975. *After the Australopithecines*. Chicago: Aldine.

Cadogan, G., and R. K. Harrison. 1978. Evidence of Tephra in soil samples from Pyrgos, Crete. In *Thera and the Aegean world, I*, ed. C. Doumas, 235−55. London: Thera and the Aegean World.

Cahen, D., and J. Moeyersons. 1977. Subsurface movements of stone artifacts and their implications for the prehistory of Central Africa. *Nature* 266:812−15.

Campbell, A. S., ed. 1971. *Geology and history of Turkey*. Tripoli: Petroleum Exploration Society of Libya.

Carlson, J. B. 1975. Lodestone compass: Chinese or Olmec primacy. *Science* 189:753−60.

Carter, G. 1977. *Archaeological chemistry II*. Advances in Chemistry Series, no. 171. New York: American Chemical Society.

Carter, G. F. 1956. On soil color and time. *Southwestern Journal of Anthropology* 12:295−324.

Carter, T. H., and R. Pagliero. 1966. Notes on mud-brick preservation. *Sumer* 22:65–76.

Clark, J. G. D. 1936. *The Mesolithic settlement of northern Europe.* Cambridge: Cambridge Univ. Press.

———. 1954. *Excavations at Starr Carr.* London: Cambridge Univ. Press.

Clark, J. D. 1970. *The prehistory of Africa.* London: Thames and Hudson.

Clark, J. D., and C. V. Haynes. 1969. An elephant butchery site at Mwanganda's village, Karonga, Malawi, and its relevance for palaeolithic archaeology. *World Archaeology* 1:390–411.

Cole, S. 1963. *The prehistory of East Aftica.* New York: Macmillan.

Coles, J. 1979. *Experimental archaeology.* London: Academic Press.

Cook, S. F., and R. F. Heizer. 1965. Studies on the chemical analysis of archaeological sites. *University of California Publications in Anthropology* 2:1–102.

Cornwall, I. W. 1958. *Soils for the archaeologist.* London: Phoenix House.

———. 1960. Soil investigations in the service of archaeology. In *The application of quantitative methods in archaeology,* ed. R. F. Heizer, and S. F. Cook, 265–84. Viking Fund Publications in Anthropology 28, Chicago.

———. 1963. Soil micromorphology and the study of prehistoric environment. *Microscope* 13:342–45.

Dakaris, S. I., E. S. Higgs, and R. W. Hey. 1964. The climate, environment and industries of Stone Age Greece: Part 1. *Proceedings of the Prehistoric Society* 30: 199–244.

Dalrymple, J. B. 1958. The application of soil micromorphology to fossil soils and other deposits from archaeological sites. *Journal of Soil Science* 9:199–209.

Dauncey, K. D. M. 1952. Phosphate content of soils on archaeological sites. *Advancement of Science* 9 (33):33–37.

Davidson, D. A. 1978. Soils on Santorini at ca. 1500 B.C. *Nature* 272:243–44.

DeLaguna, F. 1958. Geological confirmation of native traditions, Yakutat, Alaska. *American Antiquity* 23:434.

deTerra, H. 1934. Geology and archaeology as border sciences. *Science* 80:447–49.

Dimbleby, G. W., and M. C. D. Speight. 1969. Buried soils. *Advancement of Science* 26:203–05.

Dimbleby, G. W., and R. J. Bradley. 1975. Evidence of pedogenesis from a Neolithic site at Rackham, Sussex. *Journal of Archaeological Science* 2:179–86.

Engel, C. G., and R. P. Sharp. 1958. Chemical data on desert varnish. *Bulletin of the Geological Society of America* 69:487–518.

Fairbridge, R. W. 1963. Nile sedimentation above Wadi Halfa during the last 20,000 years. *Kush* 11:96–107.

Fant, J. E., and W. G. Loy. 1972. Surveying and mapping. In *The Minnesota Messenia expedition: Reconstructing a Bronze Age regional environment,* ed. W. A. McDonald, and G. Rapp, Jr., 18–35. Minneapolis: Univ. of Minnesota Press.

Farrand, W. R. 1961. Frozen mammoths and modern geology. *Science* 133:729–35.

Farrand, W. R., R. W. Redding, M. H. Wolpoff, and H. T. Wright, III. 1976. *An archaeological investigation of the Loboi Plain, Baringo District, Kenya.* Technical Reports no. 4, Museum of Anthropology, Univ. of Michigan.

Figgins, J. D. 1927. The antiquity of man in America. *Natural History* 27: (3):229–39.

Flemming, N. C., N. M. G. Czartoryski, and P. M. Hunter. 1971. Archaeological evidence for eustatic and tectonic components of relative sea level change in the South Aegean. In *Marine archaeology,* ed. D. J. Blackman, 1–66. London: Butterworth.

Franklin, A. D., J. S. Olin, and T. A. Wertime, eds. 1978. *The search for ancient tin*. Washington, DC: Smithsonian Institution.

Freeman, L. G. 1975. Acheulean sites and stratigraphy in Iberia and the Maghreb. In *After the Australopithecines: Stratigraphy, ecology, and culture change in the Middle Pleistocene*, ed. K. W. Butzer and G. L. Isaac, 661–743. Chicago: Aldine.

Frison, G. C. 1974. The application of volcanic and non-volcanic natural glass studies to archaeology in Wyoming. In *Applied geology and archaeology: The Holocene history of Wyoming*, ed. M. Wilson, 61–64. Geological Survey of Wyoming Report of Investigations no. 10.

Fry, R. E. 1972. Manually operated post-hole diggers as sampling instruments. *American Antiquity* 37:259–61.

Galanopoulos, A. G. 1960a. On the origin of the deluge of Deukalion and the myth of Atlantis. *Athenais Archaiologike Hetaireia* 3:226–31.

———. 1960b. Tsunamis observed on the coasts of Greece from antiquity to present time. *Annali di Geofisica* 13:369–86.

———. 1971. The eastern Mediterranean trilogy in the Bronze Age. In *Acta of the First International Scientific Congress on the Volcano of Thera*, ed. A. Kaloyeropoyloy, 184–97. Athens: Archaeological Services of Greece.

Galanopoulos, A. G., and E. Bacon. 1969. *Atlantis, the truth behind the legend*. London: Thomas Nelson and Sons.

Gardner, E. W., and G. Caton-Thompson. 1926. The recent geology and Neolithic industry of the northern Fayum Desert. *Journal of the Royal Anthropological Institute* 61:301–08.

Gifford, D. P. and A. K. Behrensmeyer. 1977. Observed formation and burial of a recent human occupation site in Kenya. *Quaternary Research* 8:245–66.

Goffer, Z. 1980. *Archaeological chemistry*. New York: Wiley.

Goldberg, P. S. 1974. Sediment peels from prehistoric sites. *Journal of Field Archaeology* 1:323–28.

Goldberg, P. S. and Y. Nathan. 1975. The phosphate mineralogy of et-Tabun Cave, Mount Carmel, Israel. *Mineralogical Magazine* 40:253–58.

Graham, R., C. V. Haynes, D. Johnson, and M. Kay. 1981. Kimmswick: A clovis-mastodon association in eastern Missouri. *Science* 213:1115–16.

Hall, E. T. 1971. Two examples of the use of chemical analysis in the solution of archaeological problems. In *Science and archaeology*, ed. R. H. Brill. Cambridge, MA: MIT press.

Hanson, C. B. 1980. Fluvial taphonomic processes: Models and experiments. In *Fossils in the making*, ed. A. K. Behrensmeyer and A. P. Hill, 156–81. Chicago: Univ. of Chicago Press.

Harris, J. E., and K. R. Weeks. 1973. *X-raying the pharaohs*. New York: Charles Scribner's Sons.

Harrison, W. 1971. Atlantis undiscovered—Bimini, Bahamas. *Nature* 230:287–89.

Haynes, C. V. 1964. Fluted projectile points: Their age and dispersion. *Science* 145:1408–13.

———. 1969. The earliest Americans. *Science* 166:709–15.

Haynes, C. V. Jr., 1969. Reply to Bryan: Early man in America and the late Pleistocene chronology of western Canada and Alaska: *Current Anthropology* 10:253–54.

Herrmann, G. 1968. Lapis lazuli: the early phases of its trade. *Iraq* 30:21–57.

Hey, R. W. 1962. The Quaternary and Palaeolithic of northern Libya. *Quaternaria* 6:435–49.

———. 1968. The Quaternary geology of the Jabal al-Akhdar coast. In *Geology and archaeology of northern Cyrenaica*, ed. F. T. Barr, 159–65. Tripoli: Petroleum Exploration Society of Libya.

Ho, P. T. 1969. The loess and the origin of Chinese agriculture. *American Historical Review* 75:1–36.

Hodges, H. 1970. *Technology in the ancient world*. London: Penguin Press.

Hopkins, D. M., ed. 1967. *The Bering land bridge*. Stanford, CA: Stanford Univ. Press.

Hopkins, D. M., and J. L. Giddings. 1965. The Quaternary geology and archaeology of Alaska. In *The Quaternary of the United States*, ed. H. E. Wright and D. G. Frey, 355–76. Princeton, NJ: Princeton Univ. Press.

Horowitz, A. 1979. *The Quaternary of Israel*. New York: Academic Press.

Howell, F. C. 1959. Upper Pleistocene stratigraphy and early man in the Levant. *Proceedings of the American Philosophical Society* 103:1–65.

Huffington, R. M., and C. C. Albritton. 1941. Quaternary sand on the southern high plains of western Texas. *American Journal of Science* 239:325–38.

Hunt, C. B. 1975. *Death Valley: Geology, ecology, and archaeology*. Berkeley: Univ. of California Press.

Ingersoll, D., J. E. Yellen, and W. MacDonald, eds. 1977. *Experimental archeology*. New York: Columbia Univ. Press.

Jennings, J. D. 1968. *Prehistory of North America*. New York: McGraw-Hill.

Jennings, J. D., and E. Norbeck, eds. 1964. *Prehistoric man in the New World*. Chicago: Univ. of Chicago Press.

Johnson, F. 1951. Collaboration among scientific fields with special reference to archaeology. In *Essays on archaeological methods*, ed. J. B. Griffin, 34–50. Univ. of Michigan Anthropology Papers, no. 8.

Jopling, A., W. Irving, and B. Beebe. 1981. Stratigraphic, sedimentological and faunal evidence for the occurrence of pre-Sangamonian artifacts in northern Yukon. *Arctic* 34 (1):3–33.

Karcz, I., U. Kafri, and Z. Meshel. 1977. Archaeological evidence for subrecent seismic activity along the Dead Sea—Jordan Rift. *Nature* 269:234–35.

Karcz, I., and U. Kafri. 1978. Evaluation of supposed archaeoseismic damage in Israel. *Journal of Archaeological Science* 5:237–53.

Keeley, H. C. M., G. E. Hudson, and J. Evans. 1977. Trace elements contents of human bones in various states of preservation: 1. The soil silhouette. *Journal of Archaeological Science* 4:19–24.

Keller, J. 1980. Prehistoric pumice tephra on Aegean islands. In *Thera and the Aegean world II*, ed. C. Doumas, 49–56. London: Thera and the Aegean World.

Kittleman, L. 1979. Tephra. *Scientific American* 241:160–77.

Klasner, J., and P. Calengas. 1981. Electrical resistivity and soil studies at Orendorph archaeological site, Illinois: A case study. *Journal of Field Archaeology* 8:167–74.

Komlos, G., P. Hedervari, and S. Meszaros. 1978. A brief note on tectonic earthquakes related to the activity of Santorini from antiquity to the present. In *Thera and the Aegean world*, vol. 1, ed. C. Doumas, 97–107. London: Thera and the Aegean World.

Kopper, J. 1976. Paleomagnetic dating and stratigraphic interpretation in archaeology. *MASCA Newsletter* 12 (1).

Lechtman, H., and R. Merrill, eds. 1977. *Material culture: Styles, organization and dynamics of technology*. St. Paul, MN: West Publishing Co.

Lee, R. B., and I. DeVore, eds. 1968. *Man the hunter*. Chicago: Aldine.

Lewin, S. Z. 1966. The preservation of natural stone, 1839–1965: An annotated bibliography. *Art and Archaeological Technical Abstracts, Supplement* 6 (1):185–272.

Limbrey, S. 1975. *Soil science and archaeology*. New York: Academic Press.

Lisitsina, G. N. 1976. Arid soils—the source of archaeological information. *Journal of Archaeological Science* 3:55–60.

Lobdell, J. E. 1974. The combined use of varied geologic features in bison procurement: An early middle period example from southcentral Wyoming. In *Applied geology and archaeology: The Holocene history of Wyoming*, ed. M. Wilson, 19–21. Geological Survey of Wyoming Report of Investigations, no. 10.

Lucas, A., and J. R. Harris, 1962. *Ancient Egyptian materials and industries*. 4th ed. London: Arnold.

McBurney, C. B. M. 1967. *The Haua Fteah (Cyrenaica) and the stone Age of the southeast Mediterranean*. Cambridge: Cambridge Univ. Press.

McBurney, C. B. M. and R. W. Hey. 1955. *Prehistory and Pleistocene geology in Cyrenaican Libya*. Cambridge: Cambridge Univ. Press.

McCoy, F. W. 1980. The upper Thera (Minoan) ash in deep-sea sediments: Distribution and comparison with other ash layers. In *Thera and the Aegean world II*, ed. C. Doumas, 57–78. London: Thera and the Aegean World.

McGuire, K. R. 1980. Cave sites, faunal analysis, and big-game hunters of the Great Basin: A caution. *Quaternary Research* 14:263–68.

McIntosh, R. J. 1977. The excavation of mud structures: An experiment from West Africa. *World Archaeology* 9:185–99.

MacNeish, R. S., A. Nelken-Terner, and I. W. Johnson. 1967. *The prehistory of the Tehuacán valley*. Vol. 2, *Non-ceramic artifacts*. Austin: Univ. of Texas Press.

Malde, H. E. 1964. Environment and man in arid America. *Science* 145:123–39.

Marinatos, S. P. 1971. Geology and archaeology of a volcano. In *Acta of the First International Scientific Congress on the Volcano of Thera*, ed. A. Kaloyeropoyloy, 407–12. Athens: Archaeological Services of Greece.

Martin, P. 1967. Overkill at Olduvai Gorge. *Nature* 215:212–13.

Martin, P. S., and H. E. Wright, eds. 1967. *Pleistocene extinctions: The search for a cause*. New Haven: Yale Univ. Press.

Mattingley, G. E. G., and R. J. B. Williams. 1962. A note on the chemical analysis of a soil buried since Roman times. *Journal of Soil Science* 13:254–58.

Mavor, James W., Jr. 1969. *Voyage to Atlantis*. New York: G. Putnam's Sons.

Mayer-Oakes, W. J., ed. 1967. *Life, land and water*. Occasional Papers no. 1, Department of Anthropology, Univ. of Manitoba, Winnipeg.

Meighan, C. W., and C. V. Haynes. 1970. The Borax Lake site revisited. *Science* 167:1213–21.

Meyers, J. T. 1970. Chert resources of the lower Illinois Valley. *Reports of Investigations*, no. 18, Illinois State Museum, vol. 2.

Moir, J. R. 1926. The silted-up Lake Hoxne and its contained flint implements. *Proceedings of the Prehistoric Society* 5:137–65.

Movius, H. L., Jr., and S. Judson. 1956. The rock-shelter of La Colombière. *American School of Prehistoric Research Bulletin*, no. 19. Cambridge, MA: Peabody Museum.

Mueller, J. W., ed. 1975. *Sampling in archaeology*. Tucson: Univ. of Arizona Press.

Murray, G. W. 1967. Trogodytica: The Red Sea littoral in Ptolemaic times. *Geographical Journal* 133:23–33.

North, F.J. 1937. Geology for archaeologists. *Archaeological Journal* 94:73–115.

Oakley, K. P. 1964. *Frameworks for dating fossil man*. Chicago: Aldine-Atherton.

Page, D. L. 1970. *The Santorini volcano and the desolation of Minoan Crete*. London: Society for Promotion of Hellenic Studies, London.

Parker, R. B., and H. Toots. 1980. Trace elements in bones as palaeobiological indicators. In *Fossils in the making*, ed. A. K. Behrensmeyer, and A. P. Hill, 197–207. Chicago: Univ. of Chicago Press.

Pichler, H., and W. Schiering. 1977. The Thera eruption and late Minoan-IB destructions on Crete. *Nature* 267:819–22.

Pyddoke, E., ed. 1963. *The scientist and archaeology*. London: Phoenix House.

Reagan, M., W. Dort, Jr., V. Bryant, Jr., and C. Johannsen. 1978. Flake tools stratified below paleo-Indian artifacts. *Science* 200:1272–74.

Reilly, F. A., ed. 1964. *Guidebook to the geology and archaeology of Egypt*. Petroleum Exploration Society of Libya, Sixth Annual Field Conference, Tripoli.

Rosenfeld, A. 1965. *The inorganic raw materials of antiquity*. New York: Praeger.

Said, R. 1975. The geological evolution of the River Nile. In *Problems in prehistory: North Africa and the Levant*, ed. F. Wendorf and A. E. Marks, 7–44. Dallas: SMU Press.

Said, R., F. Wendorf, and R. Schild. 1970. The geology and prehistory of the Nile Valley in upper Egypt. *Archaeologia Polona* 12:43–60.

Said, R., C. Albritton, F. Wendorf, R. Schild, and M. Kobusiewicz. 1972a. A preliminary report on the Holocene geology and archaeology of the northern Fayum Desert. In *Playa Lake symposium*, ed. C. Є. Reeves, Jr., 41–61. Lubbock, TX: Icasals.

———. 1972b. Remarks on the Holocene geology and archaeology of nothern Fayum Desert. *Archaeologia Polona* 13:7–22.

Sandford, K. S., and W. J. Arkell. 1939. *Paleolithic man and the Nile Valley in Lower Egypt, with some notes upon a part of the Red Sea littoral*. Univ. of Chicago Oriental Institute Publications, vol. 46.

Sauer, C. O. 1944. A geographic sketch of early man in America. *Geographical Review* 34:529–73.

Schild, R., and F. Wendorf. 1975. New explorations in the Egyptian Sahara. In *Problems in prehistory: North Africa and the Levant*, ed. F. Wendorf and A. E. Marks, 65–112. Dallas: SMU Press.

Sellards, E. H. 1952. *Early man in America*. Austin: Univ. of Texas Press.

Shackleton, N. J. 1973. Oxygen isotope analysis as a means of determining season of occupation of prehistoric midden sites. *Archaeometry* 15:133–41.

Shackley, M. L. 1978. The behaviour of artefacts as sedimentary particles in a fluviatile environment. *Archaeometry* 20:55–61.

———. 1979. Geoarchaeology: Polemic on a progressive relationship. *Naturwissenschaften* 66:429–33.

Sheets, P. 1981. Volcanoes and the Maya. *Natural History* 90 (8):32–41.

Shotton, F. W., and G. L. Hendry. 1979. The developing field of petrology in archaeology. *Journal of Archaeological Science* 6:75–84.

Singer, C. A., and J. E. Ericson. 1977. Quarry analysis at Bodie Hills, Mono County,

California. In *Exchange systems in prehistory*, ed. T. K. Earle and J. E. Ericson, 171–88. New York: Academic Press.

Sjoberg, A. 1976. Phosphate analysis of anthropic soils. *Journal of Field Archaeology* 3:447–54.

Slager, S., and H. T. J. Van Wetering. 1977. Soil formation in archaeological pits and adjacent loess soils in southern Germany. *Journal of Archaeological Science* 4: 259–67.

Smith, P. E. L. 1958. New investigations in the late Pleistocene archeology of the Kom Ombo Plain (Upper Egypt). *Quaternaria* 9:141–52.

Sohnge, P. G., D. J. L. Visser, and C. V. Lowe. 1937. *The geology and archaeology of the Vaal River basin*. Geological Survey Memoir, vol. 35, pt. 1 and 2, Pretoria.

Sokoloff, V. P., and G. F. Carter. 1952. Time and trace metals in archeological sites. *Science* 116:1–5.

———. 1953. Modern and ancient soils at some archaeological sites in the valley of Mexico. *American Antiquity* 19:50–55.

Solecki, R. S. 1951. Notes on soil analysis and archaeology. *American Antiquity* 16: 254–56.

Solecki, R. S., and A. Leroi-Gourhan. 1961. Paleoclimatology and archeology in the Near East. *Annals of the New York Academy of Sciences* 95:729–39.

Stafford, T., Jr. 1981. Alluvial geology and archeological potential of the Texas Southern High Plains. *American Antiquity* 46:548–65.

Stanford, D., R. Bonnichsen, and R. Morlan. 1981. The Ginsberg experiment: Modern and prehistoric evidence of a bone-flaking technology. *Science* 212:438–39.

Stross, F. H., and A. E. O'Donnell. 1972. Laboratory analysis of organic materials. *Addison-Wesley Module in Anthropology* 22:1–24.

Tate, G. P. 1910–1912. *Seistan: A Memoir on the history, topography, ruins and people*. Calcutta: Superintendent Government Printing.

Taylor, R. E., ed. 1977. *Advances in obsidian glass studies: Archaeological and geochemical perspectives*. Park Ridge, NJ: Noyes.

de Terra, H., and T. T. Patterson. 1939. *Studies on the Ice Age in India and associated human cultures*. Carnegie Institute, Publication 493, Washington, DC.

Thomas, W. L., ed. 1956. *Man's role in changing the face of the earth*. Chicago: Univ. of Chicago Press.

Thorarinsson, S. 1978. Some comments on the Minoan eruption of Santorini. In *Thera and the Aegean world*, vol. 1, ed. C. Doumas, 263–75. London: Thera and the Aegean World.

Turnbaugh, W. A. 1978. Floods and archaeology. *American Antiquity* 43:593–607.

Ucko, P. J., and G. W. Dimbleby, eds. 1969. *The domestication and exploitation of plants and animals*. Chicago: Aldine.

Van Bemmelen, R. W. 1971. Four volcanic outbursts that influenced human history: Toba, Sunda, Merapi and Thera. In *Acta of the First International Scientific Congress on the Volcano of Thera*, ed. A. Kaloyeropoyloy, 5–50. Athens: Archaeological Services of Greece.

Van Liere, W. 1966. The Pleistocene and Stone Age of the Orontes River (Syria). *Annales archéologiques arabes syriennes* 16 (2):7–30.

Vita-Finzi, C. 1978. *Archaeological sites in their settings*. London: Thames and Hudson.

Vitaliano, C. J., J. S. Fout, and D. B. Vitaliano. 1978. Petrochemical study of the Tephra sequence exposed in the Phira Quarry. In *Thera and the Aegean world*, vol. 1, ed. C. Doumas, 203–15. London: Thera and the Aegean World.

Vitaliano, D. B. 1978. Atlantis from the geologic point of view. In *Atlantis: Fact or fiction*, ed. E. S. Ramage, 137–60. Bloomington: Indiana Univ. Press.

Vitaliano, D. B., and C. J. Vitaliano. 1971. Plinian eruptions, earthquakes, and Santorin: A review. In *Acta of the First International Scientific Congress on the Volcano of Thera*, ed. A. Kaloyeropoyloy, 88–99. Athens: Archaeological Services of Greece.

Voute, C. 1957. A prehistoric find near Razzaza (Karbala Liwa): Its significance for the morphological and geological history of the Abu Dibbis depression and surrounding area. *Sumer* 33:1–14.

———. 1963. Some geological aspects of the conservation project for the Philae temples in the Aswan area. *Geologische Rundschau* 52:665–75.

Wallis, F. S. 1955. Petrology as an aid to prehistoric and medival archaeology. *Endeavour* 14:146–51.

Ward, G. K. 1974. A systematic approach to the definition of sources of raw material. *Archaeometry* 16:41–53.

Watanabe, H. 1959. The direction of remanent magnetism of baked earth and its application to chronology for anthropology and archaeology in Japan. *Journal of the Faculty of Science, University of Tokyo* 2:1–188.

Watson, P. J. 1966. Prehistoric miners of Salts Cave, Kentucky. *Archaeology* 19:237–43.

Watson, R. A., and P. J. Watson. 1969. *Man and nature*. New York: Harcourt, Brace, Jovanovich.

Weaver, M. E. 1971. A new water separation process for soil from archaeological excavations. *Anatolian Studies* 21:65–68.

Wendorf, F., ed. 1968. *The prehistory of Nubia*, vol. 1 and 2. Dallas: SMU Press.

Wendorf, F., and A. E. Marks, eds. 1975. *Problems in prehistory: North Africa and the Levant*. Dallas: SMU Press.

Wiegers, F. 1920. Diluvial Prähistorie als geologische Wissenschaft. *Abhandlungen der Preussischen Geologischen Landesanstalt*, n. F., vol. 84.

Williston, S. W. 1902. An arrowhead found with bones of *Bison occidentalis* Lucas in western Kansas. *American Geologist* (Minneapolis, MN) 30:313–15.

Wiseman, J. R. 1980. Archaeology in the future: An evolving discipline. *American Journal of Archaeology* 84:279–85.

Wood, W. R., and D. L. Johnson. 1978. A survey of disturbance process in archaeological site formation. *Advances in Archaeological Method and Theory* 1:315–81.

Wooldridge, S. W. 1957. Some aspects of the physiography of the Thames Valley in relation to the Ice Age and early man. *Proceedings of the Prehistoric Society*. 23:1–19.

Wright, H. E. 1957. Geology. In *The identification of non-artifactual archaeological materials*, 50–51. Washington, DC: National Academy of Sciences, National Research Council, Publication 565.

———. 1962. Late Pleistocene geology of coastal Lebanon. *Quaternaria* 6:525–40.

———. 1972. Vegetation history. In *The Minnesota Messenia expedition: Reconstructing a Bronze Age regional environment*, ed. W. A. McDonald and G. Rapp, Jr., 188–99. Minneapolis: Univ. of Minnesota Press.

Wright, H. E., and D. G. Frey, eds. 1965. *The Quaternary of the United States*. Princeton NJ: Princeton Univ. Press.

Wymer, J. 1957. A Clactonian flint industry at Little Thurrock, Grays, Essex. *Proceedings of the Geological Association* 68:159–77.

Yassoglou, N. J., and C. Nobeli. 1972. Soil studies. In *The Minnesota Messenia expedition: Reconstructing a Bronze Age regional environment*, ed. W. A. McDonald, and G. Rapp, Jr., 171–76. Minneapolis: Univ. of Minnesota Press.

Yassoglou, N. J., and C. F. Haidouti. 1978. Soil formation, In *Excavations at Nichoria in southwest Greece*. Vol. 1, *Site, environs, and techniques*, ed. G. Rapp, Jr., and S. E. Aschenbrenner, 31–40. Minneapolis: Univ. of Minnesota Press.

Zagwijn, W. H., H. M. VanMontfrans, and J. G. Zandstra. 1971. Subdivision of the "Cromerian" in the Netherlands: Pollen-analysis, paleomagnetism and sedimentary petrology. *Geologie en Mijnbouw* 50:41–58.

Znaczko-Jaworski, I. L. 1958. Experimental research on ancient mortars and binding materials. *Quarterly of History of Science and Technology* 3:377–407.

PALEOCLIMATOLOGY

Allchin, B., and A. Goudie. 1974. Pushkar: Prehistory and climatic change in western India. *World Archaeology* 3:358–68.

Al-Sayari, S. S., and J. A. Zotl, eds. 1978. *The Quaternary period in Saudi Arabia*. Vol. 1. New York: Springer-Verlag.

Antevs, E. 1937. Climatic changes and pre-white man. *University of Utah Bulletin* 38:168–91.

Baerreis, D. A., and R. A. Bryson. 1965. Climatic episodes and the dating of the Mississippian cultures. *Wisconsin Archaeologist* 46:203–20.

Bakker, E. van Zinderen, and J. D. Clark. 1962. Pleistocene climates and cultures in north-eastern Angola. *Nature* 196:639–42.

Brice, W. C., ed. 1978. *The environmental history of the Near and Middle East*. New York: Academic Press.

Bryan, K., and C. C. Albritton, Jr. 1943. Soil phenomena as evidence of climatic changes. *American Journal of Science* 241:469–90.

Butzer, K. W. 1958. Quaternary stratigraphy and climate in the Near East. *Bonner Geographische Abhandlungen*, vol. 24.

———. 1976. Pleistocene climates. In *Geoscience and man*. Vol. 13, *The ecology of the Pleistocene*, ed. R. C. West. Baton Rouge: Louisiana State Univ. School of Geoscience.

———. 1980. Adaptation to global environmental change. *Professional Geographer* 32:269–78.

Caton-Thompson, G., and E. W. Gardner. 1939. Climate, irrigation and early man in the Hadhramaut. *Geographical Journal* 93:18–38.

Cerling, T. E., R. L. Hay, and J. R. O'Neil. 1977. Isotopic evidence for dramatic climatic changes in East Africa during the Pleistocene. *Nature* 267:137–38.

Dimbleby, G. W. 1976. Climate, soil and man. *Philosophical Transactions of the Royal Society of London, Biological Sciences* 275:197–208.

Dort, W., et al. 1965. Paleotemperatures and chronology at archaeological cave sites revealed by thermoluminescence. *Science* 150:480–82.

Emiliani, C., L. Cardini, T. Mayeda, C. B. M. McBurney, and E. Tongiogi. 1963. Paleotemperature analysis of fossil shells of marine mollusks (food refuse) from the Arene Candide cave, Italy, and the Haua Fteah cave, Cyrenaica. In *Isotopic and cosmic chemistry*, ed. H. Craig, S. L. Miller, and G. J. Wasserburg, 133−56. Amsterdam: Elsevier.

Farrand, W. R. 1971. Late Quaternary paleoclimates of the eastern Mediterranean area. In *Late Cenozoic glacial ages*, ed. Karl K. Turekian, 529−64. New Haven and London: Yale Univ. Press.

———. 1979. Chronology and paleoenvironment of Levantine prehistoric sites as seen from sediment studies. *Journal of Archaeological Science* 6:369−92.

Flint, R. F. 1959. Pleistocene climates in East and southern Africa. *Bulletin of the Geological Society of America* 70:343−74.

Fuchs, V. E., and T. T. Paterson. 1947. The relation of volcanicity and orogeny to climatic change. *Geological Magazine* 84:321−33.

Hamilton, A. C. 1982. *Envirnomental history of East Africa: A study of the Quaternary*. New York: Academic Press.

Harmon, R. S., P. Thompson, H. P. Schwarcz, and D. C. Ford. 1975. Uranium-series dating of speleothems. *Quaternary Research* 9:54−70.

Horowitz, A. 1971. Climatic and vegetational developments in northeastern Israel during the Upper Pleistocene—Holecene times. *Pollen et spores* 13:255−78.

Huntington, E. 1914. *The climatic factor as illustrated in arid America*. Washington, D.C.: Carnegie Institution.

Irwin-Williams, C., and C. V. Haynes. 1970. Climatic change and early population dynamics in the southwestern United States. *Quaternary Research* 1:59−71.

Legge, A. J. 1972. Cave climates. In *Papers in economic prehistory*, ed. E. S. Higgs, 97−103. Cambridge: Cambridge Univ. Press.

Mather, J. R. 1954. The effect of climate on the New World migration of primitive man. *Southwestern Journal of Anthropology* 10:304−21.

Moreau, R. E. 1933. Pleistocene climatic changes and the distribution of life in East Africa. *Journal of Ecology* 211:415−35.

Murray, G. W. 1951. The Egyptian climate: An historical outline. *Geographical Journal* 117:422−34.

Sandford, K. S. 1933. Past climate and early man in the southern Libyan Desert. *Geographical Journal* 82:219−22.

Solecki, R., and A. Leroi-Gourhan. 1961. Paleoclimatology and archaeology in the Near East. *Annals of the New York Academy of Science* 95:729−39.

VanZinderen Bakker, E. M., and K. W. Butzer. 1973. Quaternary environmental changes in southern Africa. *Soil Science* 116:236−48.

Washbourn, C. K. 1967. Lake levels and Quaternary climates in the Eastern Rift Valley of Kenya. *Nature* 216:672−73.

Wright, H. E. 1960. Climate and prehistoric man in the eastern Mediterranean. In *Prehistoric investigations in Iraqi Kurdistan*, ed. R. T. Braidwood, and B. Howe, 71−97. Chicago: Univ. of Chicago Press.

———. 1961. Late Pleistocene climate of Europe: A review. *Bulletin of the Geological Society of America* 72:933−84.

———. 1964. Late Quaternary climates and early man in the mountains of Kurdistan. *Report of the Sixth International Congress on the Quaternary* (Warsaw, 1961). Vol. 2, *Paleoclimatology*, 341−48.

———. 1968. Climatic change in Mycenaean Greece. *Antiquity* 42:123−27.

PALEONTOLOGY

Aleem, A. A. 1958. Taxonomic paleoecological investigation of the diatom-flora of the extinct Fayorem Lake (Upper Egypt). *University of Alexandria Bulletin* 2:217–44.

Alteena, C. O. Van R. 1962. Molluscs and echinoderms from Paleolithic deposits in the rockshelter of Ksar Akil, Lebanon. *Zoologische Mededelingen* 38 (5).

Ascenzi, A. 1969. Microscopy and prehistoric bone. In *Science in archaeology*, rev. ed., ed. D. Brothwell and E. Higgs, 526–38. London: Thames and Hudson.

Asch, N., R. I. Ford, and D. L. Asch. 1972. *Paleoethnobotany of the Koster site: The Archaic horizons*. Illinois State Museum, Reports of Investigations, vol. 24.

Baerreis, D. A. 1973. Gastropods and archaeology. In *Variations in anthropology: Essays in honor of John C. McGregor*, ed. D. W. Lathrap and J. Douglas. Illinois Archaeological Survey no. 44, Springfield, IL.

Baker, J., and D. Brothwell. 1980. *Animal diseases in archaeology*. London: Academic Press.

Bass, W. M. 1971. *Human osteology: A laboratory and field manual of the human skeleton*. Columbia, MO: Missouri Archaeological Society Special Publication.

Bassett, E., M. Keith, G. Armelagos, D. Martin, and A. Villanueva. 1980. Tetracycline-labeled human bone from ancient Sudanese Nubia (A.D. 350). *Science* 209:1532–34.

Bate, D. M. A. 1940. The fossil antelopes of Palestine in Natufian (Mesolithic) times. *Geological Magazine* 77:418–33.

———. 1955. Vertebrate faunas of Quaternary deposits in Cyrenaica. In *Prehistory and Pleistocene geology in Cyrenaica, Libya*, ed. C. B. M. McBurney and R. W. Hey, 284–91. Cambridge: Cambridge Univ. Press.

Behrensmeyer, A. K. 1975. The taphonomy and paleoecology of Plio-Pleistocene vertebrate assemblages east of Lake Rudolf, Kenya. *Bulletin, Museum of Comparative Zoology, Harvard University* 146:473–578.

Behrensmeyer, A. K., D. Western, and D. E. Dechant-Boaz. 1979. New perspectives in vertebrate paleoecology from a recent bone assemblage. *Paleobiology* 5 (1):12–21.

Behrensmeyer, A. K., and A. P. Hill, eds. 1980. *Fossils in the making: Vertebrate taphonomy and paleoecology*. Chicago: Univ. of Chicago Press.

Binford, L. R., and J. B. Bertram. 1977. Bone frequencies—and attritional processes. In *For theory building in archaeology: Essays on faunal remains, aquatic resources, spatial analysis, and systemic modeling*, ed. L. R. Binford, 77–153. New York: Academic Press.

Bokoyni, S. 1970. Animal remains from Lepenski Vir. *Science* 167:1702–04.

Bonnichsen, R. 1973. Some operational aspects of human and animal bone alteration. In *Mammalian osteoarchaeology: North America*, ed. M. Gilbert, 9–24. Columbia, MO: Missouri Archaeological Society.

Boule, M. 1900. Étude paléontologique et archéologique sur la station paléolithique du lac Karar (Algerie). *L'Anthropologie* 11:1–21.

Boussneck, I. 1969. Osteological differences between sheep and goats. In *Science in archaeology*, 2d ed., ed. D. Brothwell and E. Higgs, 331–68. London: Thames and Hudson.

Brain, C. K. 1967. The Transvaal Museum's fossil project at Swartkrans. *South African Journal of Science* 63:378–84.

———. 1976. Some principles in the interpretation of bone accumulations associated with man. In *Human origins: Louis Leakey and the East African evidence*, ed. G. L. Isaac and B. R. McCown, 97–116. New York: W. A. Benjamin.

———. 1980. Some criteria for the recognition of bone-collecting agencies in African caves. In *Fossils in the making: Vertebrate taphonomy and paleoecology*, ed. A. K. Behrensmeyer and A. P. Hill, 107–30. Chicago: Univ. of Chicago Press.

Broom, R. 1937. On some new Pleistocene mammals from limestone caves of the Transvaal. *South African Journal of Science* 33:750–68.

Brothwell, D. R. 1977. *Digging up bones: The excavation, treatment and study of human skeletal remains*. 2d ed. Oxford: Oxford Univ. Press.

Buckland, P. C. 1976. The use of insect remains in the interpretation of archaeological environments. In *Geoarchaeology*, ed. D. A. Davidson and M. L. Shackley, 369–96. London: Duckworth.

Burleigh, R., J. Clutton-Brock, P. J. Felder, and G. de G. Sieveking. 1977. A further consideration of Neolithic dogs with special reference to a skeleton from Grimes Graves (Norfolk), England. *Journal of Archaeological Science* 4:353–67.

Callen, E. O. 1969. Diet as revealed by coprolites. In *Science in archaeology*, 2d ed., ed. D. Brothwell and E. Higgs, 235–43. London: Thames and Hudson.

Carr, C. J. 1976. Plant ecological variation and pattern in the lower Omo Basin. In *Earliest man and environments in the Lake Rudolf Basin*, ed. Y. Coppens, F. C. Howell, G. L. Isaac, and R. E. F. Leakey, 432–70. Chicago: Univ. of Chicago Press.

Carter, G. F. 1950. Plant evidence for early contacts with America. *Southwestern Journal of Anthropology* 6 (2):161–82.

Casteel, R. W. 1976. *Fish remains in archaeology and palaeo-environmental studies*. New York: Academic Press.

———. 1977. Characterization of faunal assemblages and the minimum number of individuals determined from paired elements: Continuing problems in archaeology. *Journal of Archaeological Science* 4:125–34.

Chaplin, R. E. 1971. *The study of animal bones from archaeological sites*. London: Seminar Press.

Chowdhury, K. A., and S. S. Ghosh. 1951. Plant remains from Harappa, 1946. *Ancient India* 7.

Churcher, C. S. 1972. *Late Pleistocene vertebrates from archaeological sites in the Plain of Kom Ombo, Upper Egypt*. Life Sciences Contribution, Royal Ontario Museum, vol. 82.

Churcher, C. S., and P. Smith. 1972. Kom Ombo: Preliminary report on the fauna of late Paleolithic sites in Upper Egypt. *Science* 77:259–61.

Clason, A. T. 1972. Some remarks on the use and presentation of archaeozoological data. *Helinium* 12:139–53.

———, ed. 1975. *Archaeozoological studies*. Amsterdam: North Holland.

Clason, A. T. and W. Prummel. 1977. Collecting, sieving and archaeozoological research. *Journal of Archaeological Science* 4:171–75.

Clutton-Brock, J. 1971. The primary food animals of the Jericho Tell from the Proto-Neolithic to the Byzantine period. *Levant* 3:41–55.

———. 1975. A system for the retrieval of data relating to animal remains from archaeological sites. In *Archaeozoological studies*, ed. A. T. Clason, 21–34. Amsterdam: Elsevier.

———. 1978. Early domestication and the ungulate fauna of the Levant during the prepottery neolithic period. In *The environmental history of the Near and Middle East,* ed. W. C. Brice, 29–40. London: Academic Press.

Cook, S. F. 1951. The fossilization of human bone: Calcium phosphate and carbonate. *University of California Publications in American Archaeology and Ethnology* 40: 263–80.

Cook, S. F., and R. F. Heizer. 1952. *The fossilization of human bone: Organic components and water.* University of California Archaeological Survey Department, no. 17.

Cook, S. F., S. T. Brooks, and H. E. Ezra-Cohn. 1962. Historical studies on fossil bone. *Journal of Paleontology* 36:483–94.

Cooke, H. B. S. 1952. Mammals, ape-men and stone-age men in southern Africa. *South African Archaeological Bulletin* 7:59–69.

———. 1963. Pleistocene mammal faunas of Africa with particular reference to southern Africa. In *African ecology and human evolution,* ed. F. C. Howell and F. Bourleire, 65–116. Chicago: Aldine.

Coppens, Y., and F. C. Howell. 1976. Mammalian faunas of the Omo Group: Distributional and biostratigraphic aspects. In *Earliest man and environments in the Lake Rudolf Basin,* ed. Y. Coppens, F. C. Howell, G. L. Isaac, and R. E. F. Leakey. Chicago: Univ. of Chicago Press.

Cornwall, I. W. 1956. *Bones for the archaeologist.* London: Phoenix House.

Craig, A. K., and N. P. Psuty. 1971. Paleoecology of shellmounds at Otuma, Peru. *Geographical Review* 61:125–32.

Daly, P. 1969. Approaches to faunal analysis in archaeology. *American Antiquity* 34: 146–53.

Dawson, E. W. 1969. Bird remains in archaeology. In *Science in archaeology,* ed. D. Brothwell and E. S. Higgs, 359–75. London: Thames and Hudson.

Delcourt, P. A., O. K. Davis, and R. C. Bright. 1979. *Bibliography of taxonomic literature for the identification of fruits, seeds, and vegetative plant fragments.* Environmental Sciences Division, Oak Ridge National Laboratory, Publication 1328.

Dennell, R. W. 1974. Botanical evidence for prehistoric crop processing activities. *Journal of Archaeological Science* 1:275–84.

Dimbleby, G. W. 1978. *Plants and archaeology.* 2d ed. London: John Baker.

Dimbleby, G. W., and J. G. Evans. 1974. Pollen and land-snail analysis of calcareous soils. *Journal of Archaeological Science* 1:117–33.

Dreimanis, A. 1967. Mastodons, their geologic age and extinction in Ontario, Canada. *Canadian Journal of Earth Sciences* 4 (4):663–75.

Driesch, A. von den. 1976. *A Guide to the measurement of animal bones from archaeological sites.* Peabody Museum, Bulletin 1. Cambridge, MA: Harvard Univ. Press.

Ducos, P. 1975. A new find of an equid metatarsal bone from Tel Mureibet in Syria and its relevance to the identification of equids from the early Holocene of the Levant. *Journal of Archaeological Science* 1 (1):71–73.

Evans, J. G. 1969. The exploitation of molluscs. In *The domestication and exploitation of plants and animals,* ed. P. J. Ucko and G. W. Dimbleby, 477–84. London: Duckworth.

———. 1972. *Land snails in archaeology.* London: Seminar Press.

———. 1976. Subfossil land-snail faunas from rock-rubble habitats. In *Geoarchaeology,* ed. D. A. Davidson and M. L. Shackley, 397–99. London: Duckworth.

Ewer. R. F. 1955. The fossil carnivores of the Transvaal caves: The Lycyaenas of Sterkfontein and Swartkrans, together with some general considerations of the Transvaal fossil hyaenids. *Proceedings of the Zoological Society* 124:839–57.

———. 1965. Large Carnivora. In *Olduvai Gorge*, vol. 1, ed. L. S. B. Leakey, 19–22. Cambridge: Cambridge Univ. Press.

———. 1967. The fossil hyaenids of Africa: A reappraisal. In *Background to evolution in Africa*, ed. W. W. Bishop, and J. D. Clark, 109–23. Chicago: Univ. of Chicago Press.

Flerow, C. C. 1967. On the origin of the mammalian fauna of Canada. In *The Bering land bridge*, ed. D. M. Hopkins, 271–80. Stanford, CA: Stanford Univ. Press.

Ford, R. I. 1979. Paleoethnobotany in American archaeology. *Advances in Archaeological Method and Theory* 2:285–326.

Freeman, L. G. 1973. The significance of mammalian faunas from Paleolithic occupations in Cantabrian Spain. *American Antiquity* 38 (1):3–44.

Frison, G., M. Wilson, and D. J. Wilson. 1976. Fossil bison and artifacts from an early Altithermal period arroyo trap in Wyoming. *American Antiquity* 41 (1):28–57.

Gautier, A. 1976. Assemblages of fossil freshwater mollusks from the Omo Group and related deposits in the Lake Rudolf Basin. In *Earliest man and environments in the Lake Rudolf Basin*, ed. Y. Coppens, F. C. Howell, G. L. Isaac, and R. E. F. Leakey, 383–401. Chicago: Univ. of Chicago Press.

Gibert, A. S., and B. H. Singer. 1982. Reassessing zooarchaeological quantification. *World Archaeology* 14:21–40.

Gifford, D. P. 1981. Taphonomy and paleoecology: A critical review of archaeology's sister discipline. In *Advances in archaeological method and theory*, vol. 4, ed. M. B. Schiffer, 365–438. New York: Academic Press.

Gilbert, B. M. 1973. *Mammalian osteo-archaeology: North America*. Columbia, MO: Univ. of Missouri Press.

Girling, M. A. 1979. Calcium carbonate–replaced arthropods from archaeological deposits. *Journal of Archaeological Science* 6:309–20.

Gordon, C. C., and J. E. Buikstra. 1981. Soil pH, bone preservation, and sampling bias at mortuary sites. *American Antiquity* 46:566–71.

Grayson, D. K. 1973. On the methodology of faunal analysis. *American Antiquity* 39:432–39.

———. 1979. On the quantification of vertebrate archaeofaunas. In *Advances in archaeological method and theory*, vol. 2, ed. M. B. Schiffer, 199–237. New York: Academic Press.

———. 1981. The effects of sample size on some derived measures in vertebrate faunal analysis. *Journal of Archaeological Science* 8:77–88.

Green, F. J. 1979. Phosphatic mineralization of seeds from archaeological sites. *Journal of Archaeological Science* 6:279–85.

Greenwood, P. H., and E. J. Todd. 1970. Fish remains from Olduvai. In *Fossil vertebrates in Africa*, vol. 2, ed. L. S. B. Leakey and R. J. G. Savage, 225–41. London: Academic Press.

Greig, J. R. A., and J. Turner. 1974. Some pollen diagrams from Greece and their archaeological significance. *Journal of Archaeological Science* 1:177–94.

Grigson, C. 1980. The craniology and relationships of four species of *Bos*: 5. *Bos indicus* L. *Journal of Archaeological Science* 7:3–32.

Haas, G. 1966. *On the vertebrate fauna of the Lower Pleistocene site 'Ubeidiya.* Jerusalem: Israel Academy of Sciences and Humanities.

Hall, S. 1980. Snails from Quaternary valley fill at Chaco Canyon, New Mexico. *Nautilus* 94:60–63.

Hallam, J. S., J. N. Edwards, B. Barnes, and A. J. Stuart. 1973. A Late Glacial elk with associated barbed points from High Furlong, Lancs. *Proceedings of the Prehistoric Society* 39:100–28.

Harcourt, R. A. 1974. The dog in prehistoric and early historic Britain. *Journal of Archaeological Science* 1:151–75.

Hare, P. E. 1980. Organic geochemistry of bone and its relation to the survival of bone in the natural environment. In *Fossils in the making*, ed. A. K. Behrensmeyer and A. P. Hill, 208–19. Chicago: Univ. of Chicago Press.

Harris, J. M. 1976. Bovidae from the East Rudolf succession. In *Earliest man and environments in the Lake Rudolf Basin*, ed. Y. Coppens, F. C. Howell, G. L. Isaac, and R. E. F. Leakey, 293–301. Chicago: Univ. of Chicago Press.

Harris, J. M., and T. D. White. 1977. Suid evolution and correlation of African hominid localities. *Science* 198:13–21.

Heizer, R. F., and L. K. Napton. 1969. Biological and cultural evidence from prehistoric human coprolites. *Science* 165:563–68.

Helbaek, H. 1969. Palaeo-ethnobotany. In *Science in archaeology*, ed. D. R. Brothwell and E. S. Higgs, 206–14. London: Thames and Hudson.

Henry, D. 1975. Fauna in Near Eastern archaeological deposits. In *Problems in prehistory: North Africa and the Levant*, ed. F. Wendorf and A. E. Marks, 379–85. Dallas: SMU Press.

Herre, W. 1969. The science and history of domestic animals. In *Science in archaeology*, 2d ed., ed. D. R. Brothwell and E. S. Higgs, 257–72. London: Thames and Hudson.

Higgs, E. S. 1961. Some Pleistocene faunas of the Mediterranean coastal areas. *Proceedings of the Prehistorical Society* 27:144–54.

Higham, C. F. W., A. Kijngam, and B. F. J. Manly. 1980. An analysis of prehistoric canid remains from Thailand. *Journal of Archaeological Science* 7:149–65.

Hooijer, D. 1961. *The fossil vertebrates of Ksar Akil, a paleolithic rockshelter in Lebanon*. Zoologische Verhandelingen, vol. 49, Leiden.

Jaeger, J. J., and H. B. Wesselman. 1976. Fossil remains of micromammals from the Omo Group deposits. In *Earliest man and environments in the Lake Rudolf Basin*, ed. Y. Coppens, F. C. Howell, G. L. Isaac, and R. E. F. Leakey, 351–59. Chicago: Univ. of Chicago Press.

Jansma, M. J. 1977. Diatom analysis of pottery. *ExHorreo* 4:78–85.

Jarman, H. N., A. J. Legge, and J. A. Charles. 1972. Retrieval of plant remains from archaeological sites by froth flotation. In *Papers in economic prehistory*, ed. E. S. Higgs, 39–48. London: Cambridge Univ. Press.

Johnson, D. 1980. Problems in the land vertebrate zoogeography of certain islands and the swimming power of elephants. *Journal of Biogeography* 7:1–17.

———. 1981. More comments on the northern Channel Islands' mammoths. *Quaternary Research* 15:105–06.

Johnson, D. C., M. Splengaer, and N. T. Boaz. 1976. Paleontological excavations in the Shungura formation, lower Omo Basin, 1969–73. In *Earliest man and en-*

vironment in the Lake Rudolf Basin, ed. Y. Coppens, F. C. Howell, G. L. Isaac, and R. E. F. Leakey, 402–20. Chicago: Univ. of Chicago Press.

Keeley, H. C. M. 1978. The cost-effectiveness of certain methods of recovering macroscopic organic remains from archaeological deposits. *Journal of Archaeological Science* 5:179–85.

Keepax, C. 1977. Contamination of archaeological deposits by seeds of modern origin with particular reference to the use of flotation machines. *Journal of Archaeological Science* 4:221–29.

Kenward, H. K. 1975. The biological and archaeological implications of the beetle *Algenus brunneus* (Gyll.) in ancient faunas. *Journal of Archaeological Science* 2: 63–69.

Kiszely, I. 1974. On the possibilities and methods of the chemical determination of sex from bones. *Ossa* 1:51–62.

Kitching, J. W. 1963. *Bone, tooth and horn tools of palaeolithic man*. Manchester. England: Manchester Univ. Press.

Klein, R. 1979. Stone Age exploitation of animals in southern Africa. *American Scientist* 67:151–60.

Klein, R. G. 1980. The interpretation of mammalian faunas from Stone Age archaeological sites, with special reference to sites in the southern Cape Province, South Africa. In *Fossils in the making*, ed. A. K. Behrensmeyer and A. Hill, 223–46. Chicago: Univ. of Chicago Press.

Kochetkova, V. I. 1970. Reconstruction de l'endocrane de l'*Atlanthropus mauritanicus* et de l'*Homo habilis*. *Proceedings of the Eighth International Congress of Anthropological and Ethnological Sciences* 1:102–04.

Koike, H. 1979. Seasonal dating and the valve-pairing technique in shell-midden analysis. *Journal of Archaeological Science* 6:63–74.

Kurten, B. 1957. Mammal migrations, Cenozoic stratigraphy, and the age of Peking man and the Australopithecines. *Journal of Paleontology* 31:215–27.

———. 1965. The Carnivora of the Palestine caves. *Acta Zoologica Fennica* (Helsinki) 107:1–74.

Lance, J. F. 1959. Faunal remains from the Lehner mammoth site. *American Antiquity* 25:35–42.

Lewin, R. 1981. Protohuman activity etched in fossil bones. *Science* 213:123–24.

Lyon, M. W. 1906. Mammal remains from two prehistoric village sites in New Mexico and Arizona. *Proceedings of the U.S. National Museum* (Washington, DC) 31:647–49.

Maglio, V. J. 1972. Vertebrate faunas and chronology of hominid-bearing sediments east of Lake Rudolf, Kenya. *Nature* 239:379–85.

Marquarat, W. H. 1974. A statistical analysis of constituents in human paleofecal specimens from Mammoth Cave. In *Archaeology of the Mammoth Cave area*, ed. P. J. Watson, 193–202. New York: Academic Press.

Martin, P. S. 1967. Prehistoric overkill. In *Pleistocene extinctions: The search for a cause*, ed. P. S. Martin and H. E. Wright, Jr., 75–120. New Haven: Yale Univ. Press.

Matteson, M. R. 1960. Reconstruction of prehistoric environments through the analyses of molluscan collections from shell middens. *American Antiquity* 26:117–20.

Meadow, R. A., and M. A. Zeder. 1978. Approaches to faunal analysis in the Middle East. *Peabody Museum Bulletin* 2:1–186.

Meighan, C. W. 1969. Molluscs as food remains in archaeological sites. In *Science in archaeology*, ed. D. Brothwell and E. S. Higgs, 415–22. London: Thames and Hudson.

Minnis, P. E. 1981. Seeds in archaeological sites: Sources and some interpretive problems. *American Antiquity* 46:143–52.

Morgan, A. 1973. Late Pleistocene environmental changes indicated by fossil insect faunas of the English Midlands. *Boreas* 2:173–212.

Morse, D. F., and D. F. Dickson. 1956. Prehistoric pathology. *Central States Archaeological Journal* 2 (4):143–51.

Mostny, G. 1968. Association of human industries with Pleistocene fauna in central Chile. *Current Anthropology* 9:214–15.

Olsen, S. J. 1964. *Mammal remains from archaeological site*. Peabody Museum Papers 56:1–162.

Olson, E. C. 1964. *Mammal remains from archaeological sites: Part 1, Southeastern and Southwestern United States*. Papers of the Peabody Museum of Archaeology and Ethnology, vol. 56, no. 1, Harvard Univ. Cambridge, MA.

Osborne, P. J. 1971. An insect fauna from the Roman site at Alcester, Warwickshire. *Britannia* 2:156–65.

———. 1973. Insects in archaeological deposits. *Science and Archaeology* 10:4–6.

Ostergard, M. 1980. X-ray diffractometer investigations of bones from domestic and wild animals. *American Antiquity* 48:59–63.

Parmalee, P. W., and W. E. Klippel. 1974. Freshwater mussels as a prehistoric food resource. *American Antiquity* 39:421–34.

Payne, S. 1969. A metrical distinction between sheep and goat metacarpals. In *The domestication and exploitation of plants and animals*, ed. P. J. Ucko and G. W. Dimbleby, 295–305. London: Duckworth.

———. 1975. Partial recovery and sample bias. In *Archaeozoological studies*, ed. A. T. Clason, 7–17. Amsterdam: Elsevier.

Pearsall, D. M. 1978. Phytolith analysis of archeological soils: Evidence for maize cultivation in formative Ecuador. *Science* 199:177–78.

Perkins, D. 1964. Prehistoric fauna from Shanidar, Iraq. *Science* 144:1565–66.

Perkins, D., Jr. 1968. The Pleistocene fauna from the Yabrudian rockshelters. *Annales archéologiques arabes syriennes* 1–2:123–30.

———. 1969. Fauna of Catal Huyuk: Evidence for early cattle domestication in Anatolia. *Science* 164:177–79.

Perkins, D., and P. Daly. 1968. The potential of faunal analysis: An investigation of the faunal remains from Suberde, Turkey. *Scientific American* 219 (5):96–106.

Rackham, O. 1978. The flora and vegetation of Thera and Crete before and after the Great Eruption. In *Thera and the Aegean world*, vol. 1, ed. C. Doumas, 755–64. London: Thera and the Aegean World.

Reed, C. A. 1959. Animal domestication in the prehistoric Near East. *Science* 130:1629–39.

Renfrew, J. M. 1973. *Palaeoethnobotany: The prehistoric food plants of the Near East and Europe*. London: Methuen.

Romer, A. S. 1928. A contribution to the study of prehistoric man in Algeria, North Africa. *Bulletin of the Logan Museum* (Beloit, Wisconsin), pp. 81–133.

Rovner, I. 1971. Potential of opal phytoliths for use in paleoecological reconstruction. *Quaternary Research* 1:343–59.

Ryder, M. L. 1968. *Animal bones in archaeology*. Oxford: Basil Blackwell.

———. 1969. Remains of fishes and other aquatic animals. In *Science in archaeology*, ed. D. Brothwell and E. S. Higgs, 376–94. London: Thames and Hudson.

Sandison, A. T. 1968. Pathological changes in the skeletons of earlier populations due to acquired disease, and difficulties in their interpretations. In *The skeletal biology of earlier human populations*, ed. D. R. Brothwell, 205–43. London: Pergamon Press.

Schmid, E. 1972. *Atlas of animal bones for prehistorians, archaeologists and Quaternary geologists*. Amsterdam: Elsevier.

Schoeninger, M. J., and C. S. Peebles. 1981. Effect of mollusc eating on human bone strontium levels. *Journal of Archaeological Science* 8:391–97.

Schultz, D. B., and L. C. Eiseley. 1935. Paleontological evidence of the antiquity of the Scottsbluff basin quarry and its associated artifacts. *American Anthropology* 37: 306–19.

Schultz, D. B., and W. D. Frankforter. 1946. The geologic history of the bison in the Great Plains (a preliminary report). *Bulletin of the University of Nebraska State Museum* (Lincoln, NE) 3 (1):1–10.

———. 1951. A preliminary report on the bison remains from the Finley site (Eden Bison Quarry). In *Early man in the Eden Valley*, ed. J. H. Moss, 119–24. University Museum, Univ. of Pennsylvania, Museum Monographs, vol. 6, Philadelphia, PA.

Sellards, E. H. 1955. Fossil bison and associated artifacts from Milnesand, New Mexico. *American Antiquity* 20:336–44.

Sellards, E. H., G. L. Evans, G. E. Mead, and A. D. Krieger. 1947. Fossil bison and associated artifacts from Plainview, Texas. *Bulletin of the Geological Society of America* 58:927–54.

Shackleton, J., and T. Van Andel. 1980. Prehistoric shell assemblages from Franchthi Cave and evolution of the adjacent coastal zone. *Nature* 288:357–59.

Shackleton, N. J. 1969. Marine mollusca in archaeology. In *Science in archaeology*, ed. D. Brothwell and E. S. Higgs, 407–14. London: Thames and Hudson.

Shane, O. C., III. 1978. *The vertebrate fauna of the Mountain Lake site, Cottonwood, Minnesota*. Scientific Publications of the Science Museum of Minnesota, n.s. vol. 4, no. 2.

Shipman, P. 1981. *Life history of a fossil: An introduction to taphonomy and paleoecology*. Cambridge, MA: Harvard Univ. Press.

Sloan, R. E., and M. A. Duncan. 1978. Zooarchaeology of Nichoria. In *Excavations at Nichoria in southwest Greece*. Vol. 1, *Site, environs, and techniques*, ed. G. Rapp, Jr., and S. E. Aschenbrenner, 60–77. Minneapolis: Univ. of Minnesota Press.

Smithson, F. 1958. Grass opal in British soils. *Journal of Soil Science* 9:148–54.

Sutcliffe, A. J. 1964. The mammalian fauna. In *The Swanscombe skull*, ed. C. D. Ovey, 85–111. London: Royal Anthropological Institute.

Taieb, M., D. C. Johanson, Y. Coppens, and J. L. Aronson. 1976. Geological and paleontological background of Hadar hominid site, Afar, Ethiopia. *Nature* 260: 289–93.

Thomas, D. H. 1971. On distinguishing natural from cultural bone in archaeological sites. *American Antiquity* 36:366–71.

Turnbull, P. F., and C. A. Reed. 1974. The fauna from the Terminal Pleistocene of Palegawra Cave, a Zarzian occupation site in northeastern Iraq. *Fieldiana Anthropology* 63 (3):81–146.

Turner, J. 1978. The vegetation of Greece during prehistoric times—the palynological evidence. In *Thera and the Aegean world*, vol. 1, ed. C. Doumas, 765–73. London. Thera and the Aegean World.

Uerpmann, H. P. 1973. Animal bone finds and economic archaeology: A critical study of osteo-archaeological methods. *World Archaeology* 4:307–32.

VanZeist, W. 1974. Palaeobotanical studies of settlement sites in the coastal areas of the Netherlands. *Palaeohistoria* 16:223–371.

Volman, T. P. 1978. Early archaeological evidence for shellfish collecting. *Science* 201: 911–13.

Von den Driesch, A. 1976. *A guide to the measurement of animal bones from archaeological sites*. Peabody Museum Bulletins, Cambridge, MA.

Watling, R., and M. R. D. Seaward. 1976. Some observations on puffballs from British archaeological sites. *Journal of Archaeological Science* 3:165–72.

Watson, J. P. N. 1972. Fragmentation analysis of animal bone samples from archaeological sites. *Archaeometry* 14:221–28.

———. 1975. Domestication and bone structure in sheep and goats. *Journal of Archaeological Science* 2:375–83.

Willcox, G. H. 1977. Exotic plants from Roman water-logged sites in London: *Journal of Archaeological Science* 4:269–82.

Williams, D. 1975. Identification of waterlogged wood by the archaeologist. *Science and Archaeology* 14:3–4.

Zeimens, G., and D. N. Walker. 1974. Bell Cave, Wyoming: Preliminary archaeological and paleontological investigations. In *Applied geology and archaeology: The Holocene history of Wyoming*, ed. M. Wilson, 83–90. Geological Survey of Wyoming Report of Investigations no. 10.

Zeuner, F. E. 1963. *A history of domesticated animals*. London: Hutchinson.

PROVENANCE STUDIES

Abascal-M. R., G. Harbottle, and E. V. Sayre. 1973. Correlation between terra cotta figurines and pottery from the Valley of Mexico and source clays by activation analysis. In *Archaeological chemistry*, ed. C. Beck, 81–99. Advances in Chemistry Series, no. 138. New York: American Chemical Society.

Aburto, S., S. Cruz, R. Gomez, and M. Jimenez. 1979. Mossbauer studies of ancient Mexican pottery. *Archaeo-Physica* 10:1–7. Proceedings of the 18th International Symposium on Archaeometry and Archaeological Prospection, Bonn.

Agrawal, D. P., R. V. Krishnamurthy, and S. Kusumgar. 1979. Fresh chemical data and the cultural affiliation of the Daimabad bronzes. *Archaeo-Physica* 10:8–13. Proceedings of the 18th International Symposium on Archaeometry and Archaeological Prospection, Bonn.

Artzy, M., F. Asaro, and I. Perlman. 1973. The origin of the Palestinian bichrome ware. *Journal of the American Oriental Society* 93:446–61.

Asaro, F., M. Dothan, and I. Perlman. 1971. An introductory study of Mycenaean IIIC1 ware from Tel Ashdod. *Archaeometry* 3:169–76.

Asaro, F., and I. Perlman. 1973. Provenience studies of Mycenaean pottery employing neutron activation analysis. In *Acts of the International Archaeological Symposium,*

The Mycenaeans in the eastern Mediterranean, 213–24. Cyprus: Ministry of Communications and works, Dept. of Antiquities.

Aspinall, A. 1980. Neutron activation analysis of pottery from Thera. In *Thera and the Aegean world II*, ed. C. Doumas, 155–60. London: Thera and the Aegean World.

Aspinall, A., and S. W. Feather. 1972. Neutron activation analysis of prehistoric flint mine products. *Archaeometry* 14:41–53.

————. 1978. Neutron activation analysis of Aegean obsidians. In *Thera and the Aegean world*, vol. 1, ed. C. Doumas, 517–21. London: Thera and the Aegean World.

Attas, M., F. Widemann, P. Fontes, K. Gruel, F. Laubenheimer, J. LeBlanc, and J. Lleres. 1979. Early Bronze Age ceramics from Lerna in Greece: Radiochemical studies. *Archaeo-Physica* 10:14–28. Proceedings of the 18th International Symposium on Archaeometry and Archaeological Prospection, Bonn.

Barnes, I. L., W. R. Shields, T. J. Murphy, and R. H. Brill. 1973. Isotopic analyses of Laurion lead ores. In *Archaeological chemistry*, ed. C. Beck, 1–10. Advances in Chemistry Series, no. 138. New York: American Chemical Society.

Brill, R. H., and J. M. Wampler. 1967. Isotope studies of ancient lead. *American Journal of Archaeology* 81:63–77.

Campbell Smith, A. 1967. Source of the stone used in a mace-head from Dorchester, England. *Proceedings of the Prehistoric Society* 33:455–56.

Catling, H. W., and R. E. Jones. 1977. A reinvestigation of the provenance of the inscribed stirrup jars found at Thebes. *Archaeometry* 19:137–46.

Crusoe, D. L. 1971. A study of aboriginal trade: A petrographic analysis of certain ceramic types from Florida. *Florida Anthropologist* 24:31–43.

Farquhar, R. M., and I. R. Fletcher. 1980. Lead isotope identification of sources of galena from some prehistoric Indian sites in Ontario, Canada. *Science* 207:640–43.

Frison, G., G. Wright, J. Griffin, and A. Gordus. 1968. Neutron activation analysis of obsidian: An example of its relevance to northwestern Plains archaeology. *Plains Anthropologist* 13:209–17.

Gale, N. H. 1978. Lead isotopes and Aegean metallurgy. In *Thera and the Aegean world*, vol. 1, ed. C. Doumas, 529–45. London: Thera and the Aegean World.

————. 1979. Lead isotopes and archaic Greek silver coins. *Archaeo-Physica* 10:194–208. Proceedings of the 18th International Symposium on Archaeometry and Archaeological Prospection, Bonn.

————. 1980. Some aspects of lead and silver mining in the Aegean. In *Thera and the Aegean World II*, ed. C. Doumas, 161–95. London: Thera and the Aegean World.

Genter, W., O. Muller, G. A. Wagner, and N. H. Gale. 1978. Silver sources of archaic Greek coinage. *Naturwissenschaften* 65:273–84.

Gordus, A. A. 1968. Neutron activation analysis of almost any old thing. *Chemistry* 41:8–15.

Gordus, A., W. Fink, M. Hill, J. Purdy, and T. Wilcox. 1967. Identification of the geologic origins of archaeological artifacts: An automated method of Na and Mn neutron activation analysis. *Archaeometry* 10:87–96.

Gordus, A. A., G. A. Wright, and J. B. Griffin. 1968. Characterization of obsidian sources by neutron activation analysis. *Science* 161:382–84.

Gordus, A. A., J. B. Griffin, and G. A. Wright. 1971. Activation analysis of prehistoric obsidian artifacts. In *Science and archaeology*, ed. R. H. Brill, 222−34. Cambridge, MA: MIT Press.

Grimanis, A. P., S. E. Filippakis, B. Perdikatsis. M. Vassilaki-Grimani, N. Bosana-Kourou, and N. Yaloures. 1980. Neutron activation and X-ray analysis of "Thapsos Class" vases. An attempt to identify their origin. *Journal of Archaeological Science* 7:227−39.

Hallam, B. R., S. E. Warren, and C. Renfrew. 1976. Obsidian in the western Mediterranean: Characterization by neutron activation analysis and optical emission spectroscopy. *Proceedings of the Prehistoric Society* 42:85−110.

Hansen, B. A., M. A. Sorensen, K. Heydorn, V. Mejdahl, and K. Conradsen. 1979. Provenance study of medieval decorated floor-tiles carried out by means of neutron activation analysis. *Archaeo-Physica* 10:119−40. Proceedings of the 18th International Symposium on Archaeometry and Archaeological Prospection, Bonn.

Harbottle, G. 1970. Neutron activation analysis of potsherds from Knossos and Mycenae. *Archaeometry* 12:23.

Heizer, R. F., and H. Williams. 1963. Geologic notes on the Idolo de Coatlinchan. *American Antiquity* 29:95−98.

―――. 1965. Stones used for colossal sculpture at or near Teotihuacán. *University of California, Archaeological Research Facility, Contributions* 1:55−70.

Hester, T. R., R. F. Heizer, and R. N. Jack. 1971. Technology and geologic sources of obsidian from Cerro de las Mesas, Veracruz, Mexico, with observations in Olmec trade. *University of California, Archaeological Research Facility, Contributions* 13:133−42.

Johnson, R. A., and F. S. Stoss. 1965. Laboratory-scale instrumental neutron activation for archaeological analysis. *American Antiquity* 30:345−47.

Jones, R. E. 1978. Composition and provenance studies of Cycladic pottery with particular reference to Thera. In *Thera and the Aegean world*, vol. 1, ed. C. Doumas, 471−82. London: Thera and the Aegean World.

Kinle, J. 1962. Jadeite—its importance for the problems of Asia-America pre-Columbian relationships. *Folia Orientalia* 4:231−42.

Kowalski, B., et al. 1972. Classification of archaeological artifacts by applying pattern recognition to trace element data. *Analytical Chemistry* 44:2176−80.

Leslie, V. 1964. Notes on the petrology of chipped stone artifacts from the upper Delaware Valley. *New World Antiquity* 11:74−78.

Luedtke, B. 1978. Chert sources and trace-element analysis. *American Antiquity* 43:413−23.

McDougall, J. M., D. H. Tarling, and S. E. Warren. 1983. The magnetic sourcing of obsidian samples from Mediterranean and Near Eastern sources. *Journal of Archaeological Science* 10:441−52.

Mertz, R., W. Melson, and G. Levenbach. 1979. Exploratory data analysis of Mycenaean ceramic compositions and provenances. *Archaeo-Physica* 10:580−96. Proceedings of the 18th International Symposium on Archaeometry and Archaeological Prospection, Bonn.

Montgomery, A. 1963. The source of the fibrolite axes. *El Palacio* 70:34−48.

Morey, J. E. 1950. Petrographical identification of stone axes. *Proceedings of the Prehistoric Society* 16:191−93.

Olin, J. S., G. Harbottle, and E. V. Sayre. 1977. Elemental compositions of Spanish and Spanish-colonial Majolica ceramics in the identification of provenience. In *Archaeological chemistry II*, ed. G. Carter, 200–29. Advances in Chemistry Series, no. 171. New York: American Chemical Society.

Parks, G. A., and T. T. Tieh. 1967. Identifying the geographical sources of artifact obsidian. *Nature* 211:289–90.

Peacock, D. P. S. 1967. The heavy mineral analysis of pottery. A preliminary report. *Archaeometry* 10:97–100.

Peacock, D. P. S., and C. Thomas. 1967. Class E imported post Roman pottery: A suggested origin. *Cornish Archaeology* 6.

Renfew, C., and J. R. Dixon. 1976. Obsidian in western Asia: A review. In *Problems in economic and social archaeology*, ed. G. de G. Sieveking, I. Longworth, and K. Wilson, 137–49. London: Duckworth.

Schneider, G., B. Hoffman, and E. Wirz. 1979. Significance and dependability of reference groups for chemical determinations of provenance of ceramic artifacts. *Archaeo-Physica* 10:269–85. Proceedings of the 18th International Symposium on Archaeometry and Archaeological Prospection, Bonn.

Sieveking, G. de G., P. Bush, J. Ferguson, P. Craddock, M. Hughes, and M. Cowell. 1972. Prehistoric flint mines and their identification as sources of raw material. *Archaeometry* 14:151–76.

Stevenson, D. P., F. H. Stross, and R. F. Heizer. 1971. An evaluation of X-ray fluorescence analysis as a method for correlating obsidian artifacts with source location. *Archaeometry* 13:17–25.

Stos-Fertner, Z., and N. H. Gale. 1979. Chemical and lead isotope analysis of ancient Egyptian gold, silver and lead. *Archaeo-Physica* 10:299–314. Proceedings of the 18th International Symposium on Archaeometry and Archaeological Prospection, Bonn.

Towle, J. 1973. Jade: An indicator of trans-Pacific contact? *Yearbook of the Association of Pacific Coast Geographers*: 35:165–72.

Tylecote, R. F. 1970. The composition of metal artifacts: A guide to provenance? *Antiquity:* 44:19–25.

Walthall, J., S. Stow, and M. Karson. 1980. Copena galena: Source identification and analysis. *American Antiquity* 45:21–42.

Ward, G. K. 1974. A systematic approach to the definition of sources of raw material. *Archaeometry* 16:41–53.

Wasson, J. T., and S. P. Sedwick. 1969. Possible sources of meteoritic material from Hopewell Indian mounds. *Nature* 222:22–24.

Williams, H., and R. F. Heizer. 1965. Sources of rocks used in Olmec monuments. *University of California, Archaeological Research Facility, Contributions* 1:1–40.

Wright, F. E. 1920. A petrographic description of the material of the Copan monuments. In *The inscriptions of Copan*, ed. D. G. Morley, 463–64. Carnegie Institution Publication 219.

Wright, G. A. 1969. *Obsidian analysis and prehistoric Near Eastern trade: 7500–3500 B.C.* Anthropological Papers, Museum of Anthropology, Univ. of Michigan, vol. 37.

Wright, G. A., and A. A. Gordus. 1969. Source areas for obsidian recovered at Munhata, Beisamoun, Hazorea and el-Khiam. *Israel Exploration Journal* 19:79–88.

Wright, G. A., J. B. Griffin, A. A. Gordus. 1969. Preliminary report on obsidian samples from Veratic rockshelter, Idaho. *Tebiwa* 12:27−30.

Zwicker, U., and F. Goudarzloo. 1979. Investigation on the distribution of metallic elements in copper slag, copper matte and copper and comparison with samples from prehistoric smelting places. *Archaeo-Physica* 10:360−75. Proceedings of the 18th International Symposium on Archaeometry and Archaeological Prospection, Bonn.

ARCHAEOLOGICAL SEDIMENTS

Biek, L. 1969. Soil silhouettes. In *Science in archaeology*, ed. D. Brothwell and E. S. Higgs, 118−23. London: Thames and Hudson.

Bintliff, J. 1976. Sediments and settlement in southern Greece. In *Geoarchaeology*, 267−75. *See* Davidson and Shackley, 1976.

Butzer, K. W. 1981. Cave sediments, upper Pleistocene stratigraphy, and Mousterian facies in Cantabrian Spain. *Journal of Archaeological Science* 8:133−83.

Catt, J. A., and A. H. Weir. 1976. The study of archaeologically important sediments by petrographic techniques. In *Geoarchaeology*, 65−90. *See* Davidson and Shackley, 1976.

Davidson, D. A., and M. L. Shackley, eds. 1976. *Geoarchaeology: Earth science and the past*. London: Duckworth.

Farrand, W. R. 1979. Chronology and palaeoenvironment of Levantine prehistoric sites as seen from sediment studies. *Journal of Archaeological Science* 6:369−92.

Fedele, F. G. 1976. Sediments as palaeo-land segments: The excavation side of study. In *Geoarchaeology*, 23−48. *See* Davidson and Shackley, 1976.

Goldberg, P. 1969. Sediment analysis of Jerf'Ajla and Yabroud rock shelters, Syria. In *8th INQUA Congress, Résumé de Communications* (Paris, 1969), p. 279.

———. 1979a. Micromorphology of Pech-de-l'Aze II sediments. *Journal of Archaeological Science* 6:17−47.

———. 1979b. Micromorphology of sediments from Hayonim Cave, Israel. *Catena* 6:167−81.

Goldberg, P., and Y. Nathan. 1975. The phosphate mineralogy from el-Tabun Cave, Mt. Carmel, Israel. *Mineralogical Magazine* 40:253−58.

Graham, I. 1976. The investigation of the magnetic properties of archaeological sediments. In *Geoarchaeology*, 49−63. *See* Davidson and Shackley, 1976.

Hughes, P. J. and R. J. Lampert. 1977. Occupational disturbance and types of archaeological deposit. *Journal of Archaeological Science* 4:135−40.

Laville, H. 1976. Deposits in calcareous rock shelters: Analytical methods and climatic interpretation. In *Geoarchaeology*, 137−55. *See* Davidson and Shackley, 1976.

Laville, H., J. P. Rigaud, and J. Sackett. 1980. *Rock shelters of the Perigord: Geological stratigraphy and archaeological succession*. New York: Academic Press.

Proudfoot, B. 1976. The analysis and interpretation of soil phosphorus in archaeological contexts. In *Geoarchaeology*, 93−113. *See* Davidson and Shackley, 1976.

Schmid, E. 1969. Cave sediments and prehistory. In *Science in archaeology*, ed. D. Brothwell and E. S. Higgs, 151−66. London: Thames and Hudson.

Shackley, M. L. 1975. *Archaeological sediments: A survey of analytical methods*. London: Butterworth.

———. 1976. The Danebury project: An experiment in site sediment recording. In *Geoarchaeology*, 9–21. See Davidson and Shackley, 1976.

———. 1978. A sedimentological study of Devil's Lair, western Australia. *Journal and Proceedings of the Royal Society of Western Australia* 60:33–40.

Straw, A. 1976. Sediments, fossils and geomorphology: A Lincolnshire situation. In *Geoarchaeology*, 317–26. See Davidson and Shackley, 1976.

SITE GEOLOGY

Ahlbrandt, T. S. 1974. Dune stratigraphy, archaeology, and the chronology of the Killpecker dune field. In *Applied geology and archaeology: The Holocene history of Wyoming*, ed. M. Wilson, 51–60. Geological Survey of Wyoming, Report of Investigations no. 10.

Albanese, J. P. 1970. Geology of the Glenrock site area, Wyoming. *Plains Anthropologist Memoir 7* 15:56–66.

———. 1971. Geology of the Ruby site area, Wyoming 48CA302. *American Antiquity* 36:91–95.

———. 1974. Geology of the Casper archaeological site, Natrona County, Wyoming. In *Applied geology and archaeology: The Holocene history of Wyoming*, ed. M. Wilson, 46–50. Geological Survey of Wyoming, Report of Investigations no. 10.

Albritton, C. C. 1968. Geology of the Tushka site: 8905. In *The prehistory of Nubia*, vol. 2, ed. F. Wendorf, 856–64. Dallas: SMU press.

Black, R. F. 1959. Geology of Raddatz rockshelter, Sk 5, Wisconsin. *Wisconsin Archeologist* 40:69–82.

Black, R. F., and W. S. Laughlin. 1964. Anangula: A geologic interpretation of the oldest archeologic site in the Aleutians. *Science* 143:1321–22.

Bond, G. 1957. The geology of the Khami Stone Age sites. *Occasional Papers of the National Museum of Southern Rhodesia* 21A:44–55.

———. 1969. The geology of the Kalambo Falls prehistoric site. In *Kalambo Falls prehistoric site*, vol. 1, ed. J. D. Clark, 197–213. Cambridge: Cambridge Univ. Press.

Bullard, R. G. 1970. Geological studies in field archaeology: Tel Gezer, Israel. *Biblical Archaeologist* 32:98–132.

———. 1976. The archaeological geology of the Khirbet Shema area. *Annual of the American Schools of Oriental Research* 4:15–32.

———. 1978. The environmental geology of Roman Carthage. In *Excavations at Carthage, 1975*, vol. 2, ed. J. H. Humphrey, 3–26. Ann Arbor, MI: Kelsey Museum, Univ. of Michigan.

———. 1980. *Environmental geology of Carthage. The commercial harbor and the Tophet: Carthage, the Punic Project, second interim report*. Oriental Institute Communications, no. 141. Chicago: Univ. of Chicago Press.

Butzer, K. W. 1973. Geology of Nelson Bay Cave, Robberg, South Africa. *South African Archaeological Bulletin* 28:97–110.

————. 1978. Comments on the infillings of various old Babylonian and Kassite structures at Nippur. *Oriental Institute Communications*. 23:188–90.

Butzer, K. W., P. B. Beaumont, and J. C. Vogel. 1978. Lithostratigraphy of Border Cave, Kwa Zulu, South Africa. *Journal of Archaeological Science* 5:317–41.

Butzer, K. W., I. Miralles, and J. F. Mateau. 1983. Urban geoarchaeology in medieval Alzira (Prov. Valencia, Spain). *Journal of Archaeological Science* 10:333–50.

Cooke, H. B. S. 1938. The Sterkfontein bone breccia: A geological note. *South African Journal of Science* 35:204–08.

Crowl, G. H., and R. Stuckenrath, Jr. 1977. Geological setting of the Shawnee-Minisink Paleoindian archaeological site (36-Mr-43). *Annals of the New York Academy of Sciences* 288:218–222.

Crumlin-Pedersen, O., L. Nymark, and C. Christiansen. 1980. Kyholm 78—a joint archaeological-geological investigation around a thirteenth century wreck at Kyholm, Samso, Denmark. *International Journal of Nautical Archaeology* 9:193–216.

de Heinzelin, J. 1968. Geological observations near Latamne. In *The Middle Acheulean occupation site at Latamne, northern Syria*, ed. J. D. Clark, 1–76. *Quaternaria*, vol. 10.

Donahue, J. 1979. Geologic investigations at early Bronze sites. *Annual of the American Schools of Oriental Research* 46:137–54.

Dort, W., Jr. 1975. Archaeo-geology of Jaguar Cave, upper Birch Creek Valley, Idaho. *Tebiwa* 17:33–57.

Farrand, W. R. 1969. Geology, climate and chronology of Yabrud rockshelter I. In *Alfred-Rust Festschrift*, pp. 121–32. *Fundamenta*, series A, vol. 2.

Fitzgibbons, P. T., J. M. Adovasio, and J. Donahue. 1977. Excavations at Sparks rockshelter (15J019), Johnson County, Kentucky. *Pennsylvania Archaeologist* 47:1–58.

Folk, R. L. 1975. Geologic urban hindplanning: An example from a Hellenistic-Byzantine city, Stobi, Jugoslavian Macedonia. *Environmental Geology* 1:5–22.

Garrard, A., and C. P. D. Harvey. 1981. Environment and settlement during the Upper Pleistocene and Holocene at Jubba in the Great Nefud, northern Arabia. *Atlal* 5:137–48.

Harris, E. C. 1975. The stratigraphic sequence: A question of time. *World Archaeology* 7:109–21.

Haynes, C. V. 1965. Stratigraphy of the Lehner site, Arizona. In *Guidebook for INQUA field conference. H, Southwestern arid lands*, ed. L. A. Heindl, and E. K. Reed, 55–57. Lincoln, NE: Academy of Science.

Haynes, C. V., and D. C. Grey. 1965. The Sister's Hill site and its bearing on the Wyoming post-glacial alluvial chronology. *Plains Anthropologist* 10:196–207.

Haynes, C. V. and G. A. Agogino. 1966. Prehistoric springs and geochronology of the Clovis site, New Mexico. *American Antiquity* 31, p. 812–21.

Helgren, D. M., and A. S. Brooks. 1983. Geoarchaeology at Gi, a Middle Stone Age and Later Stone Age site in the northwest Kalahari. *Journal of Archaeological Science* 10:181–97.

Hopkins, D. M. and J. L. Giddings. 1953. *Geologic background of Iyatayet archeological site, Cape Denbeigh, Alaska.* Smithsonian Miscellaneous Collections, vol. 121, no. 11.

Issar, A. 1961. The Plio-Pleistocene geology of the Ashdod area. *Bulletin of the Research Council of Israel* 104:173–82.

Jelinek, A. J., W. R. Farrand, G. Haas, A. Horowitz, and P. Goldberg. 1973. New excavations at Tabun Cave, Mount Carmel, Israel, 1967–1972: A preliminary report. *Paleorient* 1 (2).

Judson, S. 1953a. Geology of the Hodges site, Quay County, New Mexico: *Bulletin of the Bureau of American Ethnology* 154:285–302.

———. 1953b. *Geology of the San Jon site, eastern New Mexico*. Smithsonian Miscellaneous Collections, vol. 121, no. 1.

Kittleman, L. 1977. Preliminary report on the geology of Dirty Shame rockshelter, Malheur County, Oregon. *Tebiwa* 5:1–22.

Kurten, B., and E. Anderson. 1972. The sediments and fauna of Jaguar Cave. II. The Fauna. *Tebiwa* 15:21–46.

Leighton, M. M. 1923. The geological aspects of some of the Cahokia (Illinois) mounds. In *The Cahokia mounds, a report of progress*, ed. W. K. Moorehead. University of Illinois Bulletin, vol. 21.

Mackay, J. R., W. H. Mathews, and R. S. MacNeish. 1961. Geology of the Engigstciak archaeological site, Yukon Territory. *Arctic* 14:25–52.

Malde, H. E., and A. P. Schick. 1964. Thorne Cave, northeastern Utah: Geology. *American Antiquity* 30:60–73.

Molyersons, J. 1978. The behaviour of stones and stone implements, buried in consolidating and creeping Kalahari sands. *Earth Surface Processes* 3:115–28.

Movius, H. L., and S. Judson. 1956. The rock-shelter of La Colombiere: Archaeological and geological investigation of an Upper Perigordian site near Poncin, Ain. *Bulletin of the American School of Prehistoric Research*, no. 19.

Oakley, K. P., and M. Leakey. 1937. Report on excavations at Jaywick Sands, Essex (1934), with some observations on the Clactonian industry, and on the fauna and geological significance of the Clacton Channel. *Proceedings of the Prehistoric Society* 3:217–60.

Picard, L., and U. Baida. 1966. *Geological report on the Lower Pleistocene of the 'Ubeidiya excavations*. Jerusalem: Proceedings Israel Academy of Sciences and Humanities 4:1–16.

Preece, R. C. 1980. The biostratigraphy and dating of the Tufa deposit at the Mesolithic site at Blashenwell, Dorset, England. *Journal of Archaeological Science* 7:345–62.

Price, J. C., R. G. Hunter, and E. V. McMichael. 1964. Core drilling in an archaeological site. *American Antiquity* 30:219–22.

Reed, N. A., J. W. Bennett, and J. W. Porter. 1968. Solid core drilling of Monks Mound: Technique and findings. *American Antiquity* 33:137–48.

Rothenberg, B. 1972. *Timna, valley of the Biblical copper mines*. London: Thames and Hudson.

Singer, R., J. Wymer, B. Gladfelter, and R. Wolff. 1973. Excavation of the Clactonian industry of the golf course, Clacton-on-Sea, Essex. *Proceedings of the Prehistoric Society* 39:6–74.

Stearns, C. E. 1967. Pleistocene geology of Cape Ashakar and vicinity. *Bulletin of the American School of Prehistoric Research* 22:6–35.

Stein, J. 1978. Augering archaeological sites. *Southeastern Archaeological Newsletter* 20:11–17.

Stein, J., and G. Rapp, Jr. 1978. Archaeological geology of the site. In *Excavations at Nichoria in southwest Greece*. Vol. 1, *Site, environs, and techniques*, ed. G. Rapp, Jr., and S. E. Aschenbrenner, 234–37. Minneapolis: Univ. of Minnesota Press.

Stock, C., and F. D. Bode. 1937. The occurrence of flints and extinct animals in pluvial deposits near Clovis, New Mexico. III, Geology and vertebrate paleontology of the late Quaternary. *Philadelphia Academy of Natural Sciences, Proceedings* 88:219−41.

Thorson, R. M., and T. D. Hamilton. 1977. Geology of the Dry Creek site: A stratified early man site in interior Alaska. *Quaternary Research* 7:149−76.

West, R. G., and C. B. M. McBurney. 1943. The Quaternary deposits at Hoxne, Suffolk, and their archaeology. *Proceedings of the Prehistoric Society* 20:131−54.

Wright, G. E. 1962. Archaeological fills and strata. *Biblical Archaeologist* 25:35−40.

Wright, H. E. 1951. Geologic setting of Ksar 'Akil, a Palaeolithic site in Lebanon. *Journal of Near Eastern Studies* 10:112−22.

―――. 1952. The geological setting of four prehistoric sites in northeastern Iraq. *Bulletin of the American Schools of Oriental Research* 128:11−24.

Index